Optical Sensors and Switches

MOLECULAR AND SUPRAMOLECULAR PHOTOCHEMISTRY

Series Editors

V. RAMAMURTHY
Professor
Department of Chemistry
Tulane University
New Orleans, Louisiana

KIRK S. SCHANZE
Professor
Department of Chemistry
University of Florida
Gainesville, Florida

Optical Sensors and Switches

edited by

V. Ramamurthy
Tulane University
New Orleans, Louisiana

Kirk S. Schanze
University of Florida
Gainesville, Florida

CRC Press
Taylor & Francis Group
Boca Raton London New York

CRC Press is an imprint of the
Taylor & Francis Group, an **informa** business

CRC Press
Taylor & Francis Group
6000 Broken Sound Parkway NW, Suite 300
Boca Raton, FL 33487-2742

First issued in paperback 2019

© 2001 by Taylor & Francis Group, LLC
CRC Press is an imprint of Taylor & Francis Group, an Informa business

No claim to original U.S. Government works

ISBN-13: 978-0-8247-0571-8 (hbk)
ISBN-13: 978-0-367-39708-1 (pbk)

Visit the Taylor & Francis Web site at
http://www.taylorandfrancis.com

and the CRC Press Web site at
http://www.crcpress.com

Preface

During the past decade the field of photochemistry has focused increasingly on the development of photochemical systems and materials that respond in a controllable manner to chemical and/or optical stimuli. Photoresponsive systems and materials are exceedingly useful, as exemplified by the fact that they form the basis for a variety of technologically important processes and products such as chemical sensors, photochromic optical elements, optical data storage media, photoresists, organic light emitting diodes, optical limiters, photoconductors for use in xerography, and molecular systems for photodynamic therapy. Although many useful photoresponsive systems are based on "simple" molecular photochemistry—for example, photochromic azobenzene and spiropyrans are used in optical data storage media and ruthenium and platinum complexes are useful optical oxygen sensors—solutions to many high-technology applications of current interest inevitably require complex supramolecular assemblies to achieve the desired photoresponsive function.

The present volume highlights recent work from leading chemical and materials research scientists that are developing supramolecular photochemical systems that possess unique functions as optical sensors and/or switches. The chapters in this volume span the range from fundamental studies of unique and novel new chemical systems to applications-driven research in which the objective is to develop new materials having useful properties for high-technology applications such as detecting biologically relevant target molecules and optical signal multiplexing.

Chapters 1 through 4 feature work on supramolecular and polymer systems that function as sensors for a variety of chemical and biological targets. In Chapter 1, Rudzinski and Nocera give an exciting overview of research focusing on the development of unique supramolecular sensors that combine an organic receptor unit (a bucket) with a luminescent reporter moiety. These "buckets of light" are highly useful in sensing a variety of targets using luminescence as the method for detection. In Chapter 2, de Silva and his coauthors present an authoritative overview of supramolecular optical sensors that function by competition between photoluminescence and photoinduced electron transfer. These luminescent PET sensors have been widely applied by the authors as well as other groups for ion and pH sensing in chemically and biologically relevant systems. Andersson and Schmehl, in Chapter 3, provide an introduction to the application of electrogenerated chemiluminescence (ECL) to chemical sensing. This chapter introduces the method of ECL sensing and provides several examples of its application. Chapter 4 describes a very interesting new application of supramolecular photophysics to sensing chemical and biological analytes. Whitten and coworkers from QTL Biosystems (QTL = quencher-tether-ligand) describe the remarkably sensitive fluorescence quenching of π-conjugated polyelectrolytes by positively charged species such as viologens. They have parlayed this effect into the design of supramolecular systems that can detect biologically relevant analytes at nM concentrations.

Chapters 5 through 8 describe new polymeric materials that display useful optical properties. In Chapter 5, Lees discusses the use of photoluminescent metal complexes as probes to explore the properties of polymers while in Chapters 6, 7, and 8 Wang, Wiederrecht and Sponslor, respectively, and coauthors describe the properties of unique new polymer- and liquid crystalline–based materials. In the final chapters a variety of novel polymeric and supramolecular materials that display interesting and useful photochemical and optical properties are described.

Nagamura describes in Chapter 9 the fabrication and properties of supramolecular and polymeric systems that display unique linear and nonlinear optical response. Such materials may find application in high-density information processing systems of the future. Chapter 10, by Shinkai and James, describes the properties of supramolecular photoswitchable ion receptors. Finally, in Chapter 11, Ishikawa and Ye discuss the application of state-of-the art fluorescence methods to explore the properties of polymers with nm-scale resolution.

Taken together, the contributions in this volume provide an overview of work in the field of photochemistry that is directed toward the development of state-of-the art supramolecular systems and materials that display unique optical response. The book will serve as a valuable resource and guide to scientists engaged in chemical and materials research as well as to students seeking to learn more about this exciting field of scientific research.

Kirk S. Schanze
V. Ramamurthy

Contents

Contributors

Ann-Margret Andersson   Department of Chemistry, Tulane University, New Orleans, Louisiana

Troy Bergstedt, Ph.D.   QTL Biosystems, LLC, Santa Fe, New Mexico

Liaohai Chen, Ph.D.   Los Alamos National Laboratory, Los Alamos, New Mexico

A. Prasanna de Silva, Ph.D.   School of Chemistry, Queen's University of Belfast, Belfast, Northern Ireland

David B. Fox   School of Chemistry, Queen's University of Belfast, Belfast, Northern Ireland

Peter Heeger, M.D.   Case Western Reserve University, Cleveland, Ohio

Mitsuru Ishikawa   Joint Research Center for Atom Technology (JRCAT) and Angstrom Technology Partnership (ATP), Ibaraki, Japan

Tony D. James, Ph.D.   Department of Chemistry, University of Birmingham, Birmingham, United Kingdom

Robert Jones, Ph.D.   QTL Biosystems, LLC, Santa Fe, New Mexico

Alistair J. Lees, Ph.D.   Department of Chemistry, State University of New York at Binghamton, Binghamton, New York

Duncan McBranch, Ph.D. QTL Biosystems, LLC, Santa Fe, New Mexico

Thomas S. Moody School of Chemistry, Queen's University of Belfast, Belfast, Northern Ireland

Toshihiko Nagamura, Ph.D. Molecular Photonics Laboratory, Research Institute of Electronics, Shizuoka University, Hamamatsu, Japan

Man-Kit Ng, Ph.D. Department of Chemistry and The James Franck Institute, University of Chicago, Chicago, Illinois

Daniel G. Nocera, Ph.D. Department of Chemistry, Massachusetts Institute of Technology, Cambridge, Massachusetts

Christina M. Rudzinski Department of Chemistry, Massachusetts Institute of Technology, Cambridge, Massachusetts

Russell H. Schmehl, Ph.D. Department of Chemistry, Tulane University, New Orleans, Louisiana

Seiji Shinkai, Ph.D. Department of Chemistry and Biochemistry, Graduate School of Engineering, Kyushu University, Fukuoka, Japan

Michael B. Sponsler, Ph.D. Department of Chemistry and W. M. Keck Center for Molecular Electronics, Syracuse University, Syracuse, New York

Liming Wang, Ph.D. Department of Chemistry and The James Franck Institute, University of Chicago, Chicago, Illinois

Qing Wang, Ph.D. Department of Chemistry and The James Franck Institute, University of Chicago, Chicago, Illinois

Sheenagh M. Weir School of Chemistry, Queen's University of Belfast, Belfast, Northern Ireland

David Whitten, Ph.D. QTL Biosystems, LLC, Santa Fe, New Mexico

Gary P. Wiederrecht, Ph.D. Chemistry Division, Argonne National Laboratory, Argonne, Illinois

Jing Yong Ye Joint Research Center for Atom Technology (JRCAT) and Angstrom Technology Partnership (ATP), Ibaraki, Japan

Luping Yu, Ph.D. Department of Chemistry and The James Franck Institute, University of Chicago, Chicago, Illinois

Contents of Previous Volumes

1

Buckets of Light

**Christina M. Rudzinski and
Daniel G. Nocera**
Massachusetts Institute of Technology,
Cambridge, Massachusetts

I. INTRODUCTION

Chemosensors comprise molecular-scale structures that recognize and signal the presence of analytes [1–6]. Most chemosensor designs are based on a "3R scheme"—recognize, relay, and report. As shown in Fig. 1, a noncovalent molecular recognition event at the receptor site is communicated, by physical or chemical means, to a reporter site, which produces a measurable signal. A rapid equilibrium between the analyte and receptor site affords the chemosensor a real-time response that varies with the concentration of analyte. Sensitivity and selectivity are the two most important parameters of any sensing application. For the chemosensor platforms of the type depicted in Fig. 1, the "lock and key" fit of analyte to the recognition site and the strength of this interaction (described by the association constant) are crucial determinants of the overall sensitivity and selectivity. The sensitivity of the chemosensor is further augmented by the efficiency of the relay mechanism in communicating the binding event to a receptor site and by the ability of the receptor site to produce a signal.

The greatest advances in the elaboration of Fig. 1 in recent years have come about from the use of supramolecules as the basic building block of the chemosensor architecture. Although only 10 to 100 angstroms in dimen-

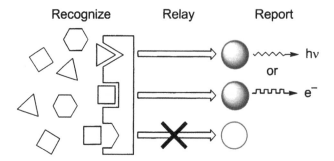

Figure 1 The "3R" chemosensor strategy. An electrical or optical signal reports the noncovalent binding of analyte to a receptor site via a relay mechanism.

sion, supramolecules have many familiar shapes including balls, bowls, boxes, bracelets, buckets, chains, clamps, cubes, ladders, lariats, pagodas, saddles, starbursts, and tweezers [7–9] to name a few. But more than whimsical reflections of chemists' imaginations, these supramolecule frameworks provide a noncovalent receptor site for analyte and a scaffold on which to attach a reporter site [1,3,4,8,10]. Signaling from the reporter site may be accomplished by electrical or optical methods, although the latter have come to the forefront owing to their sensitivity and ease of implementation [11,12], especially when the optical signal is derived from luminescence (*vide infra*). A conformational change, energy transfer, electron transfer, or some combination of these processes normally provides the relay mechanism by which the noncovalent recognition of analyte, under dynamic equilibrium with the supramolecular binding site, is communicated to the reporter site.

II. SCOPE OF CHAPTER

Of the diverse supramolecule architectures considered for chemosensor design, we emphasize one—a miniature bucket. Supramolecular buckets maintain a cylindrical cavity with structurally and chemically well-defined upper and lower rims. By virtue of its molecular shape, the bucket is an intrinsic receptor site. A supramolecular bucket alone, however, is an inadequate chemosensor because a signal cannot be produced. In the simplest chemosensing constructs, the analyte generates a signal upon association with the bucket whereas in more elaborate designs, functionality at the rim of the bucket offers sites to attach a discrete reporter site. The thematic focus of this chapter is bucket chemosensors that operate by the 3R scheme shown in Fig. 2, namely, a measurable change in a

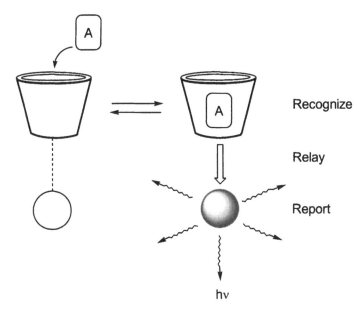

Figure 2 The 3R scheme that is the focus of this chapter. The supramolecule is a bucket receptor site, which can be filled with analyte A. Recognition of analyte produces a measurable change in a luminescence signal from a reporter site.

luminescence signal (increase or decrease) is produced when the bucket is filled with analyte.*

The overall structure of the chapter is as follows. Basic photophysics of luminescent excited states are presented, followed by a survey of the different types of luminescent excited states that are used as reporter sites for chemosensors featuring bucket motifs. The supramolecular buckets that are most commonly used in chemosensor design are then presented followed by a discussion of the important mechanisms by which analyte binding is transduced into a luminescence signal. With this general background in place, specific luminescent bucket chemosensors are examined. The chapter concludes with brief discussions of new avenues of exploration for the further development of 3R schemes and of new technologies, which when interfaced to the supramolecules described here, have the potential to lead to new sensors possessing unprecedented properties and function.

---

*In this chapter, a gray-scaled object indicates luminescence; thus Fig. 2 depicts a signal that is turned on.

III. BACKGROUND

A. Electronic Structure and Photophysics of Reporter Sites

A light emitting signal is superior for chemosensor design because it can report on nanometer length-scales [13,14] with nanosecond time responses [15], permit analytes and their influences to be monitored continuously in real time and *in situ* [16–19], possess an inherently large bandwidth (and hence information capacity), feature intrinsic selectivity owing to flexible choices of wavelength and polarization, achieve sensitivity down to the single molecule limit [20–33], and be married to a variety of imaging technologies including optical wave guides, the most important of which is optical fibers. These latter two issues have been especially prominent driving forces behind the emergence of optical sensing schemes in recent years [35–39]. With regard to the 3R scheme of Fig. 2, this chapter illustrates one other significant advantage of luminescence-based signaling: a sound knowledge of the photophysical properties that govern a particular luminescence event enables the investigator to readily tailor the signal transduction mechanism. In this manner, novel optical detection schemes involving chemosensor active sites may be elaborated.

Figure 3a depicts the intramolecular processes that govern the decay of an isolated molecule in an electronic excited state (M^*) to its ground state (M). Competing thermal relaxation and photon emission pathways are described by the nonradiative (k_{nr}) and radiative (k_r) rate constants, respectively. k_r is an intrinsic property of the molecule and it reflects the probability for emission of a photon of a given frequency [40–42]. k_{nr} encompasses all intramolecular decay pathways that do not lead to emission. Nonradiative decay most typically entails the conversion of an excited state's electronic energy into high-energy vibrations of the ground-state molecule [43–45]. Vibrational relaxation to the equilibrated ground-state molecule (M), or in the case of a photochemical process, to product molecule (P), is accompanied by the concomitant release of heat.

The interplay between k_r and k_{nr} determines the fundamental properties of an excited state molecule [46,47]. Luminescence intensity (I_o), which is directly proportional to the emission quantum yield (ϕ_e), is related to k_r and k_{nr} by the following expression [41,42,46],

$$I_o \approx \phi_e = \frac{k_r}{k_r + k_{nr}} = k_r \tau_o, \tag{1}$$

where τ_o $(= (k_r + k_{nr})^{-1})$ is the natural lifetime of the electronic excited state. Nonradiative relaxation pathways usually dominate (i.e., $k_{nr} \gg k_r$) and as related by Eq. (1), molecules under this condition will remain dark upon excitation. Conversely, M^* will luminesce upon excitation when thermal relaxation is inefficient with regard to photon emission (i.e., $k_r \gg k_{nr}$).

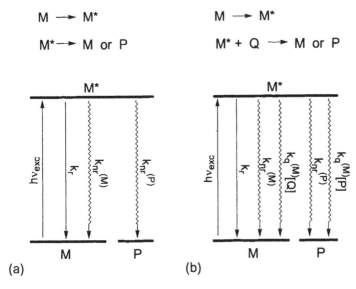

Figure 3 Excited state diagram for (a) intramolecular and (b) intermolecular decay processes of a molecule in an excited state M*. k_r and k_{nr} are the radiative and nonradiative rate constants, respectively. Superscript M refers to processes that return the excited state to ground state without chemical change. Superscript P refers to processes that result in the formation of a product, chemically distinct from M. Quenching rate constants depend on the concentration of quencher Q or P.

The physical or chemical interaction of M* with a species in its environment (usually denoted quencher, Q) will further modify luminescence intensity and lifetime (see Fig. 3b). Such bimolecular processes involving Q are nonradiative in nature and therefore Eq. (1), which only considers the intramolecular decay processes of an excited state molecule, must be modified by adding $k_q[Q]$ to the denominator [48],

$$I \approx \phi_e = \frac{k_r}{k_r + k_{nr} + k_q[Q]} = k_r\tau, \tag{2}$$

where k_q is the quenching rate constant and $\tau \ (= (k_r + k_{nr} + k_q[Q])^{-1})$ is the lifetime of the excited state in the presence of a quencher at concentration [Q]. It follows from Eq. (2) that bimolecular quenching pathways are dissipative and their presence will diminish the luminescence intensity and shorten the excited state lifetime. The Stern–Volmer relation quantitatively defines the luminescence lifetime and intensity under quenching conditions as [48],

$$\frac{I_0}{I} = \frac{\tau_0}{\tau} = 1 + \tau_0 k_q[Q], \tag{3}$$

where I_o and I, and τ_o and τ are the luminescence intensity and lifetime in the absence and presence of Q, respectively.

Emission of a given frequency can be turned on or off with changes in k_r, k_{nr}, and k_q. If these rate constants are perturbed by an analyte then a shift in the intensity or energy of luminescence will occur, thus providing a means to signal the analyte's presence. Because excited state molecules are highly reactive, it is easy to implement detection schemes in which the quencher is the analyte (Q = A). Despite their preponderance, the sensitivity and selectivity of quenching-based schemes are intrinsically limited. Electronic excited states are not very discriminating and invariably are subject to any number of quenching processes, causing quenching-based chemosensors to be prone to interferents. In addition, a detection signal produced by quenching complicates chemosensor design because a decrease in luminescence intensity must be measured against a bright background. The difficulty in detecting a signal under this condition is illustrated in Fig. 4a. The drawbacks of quenching-based detection are overcome when an analyte triggers a luminescence signal relative to a dark background (Fig. 4b). For this reason, chemosensing schemes of recent years have moved towards designs in which the analyte shifts the luminescence wavelength or prompts the appearance of a new emission signal. This preferred "turn-on" response is achieved when molecular recognition of an analyte causes an increase

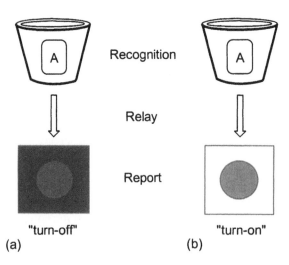

Figure 4 (a) Signal transduction by a quenching mechanism requires the detection of an attenuated emission against a bright background (light intensity scales with gray-scale intensity, black = bright). (b) A triggered signal demonstrates much greater sensitivity, as emission is detected against a dark background. Both balls in (a) and (b) have identical gray-scale fill patterns and intensities.

in the effective k_r, a decrease in k_{nr}, or a decrease in k_q. For any of these cases, ϕ_e will approach its limiting value of unity. As we discuss below, the rate constants k_r, k_{nr}, or k_q of numerous types of excited states have been manipulated to produce a luminescent signal from chemosensor reporter sites. Brief descriptions of the most common excited states used in chemosensor design are now presented.

1. $\pi\pi^*$, $n\pi^*$, and PET Excited States

The lowest energy excited state of a simple aromatic is often intensely fluorescent (Fig. 5a) [46,49–51]. With electron promotion confined to a rigid π framework, k_{nr} is not dominant and k_r prevails (i.e., $k_r \geq k_{nr}$). As delineated in Eq. (1), ϕ_e is therefore large (0.5–1.0) and τ approaches the reciprocal of k_r, which is also large (10^8–10^9 s^{-1}) owing to the sizable absorption cross-section of $^1\pi \rightarrow \pi\pi^*$ transitions. The presence of a heavy atom or the manifestation of other spin-orbit coupling mechanisms will cause intersystem crossing from $^1\pi\pi^*$ to $^3\pi\pi^*$ (Fig. 5b) [46]. The phosphorescence attendant on $^3\pi\pi^*$ decay may be exceptionally long lived (ms lifetimes are not unusual) [51] making the luminescence from this type of excited state especially susceptible to quenching. Oxygen is the most contentious quencher of phosphorescence because of its prevalence and the proclivity of its low energy singlet state to accept energy [46,52]. The problem that oxygen brings to bear on chemosensor design schemes

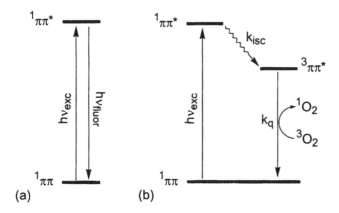

Figure 5 (a) Singlet $\pi\pi^*$ excited states typically relax radiatively (fluorescence) back to the ground state. (b) When a heavy atom is proximate to the excited state, a triplet $\pi\pi^*$ excited state may be populated by intersystem crossing. The long lifetime of triplet excited states makes them readily susceptible to quenching (e.g., by oxygen) or deactivation by other nonradiative processes. Accordingly, emission from the triplet excited state molecule (phosphorescence) is difficult to observe in the absence of environmental control.

is evident from considerations of Eq. (3). Substituting typical values for k_q (usually diffusion-controlled, $\approx 10^9$ $M^{-1}s^{-1}$) and the concentration of oxygen (10^{-3} M in solution and 10^{-2} M in air) reveals that the lifetime and intensity of luminescence from a $^3\pi\pi^*$ state of 1 ms duration will be attenuated by $> 10^3$ in oxygenated solutions and by $> 10^4$ in air. This seemingly problematic quenching reaction can be turned to an advantage in chemosensor design schemes when the phosphorescence from an efficiently quenched $^3\pi\pi^*$ excited state is recovered by analytes that physically shield the molecular phosphor from oxygen (see Sec. IV.B). In this case, the analyte will be detected by the appearance of a phosphorescence signal.

Luminescence from a $^1\pi\pi^*$ excited state may be used for signaling when an $n\pi^*$ or a charge transfer excited state is positioned energetically nearby. These latter states result from situating a lone pair proximate to a π-aromatic system [46,53–56]. Charge transfer from the lone pair to the frontier orbitals of the excited π-aromatic acceptor (a process that is commonly referred to as photoinduced electron transfer or PET) quenches luminescence. A nonemitting $n\pi^*$ or charge transfer state lying below $^1\pi\pi^*$ will therefore result in a photon silent reporter site (Fig. 6a). Analytes that energetically displace the $n\pi^*$ or charge transfer state above that of $^1\pi\pi^*$ will trigger a bright fluorescence. This state reordering, illustrated in Fig. 6b, typically occurs by raising the chemical potential of the lone pair by its association with the analyte. Chemical association mechanisms include protonation of the long pair, H-bonding, metal ion coordination, or bonding other Lewis acids to the lone pair. Chemosensors based on the principles of Fig. 6, especially those that derive their function from PET, have been reviewed extensively in recent years [1–3,5,57–62] and the reader is also referred to Chapter 2 for a more detailed overview of this area.

2. Metal-Centered Excited States

Most chemosensing applications utilizing this type of excited state are based on the emissive $^5D_J \rightarrow {}^7F_J$ transitions of terbium(III) and europium(III). The lowest energy state manifold of these ions is diagrammed in Fig. 7. Owing to the $\Delta S = 0$ spin restriction imposed on radiative transitions, lanthanide (Ln^{3+}) ions luminesce only weakly under direct irradiation [63–67]. Indirect excitation of the Ln^{3+} ion with the energy captured by a strong photon absorber can circumvent the spin impediment to ^5D_J population. As long as energy transfer from the sensitizer to the Ln^{3+} center is efficient, the induced luminescence from the metal center can be extremely intense. Although long recognized [68,69], sensitized excitation of Ln^{3+} luminescence in supramolecule constructs have been popularized only within the last decade, largely a result of the studies by Balzini, Lehn, Sabbatini, and their coworkers [70–73]. Chemosensing applications arise when the analyte functions as the sensitizer. Under the proper conditions, the analyte

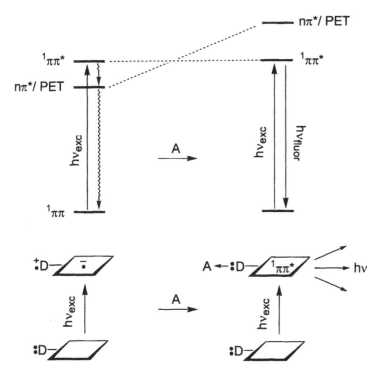

Figure 6 Sensing by a photoinduced electron transfer (PET) mechanism. A $n\pi^*$ or charge transfer excited state is repositioned relative to fluorescent $^1\pi\pi^*$ upon interaction of analyte A with a lone pair of electrons proximate to a reporter site.

may absorb the incident light and pass it on to the latent Tb^{3+} or Eu^{3+} emitting center (a process known as absorption-energy transfer emission or AETE). The presence of analyte will therefore be revealed by the appearance of a bright luminescence from the Ln^{3+} reporter site. In this scheme, k_r of the analyte, and not that of the Ln^{3+} emitting center, is relevant since it is the former that is the absorbing species [74–76]. Because $k_r = 10^{7-9}$ s^{-1} ($\varepsilon > 10^4$ M^{-1} cm^{-1}) is not uncommon for many analytes, a dramatic increase in the population of the ^5D_J emitting state may be achieved by analyte-induced AETE. Accordingly, the luminescence signal triggered by the analyte can be quite strong.

The exceptionally long lifetime of Ln^{3+} ion excited states (0.1–2 ms) might be expected to render Ln^{3+}-sensitization schemes susceptible to quenching [consider Eq. (3) for large τ_o]; however, this is not the case. Buried deep within the ion, f-centered orbitals are shielded from the external environment

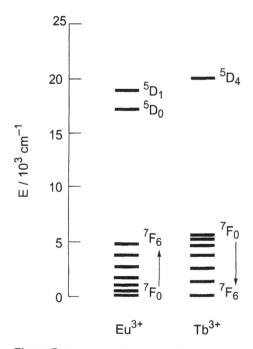

Figure 7 Energy level diagram of lowest energy excited states of europium and terbium ions. Only the emissive excited states within the ^5D_J manifold are shown.

and electronic coupling to the ^5D_J excited states is small (owing to poor overlap). A diminished k_q in Eq. (3) therefore offsets the large τ_0 and intermolecular quenching by typical reactants such as oxygen is not prevalent. On the other hand, high-energy oscillators of coordinated ligands, especially H_2O, are particularly adept at inducing efficient nonradiative decay of Ln^{3+} ion excited states [77–81]. Thus, a properly designed chemosensing supramolecule employing a Ln^{3+} reporter site must allow the analyte to access the metal center but at the same time exclude water from the primary coordination sphere.

Metal ions and complexes possessing ligand field excited states have limited application to chemosensor design because most are nonluminescent. Ligand field excited states typically involve the population of metal–ligand σ^* orbitals. Accordingly, k_{nr} is large and emission is not observed owing to dissipation of the excited state energy through the vibrations of the metal–ligand framework [82]. Metal d-centered excited states that are luminescent [83,84] include the ^2E_g excited state of Cr(III) [85–89]; the ^3E $(d^1_{xy}, (d_{xz}, d_{yz})^1)$ excited state of d^2 metal–ligand multiple bonds [90–92]; d-pπ excited states of d^8 platinum, rhodium, and iridium dithiolate complexes [93,94]; $d^8 \cdots d^8$ and $d^{10} \cdots d^{10}$

dimers [95–100]; selected $d\sigma^*$ excited states of d^7–d^7 and d^9–d^9 dimers [101–103]; multiple metal–metal bonded complexes [104–106]; the hexanuclear $(d^4)_6$ metal clusters of molybdenum, tungsten, and rhenium [107–111]; and d^{10} excited states of multinuclear copper, gold, and analogous cluster complexes [112,113]. Whereas the emission properties of these metal-centered excited states have received much attention, few [114] of them have been exploited in chemosensing applications.

3. Excimers and Exciplexes

An excimer or exciplex, E^*, is an excited state complex formed from the association of M^* with a ground state molecule that is the same or different, respectively ([MM]* = excimer; [MM′]* = exciplex) [115]. The simple considerations set forth in Fig. 8 for the highest-occupied (HOMO) and lowest-unoccupied molecular orbitals (LUMO) account for E^* formation. The filled HOMOs of two ground state molecules provide no driving force for ground state interaction (Fig. 8a). Conversely, a formal net bond arises from the interaction of the partially filled

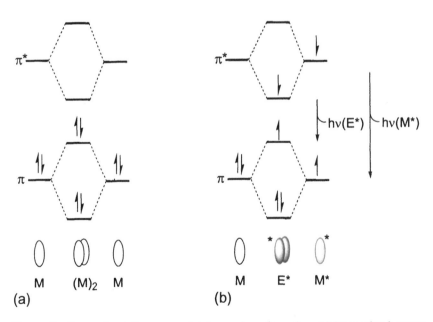

Figure 8 Energy level diagram explaining excimer formation. (a) Interaction between two ground state molecules results in no net stabilization of dimer. (b) Interaction of ground state (M) and electronically excited molecules (M^*) results in an energetically stabilized excited state complex or excimer (E^*). Emission from M^* and E^* is denoted by hν(M^*) and hν(E^*), respectively.

bonding and antibonding frontier orbitals of M* with the HOMO and LUMO of M (or M′ for an exciplex) (Fig. 8b). Decay of E* by photon emission returns the system to its individual constituents. Because the excimer or exciplex is stabilized with respect to M*, the fluorescence of E* can be red-shifted significantly (10–100 nm) from that of M*. Analyte recognition events that induce E* formation will therefore lead to the appearance of a new fluorescence signal. Excimer (exciplex) reporter sites are attractive from considerations of chemosensor design because (1) detection is facilitated by this large spectral shift and (2) the short lifetime of excimer (exciplex) fluorescence obviates the ability of oxygen and other quenchers to attenuate the luminescence signal.

4. Organic and Inorganic Charge Transfer Excited States

Red-shifted luminescence is also common to molecules that sustain intramolecular charge transfer (ICT) excited states. As shown in Fig. 9, an electron–hole

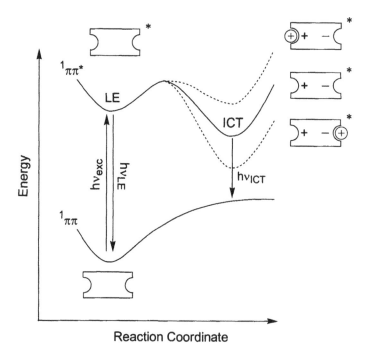

Figure 9 Intramolecular charge transfer (ICT) emission evolves from a localized excited state (LE). If a metal ion-binding site is present (indicated by the half-circle), ICT emission may be stabilized or destabilized depending on whether the metal ion (⊕) binds is at the electronegative or electropositive end of the π-aromatic system.

pair is produced from excitation of a localized excited state (LE). Owing to this charge redistribution within the π-aromatic, the energy and intensity of ICT fluorescence are very sensitive to the polarity of the environment [116]. The charge-separated electron–hole pair is resonance stabilized by electron withdrawing and electron accepting substituents (push–pull pairs), respectively, appended to or included within a π-aromatic framework. The intense fluorescence of many laser dyes and related lumophores originates from this type of excited state. Ingenious chemosensor designs feature engineered electrostatic environments that accommodate the incipient ICT electron–hole pair. For instance, consider the chemosensor schematically represented in Fig. 9 for the ICT-based detection of a metal ion. A receptor site placed at the electronegative end of the molecule will stabilize an ICT excited state upon its occupancy by a metal cation owing to the favorable $+ \cdots - +$ charge arrangement. In this instance, a red shift in the ICT fluorescence will signify the metal–ion binding event. Conversely, cation residency at a receptor site placed at the electropositive terminus of the molecule $(+ + \cdots -$ charge arrangement) will induce a blue shift of the fluorescence. The same concepts are easily applied to the detection of protons and Lewis acids and bases.

ICT luminescence is also prevalent for transition metal complexes possessing metal-to-ligand charge transfer (MLCT) excited states. Here, the hole resides at the electropositive metal center and the electron on a ligand of the primary coordination sphere [117]. This type of excited state is especially common to mononuclear d^6 metal complexes, for which *tris*(bipyridyl)ruthenium(II) is the archetype [118–124]. In this complex, the MLCT excited state is best represented chemically as $[(bpy)_2Ru^{III}(bpy\overset{\bullet}{-})]^{2+}$ where the electrons in the HOMO and LUMO levels are triplet paired. Only one polypyridyl ligand is needed to receive the electron, allowing the MLCT excited state of $[(NN)Re^{I}(CO)_3L]^x$ (NN = polypyridine, L = neutral (x = +1) or anionic (x = 0) donor ligands) complexes to also figure prominently in chemosensing applications. MLCT luminescence is typically phosphorescence based, which gives rise to large Stokes shifts and long lifetimes (hundreds of ns to tens of μs). Accordingly, most chemosensor schemes using d^6 polypyridyl complexes as reporter sites rely on detecting a quenched phosphorescence signal. For the same mechanistic reasons delineated above for organic ICT molecules, pronounced shifts in MLCT emission can be observed with the placement of a Lewis acid, Lewis base, or ion binding site proximate to the ligand framework. By monitoring the shift in MLCT luminescence, anions, cations, and protons may be readily detected.

5. Twisted Intramolecular Charge Transfer Excited States

Sometimes conformational changes are required to trap the intramolecular charge transfer events [54,125]. As represented in Fig. 10, an electron–hole pair pro-

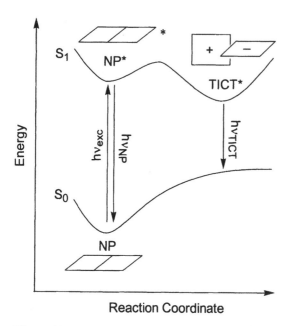

Figure 10   Potential energy diagram for emission from normal planar (NP) and twisted intramolecular charge transfer (TICT) excited states. The charge-separated state is stabilized by twisting and by the polarity of the environment.

duced from an internal charge transfer is stabilized in the twisted conformation [126–133]. Compounds possessing twisted intramolecular charge transfer (TICT) photophysics display fluorescence emission arising from both $\pi\pi^*$ (normal planar, NP) and TICT excited states (see Fig. 10), which exhibit large dipole moments. The population of TICT states may therefore be affected by the polarity of the environment and by factors that influence the twisting motion of the lumophore. Analyte binding that frees a TICT molecule from a spatially constricted (or less polar) environment will cause TICT fluorescence to increase. Conversely, analytes that promote a TICT molecule to enter a more constrained or nonpolar environment will favor NP fluorescence. Similar to ICT excited states, ions can promote TICT fluorescence, especially when the TICT molecule is appended with ion-binding sites that provide a favorable Coulombic interaction with which the TICT excited state can interact.

B.   Bucket Receptor Sites

A bucket-shaped cavity provides the consummate supramolecular architecture for analyte recognition. The most prominent buckets used in chemosensor design are cyclodextrins (CDs) and calixarenes. Both of these supramolecules enforce a cylindrical binding site that can be readily functionalized to support reporter sites and to manipulate the environment of the receptor site. Selection of guests may be achieved to a large extent by simply varying the dimensions of the bucket. Guest discrimination may be tuned further by varying the physical properties of the rim, thus modifying the entry and departure of molecules from the bucket.

In aqueous solutions, hydrogen bonding and dipole–dipole interactions provide ample driving force for guests to reside within a bucket. Electrostatic and ionic interactions also augment guest–host complexation, especially in cal-ixarenes, although the energetic degree to which these interactions contribute to analyte binding is still a matter of debate [134–136]. The construct of a bucket as a receptor site does not imply an empty interior prior to binding nor a rigid framework. Indeed, most researchers agree that the displacement of water from the bucket cavity and alleviation of ring strain are important entropic and enthalpic factors that contribute to guest association.

1. Cyclodextrins

Cyclodextrins are chiral cyclic oligosaccharides composed of six, seven, or eight D-glucose molecules (termed α, β, γ-cyclodextrin, respectively), which catenate via α-(1,4) linkages to produce the bucket-shaped cylindrical cavity of the vary-ing diameters presented in Fig. 11 [137–141]. Hydroxyl groups of the sugar rings, encircling the primary (bottom) and secondary (top) rims of the CD, impart water solubility to the bucket whereas the hydrocarbon rings of the D-glucose subunits define a hydrophobic interior suitable for binding guests. Thus the CD bucket dissolves in water but will fill itself with hydrophobic com-pounds such as aliphatic and aromatic guests and various polar compounds such as functionalized aromatics, amines, and alcohols. The countless inclusion com-plexes formed by CDs have been catalogued in numerous reviews [142,143]. Binding selectivity (and therefore selectivity in chemosensor function) at the first level of discrimination can be achieved with the CD cavity size. β-CD will bind mono- and bicyclic aromatics but not larger polyaromatics such as an-thracenes or pyrene, which show an affinity for γ-CD. Guest association to the CD may be controlled further by modifying the rims of the bucket. For example, capping a CD rim creates a more protected, hydrophobic environment. Alterna-tively, substituting the primary and secondary hydroxyls with alkoxy, ester, and a variety of other functional groups can dramatically perturb guest association constants.

The hydroxyl groups encircling the bucket rims also offer sites for the at-tachment of a reporter site. Most synthetic methodologies for site derivatization have been developed in studies aimed at promoting biomimetic transformations of substrates included within the CD cavity [144–146]. Notwithstanding, these

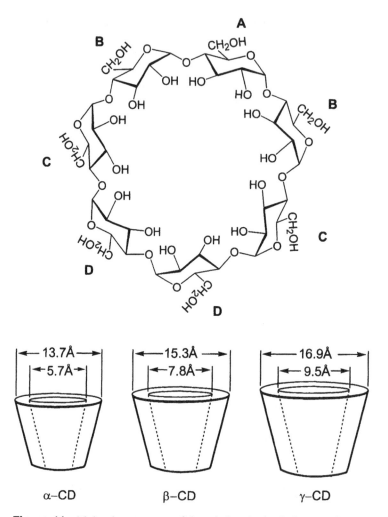

Figure 11 Molecular structure of β-cyclodextrin (top). Inner and outer wall bucket dimensions of α, β, and γ-cyclodextrins (bottom).

synthetic methodologies are readily adapted to chemosensor design. The reactivity of primary hydroxyl groups permits facile one-point modification at the bottom of the CD bucket [147]. Synthetic protocols developed by Tabushi et al. are most useful for the selective two-point functionalization of the primary hydroxyl rim [148–150]. By using rigid spacer molecules, regioselective modification at the AB, AC, and AD positions (Fig. 11) may be achieved. Methodologies to derivatize the secondary rim of the CD bucket are also well established [151], although such syntheses are less versatile and afford lower product yields owing to the diminished reactivity of the secondary hydroxyls.

2. Calixarenes

Calixarenes are formed from the cyclic catenation of 4, 6, or 8 phenolic rings [152,153]. More flexible than a CD, calixarenes can have several stable conformations; four principal conformations of calix[4]arene are shown in Fig. 12. The cone conformer most closely assumes the shape of a bucket, rigidified by strong intramolecular hydrogen bonds among the aryl hydroxyl substituents. The partial cone and alternate conformers lead to "buckets" that are flattened considerably and for these cases the supramolecule architecture is better described as a shallow bowl. Indeed, the chemosensing function of many calixarenes is derived from their ability to isomerize among cone, partial cone, and alternate conformations, all largely influenced by solvent polarity [154,155]. Acknowledging this flexibility of calixarenes, which become floppier with increasing ring size [156], a "bucket" representation is an oversimplified description of the shape of this class of supramolecules. For the sake of this chapter's simplicity and clarity, however, we generally refer to calixarenes as buckets.

As is the case for CDs, the binding affinity of calixarenes towards specific molecules can be optimized by controlling calixarene ring size and by transforming the hydroxyl groups on the rim of the bucket to other ligands (e.g., ethers, esters, amides) [152,155–158]. The oxygen ligation sphere on the lower rim (hydroxyl side) of calixarenes presents an especially effective site for binding ionic species, in particular, metal cations [159]. Enhanced binding specificity may be achieved by appending the lower rim of the calixarene with ion-selective macrocycles [156]. With the preferred binding site situated at the rim of the bucket, the analyte is seldom included within the interior of a calixarene, thus engendering yet another difference between the properties of CD-based versus calixarene-based chemosensors. The *tert*-butyl groups on the upper rim of the calixarene bucket can be removed easily under acidic conditions, providing added synthetic flexibility for the preparation of derivatized calixarenes possessing specialized functions. Recent synthetic methodologies allow calixarenes to be linked to other macrocyclic structures including CDs [160,161]. Such multireceptor designs have been used in the detection of anions and neutral guest molecules [155,162].

3. Other Bucket-Like Receptor Sites

Cyclodextrins and calixarenes are members of a general class of macrocycles commonly referred to as cyclophanes, although the naturally occurring CDs are usually considered separately from the synthetic members of the cyclophane family of compounds. The properties and guest–host complexes of cyclophanes

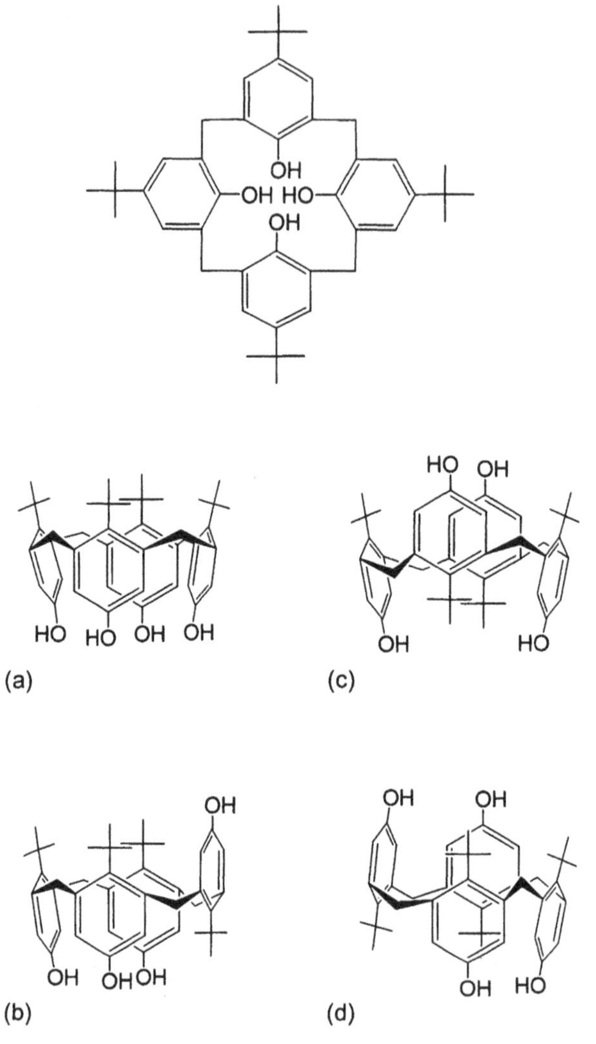

Figure 12 Molecular structure and stable conformers of p-*tert*-butycalix[4]arene: (a) cone; (b) partial-cone; (c) 1,3-alternate; (d) 1,2-alternate.

have been reviewed [134,163–171]. Many cyclophanes display bucket architectures. Cucurbituril macrocycles are composed of 6 glyoxal molecules stitched together with 12 molecules each of urea and formaldehyde [172–175]. This cyclophane is a rigid electroneutral bucket that provides an alternative structure to CDs [176–180]. Carceplexes and hemicarceplexes also present a rigid bucket possessing well-known complexation dynamics [155,181,182]. Guests often fully encapsulate within the interior of these structures. Exchange kinetics are slow and therefore these types of active sites are not optimal for reversible real-time chemosensing applications. Rather, these supramolecules are more readily adapted to applications requiring a fluorescent indicator. Chemosensing function of these receptors can result from chemical and physical alterations, such as changing the temperature of the system, to release analyte from the cavity. For the most part, however, the need for an external stimulus to drive sensory function restricts the utility of these supramolecules for chemosensor design. Many other related macromolecules demonstrate "bucket"-type interactions and it is likely that these structures can be extended to encompass the principles delineated in this chapter for CDs and calixarenes. However, in consideration of our imposed constraints on the scope of this chapter, these supramolecules are not further discussed.

C. Relay Mechanisms

Three basic types of relay mechanisms are relevant to the chemosensors of Sec. IV: a conformational change, energy transfer, and electron transfer. Operating alone or in concert, these relay mechanisms communicate the molecular recognition event of analyte binding at the receptor site to the luminescent reporter site. We now briefly discuss each of these relay mechanisms.

1. Changes in Conformation

A conformational change induced by analyte binding is often the crucial transduction step that triggers a change in the luminescence properties of a chemosensor. In many sensing schemes, analyte-induced displacement of the reporter site from the protected interior of the bucket will affect k_{nr} and k_q, resulting in a variation in the intensity of the luminescence signal or the appearance of a new luminescence signal. For example, a freed reporter can exhibit increased torsional flexibility (e.g., twisting to a TICT state) or it is able to react by electron or energy transfer with species in the external environment (e.g., solvent or ion stabilization of a ICT state, excimer/exciplex formation). The simplest designs involve the release of a luminescent reporter site from the supramolecule into an external environment where the excited state is quenched. Alternatively, a luminescence signal may be triggered if a quenching reaction, present from the start, is disrupted by a conformational change. In this case, k_{nr} (or k_q if an external reagent in the environment is quenching the reporter site) will decrease, leading to an increase in the luminescence lifetime and intensity [see Eqs. (2) and (3)]. The many different ways a conformational change is involved in a relay mechanism make it difficult to comprehensively generalize them here.

Rather, conformational effects are discussed as necessary for the specific bucket chemosensors of Sec. IV.

2. Electron Transfer

The electron transfer rate constant k_{ET} is derived from the product of Franck–Condon (FC) and electronic coupling terms (H_{DA}),

$$k_{ET}(r) = \frac{2\pi}{\hbar}H_{DA}^2 (FC). \tag{4}$$

Electron transfer has been reviewed extensively in recent years [183–188] and there is no need to reformulate the details of Eq. (4) here. Rather, we focus on how FC and H_{DA} influence signal production in chemosensor design schemes.

The FC term describes the energetics of the electron transfer event,

$$FC = \frac{\lambda}{(4\pi\lambda RT)^{1/2}}e^{-(\Delta G^o+\lambda)^2/4\lambda RT}, \tag{5}$$

where the overall reorganization energy λ contains inner- (λ_i) and outer-sphere (λ_o) terms, and ΔG^o is the free energy driving force corrected for product work term contributions. The relative sizes of the energies that contribute to the activation energy in Eq. (5) lead to the three regimes defined by Marcus as normal ($|\Delta G^o| < \lambda$), activationless ($|\Delta G^o| = \lambda$), and inverted ($|\Delta G^o| > \lambda$) [189,190]. As schematically diagrammed in Fig. 13, the electron transfer rate constant in the normal regime increases with driving force whereas in the inverted region, the opposite is true.

The most common electron transfer relay mechanism, PET [1–3,5,57], operates through the FC term. PET chemosensors incorporate a fluorophore whose emission is quenched by intramolecular electron transfer from an electron-donor group (e.g., N, S, or O lone pair) of the molecule. Analyte binding is designed to interrupt the PET process. Stabilization of the donor-electron pair by protonation, cation binding, or, in some cases, hydrogen bonding, decreases the overall driving force of the reaction. Because the encumbered donor is more difficult to oxidize PET is inefficient and an enhanced emission from the reporter site is observed.

The effect of the FC term on ICT and MLCT-based chemosensors appears when the electron transfer rate constant is generalized within the context of nonradiative decay theory [191–193]. MLCT excited states are produced directly upon excitation whereas ICT states are produced by a surface crossing from an initially prepared localized excited state (see Fig. 9). Return of the system from the charge transfer excited state to ground state has the overall form of an electron transfer recombination problem that is described by the inverted Marcus curve of Fig. 13. As described by the FC term of Eq. (5), the rate constant for

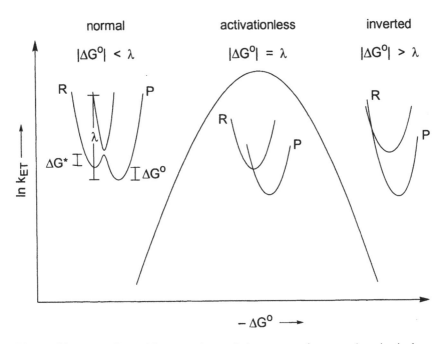

Figure 13   The different Marcus regimes of electron transfer: normal; activationless; inverted.

excited state decay will decrease as energy of the excited state (i.e., the driving force for charge recombination) increases owing to its inverted nature. In the jargon of nonradiative decay theory, k_{nr} (i.e., the recombination electron transfer rate constant) will decrease as the energy of the excited state increases. Per Eq. (1), the charge transfer excited state will exhibit a longer lifetime and greater intensity as its excited state energy increases. This trend in k_{nr} is known as the energy gap law [194–196]. In view of the profound effect that the dielectric constant of the medium can have on the yield and energy of ICT and MLCT excited states, the luminescence intensity from ICT or MLCT reporter sites will depend critically on the polarity and electrostatics of the environment.

An electron transfer relay may also operate via the electronic coupling term, H_{DA}. Mixing between the electron-donor and acceptor orbitals provides the physical basis for electron coupling. In view of the exponential decay of atomic wavefunctions and their corresponding linear combinations, it is not surprising that H_{DA} exhibits an exponential dependence on the distance between the electron-acceptor (A) and donor (D) [197],

$$H_{DA} = H_{DA}^{o} e^{[-\beta(R_{DA}-\sigma)]},$$

(6)

where H_{DA}^o is the coupling between the reactant and product electronic states at contact (i.e., $R_{DA} = \sigma$, the distance between the reactants is equal to the sum of the effective radii of the reacting partners) β is a constant that depends on the "conductivity" of the intervening medium. Electron transfer studies of biological, inorganic, and organic systems in condensed media establish $\beta = 0.82$–1.2 [193,198], although deviations from this range are observed under special circumstances [199]. For the purposes of most chemosensors, $\beta = 1$ is a good approximation. Thus the electron transfer rate constant will decrease by a factor of 10 for every 2-Å increase in the electron transfer distance. This distance dependence provides the impetus for the design of chemosensor active sites possessing donor–acceptor pairs. For illustrative purposes, let's define the donor as an emitting center (i.e., the reporter site). The proximity of the reporter site to acceptor will give rise to a quenched luminescence signal. An analyte that increases the distance between the reporter and acceptor will be detected by an increase in the luminescence signal. Conversely, a decrease in the reporter–acceptor distance upon the molecular recognition of the analyte will be reflected by attenuated luminescence intensity. In either case, chemosensors may be engineered to exhibit high sensitivity owing to the severe distance dependence of k_{ET}.

3. Energy Transfer

A lumophore can transfer energy to an acceptor according to classical or quantum mechanical mechanisms. As with electron transfer, both are sensitive to distance. In the classical, or Förster, energy transfer mechanism (also called fluorescence resonance energy transfer or FRET) [200–202], energy is exchanged by a dipole–dipole interaction. Because the potential energy of two interacting dipoles has the form of $\mu_D\mu_A/R_{DA}^3$ and the probability, and consequently the rate constant (k_{en}), is proportional to the square of the interaction energy, it follows that the distance dependence of Förster energy transfer is $1/R_{DA}^6$. The complete expression for the Förster energy transfer rate constant is [200],

$$k_{en} = \frac{9(\ln 10)\kappa^2\phi_e J}{128\pi^2 n^4 N\tau_o R_{DA}^6}. \tag{7}$$

τ_o and ϕ_e are the luminescence lifetime and quantum yield, respectively, of the donor in the absence of the acceptor, n is the refractive index of the medium, and N is Avogadro's number. Other factors important in determining the overall rate constant of energy transfer include the integrated overlap J between the emission curve of the donor and the absorption curve of the acceptor, weighted by a quadratic factor of the wavelength and the orientation angle between the donor and acceptor dipoles, κ^2.

The Dexter mechanism of energy transfer is quantum mechanical in origin and involves the double exchange of two electrons [201,203–205]. As with electron transfer, exchange requires the direct orbital overlap of the reacting species and thus the overall rate constant takes a form that is similar to the electronic coupling matrix element of Eq. (6) with the caveat that four orbitals are involved instead of two (two frontier orbitals of each of the donor and acceptor) [206]. For this reason, the overall distance dependence for energy transfer scales, all things equal, as $e^{-2\beta R_{DA}}$ as opposed to the $e^{-\beta R_{DA}}$ dependence observed for electron transfer reactions [207],

$$k_{en} = k_{en}^o J' e^{-2\beta(R_{DA}-\sigma)},$$ (8)

where k_{en}^o is the energy transfer rate constant at $R_{DA} = \sigma$ (contact). The spectral overlap J' differs from that of Förster transfer because it is normalized for the absorption cross-section of the acceptor. Hence, Dexter, unlike Förster, energy transfer can occur to orbitally and spin "forbidden" excited states as long as they are energetically accessible.

Energy transfer relay mechanisms are unique in their ability to not only affect k_{nr} (or k_q) of a reporter site but k_r as well. This situation occurs for energy transfer mechanisms effecting a sensitized luminescence as discussed in Sec. III.2. An intra- or intermolecular energy transfer process may produce an emitting excited state when a sensitizer absorbs the excitation light and passes it on to a reporter. In this case, two factors are paramount to the sensitized luminescence intensity: the absorption cross-section (and therefore k_r) of the sensitizer (since it is the absorbing species) and the efficiency of energy transfer. The k_r of the reporter site is of relevance only in determining the efficiency of emission from the reporter site after it has been excited. Accordingly, given an efficient energy transfer process, the small k_r of a weakly absorbing excited state of the reporter is circumvented by exploiting the k_r of the sensitizer's absorbing state.

Sensitized luminescence can be turned to an advantage in chemosensing applications when a reporter site is an efficient emitter but a weak absorber (i.e., k_r is small but k_{nr} is even smaller) and consequently is capable of eliciting only a weak signal under direct irradiation. If the reporter site is at a short distance from the binding site and the analyte itself is strongly absorbing, then an absorption-energy transfer-emission (AETE) process from analyte to reporter will produce a signal. AETE relay has spawned the development of many luminescent chemosensors, particularly those involving aromatic donor analytes and lanthanide ion reporter sites. Other energy transfer relay schemes are also possible. Analytes that affect the energy transfer distance between acceptor and donor sites will produce a measurable change in a luminescence signal. A strongly absorbing center situated at a long distance from the reporter site is the most desirable design element. An analyte that leads to their proximate disposition will establish a sensitized energy transfer cascade from absorbing center to luminescent reporter, thus prompting a light-emitting signal to appear.

IV. BUCKET CHEMOSENSORS

A. Analyte as the Reporter Site

The emission properties of lumophores change when included within the microenvironment of a supramolecule bucket. Nonradiative decay processes are generally curtailed within the confines of the bucket interior and luminescence intensity is therefore increased [138,208,209]. Because CDs present a more protected microenvironment than calixarenes, the binary complexes of the former supramolecule have been examined most extensively. Spurred by Cramer's pioneering observation that the spectral properties of a lumophore are perturbed by complexation within a CD [210], a large body of work has sought to define the influence of CDs on the photophysics of bound lumophores. Different factors contribute to the enhanced luminescence of 1:1 CD:lumophore complexes. These include the following.

1. Quenching processes are hindered or eliminated by physically blocking the approach of quencher to the lumophore residing within the bucket [211–213].
2. Nonradiative decay promoted by hydrophilic interactions [214] (e.g., H-bonding) or other vibrationally coupled interactions with water are obviated [215,216], although hydrogen bonds to the hydroxyls along the rim of the CD bucket may be prevalent [217–219].
3. Nonradiative decay processes that arise from conformational motion of the lumophore may be reduced in the spatially restricted environment of a bucket interior [137,220,221].
4. Hydrophobic lumophores will be solubilized in water by CD; luminescence will therefore be recovered upon the introduction of CD [222,223].
5. Photodecomposition pathways or acid-base equilibria that produce nonemissive forms of a compound may be circumvented [224–226].
6. The interior of the CD bucket offers an environment with a much smaller dielectric constant than that of aqueous solution. The microscopic polarity of the cavity has been estimated to be similar to that of dioxane, 1-octanol, isopropyl ether, and t-amyl alcohol [137]. Hence, compounds that normally exhibit intense fluorescence in organic solvents will exhibit comparable emission intensity from aqueous solutions if CD is present [136,227–230].

With regard to point 6, the inverse will be true for excited states that are stabilized by polar environments (e.g., ICT states) [231]. A molecule possessing a dipole moment that is greater in its excited state than in its ground state will be destabilized upon inclusion in the CD microenvironment. If destabilization is sufficient to allow population of a nonemissive excited state, then an abated luminescence will be observed from the lumophore residing within the CD bucket.

Many investigations have utilized the fluorescence from aromatic hydrocarbons to probe the thermodynamics and dynamics of complexation within the microenvironment of a CD bucket [135,217,231–235]. With the knowledge garnered from these studies, chemosensors have been developed according to Scheme 1 [236], which is unsurpassed for its simplicity [237]. The only requirements for chemosensing to occur are a naturally fluorescent analyte and one that exhibits an affinity for a hydrophobic environment. Chemosensors operating on the principles of Scheme 1 can detect nM concentration levels of organic and pharmaceutical solutes [238–241], toxins [242], drugs [243–245], and environmental contaminants [246–249].

The rims of the CD are often derivatized to improve sensitivity by enhancing analyte association. Hydroxypropyl modification in general has proven useful for the detection of analytes by direct fluorescence and some assays have been successfully developed [250,251]. For instance, Aaron and Coly have used both native and hydroxypropyl β-CD to detect aromatic pesticides [252]. Likewise, Zhang and Gong [253] have detected the biologically relevant substance, tryptamine, by monitoring its fluorescence within hydroxypropyl β-CD. Ethyl modification of the bucket rim was sufficient to allow de Rossi et al. [254] to detect tetracycline by direct fluorescence.

Bissel and de Silva have created a H^+ chemosensor by cleverly augmenting Scheme 1 with a PET relay mechanism. The system design is depicted in Fig. 14. The long-lived, room temperature phosphorescence of bromonaphthalene is quenched by PET from an attached amine moiety (**1**) [255]. Embedded

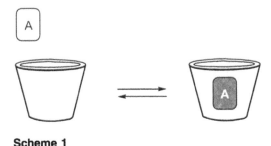

Scheme 1

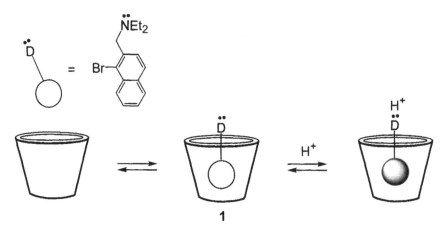

Figure 14   Modification of Scheme 1 to incorporate a PET mechanism for the detection of protons by **1**. Proton binding to amine lone pair of electrons disrupts the PET quenching process to give rise to enhanced luminescence. Within the context of Scheme 1, the protonated diethylammonium bromonaphthyl is the luminescent analyte.

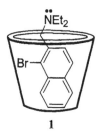

within the cavity of β-CD, the naphthalene is oriented with the hydrophilic amine group jutting outward from the bucket interior. Able to contact the aqueous solvent, protonation of the amine's lone pair at appropriate pHs disrupts PET and phosphorescence from the diethylammonium bromonaphthyl lumophore is recovered.

B.   Ternary Cyclodextrin Complexes

In many cases, the binary complex of Scheme 1 is not preferred and a third species may associate. In the most trivial mechanism, ternary complexation simply serves to modify Scheme 1 by facilitating analyte association, thus improving detection sensitivity. Luong [256] has capped CDs with tetrakis-benzoic acid porphyrin (**2**) to increase the hydrophobicity of the cavity and thus enhance the

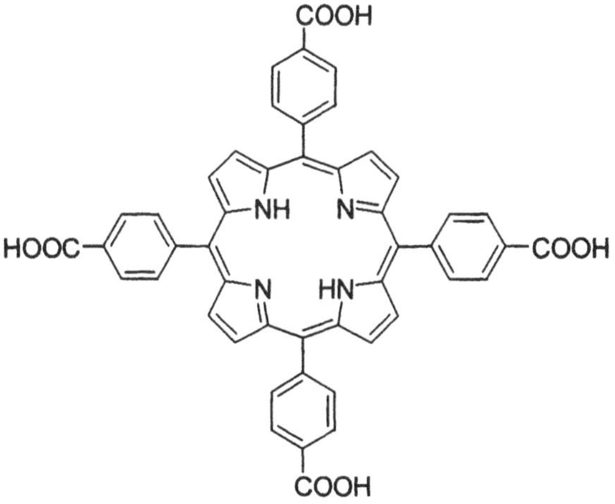

2

complexation thermodynamics of polyaromatic hydrocarbons to the CD bucket. Depending largely on chain length, a coincluded alcohol can alternatively affect analyte association by "wedging" a lumophore into the CD [257–262]. In other instances, the residency of a third species can be used to perturb the emission properties of the reporter site within the bucket.

Guest association with a supramolecule bucket depends critically on the size and shape of the lumophore. Most studies have focused on systems where size mismatches between lumophore and the CD result in reduced stability constants of the binary complex. When the lumophore is too large, 2:1 CD:lumophore complexes will form. When the lumophore is too small, the void space of the CD cavity may be occupied by a third species. Filling the void space with water is thermodynamically unfavorable [135], lending preference to other complexation agents. Alcohols have proven particularly effective in the formation of ternary CD complexes of many polyaromatic hydrocarbons (e.g., pyrene, fluorene, coronene, and naphthalene) [135,257,263–270]. When the ternary complexant is an analyte and the guest is a fluorophore, the chemosensing mechanism of Scheme 2 may be established. To date, the reporter site finding the greatest utility in Scheme 2 is pyrene. The ratio of the intensities of the first and third vibronic components of pyrene's emission band (designated the I:III ratio) increases with the polarity of the microenvironment [271,272]. While somewhat dependent on the experimental protocol, the ratio varies from 0.5 for hydrocarbon solvents to 2 for water. By measuring this variance with the addition of alcohol, the complexation stoichiometry [273–275] and the stability constants [276–278] of ternary complexes comprising CD, alcohol and pyrene may be ascertained.

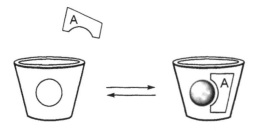

Scheme 2

Complexants other than alcohols (e.g., amines, nitriles, *tert*-butyl compounds, surfactants, and multiple fluorophores) can also be detected by the method of Scheme 2 [279–284]. Bohne and Yang have replaced alcohols with amino acids [285], although the zwitterionic charge distribution and increased steric requirements of the former lead to smaller stability constants of the ternary complex. Notwithstanding, tryptophan, leucine, and phenylalanine produce measurable changes in the I:III ratio of pyrene emission, thereby allowing for their detection at mM levels.

In Schemes 1 and 2, signal is produced by a change in the intensity of analyte luminescence with negligible shifts in wavelength. As highlighted by Fig. 4, measuring the signal against a bright background can complicate detection. This problem is eliminated when a CD imposes a heavy atom (HA) neighbor to the luminescent analyte. A proximate heavy atom can induce intersystem crossing from a singlet to a long-lived triplet excited state. Because phosphorescence is red-shifted from its parent fluorescence, emission may be detected at a distinct wavelength against a dark background. Conversion to the phosphorescent excited state by a spin–orbit coupling mechanism (H_{SO}) has a strong distance dependence [286,287],

$$H_{SO} = \sum \frac{Ze^2}{4m^2c^2} \cdot \frac{1}{r_{ij}^3} l_i s_i, \tag{9}$$

where r_{ij} is the distance between the heavy atom and lumophore and l_i and s_i are the operators for orbital and spin angular momentum, respectively. The bucket architecture provides the spatial confinement required to keep the distance between the heavy atom and incipient phosphor short so that intersystem crossing is efficient and phosphorescence is intense. Cline-Love and coworkers [288–290] and others [255,291–294] have parlayed Scheme 3 into the development of analytical assays for the detection of aromatic hydrocarbons, nitrogen heterocycles, bridged biphenyls, and drugs at subpicogram detection limits.

In some cases, room temperature phosphorescence (RTP) may be achieved in the absence of a well-defined ternary complex. A heavy atom substituent on

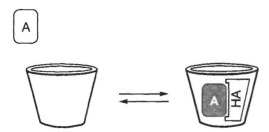

Scheme 3

the rim of the CD [295,296] can induce intersystem crossing within a polyaro-matic analyte residing in the bucket. Analytes external to the CD environment may also promote intersystem crossing. However, in view of the distance de-pendence defined by Eq. (9), the concentration of the external heavy atom must be large. For instance, a 30% w/w β-CD:NaCl mixture must be used [297,298] in the determination of selected aromatic hydrocarbons and their derivatives by RTP.

The inability of CDs to completely protect guests from external quenchers may be turned to an advantage in some chemosensing designs, especially those involving RTP from the reporter site. The long lifetimes of triplet excited states makes them susceptible to deactivation by quenching. For instance, energy trans-fer between the $^3\pi\pi^*$ excited state of 1-bromonaphthalene (1-BrNp) and oxygen is little affected by association of the lumophore with CDs; the quenching reac-tion proceeds with a rate constant near the diffusion-controlled limit. However, barriers may be designed to physically block oxygen from accessing the CD-associated lumophore. This is succinctly demonstrated by the photophysics of regioisomers **3** and **4**, which feature 1-BrNp covalently attached to the primary- and secondary-side of β-CD, respectively [299]. Whereas both molecules flu-

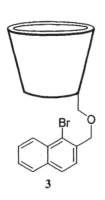

3

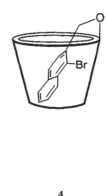

4

oresce at ca. 350 nm with nearly identical quantum efficiencies, the phosphorescence of **4** at 530 nm is $> 10^2$ times more intense than that observed for a solution of **3** at equimolar concentration. The disparate luminescence properties may be ascribed to intramolecular complex formation. The primary side of **3** is too narrow for inclusion of 1-BrNp to occur; exposed, the lumophore of **3** is quenched with efficiencies similar to those observed for 1-BrNp in solution. Conversely, the secondary side of β-CD is wide enough to permit 1-BrNp to enter the CD. The tethering arm is able to protect the electronically excited lumophore from oxygen, thus evincing RTP from oxygenated aqueous solutions of **4**.

A similar phenomenon may be observed even when BrNp is free to diffuse into and out of the CD bucket. Turro et al. have modified 4-BrNp with protecting, (n-alkyl)trimethylammonium detergent groups [300]. Although cationic phosphorescent probes **5** are quenched efficiently by oxygen and $Co(NH_3)_6^{3+}$ in aque-

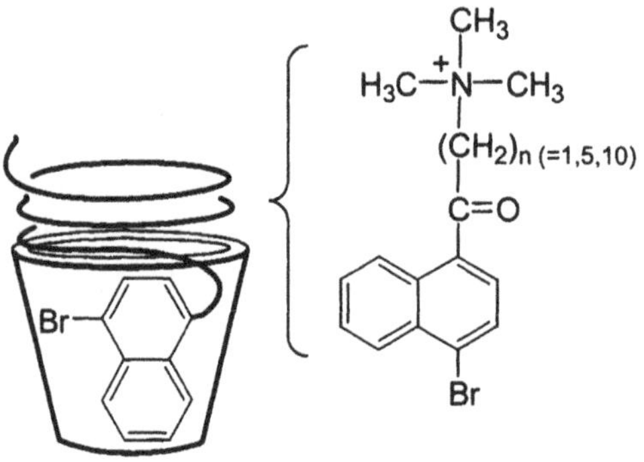

5

ous solution, their inclusion within the CD bucket preserves their intense phosphorescence. The CD-associated BrNp moiety is protected from quencher by the coiled hydrocarbon chain of the detergent appendage. Detailed kinetic measurements established that only the bromonaphthalene moiety includes within β-CD whereas in the case of the larger γ-CD, both the BrNp center and detergent chain may fill the bucket.

When alcohols or other hydrogen bonding analytes assume the protecting role embodied by the aliphatic appendages of BrNp in **4** and **5**, the chemosensing mechanism of Scheme 4 may be implemented. As we have shown, the intensity and lifetime of the green phosphorescence of 1-BrNp significantly increases when alcohols are added to aqueous solutions of CD and 1-BrNp. Exhaustive equilibria, photophysical, and kinetics studies for a series of alcohols [301] reveal that alcohol hydrogen bonds to the rim of the CD bucket. The aliphatic ends of the alcohol flip over the hydrophobic interior of the CD, thus shielding

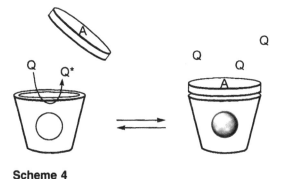

Scheme 4

1-BrNp from oxygen (see Fig. 15). A detailed investigation considering the nature of the hydrogen-bonding group, stereochemistry, and stability constants showed that the best lids of the bucket comprise a bulky *tert*-butyl or cyclohexyl group either directly connected to the alcohol (or amine) or spaced from the functional group by one methylene unit [302]. Enhancements approaching 10^5 are observed for the best lids. This approach has been extended to other analytes using the same [303] and different reporter sites [263,267].

The strategy of Scheme 4 may be extended to explore new directions in sensing that extend well beyond the world of chemicals. We have invented a new technique called molecular tagging velocimetry (or MTV) [304–307], which allows us to physically sense flow by precisely describing the velocity profile of a fluid. In the MTV experiment, the ternary complex is dissolved in water and a grid of laser lines is imposed upon a flow of interest by shining a laser beam off a grating. A glowing grid is defined by the luminescence from the ternary complex. The 10-ms luminescence lifetime of the ternary complex allows the grid to convect with flows moving as slowly as 0.02 m/s. The deformation of the grid as it moves with the flow is recorded with a CCD camera. By measuring the distance and direction each grid intersection travels and knowing the time delay between each image, the two velocity components in the grid plane may be determined (the out-of-plane velocity can be determined with a second CCD camera). Parameters important to the fluid physicist such as turbulence intensities, the Reynolds stress, and vorticity may be calculated. The solubility of the ternary complex in water has enabled us to use the MTV technique to specify fluid flows with Reynolds numbers that can be achieved in water tanks and tow facilities. Quantitative measures of swirl and tumble within the cylinder of an engine (rotational flow about and perpendicular to the cylinder axis, respectively) [308], two-phase liquid/solid flows [309,310], and other flow issues of concern at boundary layers of solid [311,312] and aerodynamic surfaces (e.g., a rotating airfoil [313]) have been characterized.

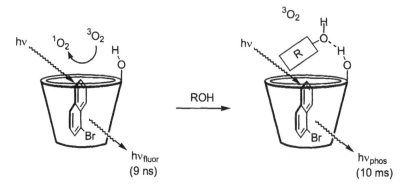

Figure 15   Addition of alcohols to a complex formed between reporter and CD provides the necessary protection to avoid O_2 quenching of room temperature phosphorescence from the reporter site. The scheme describes the operation of chemosensors described in Refs. 301 and 302.

C.   Displacement of a Reporter Site by Analyte

Some reporter sites undergo large changes in emission intensity and energy upon their displacement from the interior of a CD bucket. Prominent among these is 8-anilinonaphthalene-1-sulfonate (ANS). As compared to the modest spectral shifts of simple aromatics, ANS offers excellent signal sensitivity by displaying a ~20-fold increase in emission intensity when complexed to a CD bucket. Solvent relaxation, primarily arising from reorientation, and variations in intersystem crossing rates have been identified as the primary factors governing changes in ANS fluorescence in different media [314], although charge transfer mixing is also likely to be an important contributing factor [315–317]. Because ANS fluorescence is dramatically attenuated upon displacement of the probe from the CD interior, this system is routinely used to measure the association constants of guests to CDs in a competitive binding assay [318,319]. The overall strategy may easily be extended to allow implementation of Scheme 5. Added analyte displaces ANS from the CD bucket whereupon its emission is quenched. Aoyama et al. have exploited Scheme 5 to examine the association of sugar–oligosugar interactions in aqueous solution [320]. The measured binding constants reveal selectivity for aldopentoses; within this series of analytes, the contribution of hydrophobic interactions to sugar–oligosugar interactions is clearly apparent. Hydrogen bonding of sugar hydroxyl groups to those of CD and water provides an additional mechanism to control association and consequently tune the selectivity of the chemosensor. The ability to distinguish among sugars with extremely small binding constants (1 to 10 M^{-1}) highlights the benefits gained from choosing a reporter site that exhibits large changes in fluorescence intensity upon analyte binding.

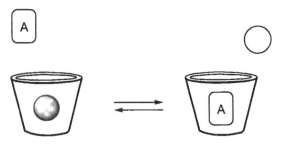

Scheme 5

D. Displacement of a Covalently Tethered Reporter Site by Analyte

Expanding on Scheme 5, a reporter site covalently attached to the bucket affords greater design flexibility since the presence of the tether prevents loss of the reporter site upon analyte recognition. The luminescence signal may increase (as shown in Scheme 6), decrease, or wavelength shift upon its displacement of the reporter site from the bucket. Analyte need not absorb or emit light, as it is not involved directly in the transduction process.

Ueno has demonstrated the utility of Scheme 6 with an assortment of reporter sites (vide infra), although dansyl (6) has been especially prominent in his studies. The underlying mechanism for luminescence signaling by a dansyl fluorescent probe arises from the considerable mixing between naphthyl $\pi\pi^*$ excited states (^1L_a and ^1L_b) and charge transfer states involving the lone pair of electrons on the nitrogen. A polar environment decreases the energy separation between the two excited states and a red-shifted but weak emission is observed owing to enhanced nonradiative decay promoted by solute–solvent vibrational relaxation [321]. Accordingly, detection is accomplished by monitoring a decrease in the fluorescence intensity of the dansyl probe upon analyte recognition.

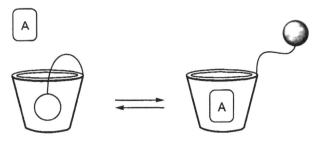

Scheme 6

The competitive association of analyte for a dansyl-included CD determines the overall sensitivity of the chemosensor. 1-Adamantanol and cyclododecanol engender significant responses from **6** (γ-CD), but smaller alcohols, such as (–)-borneol, geraniol, nerol, cyclooctanol, and cyclohexanol, give less sensitive responses. The same is true for nonfluorescent steroids such as deoxycholic acids [322]. This class of analytes fits most snugly into **6** (β-CD); association constants in excess of 10^6 M^{-1} are observed. A CD modified with the monensin cap increases the sensitivity of the chemosensor to various alcohols by increasing the hydrophobicity of the CD bucket [323]. For the same reason, a biotin cap for a dansyl bis-modified β-CD has aided in the detection of bile acids [324]. However, the addition of avidin to further modify the hydrophobic environment of the chemosensor led to poorer signal responses. Apparently, avidin protects the dansyl probe from the aqueous environment even when the reporter is displaced from the CD bucket. Because dansyl fluorescence is preserved in hydrophobic environments, the change in fluorescence for dansyl inside and outside the CD bucket is slight, thus giving rise to the inferior signal response from the avidin-treated chemosensor.

Modification of the tether will alter the properties of dansyl-derivatized CD chemosensors. For instance, glycine-linked dansyl, **7**, exhibits particular sensitivity to selected steroids although the origins of this selectivity have yet to be clarified [325]. Chemosensor performance is improved when glycine is replaced by leucine. Differences in enantiomeric configuration of the leucine engender different molecular recognition properties. NMR and binding studies of **8** show that the insertion of a leucine moiety between CD and dansyl improves the binding sensitivity of the complex to steroids relative to a complex possessing a glycine linker. In the case of **8** (β-CD), a decrease in dansyl intensity indicates that the probe molecule is displaced from the CD cavity upon analyte recognition. A dansyl moiety appended to a D-leucine tether is more deeply included within the CD than a dansyl tethered by L-leucine; consequently, higher sensitivities are observed for the former because the more protected dansyl possesses a more intense initial luminescence signal [326]. Dansyl emission of **8** (γ-CD), on the other hand, can increase or decrease depending on whether the guest enhances probe association or probe displacement from the larger cavity. Modification of the L-leucine chain serves to diminish sensitivity by stabilizing the association of the reporter to the CD bucket [327]. Thus, cavity size and the chirality and functionality of the tether can be manipulated to enhance detection sensitivity and selectivity.

A more detailed study of tether chain length has been undertaken with the ether- and amide-linked dansyls of **9** through **11** [328]. In the case of **9**, the dansyl tether is too short to allow for dansyl inclusion and the reporter is relegated to the outside of the CD bucket. No appreciable difference in fluorescence is therefore detected upon the recognition of the analyte by **9**. For **10**,

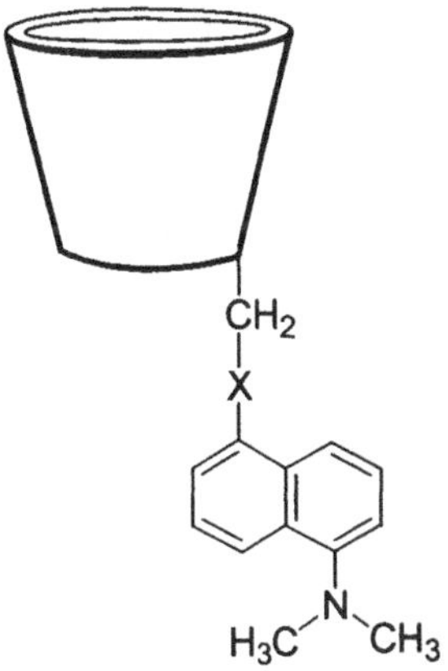

CH$_2$

X

H$_3$C$^{\diagdown}$N$^{\diagup}$CH$_3$

X =	X =
6 $-SO_3^-$	9 $-SO_2NH-$
7 $-SO_2NHCH_2CONH-$	10 $-SO_2NH(CH_2CH_2O)_n-$ $\quad$ n = 1, 3
8 $-SO_2NHCHCONH-$	11 $-SO_2NH(CH_2CH_2NH)_n-$ $\quad$ n = 1, 2
$\quad$ CH$_2$	
$\quad$ HC(CH$_3$)$_2$	

the longer tether promotes strong inclusion of the dansyl group within the CD. But dansyl association to the bucket is too strong and few analytes are able to displace it. Only when the pH of the solution is lowered, and the dimethylamino group of dansyl is protonated, can the reporter site be displaced by analyte and chemosensing function is thus achieved.

Corradini et al. have substantiated the mechanism of Scheme 6 by obtaining the crystal structure of **11** (n = 1, β-CD) [329]. The dansyl group fully encapsulates within the CD bucket, thus explaining the high fluorescence quantum yield of this chemosensor. Two-dimensional NMR ROESY studies have revealed that the length of the spacer affects the orientation of the dansyl lumophore within the CD cavity. The shorter spacer induces axial complexation and the longer spacer, allowing for more rotational freedom, induces equatorial complexation [330]. This difference in conformational flexibility of the congeners leads to very different chemosensing signals when a metal ion such as Cu^{2+} is present. The more highly constrained n = 1 spacer does not permit interaction of the amine nitrogens with Cu^{2+} and the fluorescence properties of the chemosensor is little perturbed by the presence of the metal ion. However, for the more conformationally flexible n = 2 spacer, a chelate ring is able to accommodate Cu^{2+} and dansyl fluorescence is almost completely quenched by the redox-active and Lewis acidic Cu^{2+} ion. Dansyl fluorescence was recovered when analytes, exhibiting an affinity for Cu^{2+} ion, were introduced into solution. The triggered luminescence response was especially pronounced for tryptophan and alanine, which permitted these amino acids to be conveniently detected with the Cu^{2+} complex of **11**.

Other fluorescent reporter sites exhibiting an intensity variation inside and outside the CD bucket include naphthalene, anthracene, and anthranilate. Changes in the fluorescence intensity and energy are observed for these reporters upon their displacement from the interior of β- or γ-CDs (**12** through **14**) [331–333]. The origins of these effects remain ill defined. Pyrrolinone and terphenyl reporter sites have also been appended to CD [334,335].

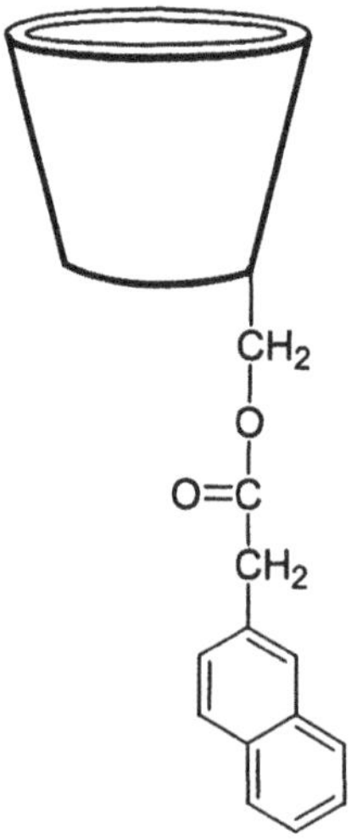

12 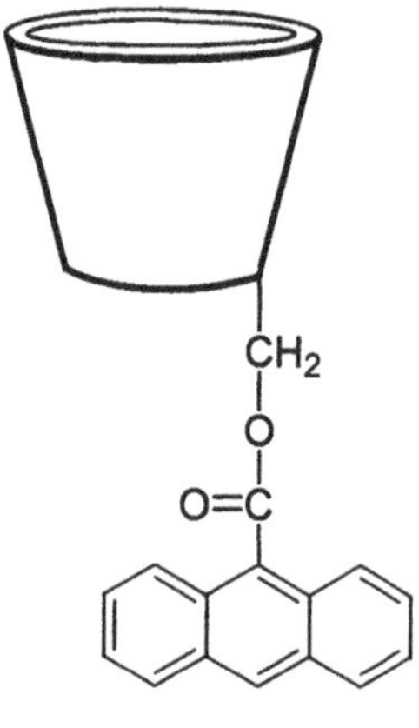   **13** 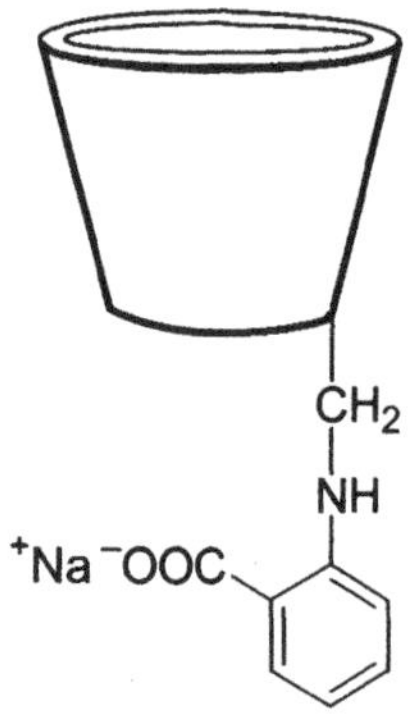   **14**

Coleman et al. [336] and Inoue et al. [337] have shown that analytes may be detected according to Scheme 6 using the peptide reporter sites of **15** and **16**. To understand the consequences of multiple probes in Scheme 6, Hamada et al. have placed bis-anthranilate moieties at various positions around the primary rim of the AB–AE glucose subunits of γ-CD (**17**) (and AB–AD glucose subunits of β-CD) [338]. Only monomer fluorescence (and no excimer emission) is observed for each of the regioisomers. Yet, substantial changes in the fluorescence intensity depend on the size of the analyte and the position of the reporter sites about the CD rim. The observed effect is due in part to one reporter site functioning as a hydrophobic cap of the bucket, thus enhancing association of its reporting rim partner with the interior of the CD bucket. Hamada et al.

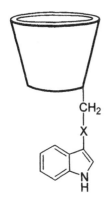

15 X = –NHCOCHCH₂–
$$X = -NHCOCHCH_2-$$

with NHAc on the substituent.

15 $X = -NHCO\overset{\overset{\displaystyle NHAc}{|}}{C}HCH_2-$

16 $X = -NH\overset{}{C}HCH_2-$ with $\overset{|}{COOH}$

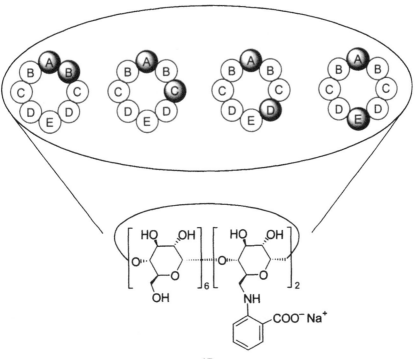

17

have elaborated their approach by placing two different fluorescent probes about the rims of β- and γ-CD [339–342]. Introduction of dansyl and tosyl groups in **18** alters and improves the sensing response as compared to the bis-dansyl

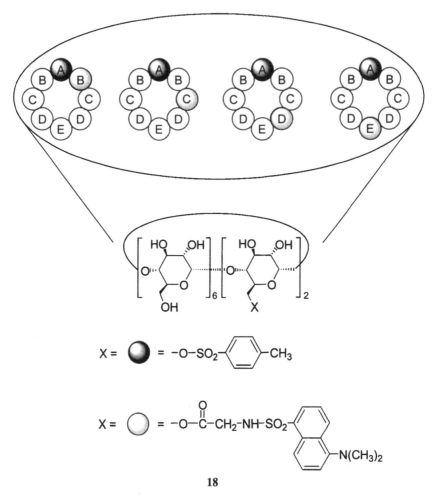

18

modified CD. Apparently, cooperatively between dansyl and tosyl groups serves to enhance analyte binding by a mechanism that has yet to be clearly elaborated.

The complexity of factors that contribute to the observed spectral shifts of Scheme 6 has been expounded by Garcia-Garibay and McAlpine in their studies of CDs appended with naphthalene reporter sites [343]. Spectroscopic investigations suggest **19** through **22** to contribute to the overall fluorescence signal of the chemosensor. Quantification of the analyte is difficult because each of the

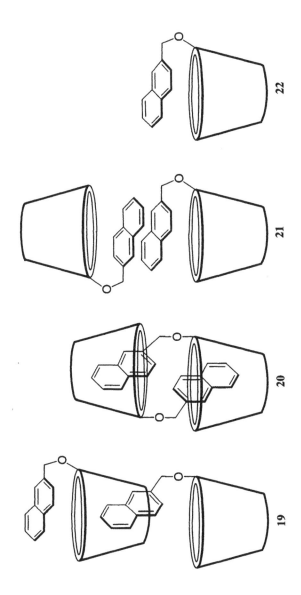

complexes exhibits different stability constants and different emission quantum efficiencies. Supramolecules **19** through **22** highlight the oversimplification of Scheme 6 and underscore the many different structures that are present for a seemingly simple chemosensor construct. The results of this study have important ramifications for future chemosensor investigations because they challenge the assumption that a fluorescence signal is produced from a single complex of known emission quantum efficiency.

The significant red-shift in the emission signal for a TICT reporter site external to the CD bucket makes them particularly attractive for implementation in Scheme 6. The appearance of the spectrally distinct luminescence signature facilitates signal detection because emitted photons are detected against a dark background. Numerous reports show that inclusion in a CD bucket enforces a normal planar conformation and consequently the intensity ratio of blue (NP) and red (TICT) emissions from probes is perturbed significantly upon inclusion of the probe within a supramolecule bucket [229,344–348]. CDs modified with p-(dimethylamino)benzoyl (DMAB) probes have been employed to detect numerous alcohols, acids, and steroids and to examine the stability of TICT probe inclusion complexes [349–351]. Ueno et al. have linked DMAB via an amide linkage to the primary rim of α-, β-, and γ-CD (**23**) [352,353]. The three

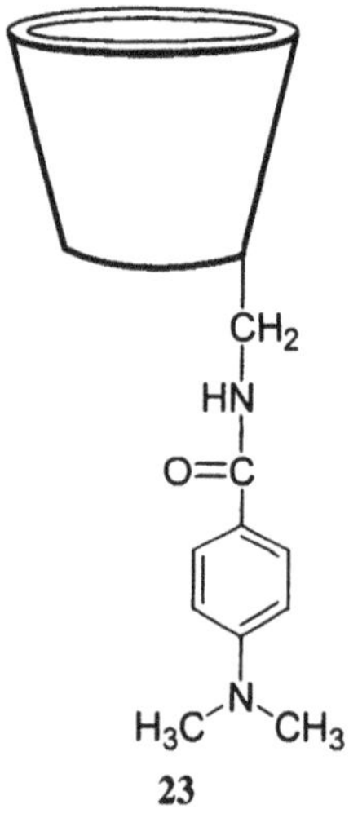

23

chemosensors exhibit different responses owing to variation in the association of the reporter site with the differently sized buckets. Minor changes in the polar environment of DMAB is suspected to account for the small differences observed between the relative intensities of NP versus TICT emission [354]. For α-CD, the bucket is too small to include DMAB, which is always located outside the bucket. In the case of γ-CD, the reporter site and small analytes such as alcohols are accommodated in the bucket. The DMAB becomes wedged within the CD and a TICT emission signal is muted since the reporter site is difficult to displace upon the addition of the analyte. The β-CD modified chemosensor displays the best chemosensing function of the triumvirate. As confirmed by circular dichroism and ^{1}H NMR studies [353], β-CD presents the ideal bucket dimension to accommodate the inside to outside conformational change required by Scheme 6. TICT reporter sites other than DMAB and its derivatives [355–357] have not been extensively utilized for Scheme 6.

Calixarenes modified with a reporter site are equally competent chemosensors for the detection of a variety of analytes according to Scheme 6, especially when the analyte is cationic. The π-cavity of a tetraanionic resorcin[4]arene readily binds cationic guests, a feature that Inouye et al. have exploited in the development of an acetylcholine chemosensor [358]. The fluorescence from a pyrene-modified N-alkylpyridinium is strongly quenched upon its association with the tetraphenolate form of resorcin[4]arene (**24**). This quenching is consis-

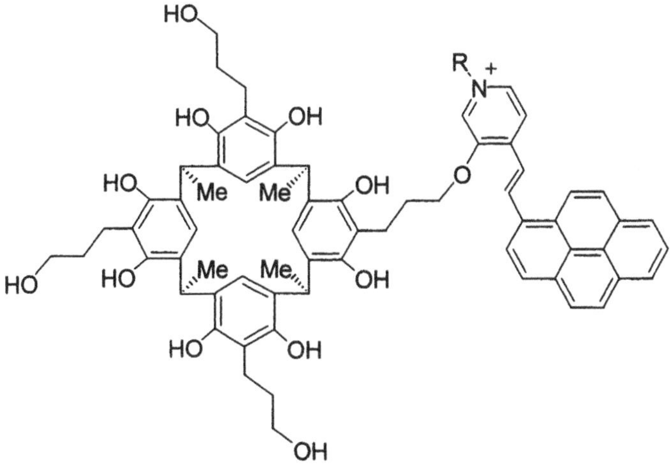

24

tent with PET from the associated anionic phenolates to the N-alkylpyridinium, which is a strong oxidant. As depicted in Fig. 16, acetylcholine displaces the reporter site from the cavity and the strong fluorescence of the pyrene lumophore is recovered due to the long distance of electron transfer between the phenolates and pyridinium. No other neurotransmitters trigger a fluorescence signal from **24**, engendering the chemosensor selective for acetylcholine. One drawback of **24** is the inability of the chemosensor to operate at neutral and acidic pHs. The deprotonated phenol, which is required for the strong association between the cationic

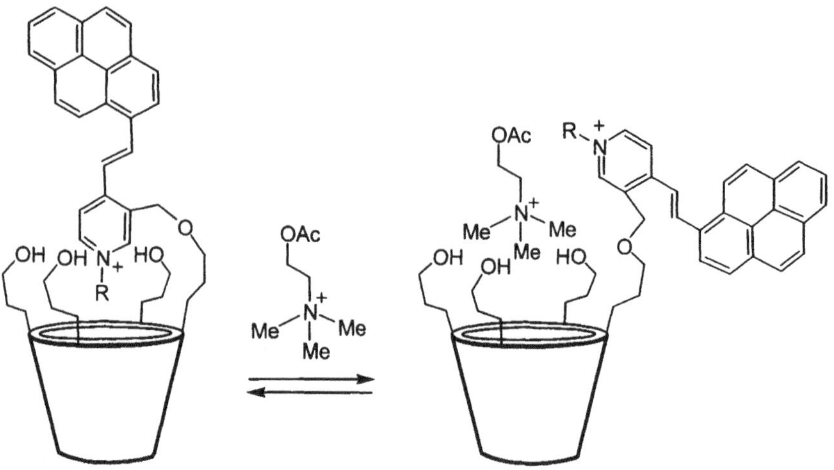

Figure 16 Acetylcholine analyte displaces a pyrene reporter from the calixarene receptor site of **24**. Strong fluorescence from the displaced pyrene is observed because of the increased distance of electron transfer from the calixarene to pyrene fluorophore.

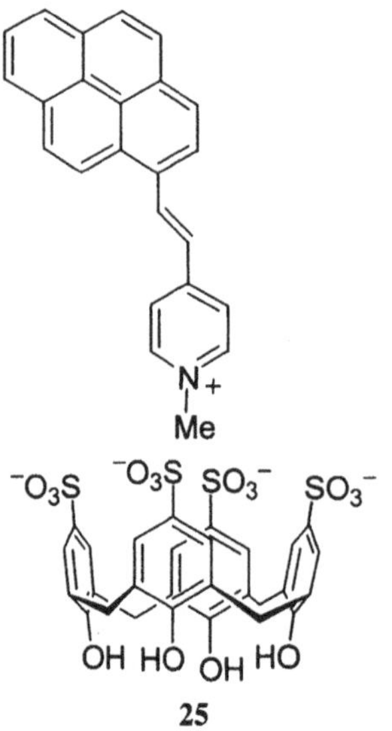

reporter site and the neurotransmitter, is prevalent only in basic solutions. With the goal of detecting neurotransmitters at physiological pH, Shinkai et al. have replaced the hydroxyl groups on the resorcin[n]arene rim (n = 4, 6) with sulfonates [359]. This modification increases the solubility of chemosensor **25** in water and provides a preformed anionic platform for the docking of the analyte and the reporter site under neutral conditions. Quenching of the reporter is preserved with sulfonate modification so the chemosensing function is maintained. The pyrene-modified N-alkylpyridinium yet remains to be covalently attached to the resorcin[n]arene (n = 4, 6) rim; a tethered reporter site will permit the practical implementation of **25**.

Reinhoudt et al. have coupled β-CD with a probe-modified calix[4]arene to detect neutral analytes in aqueous solutions [160,161]. The fluorescence capabilities of the multireceptor chemosensor arise from dansyl (**26**) and 2-naphthylamine (**27**) fluorophores attached to the rim of the calixarene. Like

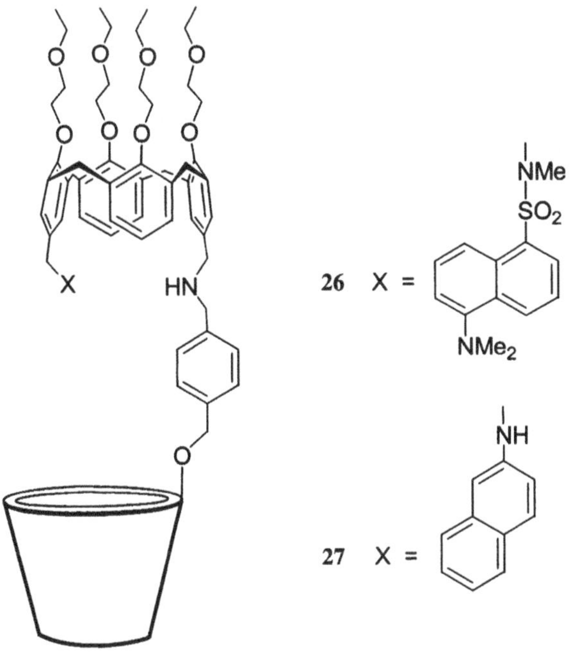

previous systems, fluorescence quenching is observed upon probe displacement from the CD cavity. The reporter sites, although separated from the CD by a calixarene spacer, are still able to associate with the CD bucket with a specific binding strength. In addition to its function as a scaffold for the fluorescent probe, the calixarene also alters the complexation properties of β-CD by expanding the dimensions of the hydrophobic cavity for the recognition of larger analytes. This was proven by examining the binding properties of noncovalently bound ANS and TNS fluorophores to the same host molecule [360]. In these studies, it is important to choose a probe that can competitively bind with a specific analyte. For instance, **26** and **27** detect steroids and terpenes. Whereas a naphthylamine moiety is easily displaced from the cavity by ethinyl-nortestosterone at μM concentration, a dansyl probe binds too strongly to permit interaction between neutral aromatic analytes and the CD.

E. Excimer and Exciplex Reporter Sites

The CD architecture is particularly fitting for excimer-based reporter sites. As long as the bucket is sufficiently large to accept two guests, monomers will cofacially organize within the interior of a CD to form excimer. Early spectroscopic studies revealed an enhanced excimer emission upon the addition of γ-CD to solutions of naphthalene [361], a result that was subsequently observed for anthracene [362] and pyrene [363,364]. Pyrene excimer formation has been examined in particular detail. Proper concentrations of pyrene and γ-CD afford 2:1 or 2:2 excimer inclusion complexes. In both cases, pyrenes interact in a cofacial manner and excimer efficiently forms. Alcohol addition can disrupt the excimer by the preference to form a 1:1:1 complex of alcohol, pyrene, and CD.

Pyrene derivatized with a benzo-15-crown-5 ionophore (**28**) provides an alternative mechanism for Yamauchi et al. to preorganize pyrene subunits [365].

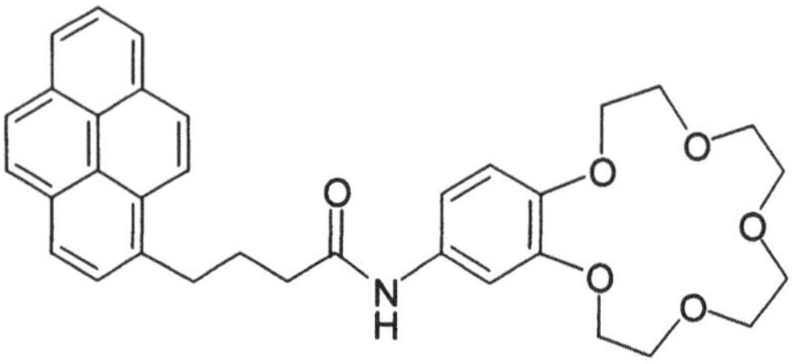

28

It is believed that two crowns bind a K^+ ion, thus leading to loosely preassembled bis-pyrene subunits. When the metal ion recognition event occurs in the presence of γ-CD, dramatic enhancements of excimer emission are observed, thus providing a method to selectively detect K^+ ion in aqueous solution. This ability to sense an ionic analyte is unusual for CD chemosensors. Modification of **28** with other ion binding sites should provide a general method for cation detection.

The observation of CD-promoted excimer formation prompted Ueno and coworkers to prepare γ-CDs monofunctionalized with pyrene (**29–31**) [366,367]. Attached to the primary rim of the CD bucket, bright excimer emission is observed for **29** and **30**, suggesting the formation of a complex in which the appended pyrenes are cofacially disposed (see Scheme 7). The formation of the bis pyrene:CD complex in Scheme 7 is supported by absorption, emission, and circular dichroism spectroscopy. A guest responsive decrease in excimer

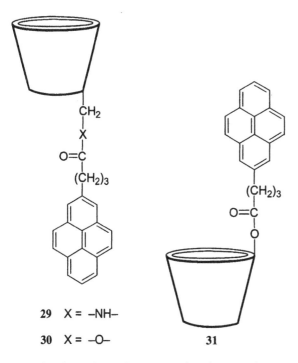

29   X = –NH–

30   X = –O–   31

emission intensity and correspondent increase in pyrene fluorescence signals the breakup of the association dimer into a 1:1 complex comprising modified CD and guest. Accordingly, analyte detection may be accomplished by the strategy of Scheme 7, where ovals represent the monomers of the excimer pair. The decrease of excimer fluorescence and attendant increase of monomer fluorescence reflects the affinity of the guest for the γ-CD bucket. On this basis, a host of

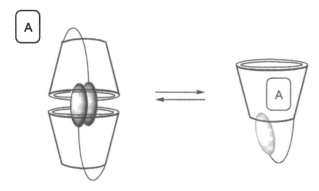

Scheme 7

organic compounds, including alcohols, amines, steroids, and other biologically important molecules may be detected by intensity ratioing the two fluorescent signals. As expected, **30** is an inadequate chemosensor when the bucket is β-CD because the cavity is too small to accommodate two pyrenes. The attachment of pyrene to the primary rim of the CD bucket in **30** also appears to be crucial to the chemosensing function of Scheme 7. Excimer emission is markedly depressed from **31**, which has pyrene attached to the secondary rim of a γ-CD bucket. It should be noted that the shortened tether length (by one methylene unit) of **31** as compared to **30** might also contribute to the decreased stability constant of the excimer complex of Scheme 7.

The utilization of monosubstituted CDs in Scheme 7 presents the drawback that excimer formation depends not only on the concentration of analyte, but also on the bimolecular assembly of the reporter site. The overall signal response of chemosensors of this type therefore depends on the concentration of the analyte *and* receptor. To alleviate the complication of assembling the reporter site, Ueno and coworkers have appended two monomers to the rim of the CD bucket. The preferred design entails the inclusion of one monomer in the CD bucket. Analyte recognition frees the included monomer, enabling it to interact with its partner tethered to the external rim of the bucket.

Displacement of monomer from the interior of the CD bucket in Scheme 8 does not ensure that excimer will form. Chemosensor design must consider the length of the tethers, points of monomer attachment to the glucosyl subunits, and the size of the bucket. For example, the γ-CDs bearing two naphthyl monomers attached to the (A,E) glucose positions via the long tether of **32** exhibit excimer emission in the absence of analyte [368]. No significant change in the excimer profile is observed upon the 1:1 complexation of analyte, which was independently confirmed by circular dichroism absorption spectroscopy. The re-

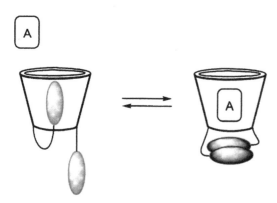

Scheme 8

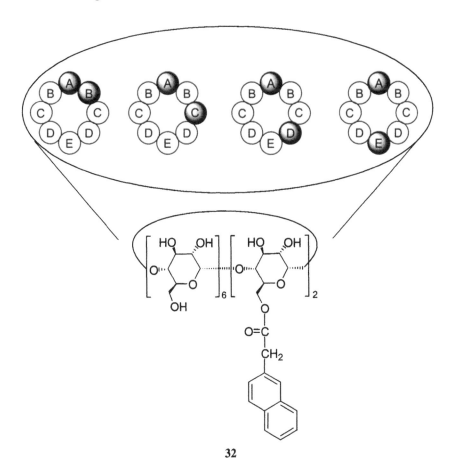

32

sults suggest that excimer is formed inside and outside the bucket, obviating the utility of **32** as a chemosensor. The shortened linkers of **33** and **34** on the other hand prevent the association of two monomers within γ-CD, and accordingly these chemosensors exhibit an analyte-induced change in the intensity of monomer and excimer emission, thus providing a basis for detection of various analytes by Scheme 8 [369–371]. In these studies, γ-CD bearing monomers at (A,B), (A,C), (A,D), and (A,E) positions permit the importance of monomer regiochemistry on excimer formation to be evaluated. Of course, the problem associated with initially preorganized excimers within the CD bucket is completely circumvented when β-CD is employed [369,372], since the smaller size of the cavity precludes occupation by two monomers, even when the tether is long.

Osa et al. have further elaborated the disubstituted CD construct by placing two naphthyl molecules on the same tether [373]. The branched arm of

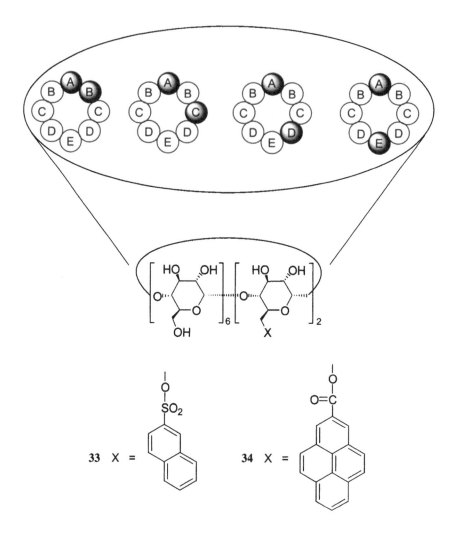

33 X =   34 X =

35, however, was found to be an ineffective chemosensing design element. Apparently, one naphthyl enters the bucket with the other naphthyl acting as a hydrophobic cap. The stability of this complex makes it difficult for analyte to displace the included monomer from the CD bucket. Thus, significant excimer emission from **35** is not observed upon the addition of analyte.

Valeur et al. have modified every glucosyl subunit of the primary rim of β-CD with naphthoate donor groups [374]. At pH > 8, the emission spectrum of **36** is composed of naphthoate monomer and excimer fluorescence, the ratio of which readily allows surfactant concentrations to be ascertained. Ionic strength and competition binding experiments strongly point toward a micelle complex rather than a 1:1 inclusion complex. Aggregates between **36** and cationic sur-

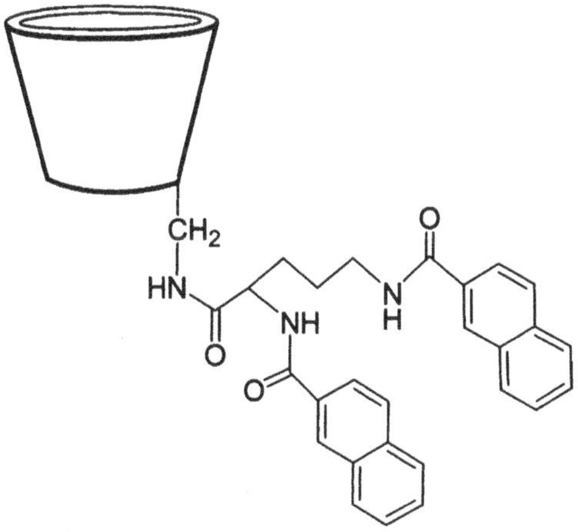

35

factants form with large stability constants, which arise from the large negative charge of the bucket. The advantage of using CD as a scaffold for the naphthoate reporter sites is apparent when referenced to control studies. Luminescence from a simple methoxy-substituted naphthoate is little affected by the presence of surfactants. In the case of redox-active surfactants, such as cetyltrimethylammonium, quenching of the naphthoate fluorescence by PET provides an additional signal transduction mechanism for the detection of surfactant. The high affinity of cationic surfactants for **36** enables their detection to µM concentration levels.

Calixarenes may also bear excimer reporter sites, although the details of signal transduction are inherently different from those of CD-based excimer chemosensors. First, unlike Schemes 7 and 8, the fluorescent monomer units are not included within the calixarene bucket. Rather, calixarene functions as only a scaffold, bearing monomer subunits. Second, analyte seldom enters the calixarene bucket. The most common design for this type of chemosensor is shown in Scheme 9. A preformed excimer is disrupted by analyte binding at the

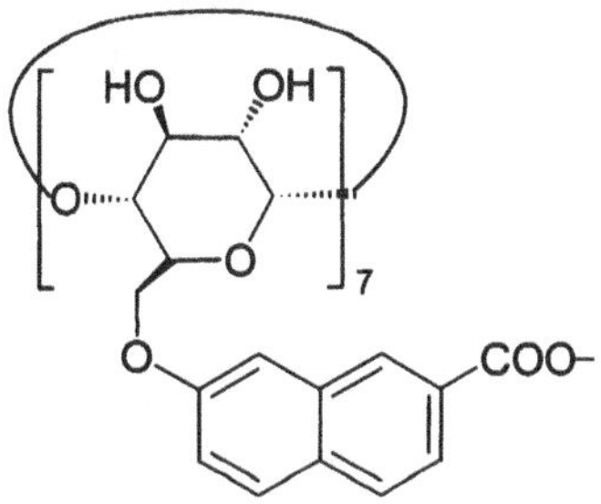

36

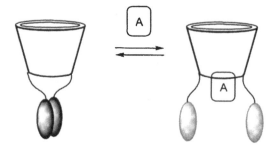

Scheme 9

oxygens around the rim of the bucket. Third, in contrast to the conformationally restricted CDs, the phenyl rings composing the bucket walls of calixarenes can rotate about their connecting arms to give rise to the cone, partial cone, and alternate isomers of Fig. 12. Some of these isomers are flattened considerably. These different conformations can be exploited to enhance or suppress interactions between monomer subunits attached to the rims of the calixarene [375].

Shinkai and coworkers have capitalized on this conformation flexibility in their designs of several pyrene-derivatized calixarene chemosensors. The rotated phenyl ring of the partial cone conformer of **37** allows two pyrene units to more easily interact by decreasing steric hindrance at the lower rim [376]. Addition of Li^+, Na^+, and K^+ ions enforces cone formation and the disruption of the initially formed excimer. Accordingly, the metal ions are detected by a decrease in pyrene excimer emission and concomitant increase in the pyrene fluorescence.

Following this same approach, Jin and coworkers have prepared a nearly identical pyrene-modified calixarene, **38**, where the methoxy groups are replaced by ester functionalities [377]. Na^+ and K^+ ions perturb the relative intensities of excimer and monomer emission in much the same manner as **37**. Unlike **37**, however, **38** is not sensitive to Li^+ ion.

Shinkai et al. have further examined conformational effects by replacing the methoxy groups on the phenyl rings with larger propyl derivatives (**39**). When the R group is made large enough, rotation of phenyl subunits is prevented and cone formation is preserved upon the addition of analytes. Despite the conformational restriction imposed by the larger propyl functionality, Shinkai et al. report that the emission spectra of both **37** and **39** are similar. Association of hydrogen-bonding analytes such as trifluoroacetic acid with the ether and ester oxygens of the tethering appendages disrupts excimer formation; thus, **37** and **39** may be used to detect a variety of acids [378].

The unusual C_3-symmetric calix[3]arene **40** has been found to selectively recognize primary ammonium ions [379]. In this instance the symmetry was

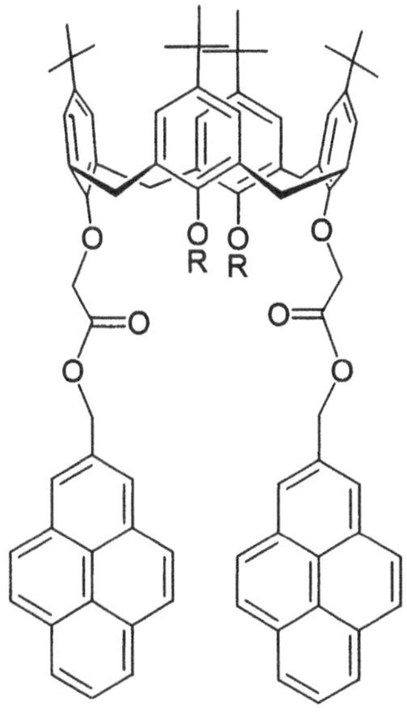

37   R = Me

38   R = CH$_2$CO$_2$Et

39   R = CH$_2$CH$_2$Me

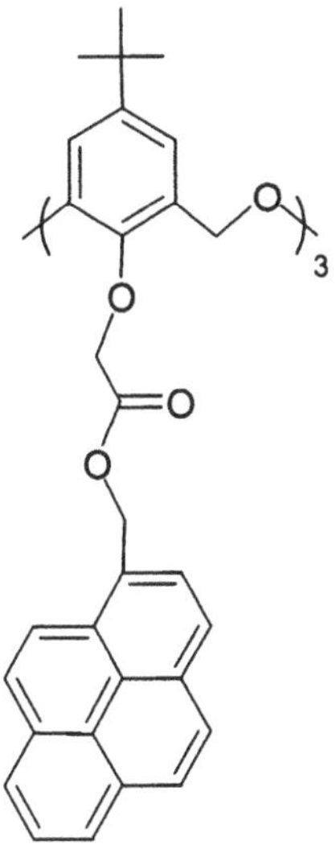

40

presumed to play a role in the complexation stability by optimizing the hydrogen bonding interactions between analyte and host. Association with the cone isomer is four to five times greater than that of the partial cone isomer, a result of the greater number of hydrogen bonding interactions presented by three pyrene ester arms on the same side of the calixarene rim.

The facile functionalization of calixarene rims with ionophores allows for the successful discrimination of different metal ions. Matsumoto and Shinkai have prepared calix[4]crown **41**, which bears a crown ionophore on the lower rim of a calixarene and two pyrenes on the upper rim [380]. In the absence of metal ions, the phenyls bearing the pyrenes assume a parallel disposition and an excimer forms. A metal ion residing in the crown-binding site pulls the ether oxygens of the pyrene-bearing phenyl groups inward. Rotation of the phenyl rings towards the interior of the bucket bottom causes the pyrenes on the upper rim to splay outward, thereby disrupting the excimer. Scheme 10 depicts the overall mechanism for analyte detection by **41**. This inventive chemosensor achieves ion selectivity through the size of the crown ether-binding site, which is large enough to accommodate Li$^+$ and Na$^+$, but not K$^+$.

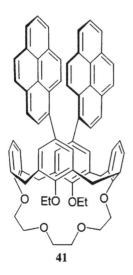

41

F. Supramolecules That are Rigidified by Analyte Association

The conformational flexibility of several bucket chemosensors is reduced upon analyte binding. In the rigidified state, the luminescence from the reporter site is enhanced. Scheme 11 pictorially represents the transduction process. The precise mechanism by which nonradiative decay is disrupted in chemosensors of this type is usually ill defined.

Many chemosensors designed by Shinkai and coworkers are based on modification of the lower rim of a calix[4]arene with a fluorophore, whose emission properties change upon association of alkali metal cations. Fluorescence from the biphenyl subunit of **42** increases markedly upon the association of Na$^+$

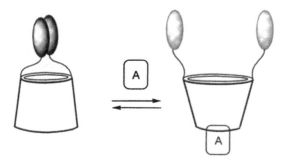

Scheme 10

42   43

with the tetraester-binding site of the lower rim [381]. Comparison to model compounds suggests that "rigidification" of the calix[4]arene subunit suppresses molecular motion that is detrimental to fluorescence. In the design of **43**, Shinkai and coworkers use benzothiazole as the reporter site [382]. Whereas the 391-nm fluorescence band is little affected by the presence of Na^+, K^+, Rb^+, and Cs^+ ions, Li^+ ions induce the appearance of a new fluorescence band at 422 nm. ^1H NMR experiments show that a bound Li^+ ion rigidifies the calix[4]arene in the partial cone conformer.

Diamond and coworkers attribute an increase in fluorescence from the anthracene reporter sites of **44** to the increased rigidity induced by complexation of Li^+, Na^+, and K^+ to the calixarene's tetraester cleft. The tetraamide derivative, **45**, shows an especially selective response to Na^+ ion [383]. Restricted motion of the calix[4]arene is believed to lead to the enhanced luminescence response. This contention is supported by ^1H NMR studies, which show metal ions to confer significant order on the calix[4]arene receptor.

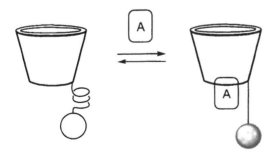

Scheme 11

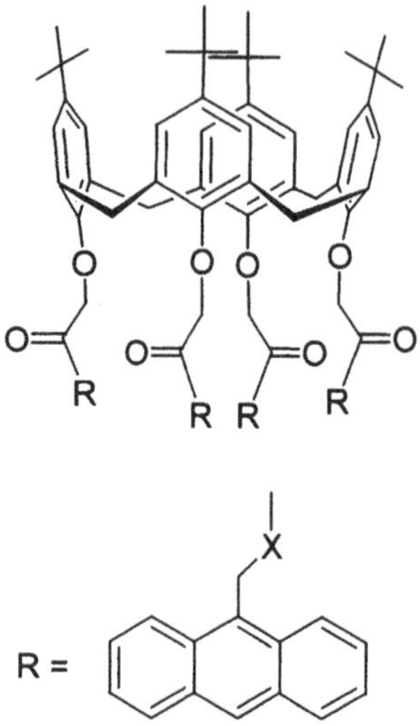

44   X = –O–

45   X = –N–
                |
              CH₃

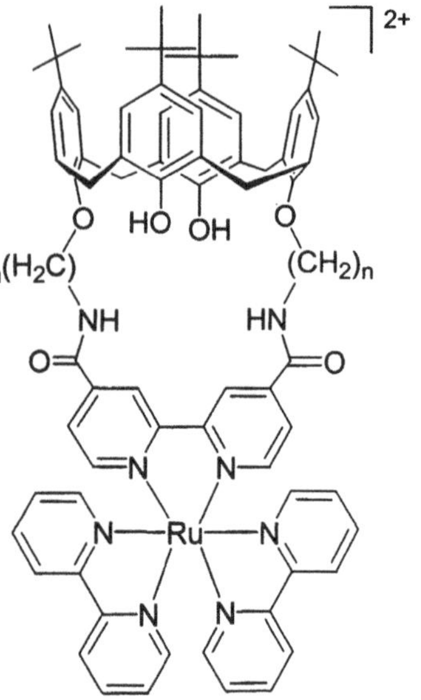

46   n = 2

47   n = 3

The detection of anionic analytes has been addressed by Beer and coworkers [384] who have affixed a $[Ru(bpy)_3^{2+}]$ reporter site to two straps extending from the lower rim of a calix[4]arene. Network closure by the metal complex in **46** and **47** produces a macrocyclic-binding site for anions. ^1H NMR titration experiments were undertaken to study the complexation properties of **46** and **47** with anions in DMSO. The stability constants of anion adducts of both complexes are quite high (10^2–10^4 M^{-1}) and follow the trend $H_2PO_4^- > Cl^- > Br^-$. Chemosensor **46**, containing the smaller and more discriminating host cavity, exhibits greatest selectivity. An X-ray crystal structure of the (Fig. 17). $H_2PO_4^-$ adduct of **46** emphasizes the importance of hydrogen bonding to the overall complexation process. Hydrogen bonds to the two amides and one hydroxy of the lower rim anchor the anion to the macrocyclic cavity. The presence of an anion within the macrocycle causes a hypsochromic shift and increase in intensity of the MLCT emission of the $[Ru(bpy)_3^{2+}]$ reporter site. The observed blue shift is likely a result of placing the negative charge of the anion near the bpy ligands, which receive the MLCT-excited electron. The large increase in emission intensity is tentatively attributed to the inhibition of vibrational relaxation brought about by rigidification of the receptor upon anion binding.

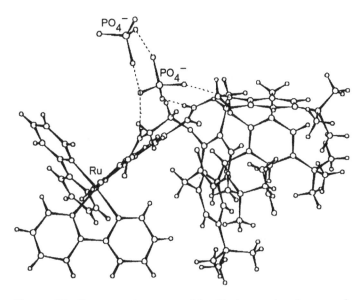

Figure 17 X-ray crystal structure of the dihydrogen phosphate complex of **46**. Thermal ellipsoids are given at 50% probability. (Reproduced with permission from the American Chemical Society.)

Beer and coworkers have modified their approach by derivatizing the upper rim of calix[4]arene with two (**48**, Tos = tosylate) and four (**49**) Ru(II) tris-bipyridyl reporter sites. The amide linkers of **48** and **49** form a hydrogen-bonding cleft for anion occupation. A preliminary report [385] indicates that MLCT luminescence from the Ru(II) centers is sensitive to anion association.

48

49

G. Relay Mechanisms Involving Electron and Energy Transfer

Many chemosensors exploit electron transfer as the signal transduction mechanism. Owing to different design constraints, we distinguish among systems whose function is derived from PET as described by Fig. 6 versus those systems whose function relies on electron (or energy) transfer between discrete donor and acceptor sites.

1. Photoinduced Electron Transfer (PET)

A PET reporter site tethered to the rim of the bucket allows the chemosensing strategy of Scheme 12 to be developed. In this construct, the bucket simply acts as a scaffold for the PET reporter site. The receptor is not required for analyte recognition and large changes in conformation are not required for signal transduction. Rather, electron transfer from a lone pair to the frontier orbitals of the excited reporter quenches luminescence, which is recovered by the interaction of the analyte with the lone pair (see Fig. 6).

Because they are good Lewis acids, protons are particularly well suited for detection by Scheme 12. Grigg has modified *tert*-butyl calix[4]arene with one and two $[Ru(bpy)_3]^{2+}$ subunits and examined the chemosensing properties of **50** through **52** [386]. At pH > 8–9, formation of the phenolate anion

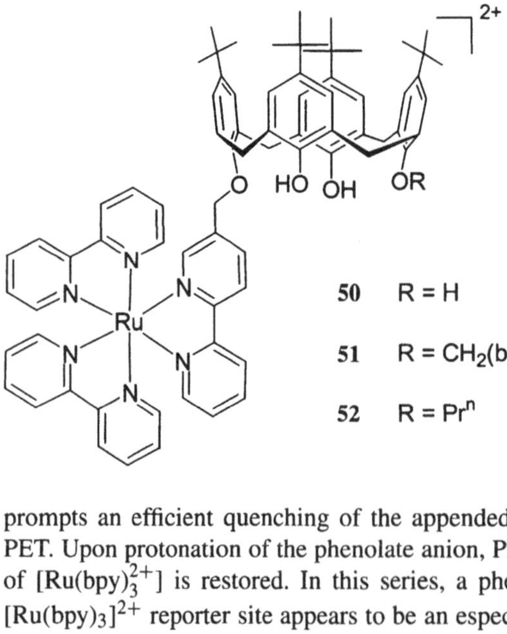

50	R = H
51	R = $CH_2(bpy)Ru(bpy)_2^{2+}$
52	R = Pr^n

prompts an efficient quenching of the appended $[Ru(bpy)_3^{2+}]$ luminescence by PET. Upon protonation of the phenolate anion, PET is averted and luminescence of $[Ru(bpy)_3^{2+}]$ is restored. In this series, a phenol diametrically opposite the $[Ru(bpy)_3]^{2+}$ reporter site appears to be an especially effective quencher; disub-

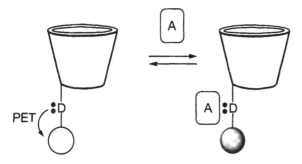

Scheme 12

stituted calix[4]arene **51** shows a marked decrease in sensitivity as compared to **50** while **52** is completely insensitive to pH.

Bodenant and coworkers [387] have taken advantage of the affinity of protons for the nitrogen lone pair of hydroxamate in the design and synthesis of **53** and **54**. Two hydroxamic acids, each bearing a pyrene, were covalently attached at the 1,3-positions of the lower rim of calix[4]arene. Methanol and diethyl ether solutions of **53** exhibit signatures of both pyrene monomer and excimer fluorescence. As is the case for **37** through **40**, the intensity ratio of

53 R = H

54 R = benzoyl

the excimer and monomer fluorescence signals depends on the extent of cone versus partial cone formation, which was carefully characterized by NMR spectroscopy. The protonation constants for the two hydroxamic acids were measured to be 9.5 and 10.2. Because the hydroxamato species efficiently quenches pyrene fluorescence by PET, a fluorescent chemosensor is thus obtained with an equivalence point of ~10. Selected metal ions cause the equivalence point to shift owing to their ability to compete for the hydroxamate-binding site. Buffered methanol/water solutions of **53** containing one equivalent of Cu(II) and Ni(II) ions exhibit pH equivalence points of ~5.5 and ~7.5, respectively. Fluorimetric titrations confirm the association of a single Cu(II) and Ni(II) ion to the bis-hydroxamato ligand sphere. The shift to more acidic pH by Cu(II) is consistent with the greater stability of copper hydroxamate complexes. No shift in the pH equivalence point of **53** is observed for divalent metal ions that do not bind the hydroxamate ligands.

A Cu(II)-induced perturbation of pyrene fluorescence has been utilized to create a sensor for glutamate [388]. A 2:2:1 Cu^{2+}:β-CD:pyrene complex is formed by the noncovalent assembly of the constituents; the site of Cu(II) binding is unknown. The pyrene emission resulting from complexation of the lumophore to β-CD is effectively quenched by the addition of Cu(II). A ~500-fold enhancement in pyrene intensity is observed upon the addition of 1.87 M glutamate, which is presumed to extract Cu(II) from the 2:2:1 complex. The precise nature of the quenching and restoration mechanisms is currently unknown.

Dabestani et al. have constructed chemosensors **55** through **57** for the detection of Cs^+ by covalently attaching one [389] and two [390] 9-cyano-10-

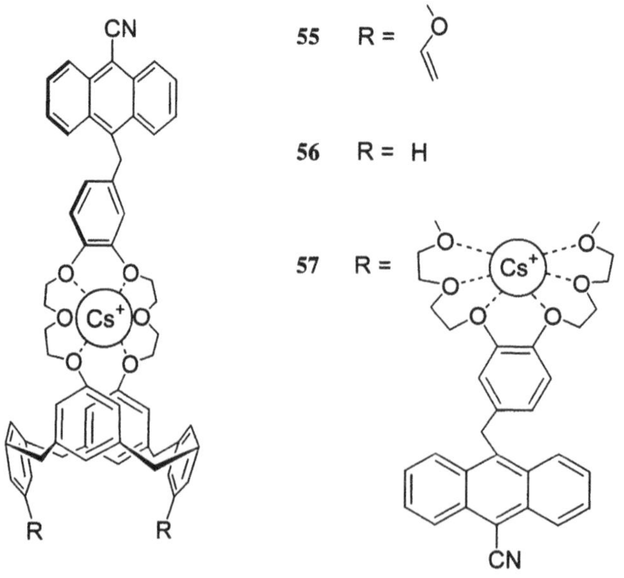

anthrylmethyl benzocrown-6 moieties to the 1–3 positions of calix[4]arene. The benzo–oxygen lone pairs effectively quench cyanoanthracene fluorescence by PET, which may be suppressed by metal association to the crown macrocycle. Fluorometric titration reveals the stepwise complexation of Cs^+ ion. Selectivity is achieved by the preference of the macrocycle to bind Cs^+ over other metal ions in both acidic and alkaline media. Of the three systems, **55** exhibits the largest fluorescence enhancement upon Cs^+ complexation ($\times 54$ as compared to $\times 5$–8 for **56** and **57**). The oxygen lone pair of **55** is believed to be more encumbered as a result of a stronger interaction between Cs^+ ion and the benzo–oxygens of the crown macrocycle. In support of this contention, X-ray crystal structure analysis shows that the Cs–O (benzo–oxygen) bond distance in **55** is 0.15 Å shorter than that in **56**. The sensitivity of **55** towards Cs^+ ion holds promise for its use in the sensing, monitoring, and remediation of ^{137}Cs, which is a fission product present in wastes generated during the reprocessing of irradiated nuclear fuels.

A suitable binding site for metal ions can often be prepared by functionalization of the lower rim of calixarenes with simple ligands. For instance, calix[4]arene tetraesters are good complexants of mono- and divalent ions [391]. Valeur et al. have exploited this property in the synthesis of **58**, which has the

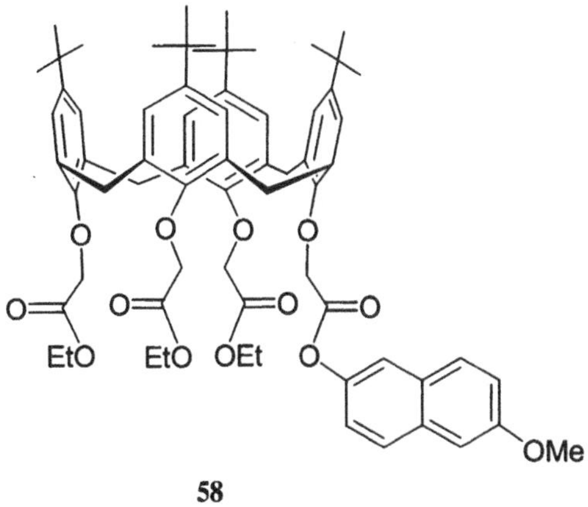

58

additional modification of a 6-acyl-2-methoxynaphthalene reporter site on one of the ester pendant arms [392]. Binding of K^+, Na^+, Li^+, and Ca^{2+} to the tetraester-modified calixarene rim of **58** causes the fluorescence quantum yield to increase by factors of 16, 35, 250, and 550, respectively. The energy diagram of Fig. 6 provides a rationalization for these large enhancements. Metal association with the carbonyl group of the ester is believed to energetically displace a weakly fluorescing $n\pi^*$ above a strongly fluorescing $\pi\pi^*$ excited state.

Bartsch et al. have observed **59** to exhibit specificity for Hg(II) cations in the presence of large excesses of a number of transition, alkali, and alkaline

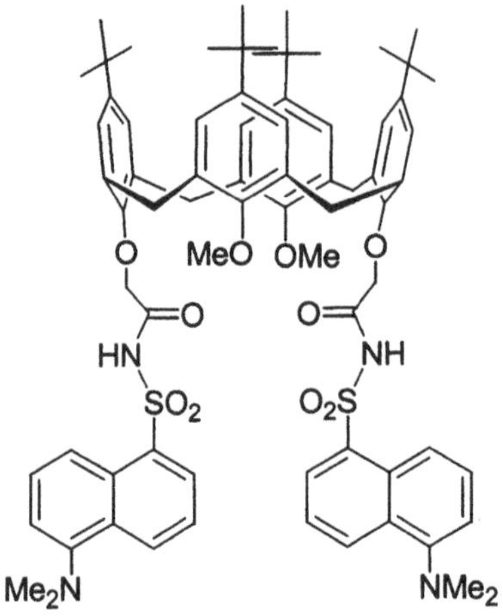

59

earth metal cations [393]. Electron transfer between dansyl and bound Hg(II) ion is thought to lead to the 89% reduction in the dansyl emission intensity. The dansyl fluorescence intensity from **59** dissolved in 25.0 mM aqueous solution of mercuric nitrate was not perturbed by the addition of a 100-fold excess of various metal ions, indicating a strong preference for the association of Hg(II) to **59**.

Suzuki et al. [394] have extended the applicability of Scheme 12 beyond proton and metal ion sensing by employing **60** to detect hydrogen-bonding substrates. The chemosensor interdisposes an azacrown between the cavity of a γ-CD bucket and pyrene reporter site. In the absence of analyte, a strong pyrene excimer emission is observed. Addition of deoxycholic acids to aqueous solutions of **60** led to a reduction in excimer fluorescence that was directly related to the decrease in the concentration of the dimer. In this situation, a concomitant increase in pyrene fluorescence is expected. But this was not observed. **60** shows anomalous fluorescence enhancements resulting from PET involving the lone pair of the distal amino group of the azacrown (the electron donating ability of the nitrogen proximate to pyrene is deactivated by the amide functionality). The differences in monomer fluorescence for selected deoxycholic acids is believed to arise from the ability of the analyte to regulate the distance between the amino group and the pyrene moiety, although no direct measure of the PET efficiency with analyte-induced changes of distance was provided to support this contention.

2. Donor–Acceptor Distance Mediated by Analyte

The dependence of electron and energy transfer rate constants on distance presents a valuable tool in chemosensor design. Large changes in the inten-

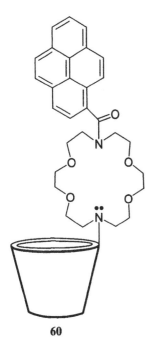

60

sity of fluorescence may be achieved by using analyte to modify the distance between a donor–acceptor pair. A proximate electron donor or acceptor will effectively quench fluorescence from a reporter site. Taking advantage of the exponential decrease in electron transfer rate with distance [see Eqs. (4) and (6)], an analyte that increases the distance between the redox sites will be detected with a large fluorescence increase from the reporter site. The general design concept is illustrated in Scheme 13 for the reductive quenching of the luminescence from a reporter site (the acceptor in this case) by an electron transfer donor (D). Conformational flexibility of the supramolecular bucket is an asset

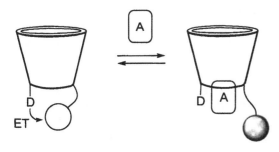

Scheme 13

to Scheme 13 because greater alteration in distance may be achieved. For this reason, Scheme 13 is most easily implemented with a calixarene receptor.

Shinkai et al. have employed nitrobenzene as the electron acceptor and pyrene as the excited state donor in the construction of **61** [395]. Addition of

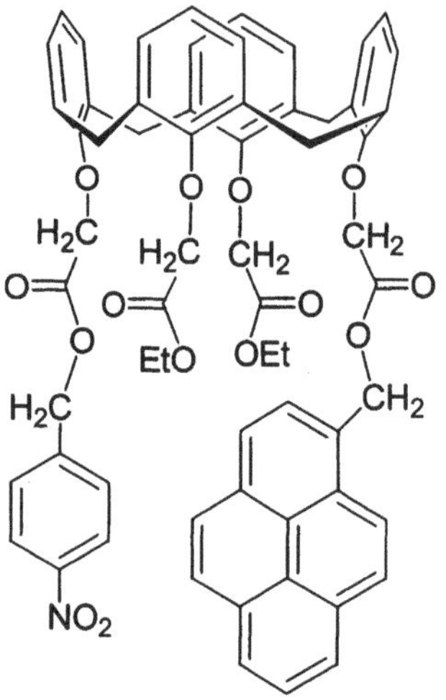

61

metal cations to diethyl ether-acetonitrile solutions of **61** causes the fluorescence intensity of pyrene to increase by a factor of ∼6. As in **58**, the tetraester cleft about a calixarene rim provides a binding site for metal cations. Binding constants of 6.3×10^{-2}, 5.0×10^{-5}, 1.2×10^{-3}, and 6.3×10^{-3} M^{-1} are measured for Li$^+$, Na$^+$, K$^+$, and Cs$^+$, respectively. Evidence for a metal ion induced conformational change of the type shown in Scheme 13 is provided by ^{1}H NMR spectroscopy. In the absence of metal cations, a large upfield shift of the protons of the nitrobenzene is ascribed to the ring current induced by the neighboring pyrene π-system. Upon addition of metal ions, the proton resonances of the nitrobenzene ring appear at their normal chemical shift values, indicating that the electron acceptor group is far removed from the electronically excited pyrene. In this conformation, electron transfer between the excited pyrene and electron accepting nitrobenzene is inefficient (large R_{DA}, small H_{DA}) and pyrene is able to fluoresce with its natural radiative lifetime.

Nolte et al. have attached a CD to each pyridine ring of a [Ru(bpy)$_3^{2+}$] core to generate the impressively large superstructures of **62** and **63** [396]. The metal center is so insulated from the external environment by the CD periphery that oxygen is unable to quench MLCT phosphorescence. **63** was the focus of chemosensor investigations because its luminescence is very intense, exceeding that of the native Ru(bpy)$_3^{2+}$ complex. Addition of the electron ac-

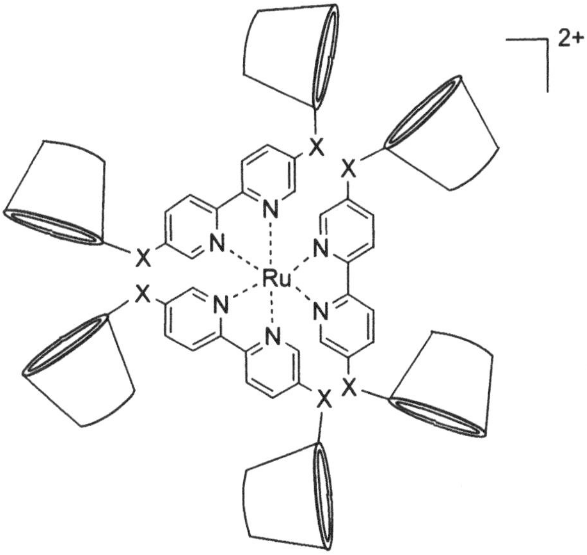

$$62 \quad X = -\overset{O}{\overset{\|}{C}}-NH-$$

$$63 \quad X = -\overset{O}{\overset{\|}{C}}-NH-(CH_2)_3-$$

ceptor, N,N'-dinonyl-4,4'-bipyridinium, to aqueous solutions of **63** efficiently quenches MLCT luminescence from the [Ru(bpy)$_3^{2+}$] core by intramolecular electron transfer. Formation of a 1:1 complex between the bipyridinium acceptor and **63** is inferred from fluorometric titration. However, this is a poor assay since quenching by the first bipyridinium is so complete that the binding of any additional bipyridiniums has no measurable effect on the luminescence intensity. Indeed, a microcalorimetric titration study clearly reveals the binding of a second bipyridinium with a comparable association constant to the first ($K_a(1:1) = 2.4 \times 10^5$, $K_a(1:2) = 4.0 \times 10^5$ M^{-1}). Displacement of bipyridinium from the CD by steroids obviates the intramolecular electron transfer quenching reaction and the MLCT luminescence of the central metal complex is recovered. Binding constants of the steroids are of similar magnitude to that of the bipyridinium quencher ($\sim 10^5$ M^{-1}). Differences in steroid binding strength allows for modest selectivities to be achieved. At a concentration of 10^{-4} M, lithocholic acid shows only a 5-fold increase in intensity whereas a 15-fold luminescence enhancement is observed for ursodeoxycholic acid at the same concentration.

In Ziessel and coworkers' design of **64** and **65**, quinones within the annulus of the calixarene receptor accept electrons from a flexibly tethered metal polypyridyl complex [397,398]. Phosphorescence from the $[Ru(bpy)_3^{2+}]$ and *fac*-[ReCl(CO)$_3$bpy] reporter sites is weak and lifetimes of the MLCT triplet states are considerably shorter (**64**: $\tau = 6$ ns, $\phi_e = 4 \times 10^{-4}$; **65**: $\tau = 5$ ns, $\phi_e = 4 \times 10^{-4}$) than those of the corresponding phenolic control compounds (**66**: $\tau = 1100$ ns, $\phi_e = 8.5 \times 10^{-2}$; **67**: $\tau = 45$ ns, $\phi_e = 3 \times 10^{-3}$). The quench-

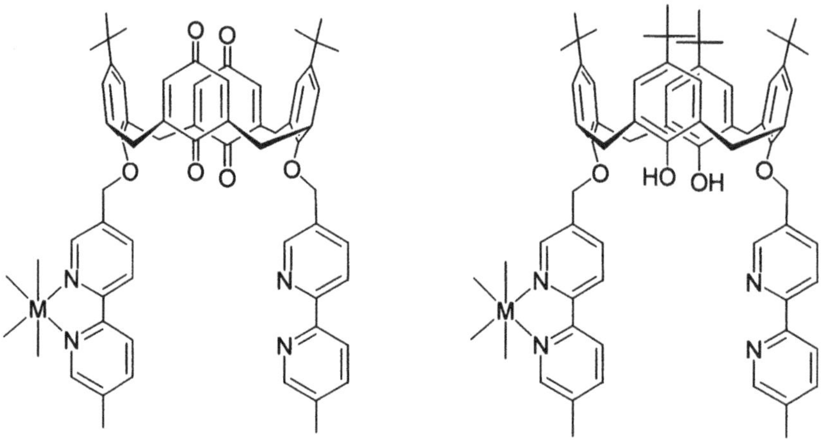

64   M = [Ru(bpy)$_2$]$^{2+}$ 2PF$_6^-$ **66**   M = [Ru(bpy)$_2$]$^{2+}$ 2PF$_6^-$

65   M = *fac*-[ReCl(CO)$_3$]   **67**   M = *fac*-[ReCl(CO)$_3$]

ing of the MLCT luminescence was ascribed to intramolecular charge transfer from the MLCT excited state of the pendant metal complex to the quinone subunit of the calix[4]diquinone. The luminescence properties of **64** and **65** in acetonitrile are markedly affected by the presence of cations, which have no affect on the photophysical properties of the control compounds **66** and **67**. Detailed NMR studies, augmented by molecular dynamics simulation, indicate that the four oxygens of the calixdiquinone and the two nitrogens of the dangling 2,2′-bipyridine hold a metal cation at the lower rim of the receptor. Association constants determined from luminescence titration experiments monotonically decrease with increasing charge of the metal cation, ranging from 200 to 400 M^{-1} for monocations (Li$^+$, Na$^+$, K$^+$), 10 to 20 M^{-1} for dications (Ca^{2+}, Ba^{2+}, Sr^{2+}, Cd^{2+}), and 0.5 to 2.0 M^{-1} for trications (La^{3+}, Gd^{3+}). The association of metal ions to the lower rim renders the quinones more easily reduced. For instance, the quinone reduction potentials of the Ba^{2+} complexes of **64** and **65** are shifted more negative by 200 and 600 mV, respectively. This shift in potential increases the driving force for electron transfer; forward and back electron transfer rate constants of $k_F = 1.1 \times 10^{10}$ s^{-1} and $k_B = 5.7 \times 10^9$ s^{-1} were determined using picosecond transient absorption spectroscopy. These enhanced electron transfer rates result in significant luminescence quenching upon metal ion association with **65**. The excited state lifetime of **65** is reduced to 85 ps and the luminescence quantum yield is $<10^{-4}$. Conversely, addition of Ba^{2+} to **64** causes an increase in luminescence intensity to values that are comparable to the emission observed from the control phenol complex, **66**. In this case, a concomitant decrease in the electron transfer rate is observed ($k_{ET} = 4.4 \times 10^5$ s^{-1}). The increase in driving force brought about by cation association with the quinone of **64** is offset by an electrostatically driven conformational change between the associated cation and charged pendant metal complex. The bound cation repels the positively charged metal complex, forcing the receptor into a cone conformation; MD simulations suggest that the edge-to-edge distance between reactants in the fully extended conformation is 5 Å. In this elongated conformation, orbital overlap between reactants is curtailed and the electron transfer from the photoexcited complex to the quinone is constrained by a small H_{DA}. The drop in k_{ET} for the metal ion complexes of **64** varies logically with the charge of the bound metal. Monocations give a 30-fold decrease in the electron transfer rate whereas a 300-fold decrease is observed for the electron transfer rate constant of dications. The average decrease of the photoinduced intramolecular electron transfer rate constant of the Ln^{3+} complexes is 1700-fold. A similar trend is observed in the electron transfer rate constants for the calix[4]diquinone receptor appended with two [Ru(bpy)$_3^{2+}$] metal complexes, although the overall luminescence enhancements for **68** are not as pronounced as those observed

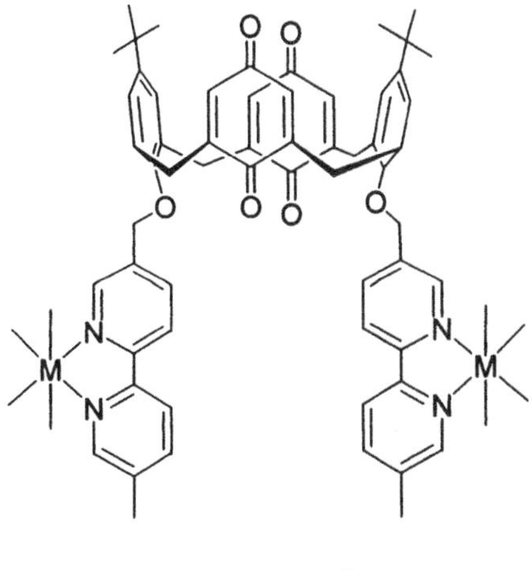

68   M = [Ru(bpy)$_2$]$^{2+}$ 2PF$_6^-$

for **64** [398]. Extremely small stability constants of the metal cation complexes ($K = 21$, 4, and 0.3 M^{-1} for K^+, Ba^{2+}, and La^{3+}, respectively) presumably result from the 4+ total charge of the two metal complexes, which electrostatically hinders the strong association of metal cations to the lower rim of the calix[4]diquinone receptor.

Employing the same calix[4]quinone receptor, Beer et al. have constructed a strap from an anion-binding macrocycle terminated with the [Ru(bpy)$_3^{2+}$] and *fac*-[ReCl(CO)$_3$bpy] reporter sites [399]. The emission quantum yields and lifetimes of **69** and **70** are significantly diminished with respect to their unoxidized calix[4]arene congeners **71** and **72** (**69**: $\tau = 30$ ns, $\phi_e = 1 \times 10^{-3}$ vs. **71**: $\tau = 510$ ns, $\phi_e = 3 \times 10^{-2}$; **70**: $\tau = 8.5$ ns, $\phi_e = 5 \times 10^{-4}$ vs. **72**: $\tau = 21$ ns,

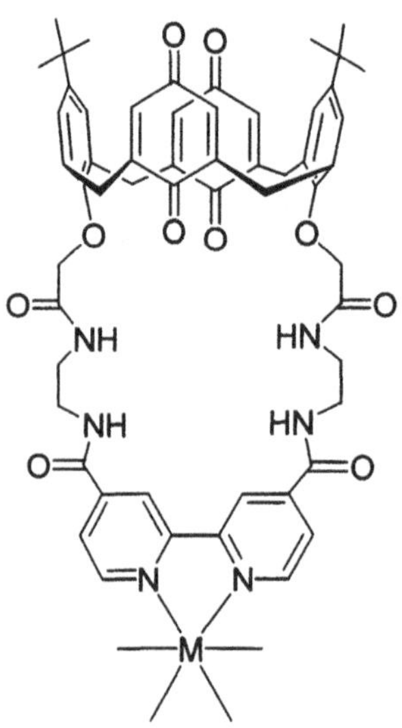

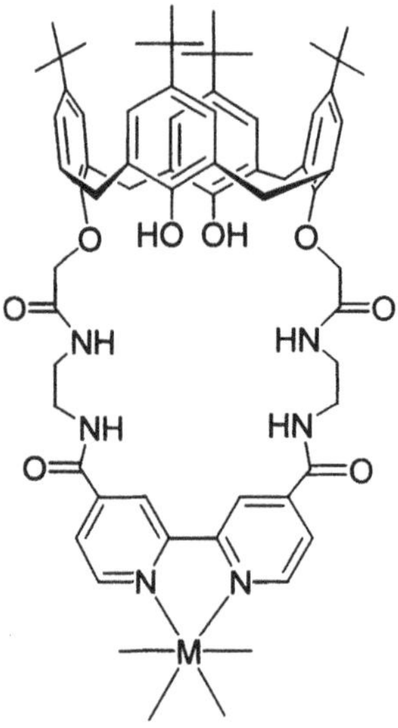

69 M = [Ru(bpy)$_2$]$^{2+}$ 2PF$_6^-$ **71** M = [Ru(bpy)$_2$]$^{2+}$ 2PF$_6^-$

70 M = *fac*-[ReCl(CO)$_3$]   **72** M = *fac*-[ReCl(CO)$_3$]

$\phi_e = 2 \times 10^{-3}$). Again, luminescence quenching is attributed to an intramolecular electron transfer from the MLCT excited state of the photoexcited metal complex to the quinones within the annulus of the calix[4]diquinone receptor. Using the lifetimes of the unoxidized receptors as a reference, intramolecular electron transfer rate constants of 3×10^7 s^{-1} and 7×10^7 s^{-1} were determined for **69** and **70**, respectively. Luminescence is recovered upon the addition of anions to DMSO and CH$_3$CN solutions of **69** and **70**; the most pronounced effect occurs for **69** with AcO$^-$, although Cl$^-$ and H$_2$PO$_4^-$ also generate a measurable luminescence response. Luminescence intensity enhancements of 500% are observed upon the addition of AcO⁻ to CH_3CN solutions of **69**. ^1H NMR titration experiments establish a 1:1 association (K = 500–1000 M^{-1} for Cl⁻ and 5000–10,000 for AcO⁻) of anion to the macrocycle of **69**. We note that the intervening macrocycle of **69** is structurally analogous to the amide macrocycle of **46**, for which anion binding has been characterized by X-ray crystallography. Presumably, anion association to the macrocycles of both complexes is similar. The anion-induced enhancement of luminescence from **69** is attributed to a decrease in the intramolecular electron transfer rate constant, suggesting that anion complexation decreases the interaction between the quinones of the calixarene annulus and the metal reporter site at the macrocyle's terminus. In view of the paucity of chemosensors for anionic analytes, the sensitivity of **69** and **70** in detecting OAc⁻ with a triggered luminescence signal is noteworthy.

Calix[4]pyrroles offer an alternative receptor site for the recognition of anions. Miyaji et al. [400] prepared a series of calixpyrrole–anthracene receptors that possess conjugated (**73**) and unconjugated bond pathways of different lengths (**74** and **75**). The stability constants (10^3–10^5 M^{-1} determined by NMR titration) for the anion complexes of each member of the series follows the trend

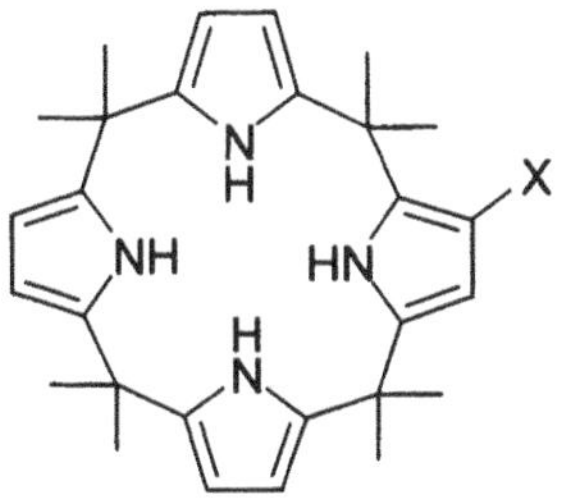

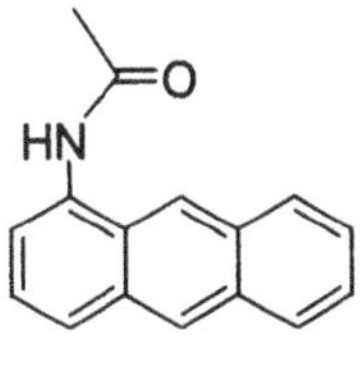

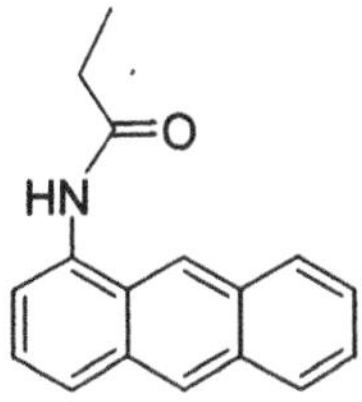

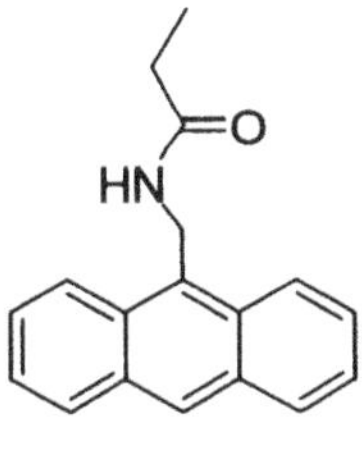

$F^- > H_2PO_4^- > Cl^- > Br^-$. The quenching efficiency of the fluorescence from the appended anthracene was observed to obey the same trend, indicating that association of the anion to the receptor determined overall selectivity. Although the mechanism for fluorescence quenching was not defined, it was noted that **73** exhibited the most sensitive response. Inasmuch as this chemosensor possesses the most strongly coupled and shortest through-bond pathway from receptor to reporter, electron transfer is suggested as the relay mechanism.

In contrast to electron transfer, if the relay mechanism is energy transfer, an enhanced luminescence response from the reporter site requires analyte to shorten the distance between the donor and acceptor [Eqs. (7) and (8)]. Energy transfer constructs of the type shown in Scheme 14 (D = energy transfer donor) have yet to be fully exploited despite their potential utility in chemosensor design. In one of the few studies of this type, Jin has tethered a pyrene donor and anthroyloxy acceptor to the 1,3-positions of *tert*-butyl calix[4]arene [401]. The emission intensity of the anthroyloxy group of **76** increases and emission intensity of pyrene decreases with the addition of Na^+ ion. Fluorometric titration revealed that the selectivity of **76** for Na^+ was ~59 greater than for K^+ due to more efficient energy transfer. The affinities of **76** for Li^+, Rb^+, and Cs^+ were too small to be detected. To verify that intermolecular energy transfer occurred, mono-fluorophore calix[4]arenes **77** and **78** were prepared; only monomer emission was observed from these supramolecules.

3. Direct Participation of Analyte in Electron or Energy Transfer

A guest in a supramolecular bucket may participate in electron transfer with substituents at the rim of the bucket. Weidner and Pikramenou have studied di-

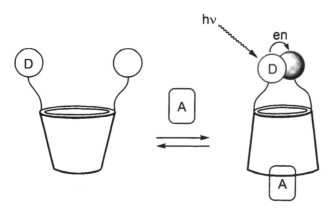

Scheme 14

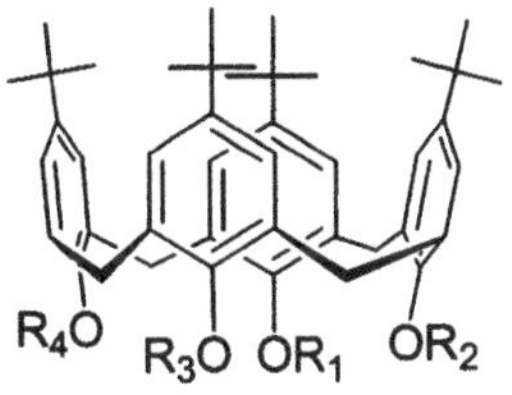

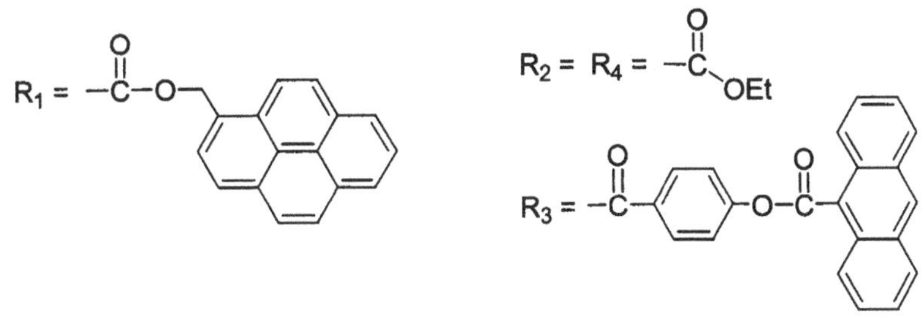

R₁ = (pyrene ester structure)

R₂ = R₄ = $-\overset{O}{\underset{}{C}}$-OEt

R₃ = (anthracene diester structure)

76

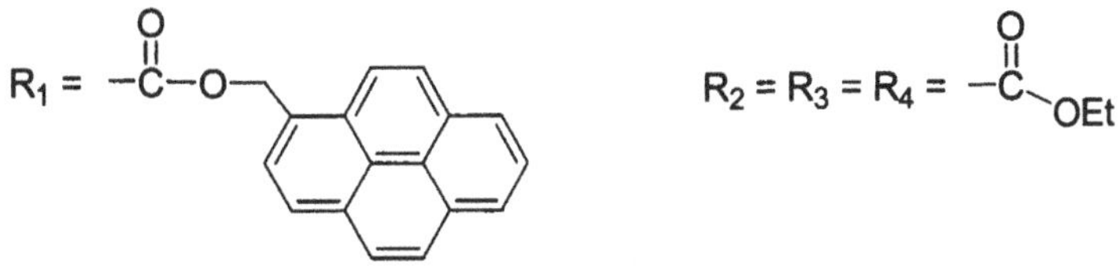

R₁ = (pyrene ester structure)

R₂ = R₃ = R₄ = $-\overset{O}{\underset{}{C}}$-OEt

77

R₁ = (anthracene diester structure)

R₂ = R₃ = R₄ = $-\overset{O}{\underset{}{C}}$-OEt

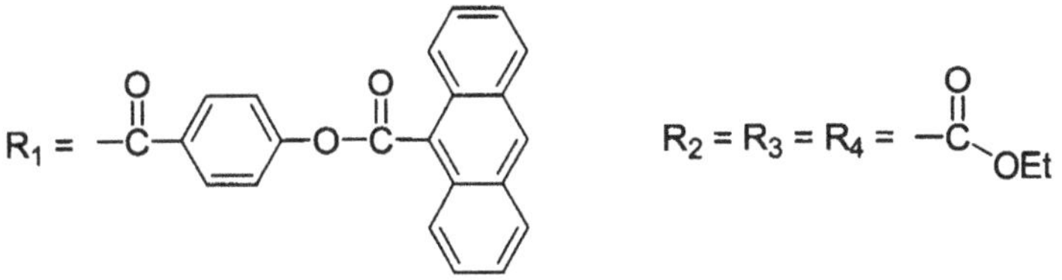

78

rectional electron transfer between a photoactive ruthenium complex appended to CD and an electron-accepting guest embedded within the cavity. Supramolecule **79** is flexible in its design because reporter sites may be introduced with facility [402]. For instance, a variety of photoactive centers may be ligated to the pendant terpyridine (tpy) upon the addition of L_3MCl_3 synthons to **79**. Using this methodology, **80** and **81** have been prepared. Consistent with the known luminescence properties of $Ru(tpy)_2^{2+}$, **80** only emits at <77 K. Excitation of

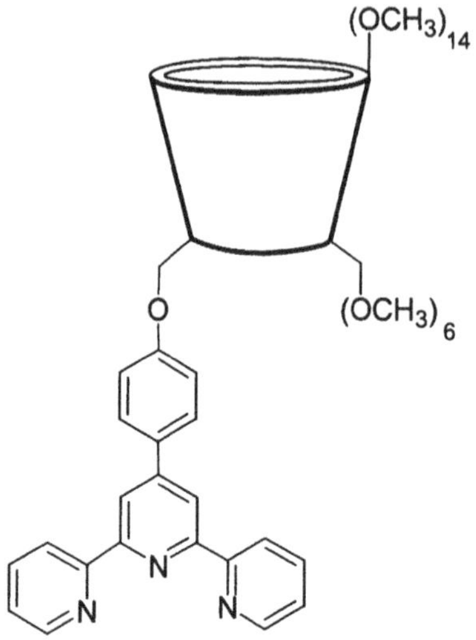

79

the MLCT transition of **81**, however, produces an easily observed room temperature phosphorescence. Addition of acceptors such as anthraquinone-2-carboxylic acid readily quenches the MLCT emission. Control experiments using $Ru(ttp)_2^{2+}$ (ttp = tolylterpyridine) establish that the electron transfer quenching reaction requires an acceptor that resides in the CD bucket.

An included guest and substituent at the rim of a bucket may also undergo directional energy transfer. Lehn and Valeur et al. have demonstrated that energy transfer processes approaching 100% efficiency may be achieved for properly designed supramolecule architecutres [403,404]. If analyte is able to assume the function of a sensitizer, then a chemosensor based on Scheme 15 may be established. The most studied relay mechanism for chemosensing of this type to date is the absorption-energy transfer-emission process described in Sec. III.C.3.

In our own work, we have recognized the benefits of Scheme 15 for the detection of mono- and bicyclic aromatic hydrocarbons. We have modified the rim of β-CD with an azamacrocycle, which ligates Eu^{3+} ion as the reporter site [405]. Tethered at only one nitrogen, the azacrown of **82** is able to swing away from the bottom of the hydrophobic interior of the CD bucket. Whereas the Eu^{3+} ion remains dark under direct irradiation of **82** with blue light, the arrival of aromatics such as benzene in the CD bucket triggers a weak red luminescence (intensity enhancement of 2 for ~440 ppm benzene). We showed the presence of an AETE process by examining the emission characteristics of the assembly as the excitation wavelength of light is scanned. The intensity of the red luminescence from an Eu^{3+} ion tracked the absorption profile of benzene

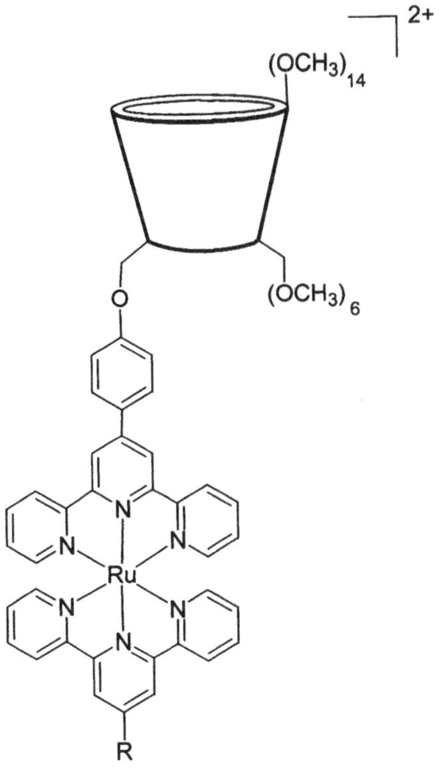

80   R = H

81   R = Ph

and not the Eu^{3+} ion itself, unequivocally establishing that the Eu^{3+} emission is excited by light passing through benzene.

The weak signal from **82** is due to the long distance for energy transfer from aromatic to Eu^{3+} ion. To rectify this problem, **83** was prepared [406]. By tethering the azamacrocycle at both of its nitrogens to the primary side of the CD, the aza metal-ion binding site is cradled below the CD bucket. We anticipated that the shorter distance imposed by the cradle geometry would result in a more efficient AETE relay process and accordingly a more intense signal response would be obtained. Surprisingly, a triggered luminescent response was not observed upon the addition of benzene or any other mono- or bicyclic aromatic hydrocarbon to aqueous solutions of **83**. Competitive binding experiments revealed that the +3 charge of the reporter site at the bottom of the bucket undermines the hydrophobic environment of the receptor site and hence

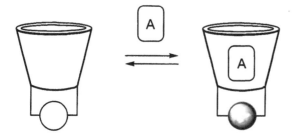

Scheme 15

decreases the association of benzene and other hydrophobic aromatics to the CD bucket (e.g., K_a for the **83**(Eu^{3+})·benzene complex is <10 M^{-1}) [407]. This diminished association of aromatics to the CD limits the overall optical response from the supramolecules even though the intrinsic energy transfer process of **83** should be more efficient than in **82**.

The results of **83** highlighted the importance of designing a cradle-binding site in which the +3 charge of the lanthanide ion was neutralized. The diethylenetriamine pentaacetic acid (DTPA) cradle binding site of **84** meets this ob-

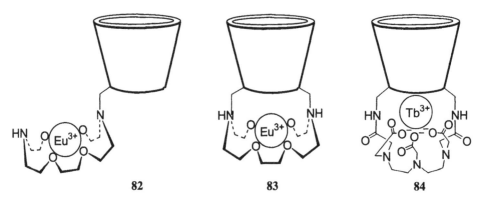

jective [408]. The choice of the nonreducible, green-emitting, Tb^{3+} ion as the reporter site for this chemosensor avoids the interference of carboxylate-based ligand-to-metal charge transfer excited states (which are prevalent for the reducible Eu^{3+} ion) in the overall AETE process. A bright green luminescence is triggered upon the introduction of aromatic hydrocarbons to aqueous solutions of **84**. The signal intensity increases monotonically with the concentration of aromatic hydrocarbon, reaching an asymptotic limit that is specific to the analyte. The smaller triggered luminescence response of **84** towards benzene (10% intensity enhancement for 200 ppm of benzene) as compared to biphenyl

(~4000% intensity enhancement for 10 ppm of biphenyl) reflects the smaller stability constant of the benzene complex and the smaller absorption cross-section
of the monocyclic aromatic.

The AETE relay mechanism of **84** was thoroughly investigated using time-
resolved spectroscopy [409]. The excitation energy captured by biphenyl included in **84** appears at the $^5D_4$ state of Tb^{3+} in 12 μs. The green luminescence
of the Tb^{3+} ion subsequently decays with a lifetime of 1.6 ms. Further insight
into the photophysics of the relay mechanism was provided by investigating the
time-resolved spectroscopy of **84**(Gd^{3+}). Because Gd^{3+} excited states are too
high in energy, the **84**(Gd^{3+}) complex provides a reference for the supramolecule
photophysics in the absence of energy transfer to the Ln^{3+} ion. A comparison
of the luminescence decay kinetics for biphenyl associated with **84**(Tb^{3+}) and
84(Gd^{3+}) leads to the model depicted in Fig. 18. The AETE process is initiated
by the absorption of an incident photon to produce the $^1\pi\pi^*$ excited state of
biphenyl. The lanthanide facilitates intersystem crossing to the triplet whereupon
energy transfer occurs to produce the $^5D_4$ emissive excited state of the Tb^{3+} ion.
Negligible spectral overlap precludes energy transfer by a Förster mechanism.
Calculations based on the prevailing Dexter mechanism yield a donor(biphenyl)–
acceptor(Tb^{3+}) distance of ~ 5 Å. This result is consistent with the distance
measured from energy minimized molecular models of the **84**(Tb^{3+})•biphenyl
complex. To further establish the role of Dexter energy transfer, the distance
between the analyte and Tb^{3+} reporter site was increased with our recent preparation of **85**. A Dexter treatment of time-resolved energy transfer kinetics data

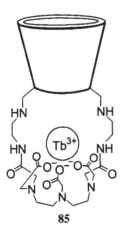

85

predicts the donor–acceptor distance in **85** to increase by ~2 Å, a distance that
is concordant with the distance measurements from energy minimized molecular
modeling calculations.

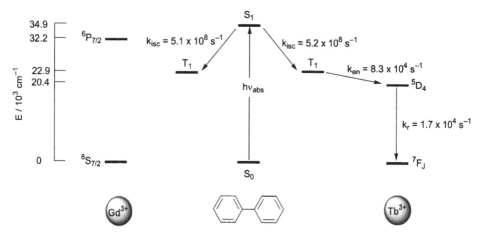

Figure 18 Photophysics scheme with measured rate constants for the AETE process of the binary complexes formed between **84**(Tb^{3+}) and **84**(Gd^{3+}) with biphenyl.

The photophysical scheme depicted in Fig. 18 is not unique to **84**(Tb^{3+}). That the triplet excited state represents a staging area for energy transfer is an emerging trend in the photophysics of many other Ln^{3+} ion supramolecules. For instance, the Tb^{3+} and Eu^{3+} complexes of modified calix[4]arenes also undergo AETE via the triplet excited state of donors [410–412]. These results, taken together with the systems described here, serve to emphasize the role of the lanthanide ion to open a conduit for energy flow in supramolecular assemblies. In addition to the obvious importance of the Ln^{3+} ion as the emitting center in the AETE process, the Ln^{3+} ion also provides a mechanism to produce a long-lived donor excited state from which energy transfer may occur. In the absence of this heavy atom effect, the singlet excited state of the aromatic would return to its ground state before energy transfer could occur. By channeling the singlet to a long-lived triplet, ample time is provided for the slower energy transfer process to effectively compete with the fast natural radiative and nonradiative processes of the analyte.

V. CONCLUDING REMARKS

As the impact of our advanced technologies on the global community becomes understood demands are ever increasing for chemosensors to selectively detect in real time and space a variety of biological and chemical substances at ultratrace levels. Within this context, by enabling dramatic new powers of observation, chemosensor science is poised at the front and center of the research enterprise. Several frontiers await exploration.

The explosion of supramolecule architectures makes the elaboration of new receptors an obvious target of future investigation. Buckets are endpoints of a continuum of supramolecules based on rings and other cylindrical structures [413]. These related macromolecules demonstrate bucket-type, "lock-and-key" interactions that are similar to those of CDs and calixarenes. The unique physical and chemical properties of the different supramolecule reporter sites enable the selectivity and sensitivity of analyte recognition to be exquisitely tuned. Depending on the analyte and the application, it is likely that new supramolecule architectures will be the cornerstones of many new chemosensor platforms.

Advances can also be expected in the development of new reporter sites. To date, most chemosensors rely on *molecular* emissive centers. However, recent years have witnessed the emergence of materials-based lumophores that display unmatched stabilities and quantum efficiencies. Prominent among these materials are quantum dots (QDs) [31,33], which are the zero-dimensional analogue of the two-dimensional quantum wells. Quantum dots offer the ultimate in quantum confinement with electronic excitations restricted in all three directions by the boundary of the dot. As a result of the quantum-size effect, the electronic states of QDs are discrete and tunable with the size of the crystallite [33,414]. There are now a number of approaches to making nanometer-size "boxes" of semiconductors. Sensory function with QDs is achieved by manipulating their luminescence intensity and spectral distribution [415,416]. Although yet to be explored, QDs should find a wide range of applications as reporter sites whose emission is triggered by energy transfer or electron charge injection into the QD. In one of many potential sensory schemes, supramolecule receptor sites can be chemically attached to the surface of the dots. Ionized, a QD is photon silent. Molecular recognition of a donor analyte capable of transferring charge back to the dot will recover the QD luminescence. Because it is possible to observe single QDs turning on and off [31–33], this application could result in sensors that detect to the single molecule limit.

On the final front of the 3R scheme, greater sensitivity and selectivity will be achieved with a sound and quantitative knowledge of the photophysics and kinetics of the relay mechanism. It is remarkable that the bucket chemosensors of Sec. IV have been largely developed with only a phenomenological knowledge of the electronic structure and relay kinetics. New opportunities abound with the advent of ultrafast and easily tunable lasers (e.g., optical parametric oscillators). The precise route of energy transduction within the chemosensor can be ascertained by following the relay event in real time. In this manner, pathways that lead the excitation energy into thermally dissipative regions of the supramolecule architecture can be uncovered. These nonradiative sinks can be eliminated by rational chemical redesign, leading to chemosensors with extremely efficient relay mechanisms. The benefit of this approach is clearly demonstrated by the evolutionary design of chemosensors **82** to **84** for the detection of polyaromatic hydrocarbons.

Extending beyond the 3R construct of chemosensing, arguably the most crucial issue at hand is signal amplification. For all the chemosensors described in this chapter, the signal is afforded by direct detection of the photon emitted from the reporter site. New transduction schemes seek to enhance the sensitivity of chemosensors by amplifying the initial photon from the reporter site. One approach is to attach the reporter site to a gain medium such as a conducting polymer [6,417]. The overall strategy is shown in Fig. 19. Individual receptor sites of monomer subunits are polymerized to produce a conducting polymer appended with N receptor sites. Absorption of a photon creates excitons that migrate along the polymer backbone. Analyte binding to a receptor produces a trapping site at which the exciton is effectively deactivated by energy or electron transfer quenching. If energy migration along the polymer backbone can sample all N sites, then the association of an analyte with one of the receptor sites will effectively lead to a signal response. Amplification is derived from the additivity of binding constants for receptors connected in series. In other terms, because the probability for binding is N times greater than for a single molecule receptor,

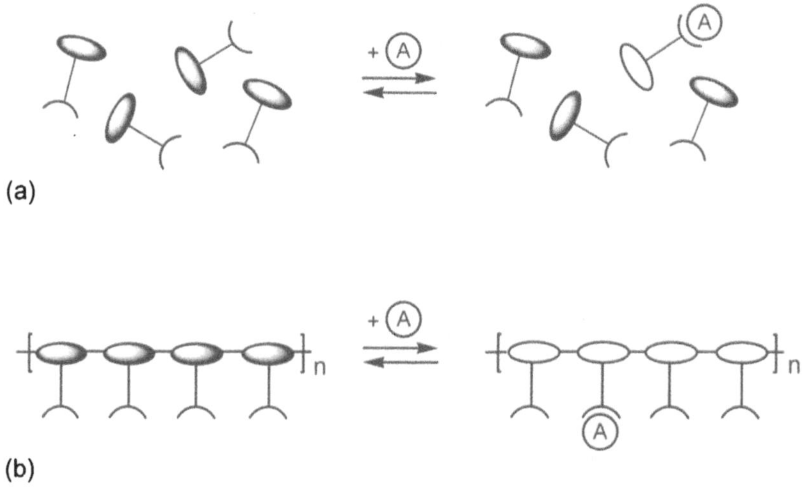

(a)

(b)

Figure 19 Signal amplification produced from chemosensors wired in series. The method has been most thoroughly elaborated for signal transduction derived from a quenching process. (a) Luminescence quenching from individual monomer subunits upon molecular recognition. (b) Signal response from receptors wired in series. Only fractional occupancy is required to quench luminescence from polymer backbone. (Adapted from Ref. 6.)

the association constant of the "wired" chemosensor is increased by a factor of N, which also represents the gain in sensitivity. Since upwards of 10^6 receptor sites may be polymerized on a single strand of conducting polymer, impressive signal gains may be achieved by wiring chemosensors in series. Although most studies to date have emphasized signal transduction mechanisms based on quenching, analyte binding to a receptor can cause a wavelength shift of the polymer's fluorescence. For example, the fluorescence from the poly(phenylene bithiophene) backbone of polymer **86** blue-shifts by 50 nm upon Na^+ association to the calix[4]arene [418]. In this instance, a signal is effectively triggered against a low intensity or dark background, greatly expanding the capabilities of the molecular wire approach to chemosensor design. This approach to chemo- and bio-sensing is developed thoroughly in Chapter 4.

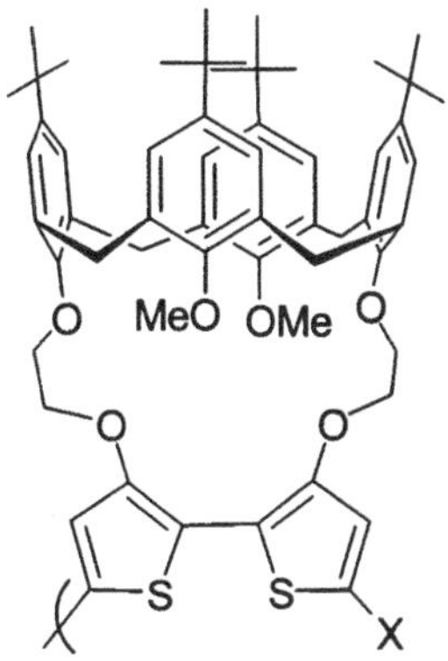

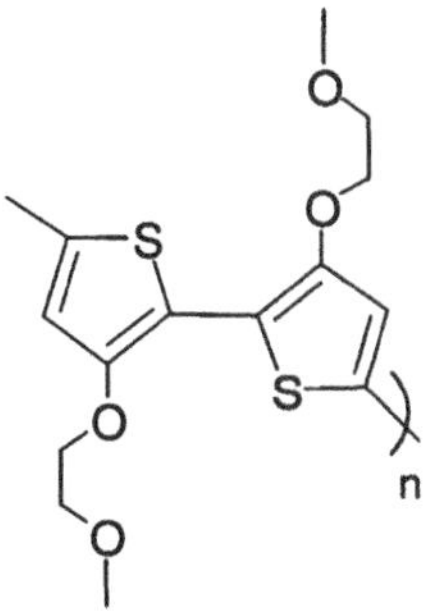

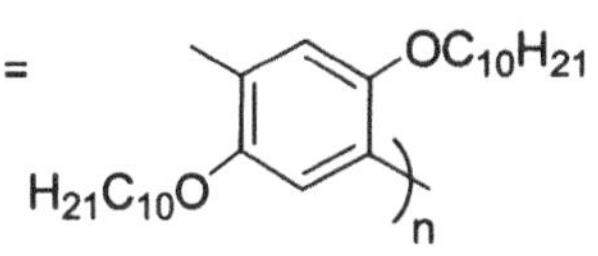

The versatility of the sensory amplification approach is expanded considerably by the ease with which it may be adapted to conductimetric detection methods. By way of example, consider sensory polymer **87**, which is produced from the polymerization of thiophene subunits appended with a calix[4]arene receptor site [419]. The polythiophene backbone is capable of transporting electrical current. The calix[4]arene endows the conducting polymer with an affinity for Na^+ ions. Analyte binding produces a resistive element in the polythiophene wire, which is experienced by all of the electrons flowing through it. The conductivity in the Na^+-free state, which is large and easily detected, is nearly absent (>99% reduced) after treatment with 0.5 mM of Na^+. Considering that high-precision resistance measurements can be readily made with simple electronics, the combination of optical and electrical transduction methods in a single sensory polymer presages the development of materials displaying ultrasensitive signal response.

Finally, bucket chemosensors have the potential to lead to advanced-age sensors possessing unprecedented performance capabilities when they are interfaced with emerging engineering technologies. For instance, the optical chemosensing buckets are ideally suited for miniaturization with their introduction onto microfluidic platforms. Collection, separation, transport, processing, and analysis of gases and aerosols can all now be carried out on lithographically patterned structures to allow sensing to be performed on a chip [420,421]. The miniaturization offered by microfluidic devices translates into rapid sampling and measurement times on microliter samples as well as convenient packaging. As we have recently shown, microfluidic optical chemosensors may be realized by incorporating bucket chemosensors into a matrix that is compatible with lithographic patterning protocols [422]. As research continues to proceed along these lines, one may envisage the creation of new technologies to benefit individuals from many different walks of life. Consider a "wristwatch" sensor for the soldier in the battlefield who needs to detect chemical and biowarfare agents, or the asthmatic wishing to know the allergens present in her or his immediate environment. The design and implementation of devices such as these is one of the formidable challenges confronting chemosensor research in the near future.

ACKNOWLEDGMENTS

The Air Force Office of Scientific Research, Air Force Material Command, USAF, under grant number F49620-98-1-0203 has supported the research presented in this review from our group. The U.S. Government is authorized to reproduce and distribute reprints for governmental purposes notwithstanding any copyright notation thereon. Partial financial support for this work was also provided by the Center for Sensor Materials, a NSF-sponsored Materials Research Science and Engineering Center (DMR-9400417).

REFERENCES

1. AP de Silva, HQN Gunaratne, T Gunnlaugsson, AJM Huxley, CP McCoy, JT Rademacher, TE Rice. Chem Rev 97:1515, 1997.
2. J-P Desvergne, AW Czarnik, eds. Chemosensors of Ion and Molecular Recognition. NATO ASI Series C, vol 492, Dordrecht, The Netherlands: Kluwer Academic, 1997.
3. AW Czarnik, ed. Fluorescent Chemosensors for Ion and Molecule Recognition. ACS Symposium Series 538, Washington, DC: American Chemical Society, 1993.
4. CM Rudzinski, WH Hartmann, DG Nocera. Coord Chem Rev 171:115, 1998.
5. AW Czarnik. Acc Chem Res 27:302, 1994.
6. TM Swager. Acc Chem Res 31:201, 1998.
7. J-M Lehn. Supramolecular Chemistry. Weinheim: VCH, 1995.
8. J-M Lehn. Science 260:1762, 1993.
9. J-M Lehn. J Incl Phenom 6:351, 1988.
10. PD Beer, PA Gale, GZ Chen. Coord Chem Rev 186:3, 1999.
11. J Janata. Principles of Chemical Sensors. New York: Plenum, 1989.
12. W Göpel, J Hesse, JN Zemel, eds. Sensors: A Comprehensive Survey. Fundamentals and General Aspects, vol 1. New York: VCH, 1988.
13. HJ Verhey, B Gebben, JW Hofstraat, JW Verhoeven. J Polym Sci A 33:399, 1995.
14. W Tan, ZY Shi, S Smith, D Birnbaum, R Kopelman. Science 29:607, 1992.
15. SL Sharp, RJ Warmack, JP Goudonnet, I Lee, TL Ferrell. Acc Chem Res 26:377, 1993.
16. M Ghodrati. Soil Sci Soc Amer J 63:471, 1999.
17. KP O'Connell, JJ Valdes, NL Azer, RP Schwartz, J Wright, ME Eldefrawi. J Immunol Meth. 225:157, 1999.
18. CK Guay, GP Klinkhammer, KK Falkner, R Benner, PG Coble, TE Whitledge, B Black, FJ Bussell, TA Wagner. Geophys Res Lett 26:107, 1999.
19. D Eastwood, RL Lidberg, SJ Simon, T Vo-Dinh. In L Pawlowski, ed. Chemistry for the Protection of the Environment. New York: Plenum, 1991.
20. AS Xie. Acc Chem Res 29:598, 1996.
21. R Kopelman, WH Tan. Science 262:1382, 1993.
22. T Ha, XW Zhuang, HD Kim, JW Orr, JR Williamson, S Chu. Proc Nat Acad Sci USA, 96:9077, 1999.
23. WE Moerner, M Orrit. Science 283:1670, 1999.
24. S Weiss. Science 283:1676, 1999.
25. AD Mehta, M Rief, JA Spudich, DA Smith, RM Simmons. Science 283:1689, 1999.
26. T Funatsu, Y Harada, M Tokunaga, K Saito, T Yanagida. Nature 374:6522, 1995.
27. TJ Ha, AY Ting, J Liang, WB Caldwell, AA Deniz, DS Chemla, PG Schultz, S Weiss. Proc Nat Acad Sci USA 96:893, 1999.
28. PM Goodwin, WP Ambrose, RA Keller. Acc Chem Res 29:607, 1996.
29. WE Moerner. Acc Chem Res 29:563, 1996.
30. VI Klimov, AA Mikhailovsky, DW McBranch, CA Leatherdale, MG Bawendi. Science 287:1011, 2000.

31. AP Alivisatos. Science 271:933, 1996.
32. SA Empedocles, R Neuhauser, K Shimizu, MG Bawendi. Adv Mater 11:1243, 1999.
33. SA Empedocles, MG Bawendi. Acc Chem Res 32:389, 1999.
34. NH Bonadeo, J Erland, D Gammon, D Park, DS Katzer, DG Steel. Science 282:1473, 1998.
35. OS Wolfbeis, ed. Fiber-Optic Chemical Sensors and Biosensors, vol 1. Boca Raton: CRC Press, 1991.
36. E Wagner, R Dändliker, K Spenner, eds. Sensors: A Comprehensive Survey. Optical Sensors, vol 6. New York: VCH, 1991.
37. AJ Rogers. In W Göpel, J Hesse, JN Zemel, eds. Sensors: A Comprehensive Survey. Fundamentals and General Aspects, Optical Sensors, vol. 6. New York: Plenum, 1992, Ch 15.
38. JR Lakowicz, ed. Topics in Fluorescence Spectroscopy, vol. IV. New York: Plenum, 1994, Ch 7.
39. DL Wise, LB Wingard, eds. Biosensors with Fiberoptics. Clifton: Humana, 1991.
40. V Balzani, V Carassiti. Photochemistry of Coordination Compounds. London: Academic, 1970.
41. LS Forster. In AW Adamson, PD Fleischauer, eds. Concepts of Inorganic Photochemistry. New York: Wiley-Interscience, 1975.
42. RP Wayne. Principles and Applications of Photochemistry. Oxford: Oxford University Press, 1980.
43. J Jortner, SA Rice, RM Hochstrasser. Adv Photochem 7:149, 1969.
44. KF Freed. Acc Chem Res 11:74, 978.
45. SH Lin. Radiationless Transitions. New York: Academic, 1980.
46. NJ Turro. Modern Molecular Photochemistry. Menlo Park: Benjamin/Cummings, 1978.
47. GJ Ferraudi. Elements of Inorganic Photochemistry, New York: Wiley-Interscience, 1988, Ch 1.
48. V Balzani, L Moggi, MF Manfrin, F Bolletta. Coord Chem Rev 15:321, 1975.
49. JG Calvert, JN Pitts Jr. Photochemistry. Chichester: Wiley, 1966.
50. JB Birks. Photophysics of Aromatic Molecules. New York: Wiley-Interscience, 1970.
51. SL Murov. Handbook of Photochemistry. 2nd ed. New York: Marcel Dekker, 1993.
52. JA Jackson, C Turró, MD Newsham, DG Nocera. J Phys Chem 94:4500, 1990.
53. CG Guilbault. Practical Fluorescence. 2nd ed. New York: Marcel Dekker, 1990.
54. EL Wehry, ed. Modern Fluorescence Spectroscopy, vols 1–4. New York: Plenum, 1976–1981.
55. J Kopecky. Photochemistry. A Visual Approach. New York: VCH, 1992.
56. JA Barltrop, JD Coyle. Excited States in Organic Chemistry. London: Wiley-Interscience, 1975.
57. RA Bissel, AP de Silva, HQN Ganaratne, PLM Lynch, GEM Maguire, CP McCoy, KRAS Sandanayake. Top Curr Chem 168:223, 1993.
58. RA Bissel, AP de Silva, HQN Ganaratne, PLM Lynch, GEM Maguire, KRAS Sandanayake. Chem Soc Rev 187, 1992.
59. AP de Silva, T Gunnlaugsson, TE Rice. Analyst 121:1759, 1996.

60. L Fabbrizzi, M Licchelli, P Pallavicini, D Sacchi, A Taglietti. Analyst. 121:1763, 1996.
61. L Fabbrizzi, A Poggi. Chem Soc Rev 24:302, 1995.
62. L Fabbrizzi, M Licchelli, P Pallavicini, A Perotti, A Taglietti, D Sacchi. Chem Eur J 2:75, 1996.
63. FH Richardson. Chem Rev 82:541, 1982.
64. SP Sinha. In SP Sinha, ed. Systematics and the Properties of the Lanthanides, NATO ASI Series 109. Dordrecht: Reidel, 1983, p 451.
65. N Sabbatini, M Guardigli, J-M Lehn. Coord Chem Rev 123:201, 1993.
66. N Sabbatini, M Guardigli, I Manet, R Ungaro, A Casnati, R Ziessel, G Ulrich, Z Asfari, J-M Lehn. Pure Appl Chem 67:135, 1995.
67. J-CG Bünzli. In J-CG Bünzli, GR Choppin, eds. Lanthanide Probes in Life, Chemical and Earth Sciences. Amsterdam: Elsevier, 1988, Ch 7.
68. SI Weissman. J Chem Phys 10:214, 1942.
69. A Heller, E Wasserman. J Chem Phys 42:949, 1965.
70. V Balzani, F Scandola. Supramolecular Photochemistry. West Sussex, UK: Ellis Horwood, 1991.
71. V Balzani. Pure Appl Chem 62:1099, 1990.
72. J-M Lehn. Angew Chem Int Ed Engl 27:89, 1988.
73. N Sabbatini, M Guardigli, J-M Lehn. Coord Chem Rev 121:201, 1993.
74. J Georges. Analyst 118:1481, 1993.
75. M Elbanowski, B Makowsaka. J Photochem Photobiol A 99:85, 1996.
76. IM Warner, SA Soper, LB McGown. Anal Chem 68:R73, 1996.
77. WDW Horrocks Jr, DR Sudnick. J Am Chem Soc 101:334, 1979.
78. A Heller. J Am Chem Soc 88:2058, 1966.
79. Y Haas, G Stein. J Phys Chem 75:3677, 1971.
80. JL Kropp, MW Windsor. J Phys Chem 71:477, 1967.
81. WDW Horocks Jr. In SJ Lippard, ed. Progress in Inorganic Chemistry, vol 30. New York: Wiley, 1984, p 1.
82. GL Geoffrey, MS Wrighton. Organometallic Photochemistry. New York: Academic, 1979.
83. DM Roundhill. Photochemistry and Photophysics of Metal Complexes. New York: Plenum, 1994.
84. AJ Lees. Chem Rev 87:711, 1987.
85. LS Forster. Chem Rev 90:331, 1990.
86. E Zinato, P Riccieri. Coord Chem Rev 125:35, 1993.
87. JF Endicott, T Ramasami, R Tamilarasan, RB Lessard, CK Ryu, GR Brubaker. Coord Chem Rev 77:1, 1987.
88. E Zinato. Coord Chem Rev 129:195, 1994.
89. AD Kirk. Chem Rev 99:1607, 1999.
90. JR Winkler, HB Gray. Comm Inorg Chem 1:257, 1981.
91. KS Heinselman, MD Hopkins. J Am Chem Soc 117:12340, 1995.
92. GA Neyhat, KJ Seward, J Boaz, BP Sullivan. Inorg Chem 30:4486, 1991.
93. W Paw, SD Cummings, MA Mansour, WB Connick, DK Geiger, R Eisenberg. Coord Chem Rev 171:125, 1998.

94. CE Johnson, R Eisenberg, TR Evans, MS Burberry. J Am Chem Soc 105:1795, 1983.
95. DM Roundhill, HB Gray, C-M Che. Acc Chem Res 22:55, 1989.
96. DC Smith, HB Gray. Coord Chem Rev 100:169, 1990.
97. DC Smith, HB Gray. In DR Salahub, MC Zerner, eds. The Challenge of d and f Electrons, ACS Symposium Series 394. Washington, DC: American Chemical Society, 1989, p 356.
98. KH Leung, DL Phillips, M-C Tse, C-M Che, VM Miskowski. J Am Chem Soc 121:4799, 1999.
99. JM Forward, JP Fackler, Z Assefa. In DM Roundhill, JP Fackler, eds. Optoelectronic Properties of Inorganic Compounds. New York: Plenum, 1999.
100. Z Assefa, BG McBurnett, RJ Staples, JP Fackler, B Assmann, K Angermaier, H Schmidbaur. Inorg Chem 34:75, 1995.
101. YK Shin, VM Miskowski, DG Nocera. Inorg Chem 29:2308, 1990.
102. J Kadis, YK Shin, JI Dulebohn, DL Ward, DG Nocera. Inorg Chem 35:811, 1996.
103. AE Stiegman, VM Miskowski. J Am Chem Soc 110:4053, 1988.
104. MD Hopkins, HB Gray, VM Miskowski. Polyhedron 6:705, 1987.
105. MD Hopkins, HB Gray. J Am Chem Soc 106:2468, 1984.
106. AM Macintosh, DG Nocera. Inorg Chem 35:7134, 1996.
107. AW Maverick, JS Najdzionek, D MacKenzie, DG Nocera, HB Gray. J Am Chem Soc 105:1878, 1983.
108. JA Jackson, C Turró, MD Newsham, DG Nocera. J Phys Chem 94:4500, 1990.
109. RD Mussell, DG Nocera. J Am Chem Soc 110:2674, 1988.
110. N Prokopuk, DF Shriver. Adv Inorg Chem 46:1, 1999.
111. TG Gray, CM Rudzinski, DG Nocera, RH Holm. Inorg Chem 38:5932, 1999.
112. PC Ford, A Vogler. Acc Chem Res 26:220, 1993.
113. VWW Yam, KKW Lo, WKM Fung, CR Wang. Coord Chem Rev 171:17, 1998.
114. VWW Yam, KKW Lo. Coord Chem Rev 184:157, 1999.
115. M Gordon, WR Ware, eds. The Exciplex. New York: Academic, 1975.
116. PF Barbara, W Jarzeba. Adv Photochem 15:1, 1990.
117. A Vlcek. Coord Chem Rev 177:219, 1998.
118. K Kalyanasundaram. Photochemistry of Polypyridine and Porphyrin Complexes. London: Academic, 1982, Ch 6.
119. E Krausz, J Ferguson. Prog Inorg Chem 37:293, 1989.
120. H Yersin, W Humbs, J Strasser. Coord Chem Rev 159:325, 1997.
121. SI Gorelsky, ES Dodsworth, ABP Lever, AA Vlcek. Coord Chem Rev 174:469, 1998.
122. JR Schoonover, CA Bignozzi, TJ Meyer. Coord Chem Rev 165:239, 1997.
123. A Juris, V Balzani, F Barigelletti, S Campagna, P Belser, A von Zelewsky. Coord Chem Rev 84:85, 1988.
124. GL Crosby. Acc Chem Res 8:231, 1975.
125. BM Krasovitski, BM Bolotin. Organic Luminescent Materials. Weinheim: VCH, 1989.
126. W Rettig. Top Curr Chem 169:253, 1994.
127. YV Ilichev, W Kuhnle, KA Zachariasse. J Phys Chem A 102:5670, 1998.

128. KA Zachariasse, M Grobys, T von der Haar, A Hebecker, YV Ilichev, YB Jiang, O Morawski, W Kuhnle. J Photochem Photobiol A 102:59, 1996.
129. U Leinhos, W Kuhnle, KA Zachariasse. J Phys Chem 95:2013, 1991.
130. G Schopf, W Rettig, J Bendig. J Photochem Photobiol A 84:33, 1994.
131. K Bhattacharyya, M Chowdhury. Chem Rev 93:507, 1993.
132. ZR Grabowski. Pure Appl Chem 64:1249, 1992.
133. ZR Grabowski, J Dobkowski. Pure Appl Chem 55:245, 1983.
134. J Rebek. Acc Chem Res 32:278, 1999.
135. KA Connors. Chem Rev 97:1325, 1997.
136. J Szejtli. Cyclodextrins and Their Inclusion Complexes. Budapest: Akademiai Kiado, 1982.
137. J Szejtli. Cyclodextrin Technology. Boston: Kluwer Academic, 1988.
138. J Szejtli, T Osa, eds. Comprehensive Supramolecular Chemistry, vol 3. New York: Elsevier, 1996.
139. ML Bender. Cyclodextrin Chemistry, vol 6. New York: Springer-Verlag, 1978.
140. J Szejtli. Chem Rev 98:1743, 1998.
141. W Saenger, J Jacob, K Gessler, T Steiner, D Hoffman, H Sanbe, K Koizumi, SM Smith, T Takaha. Chem Rev 98:1787, 1998.
142. H-J Schneider, F Hacket, V Rudiger, H Ikeda. Chem Rev 98:1765, 1998.
143. MV Rekharsky, Y Inoue. Chem Rev 98:1875, 1998.
144. R Breslow, SD Dong. Chem Rev 5:1997, 1998.
145. E Rizzarelli, G Vecchio. Coord Chem Rev 188:343, 1999.
146. K Takahashi. Chem Rev 98:2035, 1998.
147. G Wenz. Angew Chem Int Ed Engl 33:803, 1994.
148. I Tabushi, Y Kuroda, A Mochizuki. J Am Chem Soc 102:1152, 1979.
149. I Tabushi. Acc Chem Res 15:66, 1982.
150. I Tabushi, K Yamamura, T Nabeshima. J Am Chem Soc 106:5267, 1984.
151. AR Kahn, P Forgo, KJ Stine, VT D'Souza. Chem Rev 98:1977, 1998.
152. J Vicens, V Böhmer. In JED Davies, ed. Calixarenes. A Versatile Class of Macrocyclic Compounds, vol. 3. Boston: Kluwer Academic, 1991.
153. A Ikeda, S Shinkai. Chem Rev 97:1713, 1997.
154. K Iwamoto, A Ikeda, K Araki, T Harada, S Shinkai. Tetrahedron 49:609, 1993.
155. A Ikeda, S Shinkai. Chem Rev 98:1713, 1997.
156. AF Danil de Namor, RM Cleverly, ML Zapata-Ormachea. Chem Rev 98:2495, 1998.
157. M Takeshita, S Shinkai. Bull Chem Soc Jpn 68:1088, 1995.
158. S Shinkai. Tetrahedron 49:8933, 1993.
159. RM Izatt, JS Bradshaw, K Pawlak, RL Bruening, BJ Tarbet. Chem Rev 92:1261, 1992.
160. J Bügler, JFJ Engbersen, DN Reinhoudt. J Org Chem 63:5339, 1998.
161. J Bügler, N Sommerdijk, A Visser, A van Hoek, R Nolte, J Engbersen, DN Reinhoudt. J Am Chem Soc 121:28, 1999.
162. D Diamond, MA McKervey. Chem Soc Rev 15, 1996.
163. C Seel, F Vögtle. Angew Chem Int Ed Engl 31:528, 1992.
164. F Diederich. Angew Chem Int Ed Engl 27:362, 1988.
165. B Konig. Top Curr Chem 196:91, 1998.

166. PR Ashton, R Ballardini, V Balzani, SE Boyd, A Credi, MT Gandolfi, M Gomez Lopez, S Iqbal, D Philp, JA Preece, L Prodi, HG Ricketts, JF Stoddart, MS Tolley, M Venturi, AJP White, DJ Williams. Chem Eur J 3:152, 1997.
167. S Akine, K Goto, R Okazaki. Chem Lett 7:681, 1999.
168. CJ Jones. Chem Soc Rev 27:289, 1998.
169. AP Bisson, VM Lynch, MKC Monahan, EV Anslyn. Angew Chem Int Ed Engl 36:2340, 1997.
170. FY Chu, LS Flatt, EV Anslyn. J Am Chem Soc 116:4194, 1994.
171. K Niikura, AP Bisson, EV Anslyn. J Chem Soc Perkin Trans 2 6:1111, 1999.
172. WL Mock. Top Curr Chem 175:1, 1995.
173. BD Wagner, AI MacRae. J Phys Chem B 103:10114, 1999.
174. WA Freeman, WL Mock, N-Y Shih. J Am Chem Soc 103:7367, 1981.
175. WL Mock, N-Y Shih. J Org Chem 51:4440, 1986.
176. D Whang, J Heo, JH Park, K Kim. Angew Chem Int Ed Engl 37:78, 1998.
177. J Heo, SY Kim, D Whang, K Kim. Angew Chem Int Ed Engl 38:641, 1999.
178. C Meschke, HJ Buschmann, E Schollmeyer. Themochim Acta 297:43, 1997.
179. HJ Buschmann, K Jansen, E Schollmeyer. Themochim Acta 317:95, 1998.
180. HJ Buschmann, E Schollmeyer. J Incl Phenom Mol Recog 29:167, 1997.
181. A Jasat, JC Sherman. Chem Rev 99:931, 1999.
182. RG Chapman, JC Sherman. Tetrahedron 53:15911, 1997.
183. RA Marcus, N Sutin. Biochim Biophys Acta 811:265, 1985.
184. DN Beratan, SS Skourtis. Curr Opin Chem Biol 2:235, 1998.
185. M Bixon, J Jortner. Adv Chem Phys 106:35, 1999.
186. BS Brunschwig, N Sutin. Coord Chem Rev 187:233, 1999.
187. AB Myers. Chem Rev 96:911, 1996.
188. M Bixon, J Jortner. J Chem Phys 107:5154, 1997.
189. RA Marcus. Ann Rev Phys Chem 15:155, 1964.
190. RA Marcus. J Chem Phys 43:2654, 1965.
191. PY Chen, TJ Meyer. Chem Rev 98:1439, 1998.
192. D Graff, JP Claude, TJ Meyer. Adv Chem Ser 253:183, 1997.
193. PF Barbara, TJ Meyer, MA Ratner. J Phys Chem 100:13148, 1996.
194. JV Caspar, TJ Meyer. J Am Chem Soc 105:5583, 1983.
195. JV Caspar, EM Kober, BP Sullivan, TJ Meyer. J Am Chem Soc 104:630, 1982.
196. TJ Meyer. Prog Inorg Chem 30:389, 1983.
197. VG Levich. In H Henderson, W Yost, eds. Physical Chemistry—An Advanced Treatise, vol 9B. New York: Academic, 1970.
198. KV Mikkelson, MA Ratner. Chem Rev 87:113, 1987.
199. FD Lewis, TF Wu, YF Zhang, RL Letsinger, SR Greenfield, MR Wasielewski. Science 277:673, 1997.
200. BW Van der Meer, G Coker, S-YS Chen, eds. Resonance Energy Transfer Theory and Data. New York: VCH, 1994.
201. T Förster. In O Sinanoglu, Ed. Modern Quantum Chemistry, vol 3. New York: Academic, 1965.
202. DL Andrews. Resonance Energy Transfer. New York: Wiley Interscience, 1999.
203. DL Dexter. J Chem Phys 21:836, 1953.
204. T Miyakawa, DL Dexter. Phys Rev B 1:2961, 1970.

205. A Lamola. Energy Transfer and Organic Photochemistry. New York: Interscience, 1964.

206. NJ Turro. Modern Molecular Photochemistry. Menlo Park: Benjamin/Cummings, 1978, Ch 9.

207. GL Closs, P Piotrowiak, JM Macinnis, GR Fleming. J Am Chem Soc 110:2652, 1988.

208. A Ueno. In V Ramamurthy, ed. Photochemistry in Organized and Constrained Media. New York: VCH, 1991.

209. K Kalyansundaram. Photochemistry in Microheterogeneous Systems. Orlando: Academic, 1987.

210. F Cramer, W Saenger, H-C Spatz. J Am Chem Soc 89:14, 1967.

211. NJ Turro, GS Cox, X Ki. Photochem Photobiol 36:149, 1983.

212. CD Tran, JH Fendler. J Phys Chem 88:2167, 1984.

213. K Kano, I Takenoshita, T Ogawa. J Phys Chem 86:1833, 1982.

214. S Nigam, G Durocher. J Phys Chem 100:7135, 1996.

215. S Scypinski, LJ Cline-Love. Am Lab 55, 1984.

216. M Hoshino, M Imamura, K Ikehara, Y Hama. J Phys Chem 85:1820, 1981.

217. RS Murphy, TC Barros, J Barnes, B Mayer, G Marconi, C Bohne. J Phys. Chem A 103:137, 1999.

218. S Monti, G Kohler, G Grabner. J Phys Chem 97:13011, 1993.

219. L Biczok, L Jicsinszky, H Linschitz. J Incl Phenom Mol Recog 18:237, 1994.

220. P Bortolus, G Grabner, G Kohler, S Monti. Coord Chem Rev 125:261, 1993.

221. A Sarkar, S Chakravorti. J Lumin 69:161, 1996.

222. A Sarkar, S Chakravorti. J Lumin 78:205, 1998.

223. WG Herkstroeter, PA Martic, S Farid. J Am Chem Soc 112:3583, 1990.

224. DF Eaton. Tetrahedron 43: 1551, 1987.

225. M Sbai, SA Lyazidi, DA Lerner, B Delcastillo, MA Martin. Anal Chim Acta 303:47, 1995.

226. MC Rath, DK Palit, T Mukherjee. J Chem Soc Faraday Trans 94:1189, 1998.

227. BD Wagner, PJ MacDonald. J Photochem Photobiol A 114:151, 1998.

228. KA Al-Hassan, UKA Klein, A Suqaiyan. Chem Phys Lett 212:581, 1993.

229. GS Cox, PJ Hauptman, NJ Turro. Photochem Photobiol 39:597, 1984.

230. KA Al-Hassan. Chem Phys Lett 227:527, 1994.

231. P Karunanithi, P Ramamurthy, VT Ramakrishnan. J Incl Phenom Mol Recog 34:105, 1999.

232. MH Kleinman, C Bohne. In V Ramamurthy, KS Schanze, eds. Molecular and Supramolecular Photochemistry, vol 1. New York: Marcel Dekker, 1997, p 391.

233. WY Xu, JN Demas, BA DeGraff, M Whaley. J Phys Chem 97:6546, 1993.

234. A Nakamura, S Sato, K Hamasaki, A Ueno, F Toda. J Phys Chem 99:10952, 1995.

235. M Milewski, A Maciejewski, W Augustyniak. Chem Phys Lett 272:225, 1997.

236. A Ueno. New Funct Mater C 521, 1993.

237. S Li, WC Purdy. Chem Rev 92:1457, 1992.

238. ML Pola, M Algarra, A Becerra, M Hernandez. Anal Lett 33:891, 2000.

239. CC Wang, CI Li, YH Lin, LK Chau. Appl Spectrosc 54:15, 2000.

240. D Demore, A Kasselouri, O Bourdon, J Blais, G Mahuzier, P Prognon. Appl Spectrosc 53:523, 1999.

241. RP Frankkewich, KN Thimmaiah, WL Hinze. Anal Chem 63:2924, 1991.
242. J Wei, E Okerberg, J Dunlap, C Ly, JB Shear. Anal Chem 72:1360, 2000.
243. G Escandar. Analyst 124:587, 1999.
244. JA Arancibia, GM Escandar. Analyst 124:1833, 1999.
245. SM Shuang, SY Guo, MY Cai, JH Pan. Anal Lett 31:1357, 1998.
246. R Badia, ME Diaz-Garcia. J Agric Food Chem 47:4256, 1999.
247. M Del Olmo, A Zafra, A Gonzalez-Casado, JL Vilchez. Int J Environ Anal Chem 69:99, 1998.
248. AV Veglia. Molecules 5:437, 2000.
249. A Coly, J-J Aaron. Talanta 46:815, 1998.
250. M Valero, SMB Costa, JR Ascenso, MM Velazquez, LJ Rodriguez. J Incl Phenom Macro Chem 35:663, 1999.
251. SE Castrejon, AK Yatsimirsky. Talanta 44:951, 1997.
252. A Coly, J-J Aaron. Anal Chim Acta 360:129, 1998.
253. Z Gong, Z Zhang. Anal Chim Acta 351:205, 1997.
254. RE Galian, AV Veglia, RH de Rossi. Analyst 123:1587, 1998.
255. RA Bissel, AP de Silva. J Chem Soc Chem Commun 1148, 1991.
256. JHT Luong. In J-P Desvergne, AW Czarnik, eds. Chemosensors of Ion and Molecule Recognition, NATO ASI Series C, vol 492. Dordrecht, The Netherlands: Kluwer Academic, 1997.
257. J Van Stam, S De Feyter, FC De Schryver, CH Evans. J Phys Chem 100:19959, 1996.
258. Y Liao, C Bohne. J Phys Chem 100:734, 1996.
259. G Patonay, K Fowler, A Shapira, G Nelson, IM Warner. J Incl Phenom 5:717, 1987.
260. G Nelson, G Patonay, IM Warner. Anal Chem 60:274, 1988.
261. A Nakajima. Bull Chem Soc Jpn 57:1143, 1984.
262. A Ueno, K Takahashi, Y Hino, T Osa. J Chem Soc Chem Commun 194, 1981.
263. M Milewski, W Augustyniak, A Maciewjewski. J Phys Chem A 102:7427, 1998.
264. S Hamai, T Ikeda, A Nakamura, H Ikeda, A Ueno, F Toda. J Am Chem Soc 114:6012, 1992.
265. JF Huang, GC Catena, FV Bright. Appl Spectrosc 46:606, 1992.
266. A Muñoz de la Peña, TT Ndou, JB Zung, KL Greene, DH Live, IM Warner. J Am Chem Soc 113:1572, 1991.
267. GM Escandar. Spectrochim Acta A 55:1743, 1999.
268. XK Chen, L Mou, LD Li, AJ Tong. Chin J Anal Chem 27:125, 1999.
269. S Hamai. J Am Chem Soc 111:3984, 1989.
270. CH Evans, N Prud'homme, M King, JC Scaiano. J Photochem Photobiol A 121:105, 1999.
271. K Kalyanasundaram, JK Thomas. J Am Chem Soc 99:2039, 1977.
272. DC Dong, MA Winnik. Photochem Photobiol 35:17, 1982.
273. A Muñoz de la Peña, T Ndou, JB Zung, IM Warner. J Phys Chem 95:3330, 1991.
274. Y Kusumoto. Chem Phys Lett 136:535, 1987.
275. W Xu, JN Demas, BA DeGraff, M Whaley. J Phys Chem 97:6546, 1993.
276. JM Schuette, TT Ndou, A Munoz de la Pena, S Mukundan Jr, IM Warner. J Am Chem Soc 115:292, 1993.

277. JM Schuette, AY Will, RA Agbaria, IM Warner. Appl Spectrosc 48:581, 1994.
278. JB Zung, A Muñoz de la Peña, T Ndou, IM Warner. J Phys Chem 95:6701, 1991.
279. S Hashimoto, JK Thomas. J Am Chem Soc 107:4655, 1985.
280. XZ Du, Y Zhang, YB Jiang, LR Lin, XZ Huang, GZ Chen. J Photochem Photobiol A 112:53, 1998.
281. K Kano, I Takenoshita, T Ogawa. J Phys Chem 86:1833, 1982.
282. VC Anigbogu, IM Warner. Appl Spectrosc 50:995, 1996.
283. S Hamai. J Phys Chem 93:2074, 1989.
284. S Hamai. Bull Chem Soc Jpn 62:2763, 1989.
285. H Yang, C Bohne. J Phys Chem 100:14533, 1996.
286. AK Chandra, NJ Turro, AL Lyons Jr, P Stone. J Am Chem Soc 100:4964, 1978.
287. V Ramamurthy, JV Caspar, DR Corbin, BD Schyler, AH Maki. J Phys Chem 94:3391, 1990.
288. S Scypinski, LJ Cline-Love. Anal Chem 56:322, 1984.
289. S Scypinski, LJ Cline-Love. Anal Chem 56:331, 1984.
290. LJ Cline-Love, ML Grayeski, J Noroski, R Weinberger. Anal Chim Acta 170:3, 1985.
291. MH Lopez, MA Gonzalez, MIL Molina. Talanta 49:679, 1999.
292. SA Soper, IM Warner, LB McGown. Anal Chem 70:477R, 1998.
293. GM Escandar, A Muñoz de la Peña. Anal Chim Acta 370:199, 1998.
294. A Muñoz de la Peña, MC Mahedero, A Espinosa-Mansilla, AB Sanchez, M Reta. Talanta 48:15, 1999.
295. S Hamai, T Kudou. J Photochem Photobiol A 113:135, 1998.
296. S Hamai. J Phys Chem B 101:1707, 1997.
297. MD Richmond, RJ Hurtubise. Anal Chem 61:2643, 1989.
298. R Temia, S Scypinski, LJ Cline-Love. J Environ Sci Technol 19:155, 1985.
299. MA Mortellaro, WK Hartmann, DG Nocera. Angew Chem Int Ed Engl 35:1945, 1996.
300. NJ Turro, T Okube, C-J Chung. J Am Chem Soc 104:1789, 1982.
301. A Ponce, PA Wong, JJ Way, DG Nocera. J Phys Chem 97:11137, 1993.
302. WK Hartmann, MHB Gray, A Ponce, DG Nocera, PA Wong. Inorg Chim Acta 243:239, 1996.
303. XZ Du, Y Zhang, YB Jiang, LR Lin, XZ Huang, GZ Chen. J Photochem Photobiol A 112:53, 1998.
304. CP Gendrich, MM Koochesfahani, DG Nocera. Exp Fluids 23:361, 1997.
305. MA Mortellaro, DG Nocera. Chem Tech 26:17, 1996.
306. DG Nocera. New Scientist 149:24, 1996.
307. MM Koochesfahani, RK Cohn, CP Gendrich, DG Nocera. In RJ Adrian, ed. Developments in Laser Techniques in Fluid Mechanics. Berlin: Springer-Verlag, 1997.
308. B Stier, MM Koochesfahani. Exp Fluids 26:297, 1999.
309. RE Falco, DG Nocera. In MC Roco, ed. Particulate Two Phase Flows. Boston: Butterworth-Heinemann, 1992, Ch 3.
310. RE Falco, DG Nocera. Liquid-Solid Flows 118:143, 1991.
311. D Maynes, J Klewicki, P McMurtry. J Fluid Mech 388:49, 1999.
312. JC Klewicki, RB Hill. J Fluids Eng Trans ASME 120:772, 1998.

313. MM Koochesfahani. AIAA 99:3768, 1999.
314. A Upadhyay, T Bhatt, HB Tripathi, DD Pant. J Photochem Photobiol A 89:201, 1995.
315. SK Chakrabarti, WR Ware. J Chem Phys 55:5494, 1971.
316. CJ Seliskar, L Brand. J Am Chem Soc 93:5414, 1971.
317. CJ Seliskar, L Brand. Science 171:799, 1971.
318. JJ Inestal, F Gonzalezvelasco, A Ceballos. J Chem Educ 71:A297, 1994.
319. K Takahashi, Y Ohtsuka, S Nakada, K Hattori. J Incl Phenom Mol Recog Chem 10:63, 1991.
320. Y Aoyama, Y Nagai, J Otsuki, K Kobayasha, H Toi. Angew Chem Int Ed Engl 31:745, 1992.
321. Y-H Li, L-M Chan, L Tyer, RT Moody, CM Himel, DM Hercules. J Am Chem Soc 97:3118, 1975.
322. Y Wang, T Ikeda, A Ueno, F Toda. Chem Lett 863, 1992.
323. M Nakamura, A Ikeda, N Ise, T Ikeda, H Ikeda, F Toda, A Ueno. J Chem Soc Chem Commun 721, 1995.
324. T Ikunaga, H Ikeda, A Ueno. Chem Eur J 5:2698, 1999.
325. A Ueno, S Minato, I Suzuki, M Fukushima. Chem Lett 605, 1990.
326. H Ikeda, M Nakamura, N Ise, N Oguma, A Kakamura, T Ikeda, F Toda, A Ueno. J Am Chem Soc 118:10980, 1996.
327. A Ueno, A Ikeda, H Ikeda, T Ikeda, F Toda. J Org Chem 64:382, 1999.
328. HFM Nelissen, F Venema, RM Uittenbogaard, MC Feiters, RJM Nolte. J Chem Soc Perkin Trans 2 10:2045, 1997.
329. R Corradini, A Dossena, R Marchelli, A Panagia, G Sartor, M Saviano, A Lombardi, V Pavone. Chem Eur J 2:373, 1996.
330. R Corradini, A Dossena, G Galaverna, R Marchelli, A Panagia, G Sartor. J Org Chem 62:6283, 1997.
331. F Hamada, Y Kondo, K Ishikawa, H Ito, I Suzuki, T Osa, A Ueno. J Incl Phenom Mol Recog Chem 17:267, 1994.
332. F Hamada, Y Kondo, R Ito, I Suzuki, T Osa, A Ueno. J Incl Phenom Mol Recog Chem 15:273, 1993.
333. A Ueno, F Moriwaki, T Osa, F Hamada, K Murai. J Am Chem Soc 110:4323, 1988.
334. M Narita, S Koshizaka, F Hamada. J Incl Phenom Macro Chem 35:605, 1999.
335. S Ito, M Narita, F Hamada. Int J Soc Mater Eng Resour 7:156, 1999.
336. M Eddaoudi, H Parrot-Lopez, SF de Lamotte, D Ficheux, P Prognon, AW Coleman. J Chem Soc Perkin Trans 2:1711, 1996.
337. Y Liu, B-H Han, S-X Sun, T Wada, Y Inoue. J Org Chem 64:1487, 1999.
338. M Narita, F Hamada, I Suzuki, T Osa. J Chem Soc Perkin Trans 2:2751, 1998.
339. M Narita, F Hamada. J Chem Soc Perkin Trans 2:823, 2000.
340. M Narita, N Ogawa, F Hamada. Anal Sci 16:37, 2000.
341. M Sato, M Narita, N Ogawa, F Hamada. Anal Sci 15:1199, 1999.
342. M Narita, F Hamada, M Sato, I Suzuki, T Osa. J Incl Phenom Macro Chem 34:421, 1999.
343. SR McAlpine, MA Garcia-Garibay. J Am Chem Soc 120:4269, 1998.
344. S Kundu, N Chattopadhyay. J Photochem Photobiol A 88:105, 1995.

345. A Nag, K Bhuttacharyya. J Chem Soc Faraday Trans 86:53, 1990.
346. A Nag, R Dutta, N Chattopadhyay, K Bhattacharyya. Chem Phys Lett 157:83, 1989.
347. G Krishnamoorthy, SK Dogra. J Photochem Photobiol A 123:109, 1999.
348. HS Banu, K Pitchumani, C Srinivasan. J Photochem Photobiol A 131:101, 2000.
349. S Kundu, SC Bera, N Chattopadhyay. Ind J Chem 37A:102, 1998.
350. Y Matsushita, T Hikida. Chem Phys Lett 290:349, 1998.
351. T Tanabe, S Usui, A Nakamura, A Ueno. J Incl Phenom Macro Chem 36:79, 2000.
352. S Usui, K Hamasaki, T Kuwabara, A Nakamura, T Ikeda, H Ikeda, A Ueno, F Toda. Supramol Chem 9:57, 1998.
353. K Hamasaki, H Ikeda, A Nakamura, A Ueno, F Toda, I Suzuki, T Osa. J Am Chem Soc 115:5035, 1993.
354. K Hamasaki, A Ueno, F Toda, I Suzuki, T Osa. Bull Chem Soc Jpn 67:516, 1994.
355. VJP Srivatsavoy. J Luminesc 82:17, 1999.
356. S Hamai. Chem Phys Lett 267:515, 1997.
357. L Jianzhong, S Jiang, Y Xu, J Wei. Anal Chim Acta 349:17, 1997.
358. M Inouye, K Hashimoto, K Isagawa. J Am Chem Soc 116:5517, 1994.
359. KN Koh, K Araki, A Ikeda, H Otsuka, S Shinkai. J Am Chem Soc 118:755, 1996.
360. E van Dienst, BHM Snellink, I Von Piekartz, JFJ Engbersen, DN Reinhoudt. J Chem Soc Chem Commun 1151, 1995.
361. A Ueno, K Takahashi, T Osa. J Chem Soc Chem Commun 921, 1980.
362. A Ueno, F Moriwaki, T Osa, F Hamada, K Murai. J Am Chem Soc 110:4323, 1988.
363. K Kano, I Takenoshita, T Ogawa. Chem Lett 321, 1982.
364. WG Herkstroeter, PA Martic, TR Evans, S Farid. J Am Chem Soc 108:3275, 1986.
365. A Yamauchi, T Hayashita, S Nishizawa, M Watanabe, N Teramae. J Am Chem Soc 121:2319, 1999.
366. A Ueno, I Suzuki, T Osa. J Am Chem Soc 111:6391, 1989.
367. A Ueno, I Suzuki, T Osa. Anal Chem 62:2461, 1990.
368. A Ueno. In J-P Desvergne, AW Czarnik, eds. Chemosensors of Ion and Molecule Recognition, NATO ASI Series C, vol 492. Dordrecht, The Netherlands: Kluwer Academic, 1997.
369. S Minato, T Osa, M Morita, A Nakamura, H Ikeda, F Toda, A Ueno. Photochem Photobiol 54:539, 1991.
370. I Suzuki, M Ohkubo, A Ueno, T Osa. Chem Lett 269, 1992.
371. S Minato, T Osa, A Ueno. J Chem Soc Chem Commun 107, 1991.
372. A Ueno, S Minato, T Osa. Anal Chem 64:2562, 1992.
373. I Suzuki, M Ito, T Osa. Chem Pharm Bull 45:1073, 1997.
374. P Choppinet, L Jullien, B Valeur. J Chem Soc Perkin Trans 2:249, 1999.
375. C Perez-Jimenez, SJ Harris, D Diamond. J Chem Soc Chem Commun 480, 1993.
376. I Aoki, H Kawabata, K Nakashima, S Shinkai. J Chem Soc Chem Commun 1771, 1991.
377. T Jin, K Ichikawa, T Koyama. J Chem Soc Chem Commun 499, 1992.
378. I Aoki, T Sakaki, S Tsutsui, S Shinkai. Tetrahedron Lett 33:89, 1992.
379. M Takeshita, S Shinkai. Chem Lett 125, 1994.

380. H Matsumoto, S Shinkai. Tetrahedron Lett 37:77, 1996.
381. H Matsumoto, S Shinkai. Chem Lett 2431, 1994.
382. K Iwamoto, K Araki, H Fujishima, S Shinkai. J Chem Soc Perkin Trans 1:1885, 1992.
383. C Perez-Jimenez, SJ Harris, D Diamond. J Mater Chem 4:145, 1994.
384. F Szemes, D Hesek, Z Chem, SW Dent, MGB Drew, AJ Goulden, AR Graydon, A Grieve, RJ Mortimer, T Wear, JS Weightman, PD Beer. Inorg Chem 35:5868, 1996.
385. PD Beer. J Chem Soc Chem Commun 689, 1996.
386. R Grigg, JM Holmes, SK Jones, WDJA Norbert. J Chem Soc Chem Commun 185, 1994.
387. B Bodenant, T Weil, M Businelli-Puorcel, F Fages, B Barbe, I Pianet, M Laguerre. J Org Chem 64:7034, 1999.
388. S Santra, P Zhang, W Tan. J Chem Soc Chem Commun 1301, 1999.
389. H-F Ji, R Dabestani, GM Brown, RA Sachleben. J Chem Soc Chem Commun 122:833, 2000.
390. H-F Ji, GM Brown, R Dabestani. J Chem Soc Chem Commun 609, 1999.
391. F Arnaud-Neu, EM Collins, M Deasy, G Ferguson, SJ Harris, B Kultner, AJ Lough, MA McKervey, E Marques, BL Ruhl, MJ Schwing-Weil, EM Seward. J Am Chem Soc 111:8681, 1989.
392. I Leray, F O'Reilly, J-LH Jiwan, J-P Soumillion, B Valeur. J Chem Soc Chem Commun 795, 1999.
393. GG Talanova, NSA Elkarim, V Talanov, RA Bartsch. Anal Chem 71:3106, 1999.
394. I Suzuki, M Ito, T Osa, J-I Anzai. Chem Pharm Bull 47:151, 1999.
395. I Aoki, T Sakaki, S Shinkai. J Chem Soc Chem Commun 730, 1992.
396. HFM Nelissen, AFJ Schut, F Venema, MC Feiters, RJM Nolte. J Chem Soc Chem Commun 577, 2000.
397. M Hissler, A Harriman, P Jost, G Wipff, R Ziessel. Angew Chem Int Ed Engl 37:3249, 1998.
398. A Harriman, M Hissler, P Jost, G Wipff, R Ziessel. J Am Chem Soc 14, 1999.
399. PD Beer, V Timoshenko, M Maestri, P Passaniti, V Balzani. J Chem Soc Chem Commun 1755, 199.
400. H Miyaji, P Anzenbacher Jr, JL Sessler, ER Bleasdale, PA Gale. J Chem Soc Chem Commun 1723, 1999.
401. T Jin. J Chem Soc Chem Commun 2491, 1999.
402. S Weidner, Z Pikramenou. J Chem Soc Chem Commun 1473, 1998.
403. L Jullien, J Canceill, B Valeur, E Bardez, J-P Lefevre, J-M Lehn, V Marchi-Artzner, R Pansu. J Am Chem Soc 118:5432, 1996.
404. MN Berberan-Santos, P Choppinet, A Federov, L Jullien, B Valeur. J Am Chem Soc 121:2526, 1999.
405. Z Pikramenou, DG Nocera. Inorg Chem 31:532, 1992.
406. Z Pikramenou, KM Johnson, DG Nocera. Tetrahedron Lett 34:3531, 1993.
407. Z Pikramenou, DG Nocera. Proceedings of the Sixth International Symposium on Cyclodextrins. Paris: Editions de Santé, 1993, p 259.
408. MA Mortellaro, DG Nocera. J Am Chem Soc 118:7414, 1996.

409. CM Rudzinski, DS Engebretson, WK Hartmann, DG Nocera. J Phys Chem A 102:7442, 1998.

410. FJ Steemers, W Verboom, DN Reinhoudt, EB van der Tol, JW Verhoeven. J Am Chem Soc 117:9408, 1995.

411. H Matsumoto, S Shinkai. Chem Lett 901, 1994.

412. N Sabbatini, M Guardigli, A Mecati, V Balzani, R Ungaro, E Ghidini, A Casnati, A Pochini. J Chem Soc Chem Commun 878, 1990.

413. JA Semlyen, ed. Large Ring Molecules. New York: Wiley Interscience, 1996.

414. BO Dabbousi, J Rodriguez-Viejo, FV Mikulec, JR Heine, H Mattoussi, R Ober, KF Jensen, MG Bawendi. J Phys Chem B 101:9463, 1997.

415. M Bruchez Jr, M Moronne, P Gin, S Weiss, AP Alivistos. Science 281:2013, 1998.

416. WCW Chan, S Nie. Science 281:2016, 1998.

417. DT McQuade, AE Pullen, TM Swager. Chem Rev 100:2537, 2000.

418. MJ Marsella, RJ Newland, PJ Carroll, TM Swager, J Am Chem Soc 117:9842, 1995.

419. KB Crawford, MB Goldfinger, TM Swager. J Am Chem Soc 120:5187, 1998.

420. E Delamarche, A Bernard, H Schmid, A Bietsch, B Michel, H Biebuyck. J Am Chem Soc 120:500, 1998.

421. E Delamarche, A Bernard, H Schmid, B Michel, H Biebuyck. Science 276:779, 1997.

422. CM Rudzinski, AM Young, DG Nocera. (Submitted for publication.)

2

Luminescent PET
Signaling Systems

A. Prasanna de Silva, David B. Fox,
Thomas S. Moody, and
Sheenagh M. Weir
Queen's University of Belfast, Belfast, Northern Ireland

I. INTRODUCTION

Since their generalization over a decade ago, luminescent PET (photoinduced electron transfer) signaling systems have grown to be a widely researched area of molecular sensing and switching devices. Beside ourselves [1–13], Bernard Valeur [14–16] (Ecole Normale Superieur de Cachan, France), Tony Czarnik [17–21] (formerly of Ohio State University), Jean-Pierre Desvergne [21,22] (Université de Bordeaux, France), Luigi Fabbrizzi [23-27] (Universita de Pavia, Italy), Seiji Shinkai [28,29] (Kyushu University, Japan), and Kanji Kubo [30] (now at Kanagawa University, Japan) have contributed to the review literature of this area. Since our 1997 review [9] tried to cover the available literature comprehensively, the present chapter discusses pre-1997 work only to set the stage for a rational discussion of more recent material. It is a particular pleasure for us to trace the conceptual roots of several modern research lines back to our efforts at the University of Colombo, Sri Lanka and the Queen's University of Belfast, Northern Ireland stretching over 15 years.

Luminescent molecular signaling systems usually serve two general needs: gathering information (preferably continuously) from small and fragile spaces, and processing information in small spaces denied to semiconductor technology. The first need has received attention with ever-increasing success. The second has been recognized for over a decade [31], but generic solutions are yet to be established.

II. PET SIGNALING SYSTEM DESIGN

Our design principle is a simple one. Continuous information gathering requires molecular sensors. These observe their targets at contact distances followed by data transmission over a longer range to their controller, that is, a "catch-and-tell" operation. Such operations would logically require the sensor molecule to possess two nearly autonomous departments held within a federation to perform the "catching" and "telling" acts. This is why the anatomy of a sensor molecule has the lumophore-spacer-receptor format [1,2]. The receptor molecule reversibly captures a representative of the analyte population if the latter is concentrated enough. The power of light absorption and emission possessed by the lumophore module enables it to perform the powered data transmission.

How is the "catching" act translated into a signal ready for transmission? We need to examine Fig. 1 to answer this question in sufficient detail. The molecular device is powered by the excitation light. It loses the excited state energy by transferring an electron from the receptor module to the lumophore.

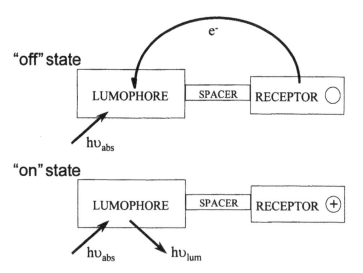

Figure 1   PET signaling system design.

The opposite direction of transfer is also perfectly exploitable if required by the designer. Such PET processes are of course responsible for green plant photosynthesis [32–36]. In the present simple instance the electron transfer is essentially reversed immediately following excitation. Such self-repair is essential following a potentially lethal PET process so that the molecular system is reasonably photostable. The important net result for us is that no luminescence is seen following excitation. This is the "off" state of the molecular switch.

The situation changes sharply the moment the receptor module within the molecular system captures the chosen analyte (which is a cation for the purposes of illustration). Again, if we consider excitation of the lumophore module, we find the analyte influences the transferring electron. At the simplest level, this is due to electrostatic charge attraction between the jumping electron and the captured analyte. So the photoinduced electron transfer is arrested and the unused energy of the excited state is dumped as a luminescence photon. This is how the analyte population is visualized. Now we have the "on" state of the molecular switch.

III. A CASE STUDY

We find **1** [37] illustrates the principles, features, and prospects of luminescent PET signaling systems rather well. As it stood, **1** was planned as a sensor for

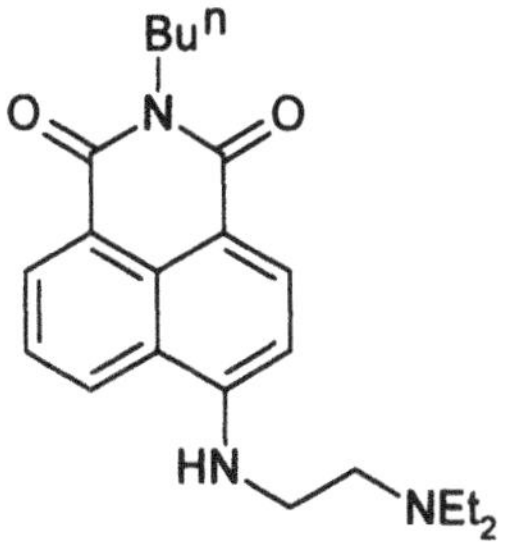

1

protons excitable at visible (blue) wavelengths. The lumophore-spacer-receptor system is composed of a 4-aminonaphthalimide lumophore, a tertiary amine receptor, and a dimethylene (C_2) spacer. Planning for PET processes requires a thermodynamic assessment, carried out three decades ago by Albert Weller at the Max Planck Institute, Gottingen, Germany [38]. Photoexcitation of the lumophore creates a vacancy in its highest occupied molecular orbital (HOMO) whose energy can be represented by its oxidation potential (+1.1 V vs. standard calomel electrode (SCE)) [37]. The nonbonding electrons on the receptor can be modeled by triethylamine whose HOMO energy is represented by its oxidation potential (+1.1 V vs. SCE) [39]. One of these nonbinding electrons transfers to the vacancy in the HOMO of the lumophore since the energies of the starting and finishing orbitals happen to be the same. In fact, the electron transfer creates a lumophore radical anion and a receptor radical cation, whose electrostatic attraction is worth 0.1 eV of stabilization in polar media [40]. So in the present case, the free energy change for the PET process (ΔG_{PET}) is −0.1 eV. The seminal research by Rudolph Marcus at the California Institute of Technology [41] shows that fast PET rates tending towards vibrational frequencies are found in this nearly isoergonic regime. So it is no surprise that PET overwhelms fluorescence as far as **1** is concerned. Of course, protonation of the tertiary amine makes its oxidation potential immeasurably high and the ΔG_{PET} value shifts to a large positive number (>+1.4 eV). Now fluorescence becomes the winner of the competition. So a large fluorescence enhancement is created upon protonation of **1**.

Concurrent measurement of fluorescence quantum yields (ϕ_F) and fluorescence lifetimes (τ_F) for the protonated and unprotonated forms of **1** (along with model **2**) permits the calculation of PET rates. **1** has a PET rate of 3.3×10^9 s^{-1}

2

when unprotonated. When protonated, the PET rate becomes a part (if any) of a total nonradiative deactivation rate of 0.03×10^9 s^{-1}.

The spacer length naturally controls the PET rate in **1** with its homologues since the electron has farther to jump in the latter cases. When **1** with its C_2 spacer is stretched into a C_3-spaced version, the PET rate falls by an order of magnitude to 0.3×10^9 s^{-1}. Further, stretching to a C_4-spaced version causes the PET rate to plummet further (by another order of magnitude) to 0.03×10^9 s^{-1}. These results are in line with older studies on distant-dependent PET [32,35]. When viewed from a signaling standpoint, the proton-induced fluorescence enhancements (FE values) are almost useless in the C_4-spaced cousin of **1**.

The mention of radical ions three paragraphs ago is part of a general truth: PET is naturally charge separating within electroneutral systems. So PET

Table 1 Solvent Dependence of the Fluorescence Quantum
Yield and Lifetime of **1**[a]

Solvent	CH	DCM	ETAC	W/M	AN
ϕ_F	0.95	0.62	0.42	0.03	0.02
τ_F (ns)	7.4	7.5	5.7	0.2	0.2

[a]CH = cyclohexane, DCM = dichloromethane, ETAC = ethyl acetate,
W/M = water/methanol (4:1, v/v), AN = acetonitrile.

rates become very sensitive to the polarity of the matrix. Since fluorescence
competes with PET for deactivation of the excited state (of **1**), ϕ_F and τ_F values
become environmentally sensitive too. The 4-aminonaphthalimide lumophore
possesses some environmental sensitivity of its own due to the push–pull nature
of its pi-electron system. In fact, this excited state has significant internal charge
transfer (ICT) from the 4-amino group to the imide unit. Previous work on **2**
from Madrid, Spain [42] and Preston, England [43] have detailed this aspect of
the lumophore. But **1** represents polarity dependences that are much larger [44]
(Table 1).

The substantial electric field enveloping the excited state of **1** is to be
expected from its ICT nature. In fact a dipole moment of 11D can be measured
[37]. Of course, such fields are vectorial and they can help or hinder electron
transfer depending on relative orientation. Such effects are clearest when the
electron transfer is not heavily biased thermodynamically. The near isoergonic
condition of **1** suits us nicely. The 4-amino group is close to the positive terminal
of the excited state dipole whereas the imide unit houses the negative end.
So, **1** fits the happy situation where the incoming photoinduced electron is
attracted towards the lumophore. On the other hand, **1**'s regioisomer **3** repulses

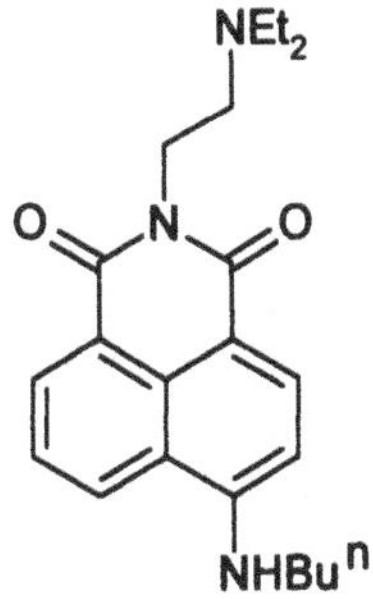

3

any electron invasion via its photogenerated electric field. Another contributory
effect to the poor PET effects in **3** is the node found on the imide nitrogen

in both frontier orbitals. Fox and Galoppini [45,46] (then at the University of Texas at Austin) also found the strong susceptibility of PET to molecular-scale electric fields, due to the α-helix polypeptides.

The fact that PET can be controlled by photogenerated molecular-scale electric fields has been beautifully established by Wasielewski's team [47] at Northwestern University, Evanston and Argonne National Laboratory. Their experiment is ingenious regarding design of the test compound as well as the measurement protocol. Rigid rod-like **4** is of the donor$_1$–acceptor$_1$–acceptor$_2$–donor$_2$ format with the modularity of all components being sterically enforced. Interspersed benzene rings reinforce the modularity at two of the three functions. Both donors are good lumophores on their own, which facilitates the assignment of their singlet energies. The aminoperyleneimide donor$_1$, and the porphyrinZn(II) donor$_2$ can be excited separately with minimal cross-talk. Each donor–receptor pair is arranged for exergonic and rapid PET, the latter being encouraged by benzenoid or virtual (C_0) spacers. Electronic energy transfer (EET), expected in bilumophore systems, is overpowered by rapid PET. A two pump-probe measurement in the picosecond domain is used such that, say, donor$_1$ is excited first. PET produces the donor$_1$$^{\cdot+}$–acceptor$_1$$^{\cdot-}$ pair. Within the lifetime of this radical ion pair, donor$_2$ is then excited by the second pump pulse. Now, PET can produce the donor$_2$$^{\cdot+}$–acceptor$_2$$^{\cdot-}$ pair, but only if the photogenerated electric field due to the preformed donor$_1$$^{\cdot+}$–acceptor$_1$$^{\cdot-}$ pair is not present. The latter field is turned off and on in a controlled manner by running the first pump pulse train (exciting donor$_1$) at half the repetition rate of the second pump pulse train (exciting donor$_2$). In this way, the preformed electric field is only present during alternate excitations of donor$_2$. Subtraction of the absorption transients from adjacent pulses of the second pump pulse train produces the signature of the species that was perturbed by the preformed electric field, that is, the donor$_1$$^{\cdot+}$–acceptor$_2$$^{\cdot-}$ pair. Actually, only the strongly absorbing acceptor$_2$$^{\cdot-}$ is experimentally observed by the probe pulse. Detailed analysis of the intensity of the absorption transient shows that every PET event is stopped by the performed electric field. More generally, this work underlines how molecular-scale photonic devices can be driven at potentially gigahertz rates [48].

The comparative signaling behavior of **1** and **3** [49] also allowed us to emulate the path-selective PET seen in the bacterial photosynthetic reaction centre [50]. The stage is set within **5** and **6**, both of which are receptor$_1$–spacer$_1$–lumophore–spacer$_2$–receptor$_2$ systems with PET originating from either receptor. Both PET paths appear feasible thermodynamically in each of **5** and **6**. The two PET paths in each molecule can be easily distinguished by the basicity of each tertiary amine group. In the event, PET is observed only along the path involving R_2 whichever tertiary amine is chosen.

Tian, Chen, and their team at the East China University of Science and Technology, Shanghai, also exploit the special features of aminonaphthalimide-

4

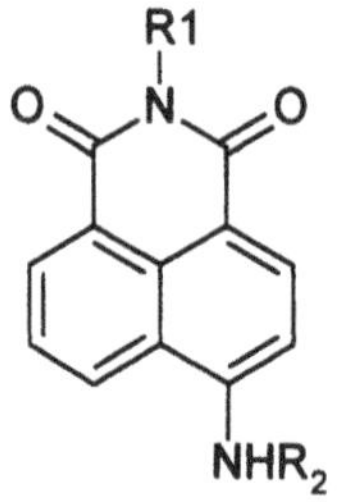

5; R_1 = diethylaminoethyl; R_2 = morpholinoethyl
6; R_1 = morpholinoethyl; R_2 = diethylaminoethyl

based PET systems for emulation of some aspects of the photosynthetic reaction center [51]. **7** is effectively of the acceptor–spacer$_1$–lumophore–spacer$_2$–receptor

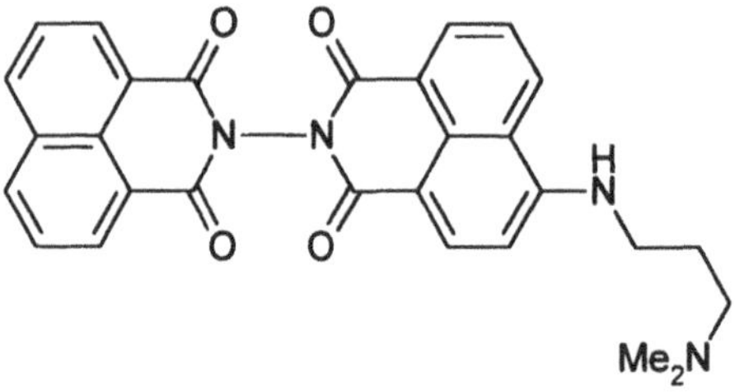

7

format, where the unswitchable electron acceptor (1.8-naphthalimide) is sterically held orthogonal to the lumophore (4-amino-1,8-naphthalimide) so that the spacer is virtual, that is, C_0. The receptor (tertiary amine) serves as the electron donor, which property can be switched off at will by protonation. So there are two PET paths involving the lumophore, one reductive (and switchable) and one oxidative. This makes **7** a particularly interesting functional model of the photosynthetic reaction center. Naturally emission will not survive in any strength if even a single PET path remains operative. The oxidative PET path is very efficient, both thermodynamically and kinetically (owing to the C_0 spacer), whereas the reductive path is efficiency limited by the length of the C_3 spacer. Still, proton-induced fluorescence enhancement factors of 3.5 are observed in suitable reference compounds. Of course the emission of **7** remains switched "off" right across the pH range.

The fact that **1** is fronting a host of molecular lumophore–spacer–receptor PET signaling systems appealed to Tusa, Leiner, and their collaborators at AVL Biosense Corporation, Atlanta, and Graz, Austria. Related sensory molecules now lie at the heart of blood electrolyte measurement in critical care units in hospitals [52]. This is perhaps the clearest endorsement to date of the device capability of luminescent PET signaling systems.

IV. PET SIGNALING SYSTEMS BASED ON ACYCLIC RECEPTORS

We begin this section by noting the commercialization of our **8** [53] and **9** [54] and relatives by Molecular Probes Inc., Eugene, OR [53], as monitors of acidic

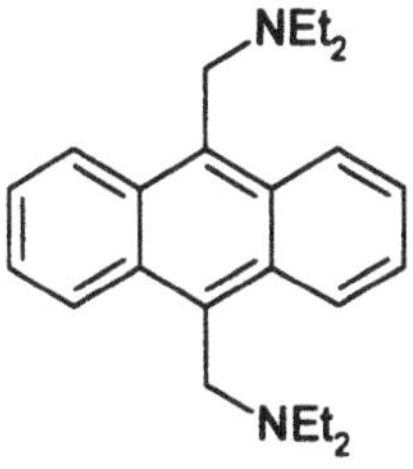

lysosomal compartments in living cells. The protonated forms of these molecules are actively transported into the lysosomes. The lysosomes show up as regions of bright luminescence and are easily tracked as they move about the cell during the ingestion and trafficking of foreign matter.

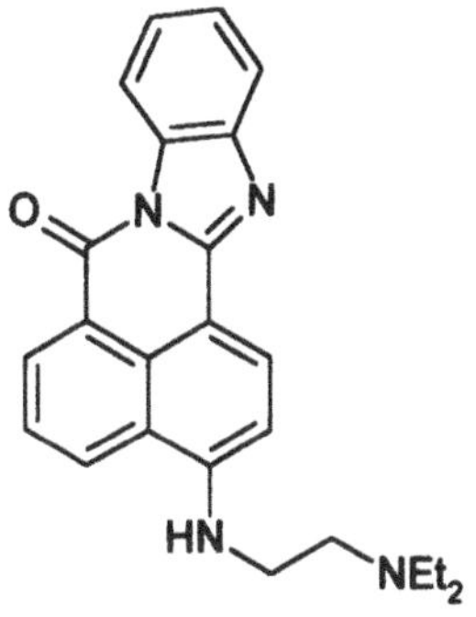

Copeland and Miller [56] at Boston College, MA have applied the classical fluorescent PET pH signaling systems to advance the field of catalyst discovery. In this powerful work, the pH sensor **10** monitors the surface regions of a

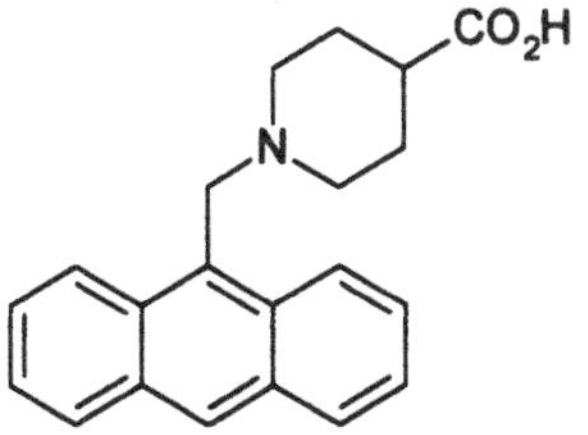

polymer bead for the presence of the reaction product. The bead carries both the catalyst and the sensor while the reactants are dispersed in the solution phase. Thus catalytic action is visualized, provided that the reaction product does not diffuse away from the point of origin within the observation period. In the example considered, alcohol **11** is acylated with acetic anhydride in toluene to produce acetic acid which switches "on" the sensor **10**, when catalyzed by **12**. The on-bead version is prepared by attaching **12** via its C-terminal. Evidently, the

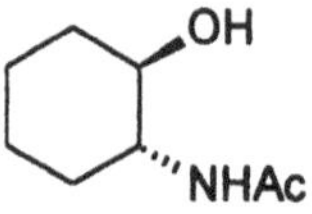

11 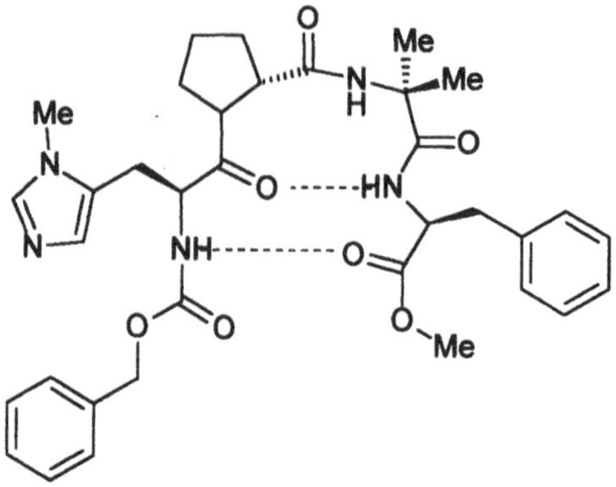   12

tertiary amine unit within the sensor does not act in a catalytic role presumably because of its steric congestion.

An early report [57] from Michael Schuster and his team at the Technical University of Munich, Germany highlighted the lumophore–spacer–receptor system **13** which showed strong fluorescence recovery with intrinsic quenchers

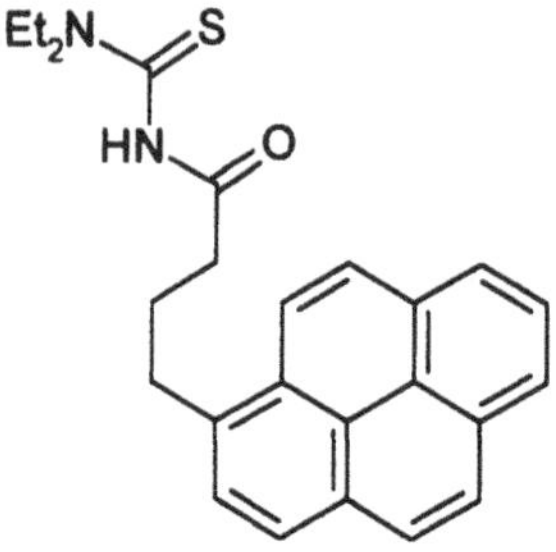

13

such as Cu^{2+}. Thiourea groups can be reasonably expected to participate in PET processes [58], hence the low fluorescence efficiency of the free ligand **13**. Cu^{2+} binding should deprotonate the acylthiourea but the complexed unit is evidently unable to launch a PET process towards the lumophore. It is important to note that such a deprotonated ligand should suppress the electron deficiency of the Cu(II) center, one of the principal contributions to the intrinsic quencher status of the latter ion.

Several structural and conceptual similarities can be seen between **13** and **14** [59]. **14** reserves its best performance for Cd^{2+} with a fluorescence enhancement factor of 6. Its lack of response to protons is no surprise owing to the lack of an undelocalized amine within the iminoylthiourea moiety. However, the true

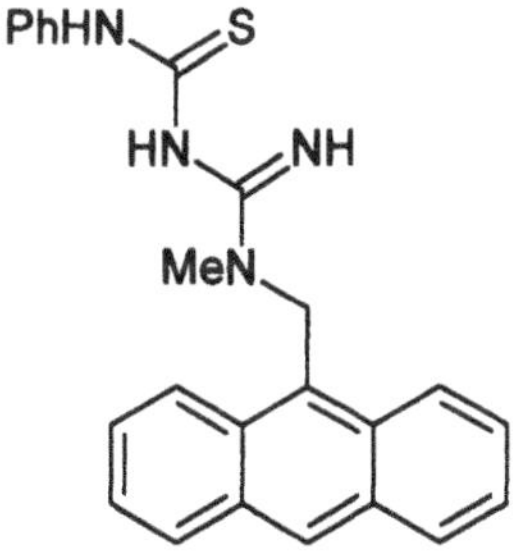

14

uniqueness of this example from Ute Resch-Genger at her Berlin laboratory and her colleagues at Leipzig, Germany lies in the fact that **14** is only one half of the story. **14** and **15** form a redox pair such that the receptor unit within the

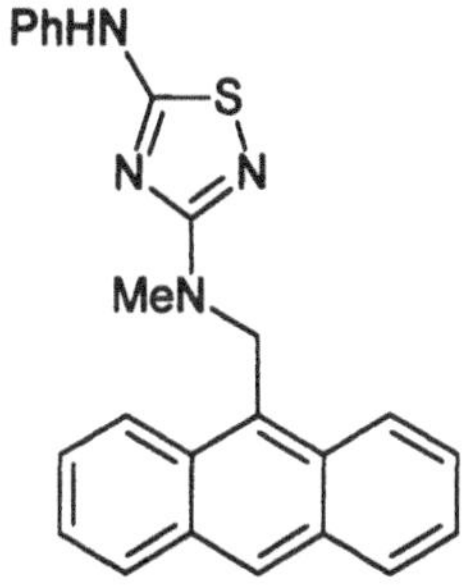

15

lumophore–spacer–receptor system can be switched at will to drastically alter the sensory capability. The aminothiadiazole receptor within **15** selects Hg^{2+} and a fluorescence enhancement factor of 44 is seen. Arguments advanced previously concerning such an unusual behavior by an intrinsic quencher are equally valid here. The slow desulfurization of **14** by Hg^{2+} (seen before in Czarnik's work [58] at Ohio State University) is the only cloud among the bright prospects offered by the device system of **14** and **15**.

The next contribution from Munich has as its target the continuing need for fluorescent sensors for various heavy metals. Heavy atom effects, intramolecular charge transfer, and magnetic interactions can often lead to fluorescence quenching in these situations. **16** [60] is capable of complexing over 25 metal ions. Nevertheless, some degree of selectivity can be achieved by varying pH, although no fluorescence can be observed at low pH values. Complexation of metal ions causes both a hypsochromic shift of emission and fluorescence quenching. The exception to this is the complexation of Cu^{2+}, which produces a bathochromic shift in emission in addition to quenching.

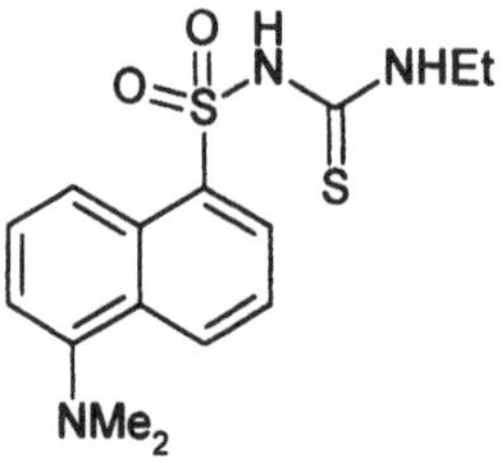

16

Schuster et al.'s latest effort [61]—a clear PET system of the lumophore–spacer–receptor format—breaks free from this quenching mold just like their 1993 case [57]. Sensor **17** is capable of binding (through the S atom to Cu^{2+} or

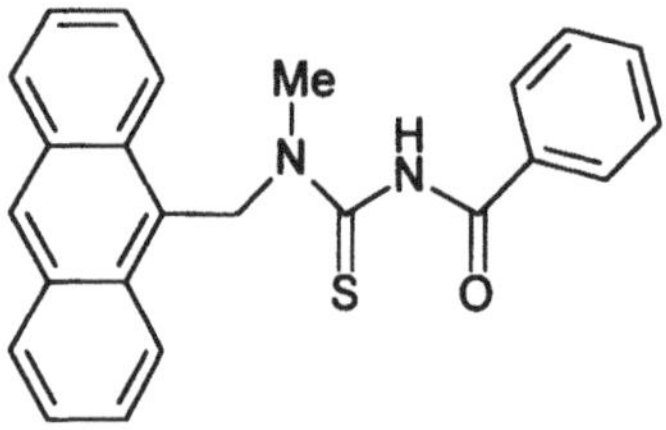

17

through the S and O to Hg^{2+}) to thiaphilic metal ions while ignoring the harder group I and II metal ions. The complexation of these ions suppresses PET from the thiourea moiety to the lumophore. Strong enhancements of fluorescence are seen. Of the metals tested, the sensor is selective for Cu^{2+} and Hg^{2+}. The interaction with Cu^{2+} is time dependent as the first step requires reduction of Cu(II) to Cu(I), a worrying feature for the stability of **17**. Furthermore, placing the sensor in micellar media (Triton X-100) shows this redox reaction enabling the sensor to be nicely selective for Hg^{2+}. It is important that signaling systems like **17** do not undergo desulfurization under the sensing conditions. This avoids the lack of reversibility seen in earlier PET-based systems [58].

The worries about the instability of sulfur-based signaling systems such as **13**, **14**, **16**, and **17** have crystallized in the following cautionary tale. Ute Resch-Genger's colleagues in Berlin and Leipzig, Germany, and Kiev, Ukraine, now find that the Cu^{2+}-induced fluorescence enhancement previously seen with **18** [62] is not due to simple complex formation but to oxidative rearrangement into **19** [63]. However, the 9-anthroyl unit in **18** seems to be the main culprit.

Just as the sulfur-based signaling systems have been successful at achieving clear fluorescence enhancements with "intrinsically" quenching transition metal ions since 1993, there is a growing body of literature on a parallel development with aliphatic amine-based compounds. Bharadwaj and coworkers

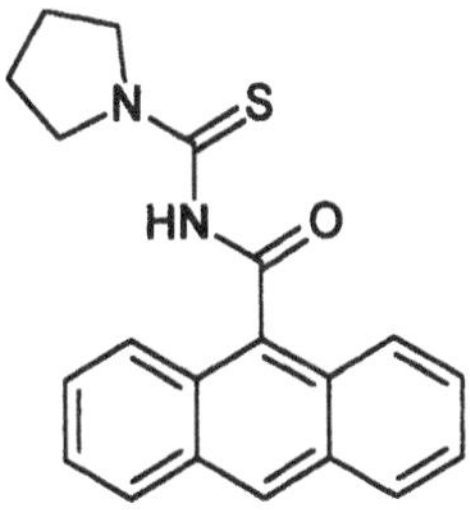

18

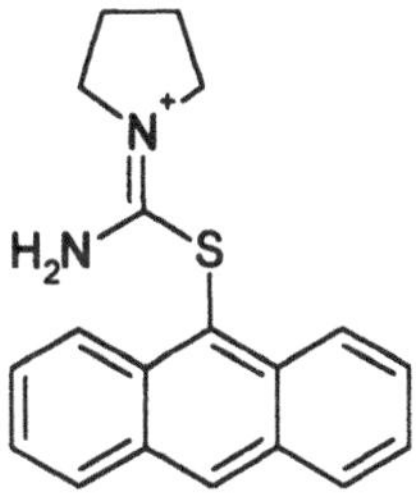

19

at The Indian Institute of Technology at Kanpur produced the first of these cases, **20** [64]. In this instance, the clear suppression of the redox activity of

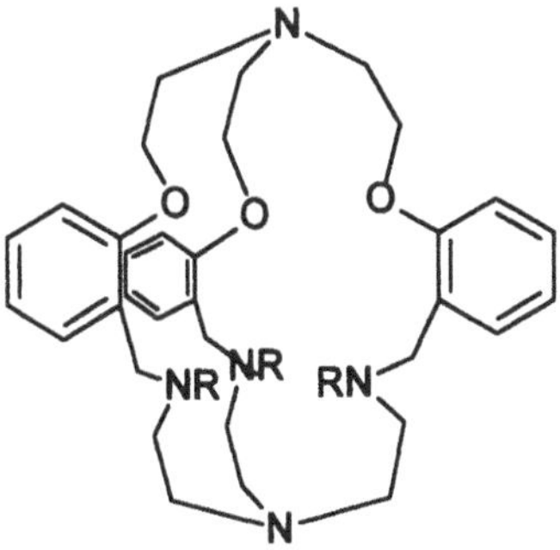

20; R = 9-Anthrylmethyl

the metal center by the cryptand environment could be a pointer to its success. The isolation of characterizable transition metal complexes of cryptands of this type is another comforting feature of this work. The same comforts are not found in several more recent reports arising from the University of Hyderabad, India, headed by Anunay Samanta [65–67]. However, the latter work is appealing because of the extreme simplicity of the systems such as **21**, **22**, and the

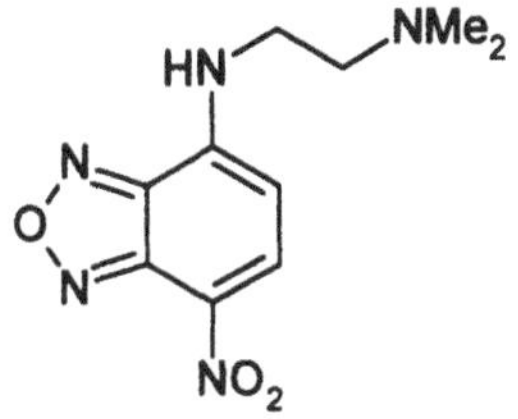

21

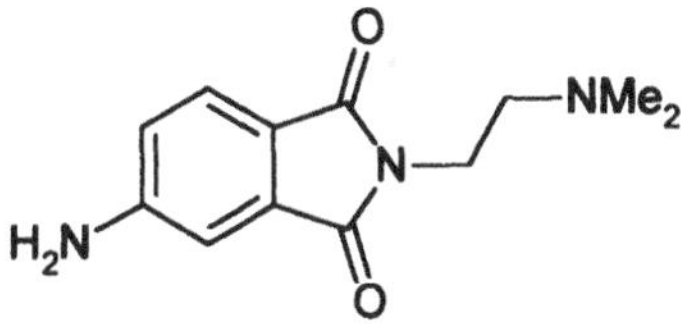

22

range of transition metal ions that produce large fluorescence enhancements. Even the extremely stubborn case of Fe^{3+} is showcased in one publication [67]. Nevertheless, some of the worries concerning the sulfur-based systems may have some relevance since the tertiary aliphatic amine-based systems such as **21** are very proton sensitive while they are not the best binders of transition metal ions due to their weakly chelating nature. The same arguments may apply to Cu^{2+}-induced fluorescence enhancement of a structural cousin of **1** [68].

An example of old PET signaling systems being applied to new purposes appears in the recent report on **23** as a boronic acid sensor by Wang and col-

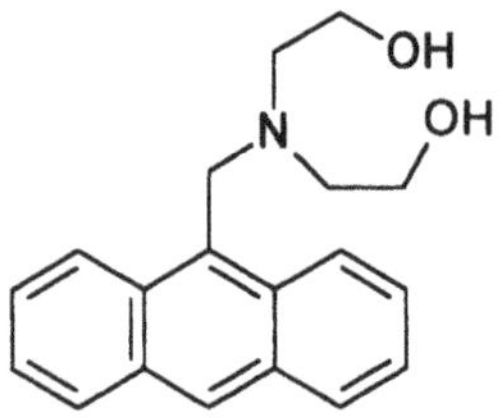

23

leagues at North Carolina State University [69]. We disclosed **23** as a member of the 9-aminomethyl anthracene family with respect to protons in 1992 [70]. Shinkai's **24** showed in 1994 how aminomethylanthracenes can produce large

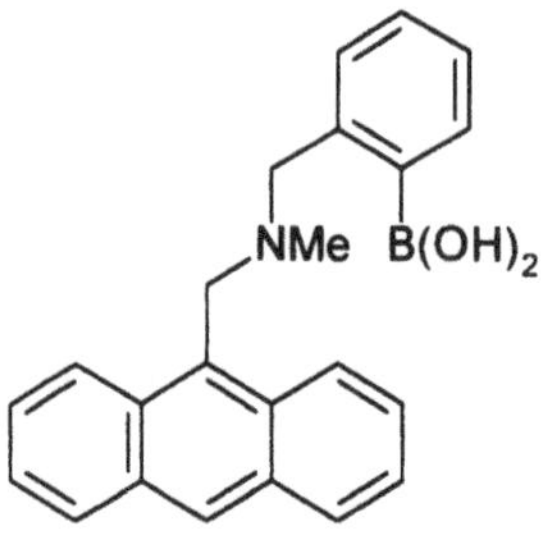

24

fluorescence enhancements when cyclic boronate esters of diols interact with the benzylic nitrogen center [71]. Wang's work considers the bor(on)ic acid as the analyte and aminodiol moiety within **23** as the responsive host whereas Shinkai looked upon the saccharide diol as the analyte and the aminoboronic acid moiety within **24** as the responsive host. Both cases involve a B−N and two B−O bonds when the fluorescence is switched "on." A further similarity in the two cases is the presence of two five-membered rings containing boron in the fluorescent state. The ever-present problem of protonation-induced fluorescence enhancement in aminomethylanthracenes such as **23** has been addressed in the present case by using another old favorite of ours, **25**, devoid of the diol features under the same experimental protocol. Bor(on)ic acids appear reluctant to

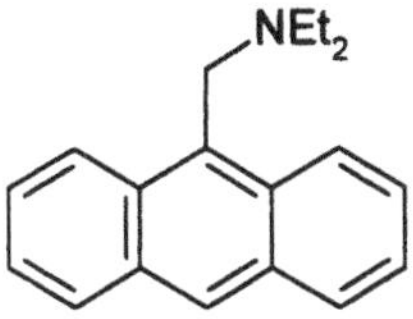

25

transfer protons to **25** in methanol solution as judged by the weaker fluorescence enhancements.

So far the receptors discussed have been largely based either on amines or on sulfur-containing groups. Sakamoto's **26** [72], from Wakayama and Nagoya

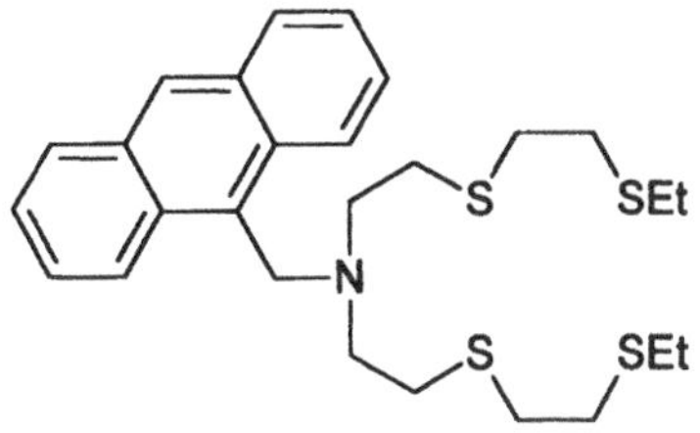

26

Universities, Japan, has an interesting receptor in that aliphatic amines and sulfur atoms are both involved. On addition of Ag^+ to **26** in acidic medium, a decrease of emission is seen. Ag^+ displaces the H^+ that had prevented the PET from the amine to the lumophore in the first place. Ag^+ can be expected to prevent PET from the amine, although not as well as H^+. This is partly because of Ag^+'s own ability to launch PET processes to itself owing to its redox activity. It follows therefore that Ag^+ does cause a significant enhancement of fluorescence in basic solution.

Most of the cases discussed above aimed for guest-induced emission enhancements, even if quenching was obtained in some cases with transition metal ion guests. But now we present the opposite situation. Such "on-off" signaling has its uses, as we show. **27** [73] from Fages' laboratory at the University of Bordeaux, France, has two ligands well disposed as a tetrahedral chelating site for Zn^{2+} and a pyrene lumophore. Complexation of Zn^{2+} results in the quenching of fluorescence, due to PET from the lumophore to the positively charged and planarized bipyridine units. The PET nature of the fluorescence quenching was checked by the revival of emission of **27**·Zn^{2+} complex in a low temperature glass. Our **28** [74] published simultaneously, while being a minimalist structure, has many similarities with **27**. These two studies along with an older one [75], confirm the "on-off" signaling achievable with pyridine receptors.

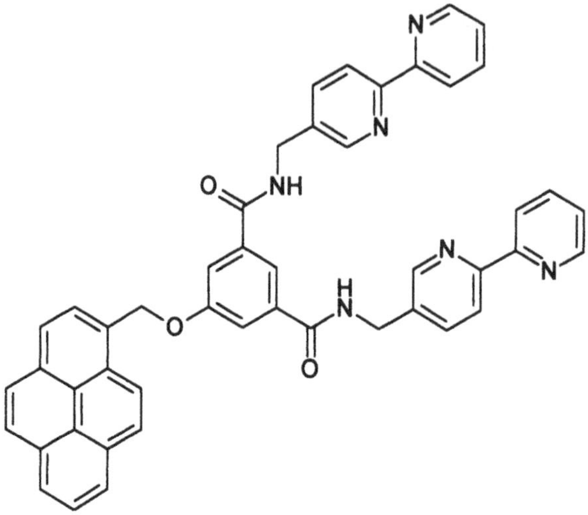

27

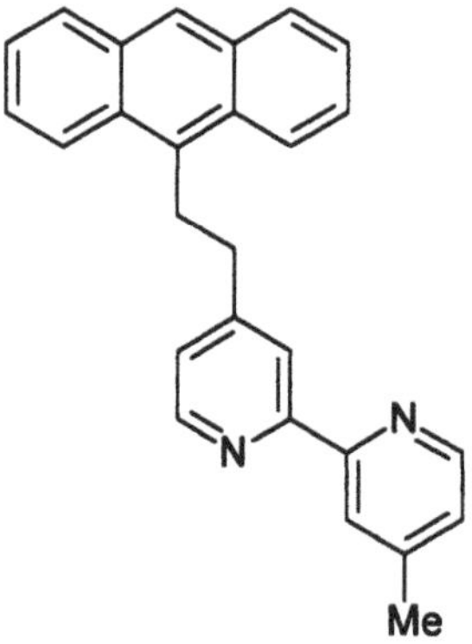

28

V. AZACROWN-BASED PET SIGNALING SYSTEMS

System **29** [76] has several things going for it. Two of its components, anthracene and monoaza-18-crown-6, are icons of photochemistry and supramolecular chemistry, respectively. The brilliant fluorescence enhancement of **29** with K$^+$ by a factor of 47 is remarkable for a structure so easily synthesized. Of course, it was an early demonstrator of luminescent PET signaling. So perhaps it is no wonder that the structure **29** has proved to be a springboard for several lines of research in the hands of several different people.

The bis-guanidium ion saxitoxin, **30**, is a poison when ingested via the consumption of shellfish. Rapid sensing of **30** could therefore be a matter of

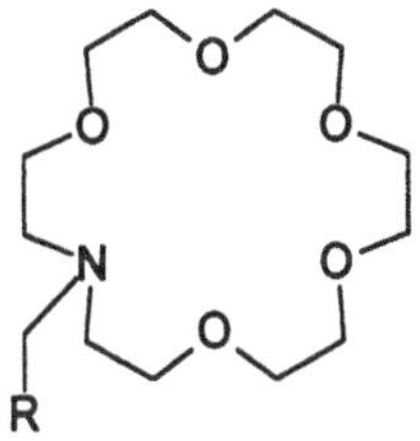

29; R = 9-Anthryl
33; R = 1-Pyrenyl

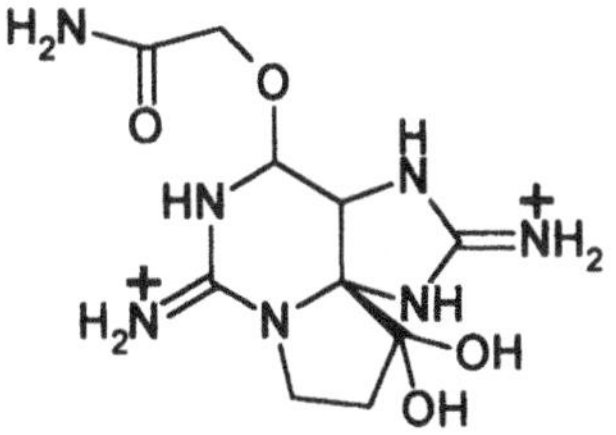

30

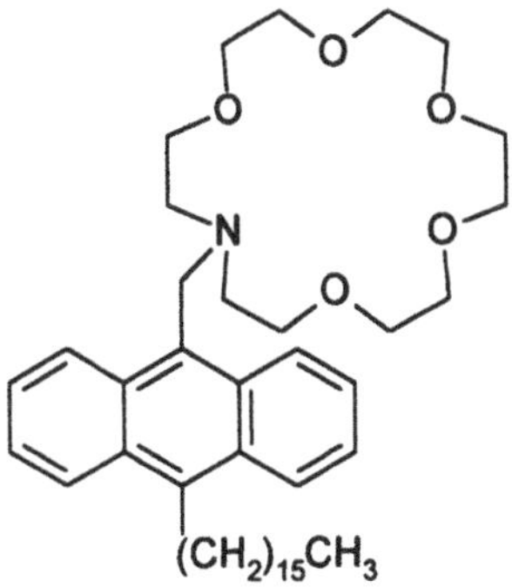

31

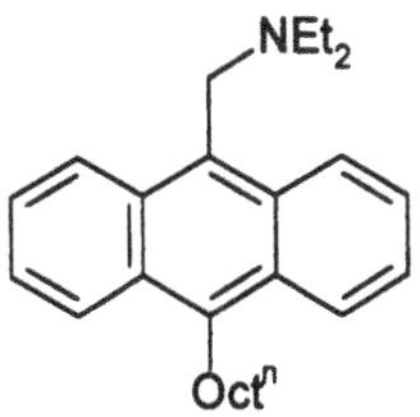

32

life and death. Now **30** is shown by Gawley and LeBlanc's team [77] at the University of Miami to bind to **29** [76] and its 15-crown-5 version [76]. Upon binding, hydrogen bond donation from **30** to the benzylic nitrogen retards the normal PET within **29** resulting in strong fluorescence enhancement of up to 30-fold. This is a new, and valuable, use for our old faithful **29**.

In an effort to carry the success of **29** forward from homogeneous solution [77] to films, the team at Miami has turned to **31** [78]. While **31** forms excellent monolayer films at the air–water interface, decent anthracenic fluorescence signatures are not produced until the films are diluted substantially with stearic acid and deposited onto hydrophobic glass. As may be anticipated, neat films of **31** only show broad excimeric emissions. Light collection problems dog the experiments on the films on water. No sensitivity of the emission towards pH is found, so experiments with saxitoxin are not reported at this stage. We note the closely related PET system **32** which shows good sensing of membrane-bounded protons in micellar media [79].

In another of their reincarnations, the archetypal PET signaling systems **29** and its smaller cousin show fluorescence intensity changes in the presence of paramagnetic metal cations such as Mn^{2+}, Co^{2+}, and Cu^{2+} [80]. Such paramagnetic metal ions quench fluorescence and can be rationalized via their unpaired electrons which can quench both singlet and triplet states of these anthrylaza-crown ethers. Fluorescence can also be recovered by the displacement of these metals from the azacrown ether by nonquencher metal ions such as Na^+.

The structural mutation of our **29** by variation of the macrocycle (via its heteroatoms and carbon bridges), the lumophore, and the spacer is obviously a rich vein to mine for luminescent PET sensors. Kubo's laboratory (formerly at Kyushu University, now at Kanagawa University, Japan) has systematically explored this line [81–84]. **33** has a straight swap of 1-pyrenyl for 9-anthryl [81]. **34** to **36** [81–83] use two nitrogens in the crown ether instead of the one in **29**,

34; R = 1-Pyrenyl
35; R = 1-Naphthyl
36; R = 9-Anthryl   37; R = 1-Naphthyl

with appropriate N-alkylation using arylmethyl groups. **37** [84] departs further from **29** by adopting the 1,5,9-triazacyclododecane receptor. **29** and **33** to **37** share the feature of significant fluorescence enhancement by s-block cations.

The C_3 symmetry of the receptor in **37** explains its selectivity towards NH_4^+ both in terms of binding and in terms of fluorescence switching. Some of these cases show weak exciplex emissions when free of cations, attesting further to their PET parentage. The well-known leanings of pyrene lumophores to engage in excimer interactions give rise to cation-induced enhancement in both the monomer and excimer emissions of **34**. Interestingly, the fluorescence enhancement selectivity pattern is clearly different for the excimer and monomer bands. Another pattern which emerges is that the systems with two lumophores (**34** through **36**) have far lower binding constants than the monolumophore cases (**29, 33**). The point that the 9-anthrylmethyl unit does not noticeably perturb the alkali cation-binding constant of **29** was made back in 1986 [76]. **34** to **36** not only have two bulky hydrocarbon moieties that can interfere with cation occupancy of the receptor, but their diazacrown receptor is intrinsically poorer at binding alkali cations than monoazacrowns. Another possible reason was recently pointed out by Gokel and his colleagues at the Washington University School of Medicine, St. Louis, after study of a whole host of dialkyl diazacrown ethers [85,86]. Pairs of aromatic C−H=O interactions are found in the X-ray crystal structures of alkali cation complexes of **35** and **36** that can decrease the affinity of the crown ether oxygens toward the alkali cation. A similar pair of interactions is also seen in the X-ray crystal structure of **34** itself, which would tend to hold the hydrocarbon moieties such that cation binding to the receptor is impeded.

A much larger structural change of **29** can be discerned in Dabestani's **38**, and also in **39** [87]. The 9-anthrylmethylazamacrobicycle clearly behaves as a

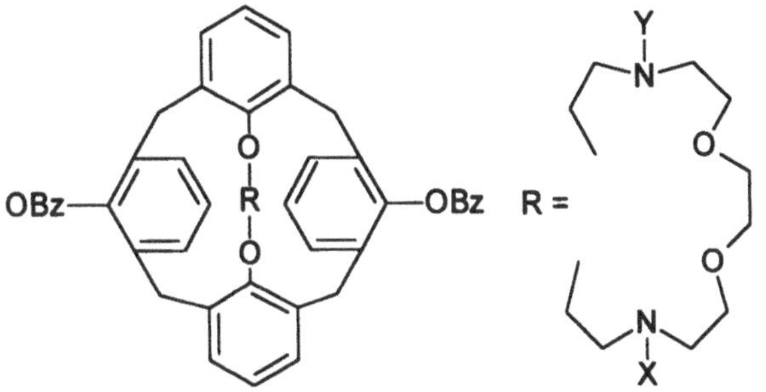

38; X = 9-Anthrylmethyl, Y=H
39; X = Y = 9-Anthrylmethyl

PET system with gradually mounting fluorescence enhancement factors as the size of the alkali cation increases. The 1,3-alternate calix[4]arene moiety within the macrobicycle is clearly the reason for this bias toward larger alkali cations.

In a Bologna–Milan collaboration [88], Maestri and his colleagues have carefully examined further structural mutations of **29** in the form of **40** and **41**.

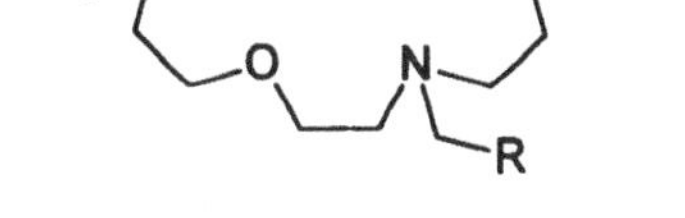

40; R = 9-Anthryl
41; R = 1-Naphthyl

The spectroscopy of ligands when free and when subjected to different extents of protonation is the subject of the present study. The influence of metal ions will doubtless be examined in later papers. Basic PET signaling is confirmed via proton-induced fluorescence enhancement, but evidence for charge-transferred excited states is also found in absorption and in emission. As seen previously during protonation studies of PET systems with multiple amine units [8], the presence of any free amines can quench fluorescence with an efficiency depending on amine–lumophore separation distance, among other things. Excimers are detected in the multilumophore cases among **40** and **41** concerning anthracenes but not naphthalenes.

The variations on the spacer module within **29** are available from Dabestani's laboratory [89]. The methylene spacer has been systematically extended to the tetramethylene case. As may be anticipated from previous studies on structurally similar aminoalkylaromatic PET systems [90–92], the fluorescence quantum yield (ϕ_F) of the ion-free sensor rises gradually with increasing spacer length as the PET rate drops. Potassium ions cause moderate to small fluorescence enhancements, but the same trend can be discerned in the ϕ_F values, though muted. As usual, protons cause the largest enhancements of all and the ϕ_F values remain constant across the series since PET has been completely killed off.

While the structures **29** and **42** only differ by a methylene group, removal of the vital spacer group completely alters the mode of signaling of **42**

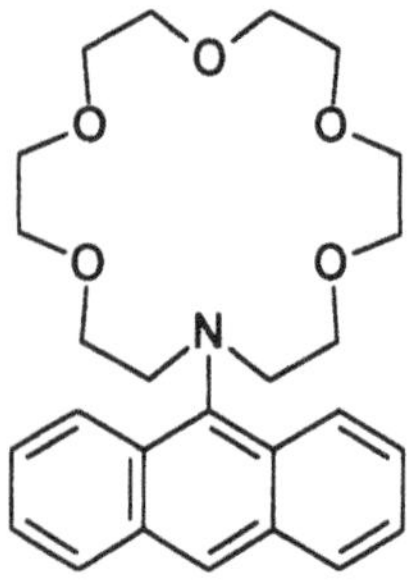

42

(c.f. **29**). **42** is the child of a Franco–German collaboration between Bouas-Laurent at Bordeaux and Witulski at Kaiserslautern [43]. Now **42** is the integrated lumophore–receptor format, with the peri-hydrogens of anthracene sterically congesting the azacrown ether moiety insufficiently to cut off delocalization of the nitrogen electron pair into the anthracene π-electron system and vice versa. The perturbation of the electronic absorption spectrum of **42** away from the anthracene model is clear evidence of this delocalization. In fact the long wavelength tail seen here reflects weaker features seen in spaced systems only by a sharp eye [88]. Such charge transfer phenomena naturally carry over into emission spectra (a common feature of internal charge transfer systems permitting cation deco-ordination in the excited state [94,96]) and the corresponding binding constants are smaller than those for **29** because **42** is effectively an N-aryl azacrown ether.

Rurack (Humboldt University, Berlin) and his collaborators provide steady state and time-resolved data for PET signaling of Ca^{2+} with **43** [95]. The isola-

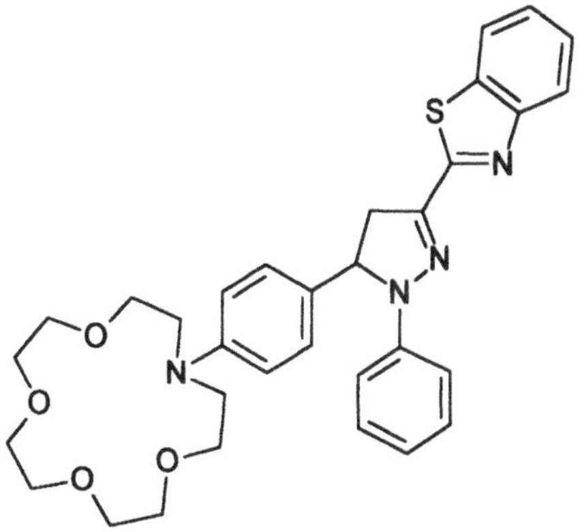

43

tion of the N-phenylazacrown receptor from the 1,3-diarylpyrazoline fluorophore in terms of their π-electron systems is clear. This sets the stage for PET across the C_1 spacer. The thermodynamic feasibility of PET is assessed by quantum chemical calculations rather than from electrochemical experiments. Indeed, a large fluorescence enhancement factor of 28 is seen when the rapid PET is stopped by calcium binding to the receptor. A corresponding increase in the fluorescence lifetime from 0.3 to 3.6 ns is also seen. A recent update [96] provides a wealth of results of **43** and several relatives. The oxidative fragility of 1,3-diarylpyrazolines is exposed during experiments with Hg^{2+} and Cu^{2+} especially in acetonitrile solvent which stabilizes species such as Cu^+.

A sensor **44** has been developed by Nagamura and coworkers [97] at the Universities of Shizuoka and Kochi, Japan, to recognize both charge and size of metal cations by transferring this information into a fluorescence signal. The designed chromophore is a good electron acceptor and marked fluorescence quenching by intramolecular PET from the nitrogen of the azacrown is observed. The changes in fluorescence quantum yields (ϕ_F) upon binding of metal ions to the azacrown are due to suppression of PET as usual. Due to their higher charge density, alkaline earth metal cations induce much larger enhancements of fluorescence intensity than the alkali metal cations. The metal ion whose size is closer to that of the cavity of the azacrown exhibits a larger fluorescence enhancement among cations of the same charge.

A deeper investigation [98] of **44** proves to be most enlightening. The model molecule **45** has an emission lifetime of 420 ps. For metal-free solutions

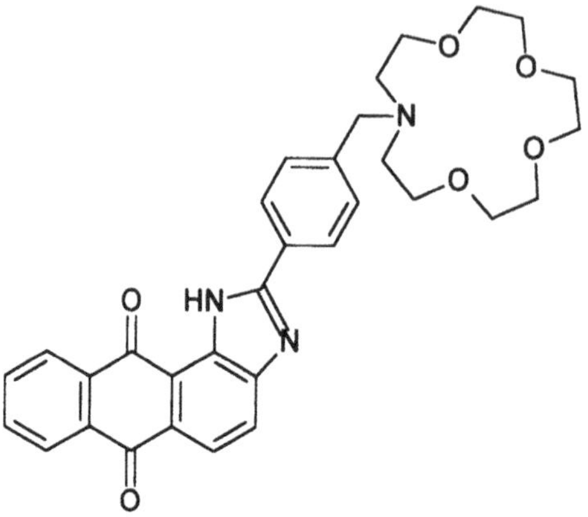

44

of **44**, the emission decay is too fast to measure, but this decay lengthens to 230 ps upon metal ion binding. However, the distinguishing feature of this work is the first demonstration of the mechanistic intermediates required of a PET signaling system—something we have looked forward to for a long time [3]. Time-resolved absorption measurements of **45** show transients at 525 and 750 nm, which are attributable to the excited singlet state. These decay giving rise to a new band at 460 nm assigned to the excited triplet. **44** shows similar behavior but decay is faster and gives rise to a new band at 610 nm which is due to the phenylimidazoanthraquinone anion radical. This is direct evidence for a photoinduced electron transfer. Binding metal ions gives a longer lifetime for the excited singlet. However, as the band at 610 nm (due to the anion radical) is still weakly observed, this is taken along with the fact that the singlet lifetime is still shorter than that of **45** as evidence that some electron transfer is still occurring.

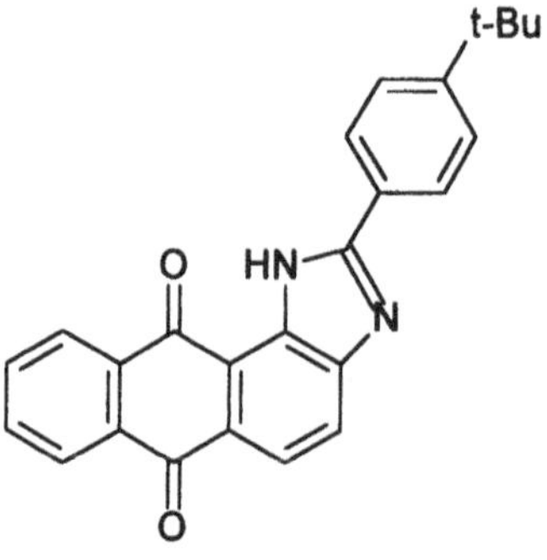

45

VI. BENZOCROWN-BASED PET SIGNALING SYSTEMS

Molecular sensing within the nuclear industry has to stand up to particularly harsh challenges. The reprocessing of irradiated material produces highly acidic and salty solutions. One of the key radioisotopes of interest is ^{137}Cs. Thus, a sensor with any realistic prospects for success must have a high degree of selectivity for cesium against sodium and protons. A rare example of a fluorescent molecular sensor that can operate in highly acid solution is our **46** [99]. It

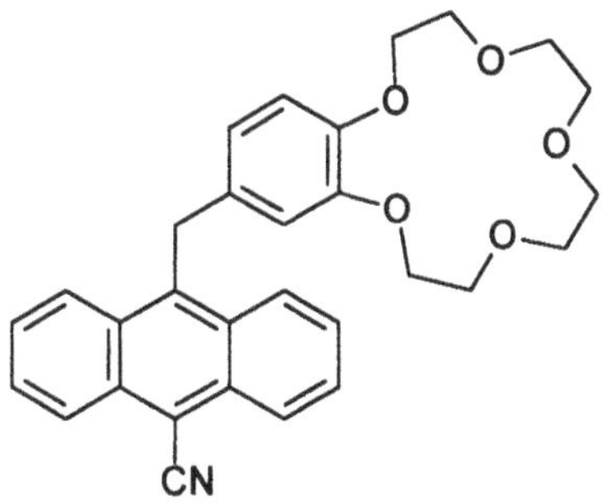

46

responds to sodium with more than an order of magnitude fluorescence enhancement. This ability of **46** to discriminate sharply against protons opens up the way to molecular logic operations (see later). In the present context, **46** provided the launchpad for Dabestani and his coworkers at Oak Ridge National Laboratory to build a useful fluorescent PET sensor for cesium [100]. The benzo-15-crown-5 ether in **46** was modified into a 1,3-alternate calix[4]benzocrown-6 in **47**. With two pi-faces available, 1,3-alternate calixarenes have a good record for selectivity binding soft cations such as Cs$^+$ [101] and Ag$^+$ [102]. The fluorescence response of simple calixarenes has been recently examined in collaboration between Prodi in Bologna and Ungaro in Parma, Italy [103]. So **47**'s receptor rejects Na$^+$ sufficiently. The nitrogen-free nature of the receptor ensures immunity against even high acidities. Still, the PET activity of the 1,2-dioxybenzene is preserved. The 9-cyanoanthracene lumophore is a good PET partner in **47** as seen within **46**. Sure enough, the fluorescence of **47** is switched "off" in the initial metal-free state. Binding of Cs$^+$ gives a FE value of 8, with Na$^+$ hardly altering the fluorescence at all. However, just as the FE selectivities, the binding selectivities could bear improvement. In fact, the binding of Cs$^+$ was only 30% more than Rb$^+$.

Such improvements came into view with the report from the Oak Ridge National Laboratory that 1,3-alternate dideoxy calix[4]benzocrown-6 was much more selective for Cs$^+$ [104]. Indeed, when this receptor was derivatized with the

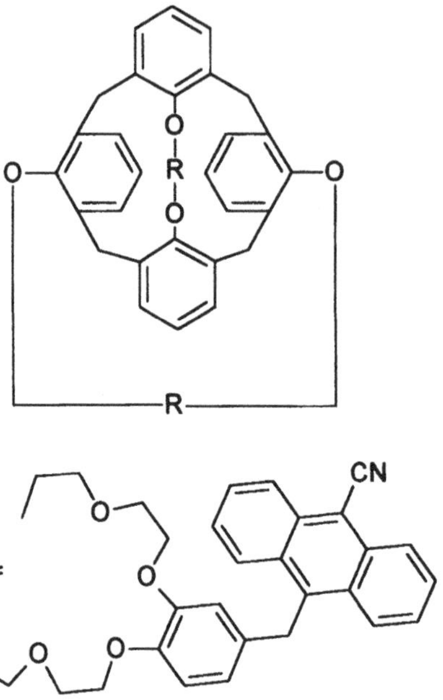

47; R =

tried and tested 9-cyanoanthryl-10-methyl lumophore, an excellent PET "off–on" sensor for Cs$^+$ was found [105]. For instance, **48**'s binding of Cs$^+$ is now 13-fold stronger than that of Rb$^+$. Equally important, the FE value for Cs$^+$ now jumps to 52. This is understandable since the dideoxygenation of the calix unit forces the Cs$^+$ to seek closer attention from the electrons of dialkoxybenzene oxygens. In fact, X-ray crystallography shows a noticeable

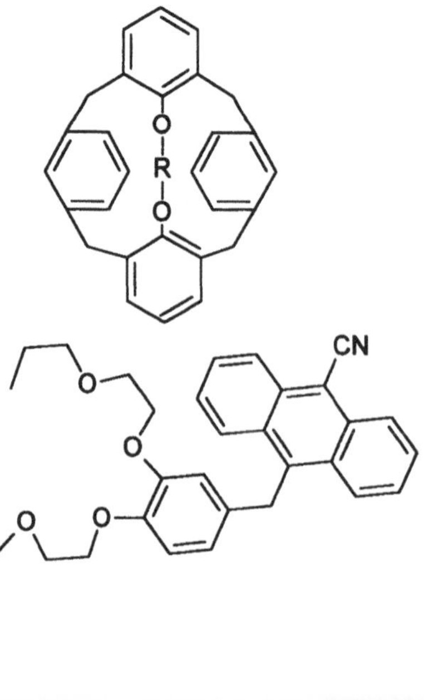

48; R =

shortening of the Cs$^+$-benzo oxygen distance. So the Cs$^+$-induced increase of the dialkoxybenzene's oxidation potential would be larger than that for **47**.

Selective Cs$^+$ sensing continues to be topical. In between the appearance of reports on **47** and **48**, Li and Schmehl at Tulane University, New Orleans produced the non-PET sensor **49** [106]. It functions by forming a 2:2 com-

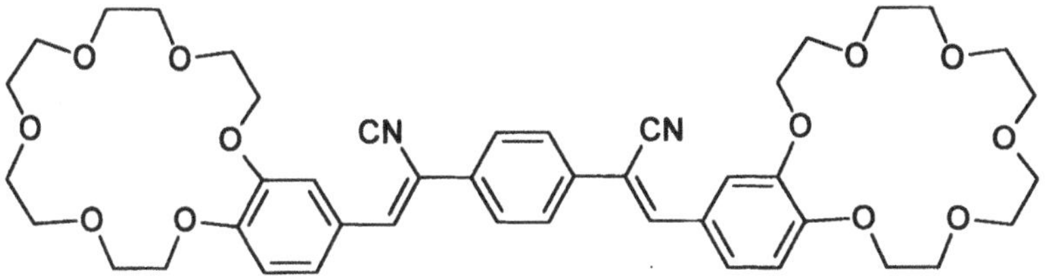

49

plex reminiscent of a baguette sandwich. Double crown sandwiching of alkali cations that exceed the hole size has been recognized previously [107]. This rigidification prevents the de-excitation channel of double bond torsion found in **49** and other stilbenoid compounds. Naturally, the smaller crown version of **49**, that is, **50** shows selectivity for K$^+$. We find it hard to avoid mentioning

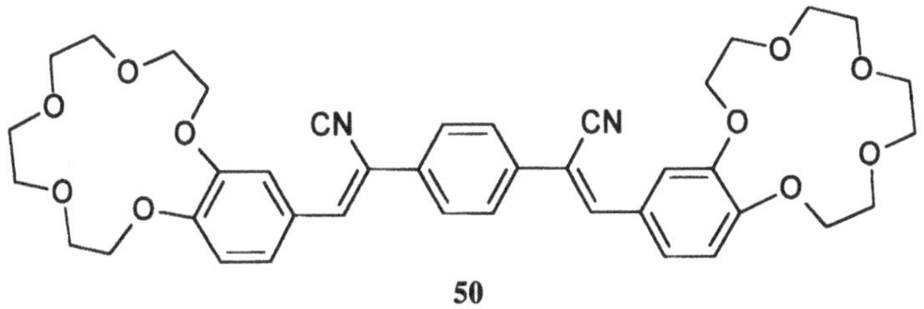

50

the related rigidification of stilbenoid [108] and cyanine [109] dyes by using terminal boronic acids to capture saccharides.

VII. CYCLOPHANE-BASED PET SIGNALING SYSTEMS

A substantial number of interesting luminescent molecular systems contain cyclophane receptor units [110] and involve PET-type processes, particularly as a result of the collaboration between the teams of Stoddart (now at the University of California at Los Angeles) and Balzani (University of Bologna, Italy) [111].

For instance, the [2]catenane, **51** [112] has been synthesized incorporating an anthracene subunit—an important electrochemical and photochemical active site which is a major building block required for switches and machines at the molecular level. The bulky anthracene stops the normal dynamic circumrotation

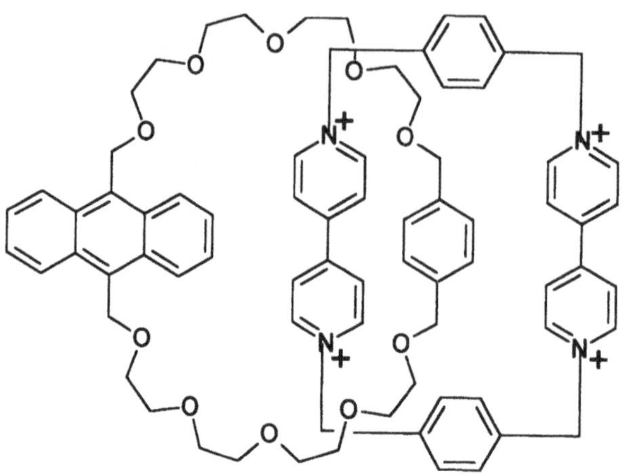

51

of the polyether macrocycle through the tetracationic cyclophane and always occupies the "alongside" position leaving the 1,4-dioxybenzene moiety always inside the cyclophane. Of course this does not stop the rotation of the cyclophane around the 1,4-dioxybenzene at ambient temperatures. The close proximity of the two rings with electron-donor and electron-acceptor components causes strong electronic interactions of the charge transfer (CT) type which are closely akin to PET processes. A new broad absorption band in the visible region can be accounted for by the contributions of two distinct CT transitions, that is, the presence of two distinct electron donor units, 1,4-dioxybenzene and 9,10-dioxyanthracene. These low energy CT excited states cause a fast radiationless decay of the 9,10-dioxyanthracene's lowest excited singlet, hence no fluorescence is observed for this [2]catenane.

It is important to note that neutral π-associated [2]catenane systems can also involve donor–acceptor interactions as seen in the work of Sanders and his team at the Cambridge Centre for Molecular Recognition, England. The correct π-donor, naphthalene diether, and π-acceptor, 1,4,5,8-naphthalenetracarboxylic diimide, favor molecular interlocking to give neutral supermolecules, **52** [113]. Fluorescence studies of these systems show only total quenching of fluorescence at low temperature due to a frozen molecular structure where each diimide subunit is locked in close proximity with a naphthalene diether. At ambient temperatures, dynamic movement of the π-donors and π-acceptors by a ring sweeping mechanism weakens their interaction and consequently their ability to quench fluorescence.

Although no signaling is involved in these two representative examples, the structures and the phenomena involved possess that potential. This is especially so because these rather large molecular systems have the capacity to spatially

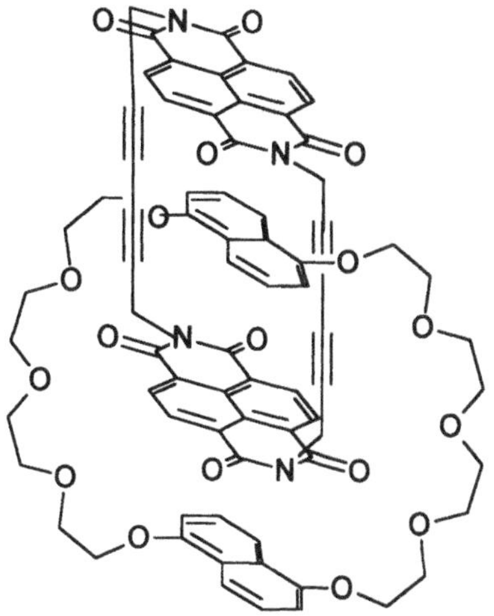

52

rearrange their components, that is, display machine-like behavior. The possibilities can be illustrated with the following case [114]. Na$^+$-induced fluorescence enhancement will probably be observed with the pseudorotaxane complex **53.54** due to electrostatic repulsion of Na$^+$-bound **54** away from tetracationic **53**. The

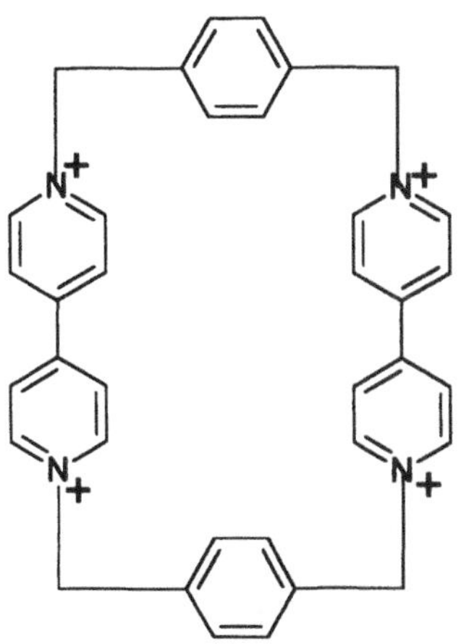

53 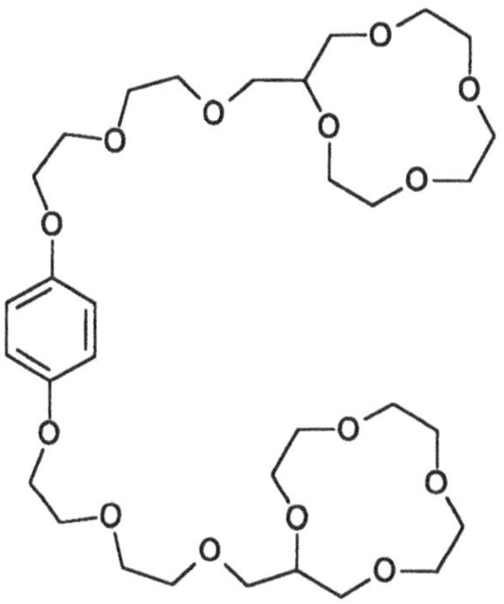   54

corresponding disappearance of the fluorescence-quenching CT interaction between **53** and **54** is clearly seen absorptiometrically when the pseudorotaxane dissociates upon Na$^+$ addition. Discussion of related work with fluorescence output is deferred until we consider logic operations.

Examples of the fruitful Portuguese–Spanish collaboration between Pina at Lisbon and Garcia-Espana at Valencia are the cyclophane compounds **55** and **56** [115] along with analogous open chain compounds where fluorescence is

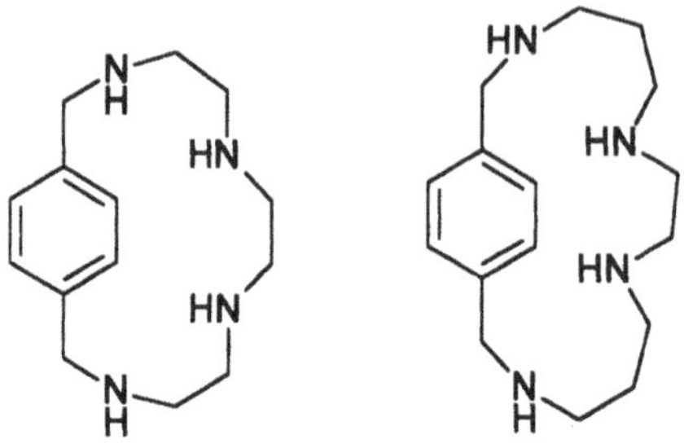

55   **56**

quenched by PET from the nitrogens to benzene. This is evidenced by fluorescence enhancement upon protonation of **55** and **56**. It is found that the stability constants for ion binding are lower compared to the open chain analogues due to the cyclophane rigidity preventing both the benzylic nitrogens from coordinating to the metal ion. This also means that upon ion (Zn^{2+}) binding it is necessary (if PET is to be prevented) for the ligand to be bound to both the ion and at least one proton. Such a species is not formed by **55** even at high acidities and hence no Zn^{2+}-induced emission enhancement is observed. **56**, however, can form $[ZnHL]^{3+}$, a species, which will be signaled by fluorescence enhancement. Cd^{2+} gives similar results but the enhancements are smaller than for Zn^{2+}. The complexation of Cu^{2+} does not result in emissive complexes. This is believed to be due to Cu^{2+} complexation resulting in the excited state being metal rather than ligand centered.

VIII. PET SIGNALING SYSTEMS WITH VIRTUAL (C_0) SPACERS

"Lumophore–receptor" systems where the two components are sterically held orthogonal can be considered to have a virtual C_0 spacer between the lumophore and receptor. Thus these systems can be approximately considered as a PET system. A deeper analysis would take twisted intramolecular charge transfer (TICT) states [116] into account whose thermodynamics is, unsurprisingly, very similar to that of PET systems.

The German collaboration between Berlin and Regensburg has produced an excellent signaling system for Hg^{2+} and Ag^+ [117]. The selectivity of **57** towards Hg^{2+} and Ag^+ (and to a lesser extent, Cu^{2+}) is arranged with the sulfur-rich thiaazacrown receptor. In the present case, as with several other related cases, a weak charge-transfer emission [116] can be observed. But the

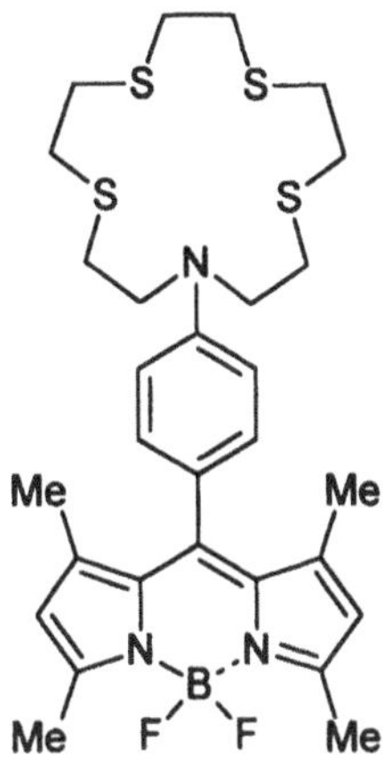

57

characteristic emission of the lumophore is essentially completely recovered upon arrival of Hg^{2+} or Ag^+ with fluorescence enhancement factors of up to 5900. Such enhancements are all the more remarkable because Hg^{2+} and Ag^+ both have a reputation as fluorescence quenchers.

The success of **57** can be understood when we note a very similar result which emerged in 1990 from Verhoeven's laboratory [118] at the Free University of Amsterdam, The Netherlands. **58** shows a large fluorescence enhancement

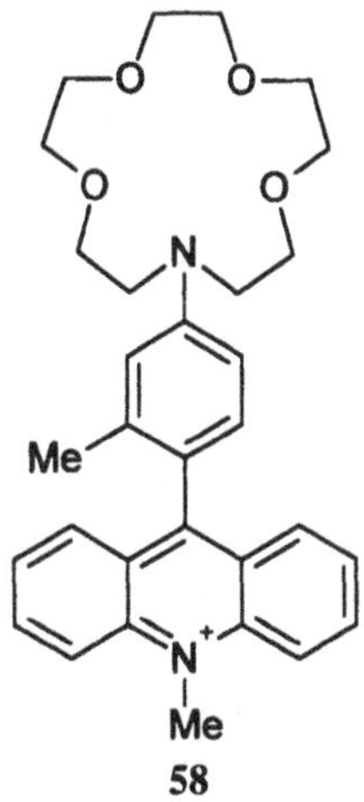

58

with Ag^+. We rationalized this result in 1993 [3] by noting that Ag^+'s ability to act as a fluorescence quencher was due to PET towards it from the lumophore. However, the lumophore within **58** is rather electron deficient, so Ag^+'s usual quenching tendency cannot be expressed. A similar argument appears reasonable for **57**.

59 [119] and **60** [120] are two other cousins of **57** to emerge from Daub's laboratory in Regensburg in collaboration with Wolfbeis. **59** responds to rather

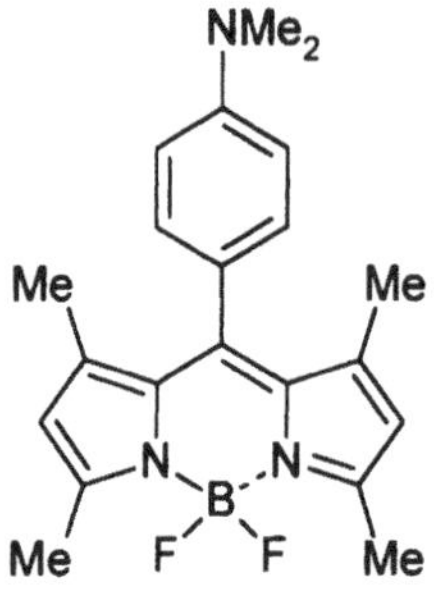

high concentrations of protons with a huge enhancement of fluorescence. **60** is highly fluorescent to start with and is almost fully quenched upon deprotonation of the phenol group. The increased electron density on the phenolate unit can easily launch a PET process across the virtual C_0 spacer whereas the phenol form cannot. Excellent optical properties characterize the borodipyromethane fluorophore, such as narrow absorption and emission bands of high intensity, with the only weakness being the small Stokes shift of ca. 10 nm.

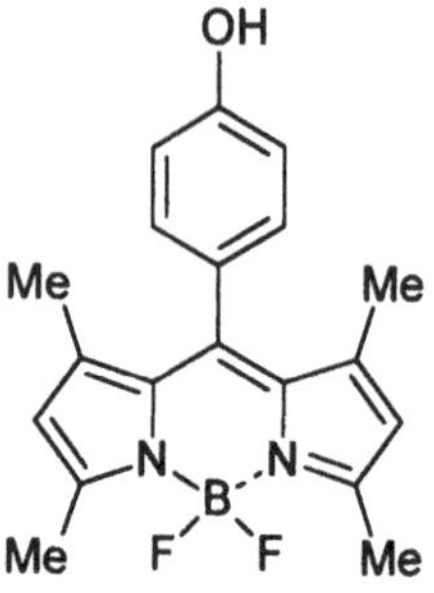

59 and **60** have some parallels in older work from Grigg and Norbert and their collaborators in the Universities of Leeds and Belfast [121,122]. They showed that the luminescence of tris(bipyridyl)Ru(II) lumophores can be quenched by intramolecular receptors such as alkylaniline or phenoxide via PET pathways. Protonated aniline or phenol units lack this ability. Thus, the pH response of **59** and **60** is mirrored, although not with the same efficiency, in these ruthenium complexes. The similarity ends there, however, since these ruthenium complexes are lumophore–spacer–receptor systems.

61 from Nagano's laboratory at Tokyo, Japan, shows excellent fluorescence enhancements with physiological levels of Zn^{2+} with the added bonus of relatively long wavelength (495 nm) excitation [123]. Zn^{2+} binding to the N-phenyl[12]aneN$_4$ macrocycle clearly increases the oxidation potential of the latter unit which should suppress the PET process from it to the lumophore. The [12]aneN$_4$ macrocycle also shows up in an offering from Kimura's laboratory [124, 125] at Hiroshima University, Japan. **62** also functions in the physiological regime but suffers from the requirement of excitation at wavelengths as low as 330 nm. The fluorescence enhancement factor of 5 is moderate but useful especially since a blue shift of 40 nm is seen. Mechanistically, **62** depends not on PET but on a Zn^{2+}-induced ionization of the sulfonamide part of the fluorophore. A still smaller luminescence enhancement (20%) is fond with Zn^{2+} and **63** [126], from David Parker's team at the University of Durham, England.

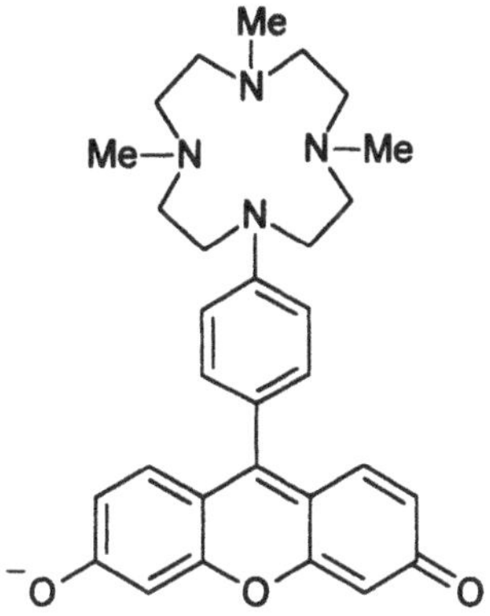

61

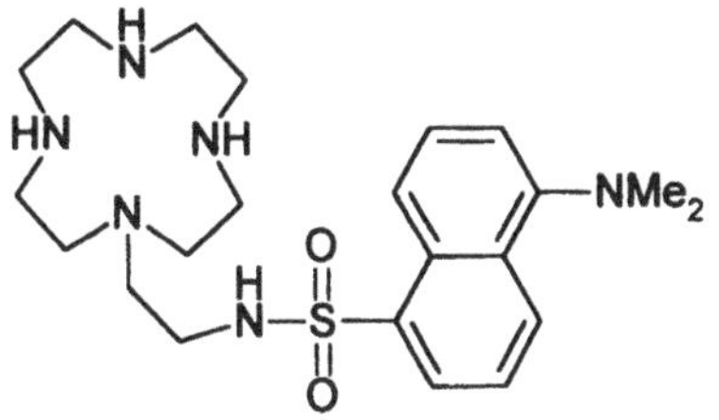

62

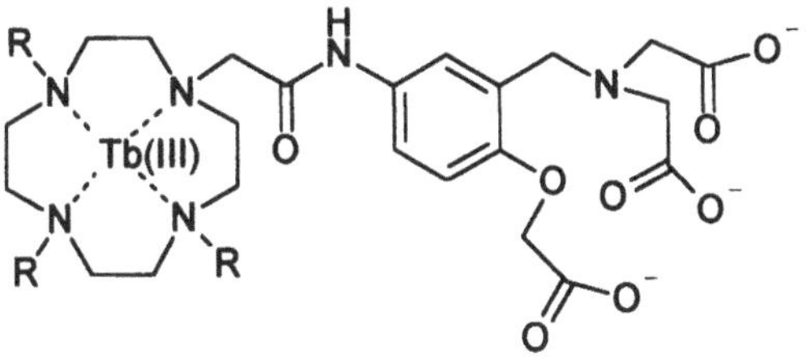

63; R = CH₂CO₂⁻

This system also needs rather short ultraviolet wavelengths for excitation but has the benefit of a time-delayed emission that allows time-gating to block out autofluorescence from the matrix. PET appears to be at work here too.

D'Souza at Wichita State University has built on his successful porphyrin-based PET sensors **64** [127] and **65** [128] for hydroquinone and benzoquinone, respectively, with **66** for catechols [129]. The structural similarities seem to be deceptive, however. The guest-induced luminescence signaling with **64** and **65** depend on hydroquinone–benzoquinone interaction via hydrogen bonding and charge transfer. In contrast **66** engages in a redox equilibrium with its catechol guests to produce the reduced form of **66**, that is, **67**. Cyclic voltammetry data show that PET is thermodynamically feasible in **66** but not **67**. So the **66** to **67** conversion results in fluorescence enhancements of up to 9. Since a reversible redox reaction rather than a recognition process is involved it is no wonder that other reducing but structurally different species such as resorcinol also give some fluorescence enhancement whereas structurally similar but less reducing species such as tetrachlorocatechol are without effect. Mechanistic relatives of **66** can be found in earlier examples of redox-switched luminescent systems, both simple [130] and complex [131–134].

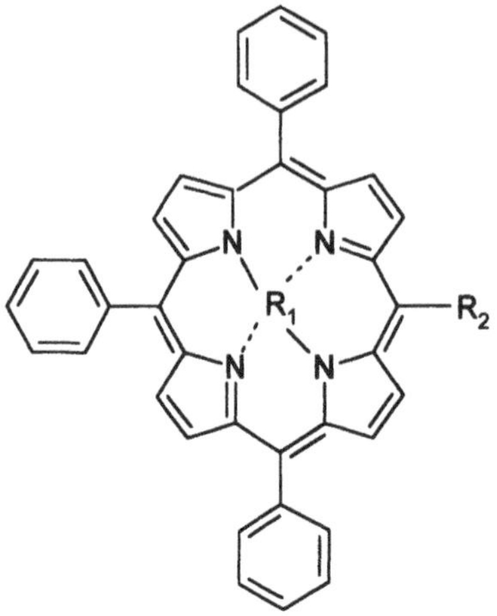

64; R₁ = H₂, R₂ = benzo-1,4-quinon-2-yl
65; R₁ = Zn(II), R₂ = 1,4-dihydroxyphen-2-yl
66; R₁ = Zn(II), R₂ = benzo-1,2-quinon-4-yl
67; R₁ = Zn(II), R₂ =1,2-dihydroxyphen-4-yl

IX. "OFF-ON-OFF" PET SIGNALING SYSTEMS

In 1996, we showed how the H^+-induced "off-on" switching typical of amino-methylaromatics [53] and the "on-off" switching observed in pyridylalkylaro-matics [73–75] could be integrated [135] within the one supramolecular system [136,137] to produce an emission-pH profile that is a combination of the parent characteristics. **68** [135] is a signaling system of the "window" type, that is, "off-on-off" switching. de Silva's team at Montclair State University, NJ, almost simultaneously published system **69** [138] with even better performance in some respects than **68**. Also, **69** is assembled even easier than **68**.

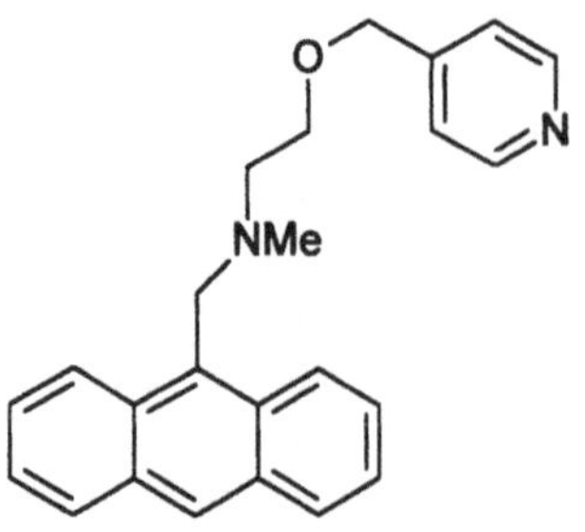

68 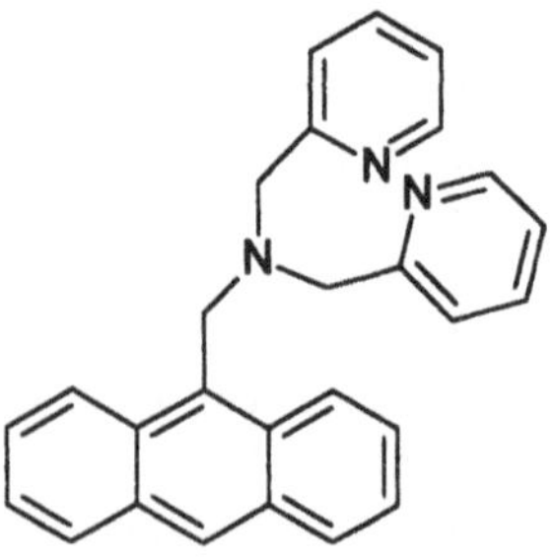   69

Fabrizzi's team at the University of Pavia saw only the beginnings of "off-on-off" switching action in the pH-responsive fluorescence of **70** and **71** [139]. Both these molecules contain the amine and pyridyl moieties that have previously

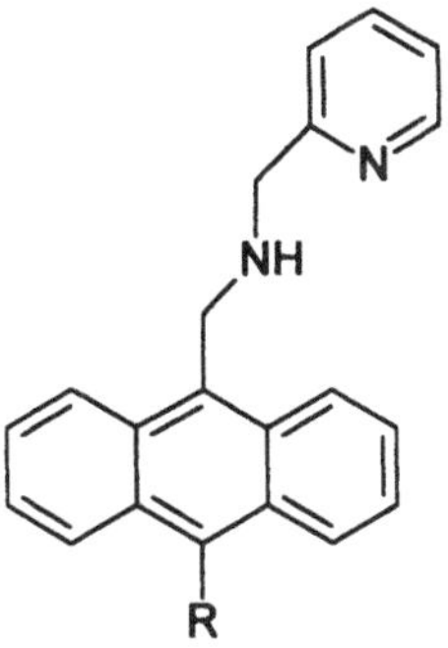

70; R = H
71; R = 2-pyridylmethylaminomethyl

led to such switching. de Silva's **69** shares the 2-pyridylmethylaminomethyl moiety with **70** and **71**. The Pavia team have also produced improved "off-on-off" systems recently [140].

A metal-based "off-on-off" system **72** comes from Ye, Ji, and collaborators at Zhongshan University, China. In fact, two metallolumophores and two

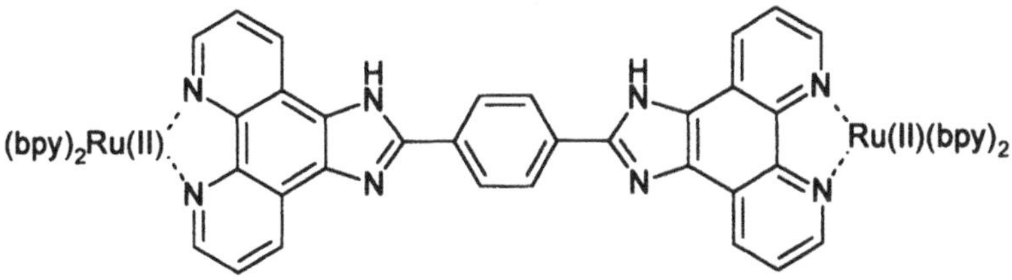

72

receptors are integrated within **72** [141]. The "off-on-off" switching occurs via protonation and deprotonation of the imidazole receptors. At low pH the imidazole units can be protonated and luminescence is quenched, though only to 40%. The protonated bisimidazole ligand can be expected to lower the energy of the MLCT excited state drastically. Emission quenching can then arise via the operation of the energy gap law and/or coupling of the imidazolium unit to the solvent water molecules. From pH 6 to 8 emission decreases and fluorescence switches "off". This is caused by the negative charge generated on the deprotonated imidazole unit which would decrease the energy gap between MLCT and metal-centered d–d excited states.

pH-Induced "off-on-off" switching behavior can also be produced by mixtures of compounds, as demonstrated recently by the Spanish–Portuguese axis of Garcia-Espana, Pina, and their colleagues [142]. **73**, on its own, displays fluores-

73

cence enhancement upon protonation seen for many 9-anthrylmethyl substituted amines. The enhancement is clearly seen at pH values below 8 and stays constant at pH <5. However, in the presence of a threefold excess of adenosine triphosphate (ATP), the fluorescence of **73** is strong in the pH range 5 to 8. The acidic branch of the fluorescence–pH profile is due to a PET-type interaction between the anthracene fluorophore in tetraprotonated **73** and the protonated adenine unit within diprotonated ATP. The registering of the tetra-ammonium and triphosphate moieties allows the pi-stacking of anthracene and protonated adenine units for a strong quenching. Such registering is so important that looping the tetra-amine chain of **73** to produce an anthracenophane completely loses this "off-on-off" effect. The genealogy of **73** can be traced back to older work by Czarnik et al. [143] and even older work by Lehn et al. [144].

The term "off-on-off" has been interpreted somewhat differently by Kubo and coworkers [145] at the Japan Science and Technology Corporation, Saitama. Even though no PET is involved here, it illustrates how concepts originating in the PET literature can take root in other research areas. "Off-on" switching in the absorbance of **74** at 734 nm is arranged with 1.5 equivalents of Ca^{2+}. The subsequent induction of "on-off" switching is done with 10 equivalents of K^+. The Ca^{2+}-induced absorption spectral changes of relatives of **74** due to stabilization of the indophenol internal charge transfer (ICT) excited state had been reported earlier [146]. They now find that K^+ binds somewhat stronger than Ca^{2+} and displaces the latter ion during the competitive phase of the experiment. However K^+ is unable to cause any significant spectral changes in **74**, apparently due to bonding more to the oligoethyleneoxy chain rather than the quinone oxygens.

There are other ways to generate "off-on-off" switching if we consider a different independent variable rather than a chemical concentration: for instance, light intensity. As mentioned previously, Wasielewski and colleagues at Argonne

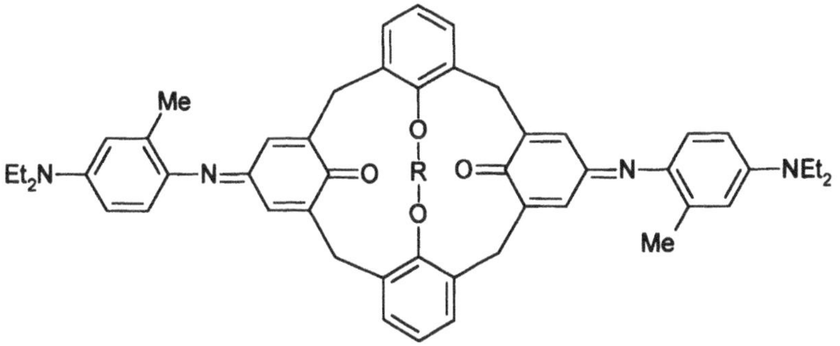

74; R = -(CH₂CH₂O)₃CH₂CH₂-

National Laboratory have optical switches like **75** based on PET that flip within picoseconds [48]. Optical pumping of the porphyrin absorption band at low intensities produces the perylenetetracarboxydiimide radical anion (and the porphyrin radical cation). The absorbance of the radical anion naturally increases with increasing pump intensity but falls at still higher pump intensities. So the absorbance versus pump intensity profiles show "off-on-off" behavior. At these higher intensities the second porphyrin unit joins in to launch a PET process into the perylenetetracarboxydiimide radical anion which pushes the latter into the dianion state. Although our main interest here is in luminescent systems, this work has much to teach us.

Although not involving PET, the "off-on-off" switching described by the Lisbon–Bologna co-operative deserves discussion here [147]. Light dose, rather than intensity, is the independent variable in this case. A mixture of **76** and $[Co(CN)_6]^{3-}$ is irradiated. If irradiated by itself in acidic solution, **76** leads to **77** [148]. On its own $[Co(CN)_6]^{3-}$ photoaquates to $[Co(CN)_5(H_2O)]^{2-}$ and the liberated cyanide picks up a proton [149]. The irradiation of the mixture at pH 3.6 leads to a buildup of **77** at first but the aquation of $[Co(CN)_6]^{3-}$ gradually asserts itself and the increasing pH strangles the path to **77**. Furthermore, the

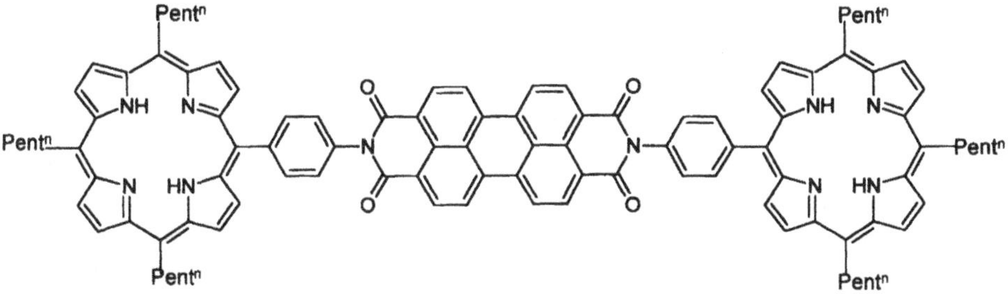

75

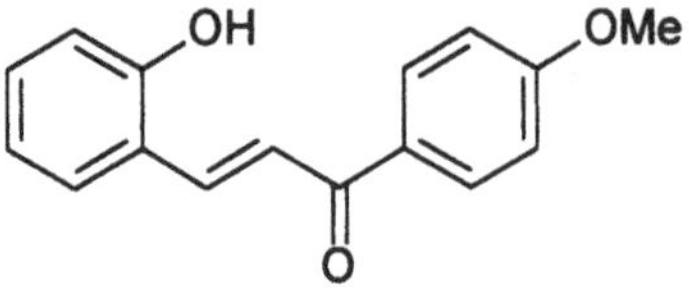

76 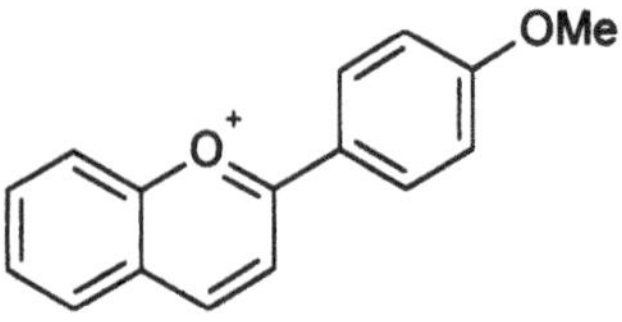   77

increasing pH causes cycloreversion of **77** to the Z isomer of **76** which promptly photoequilibriates back to **76** itself. Thus, the **77** concentration falls back to zero at long irradiation times. Fluorescence is a clear measure of the **77** concentration and serves as the output.

X. PET SIGNALING SYSTEMS WITH EET

In 1996, we described the fluorescence signaling behavior of the lumophore$_1$–spacer$_1$–receptor–spacer$_2$–lumophore$_2$ system **78** where a PET process is made

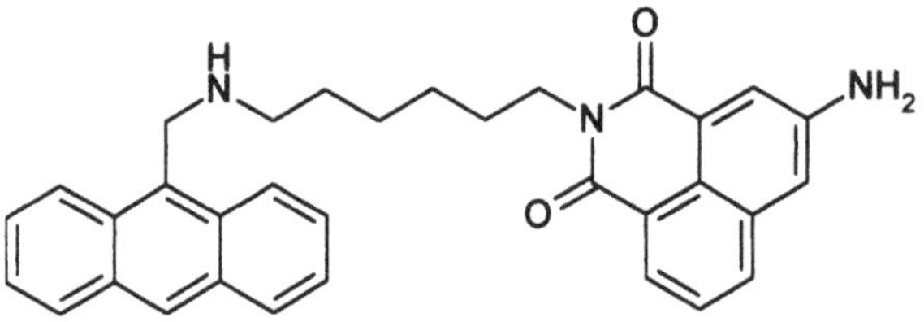

78

to compete with electronic energy transfer (EET) [150]. The anthracene moiety serves as the EET donor lumophore as well as the participant in PET originating in the amine receptor. The 3-amino-1,8-naphthalimide unit is the EET-accepting lumophore$_2$. The secondary amine serves as a receptor for H$^+$. PET is not kinetically feasible from amine to aminonaphthalimide moieties in the current configuration [37]. When applications are considered, **78** shows self-calibrated pH sensing because the anthracene-based emission is strongly pH dependent whereas the aminonaphthalimide-centered fluorescence is not. Ratioing of the two emission intensities gives a pH-dependent parameter that is largely unaffected by local concentration of **78** or quenchers or even by optical path. From a theoretical angle, **78** shows a remarkable acceleration of EET in its proton-free form assigned to an electron exchange facilitated by a "stepping stone" provided by the central secondary amine group.

Newly emerging results concerning **79** from the Bristol–Bologna collaboration among Ward, Barigelletti, and their colleagues underline several aspects of the previous paragraph from an inorganic perspective [151]. Now a

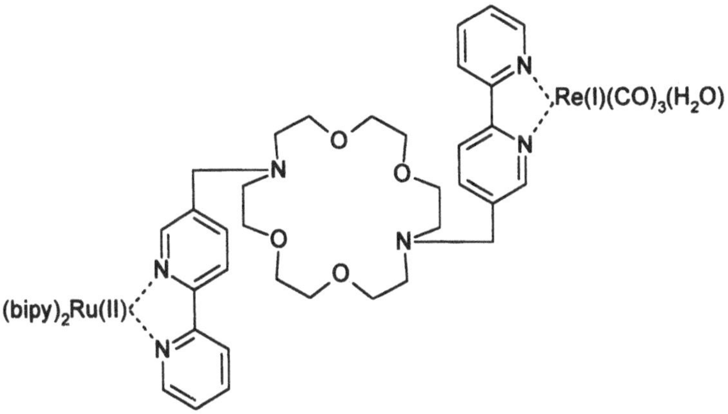

79

tris(bipyridyl)Ru(II) moiety represents lumophore$_2$, a diaza-18-crown-6 ether represents the receptor for Ba^{2+} and a (bipyridyl)(CO)$_3$(H$_2$O)Re(I) unit serves as lumophore$_1$. The amine units can engage in PET with the Re(I) lumophore$_1$ but not the Ru(II) lumophore$_2$ for thermodynamic reasons. A very fast PET (10^{10} s^{-1}) virtually kills off Re(I)-based luminescence from **79** at room temperature, with no detectable EET from the Re(I) lumophore$_1$ to the Ru(II) lumophore$_2$. Arrival of Ba^{2+} within the diazacrown ether allows recovery of the emission of Re(I) lumophore$_1$ by PET suppression. Now simultaneous emission from both lumophores is seen, without any detectable EET. A reasonable interpretation is that the dicationic Ba^{2+} repels the dicationic Ru(II) lumophore$_2$ and cationic Re(I) lumophore$_1$ termini such that the large distance between the lumophore pair minimizes the EET rate. When the solvent is frozen, PET becomes impossible even in Ba^{2+}-free **79**. Now a Re(I) to Ru(II) EET becomes visible (with a rate of 2×10^8 s^{-1}). This rate falls by a factor of 30 when Ba^{2+} is introduced. While the electronic repulsion undoubtedly contributes to this EET rate control by Ba^{2+}, the "stepping stone" facilitated electron exchange is also consistent with this observation.

Wang and Wu's Beijing laboratory [152] is the source of another nice lumophore$_1$–spacer$_1$–receptor–spacer$_2$–lumophore$_2$ system involving PET and EET [2]. Molecule **80** again features an aminomethyl anthracene which, when

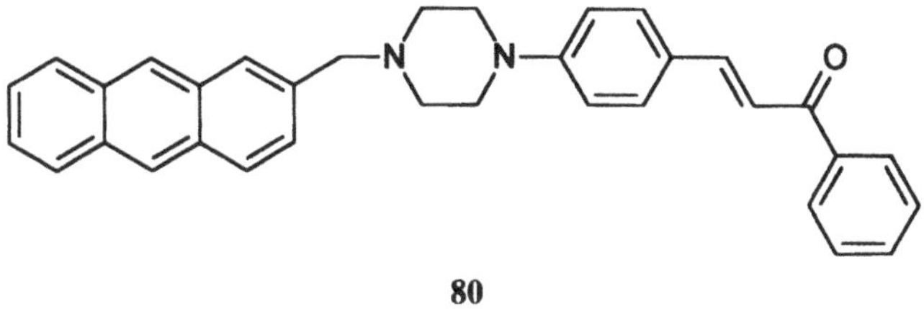

80

unprotonated, permits electron transfer from the nitrogen to quench the anthracene excited state. Protonation, however, prevents PET enabling the anthracene to transfer its excited state energy to the second chalcone-type lumophore, which can then undergo fluorescence. Thus protonation increases the quantum yield of the chalcone-type fluorescence. The additional benefit pointed out by the authors is that the flexibility offered by different lumophores can allow these systems to exhibit very large Stokes' shifts which may be useful if emission in the visible range is desired.

PET and EET also coexist in **81** [153] from Fabbrizzi's laboratory in Pavia, Italy. At low pH, there is no PET from the protonated amine sidechain to the

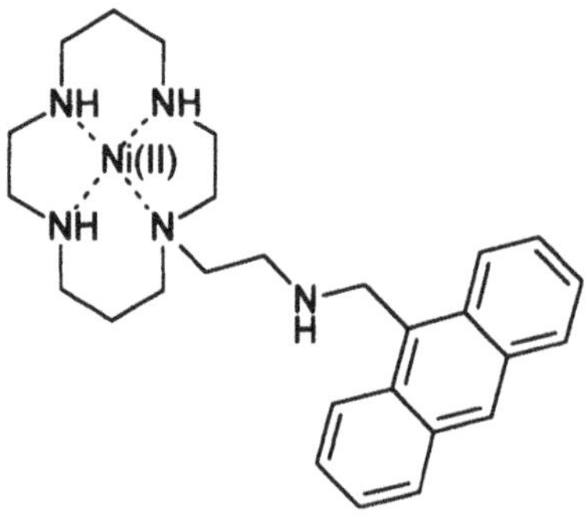

81

lumophore so no quenching is observed. The electrostatic repulsion between the Ni(II)cyclam and the protonated amine sidechain minimizes EET from the lumophore to the Ni(II) center. Between pH 4 to 8, the amine sidechain is not protonated and co-ordinates to the metal center. A water molecule co-ordinates to the opposite axial position. The Ni(II) center and the lumophore are now closer together, which causes a rather slow EET process that quenches fluorescence to 60% of the former level. Above pH 10 total quenching is observed. The co-ordinated water molecule releases a proton thus favoring access to Ni(III). Quenching is therefore due to PET from Ni(II) to the lumophore. The distinction between PET and EET is made from low temperature experiments. Besides the machine-like features of **81**, the three-level emission (high–medium–off) is also interesting in the light of recent [154] and older work [155,156].

XI. PET SIGNALING VIA IRREVERSIBLE COVALENT INTERACTIONS

The reversibility of a signaling system is essential for many applications, for example, those involving continuous monitoring. Nevertheless, such reversibility of signaling requires genuine receptors in the supramolecular [137] sense, that is, those with reversibility of analyte binding over relatively short timescales operating via noncovalent interactions. This requirement becomes a burden in many cases when we step beyond atomic analytes. The palette of suitably reversible receptors, while expanding all the time, still has not built up to the point of designating many important molecular analytes. One reason is the relatively short history of supramolecular chemistry [137] vis-à-vis inorganic coordination chemistry.

Given this situation, and the pressing need for measurement of important molecular analytes, it is no wonder that a compromise is being struck. Irreversible binding of a molecular analyte can be built into the PET signaling system because the original design logic of luminescence modulation does not concern reversibility. Additional propulsion of such efforts comes from the fact that covalent chemistry is rich in examples gathered over a long period.

We rationally launched this line by exploiting the Michael addition of thiols to enones with example **82** [157]. The π-electron system of maleimide is

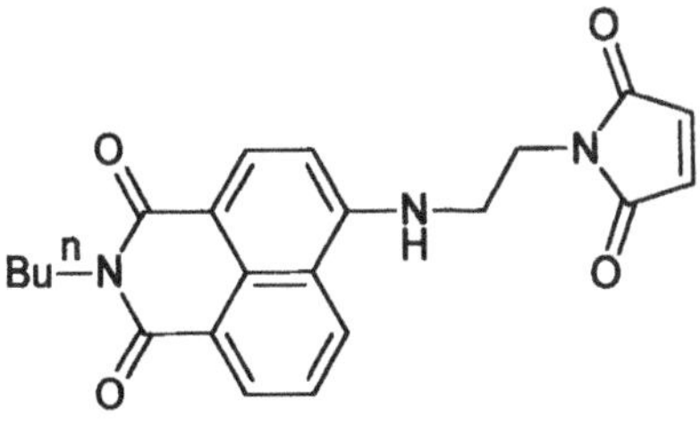

82

rather electron deficient and very suitable for being a partner in a PET process. This π-electron system loses its central double bond upon thiol addition and the electron deficiency largely disappears. We were further encouraged along this design path because thiols add to maleimides rapidly in the physiological regime. The latter point is crucial because the majority of covalent bond-forming reactions do not proceed fast enough under mild conditions for convenient signaling purposes. In the event, **82** shows a useful fluorescence enhancement upon encountering several aliphatic thiols.

The positive response of **82** offers evidence for the luminescent PET signaling of irreversible interactions. More specifically, it underpins the previously known, but empirically designed, lumophore–spacer–maleimide systems **83** [158] and **84** [159] operating at shorter communication wavelengths. In addition, the present results help to explain the larger thiol-induced fluorescence enhancements found in **85** [160] with unassigned mechanism of action. The latter are twisted lumophore–maleimide systems. Their virtual C_0 spacers ensure rapid PET rates and, subsequently, low fluorescence quantum yields in the absence of thiol.

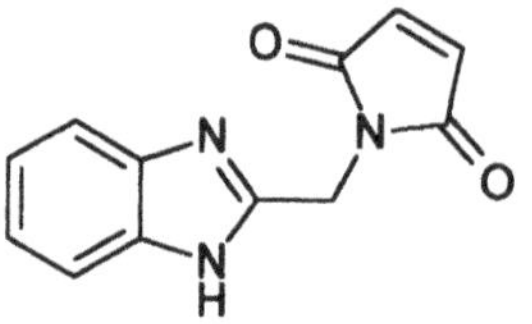

83 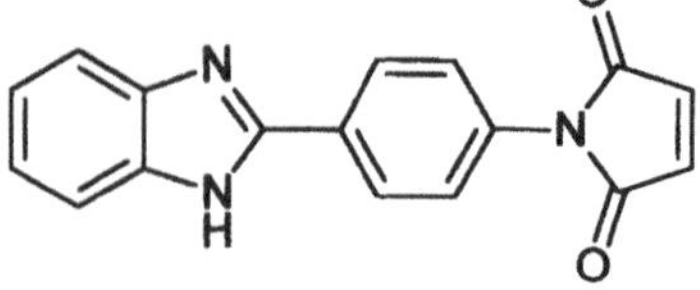   84

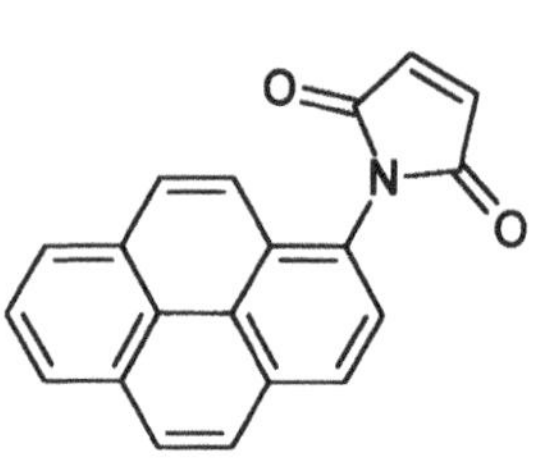

85 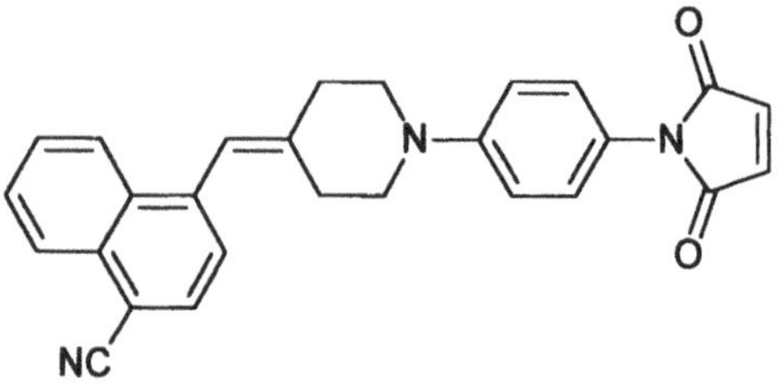   86

The ability of the maleimide unit to switch off emission is also exemplified by **86**, due to Verhoeven's coworkers [161] at the University of Amsterdam and Akzo Nobel in The Netherlands. Again the Michael reaction of the maleimide with thiols produces nicely emissive material. Solvent-sensitive emission, characteristic of these donor–acceptor systems with strongly coupling bridges, is a special feature of **86** after thiolation. An added interest of **86** stems from the occurrence of PET to the maleimide unit from the through-bond charge-transfer excited state [162], an unusual combination of photophenomena.

The tragedy at the Union Carbide plant in Bhopal, India in 1984 highlighted, among other issues, the critical need for careful monitoring of methyl isocyanate in the environment. **87** and **88** are both lumophore–spacer–secondary

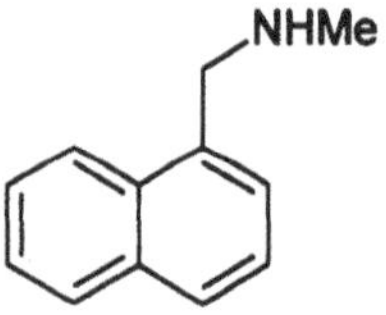

87 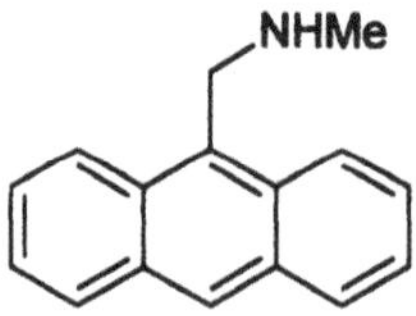   88

amine systems with good PET activity producing low fluorescence efficiencies. Secondary amines are, of course, excellent nucleophiles that smoothly attack the carbonyl carbon of the isocyanate group. The product urea possesses nitrogen atoms of far less electron density owing to the delocalization into the carbonyl group. So PET activity ceases and a large fluorescence enhancement is seen. **87** [163] and **88** [164] therefore serve very well in chromatography protocols with fluorescence detection to distinguish and measure various isocyanates.

The suddenness with which nitric oxide entered scientific discussions during the past decade [165] pressured scientists to provide sensors for this small molecule. Luminescent sensors are high on the list because they provide spatially and temporally resolved information, among other things. Luminescent sensors should also allow reversible binding so that they may not remove the very nitric oxide they are meant to measure with high fidelity. Such luminescent sensors are yet to emerge. In such a situation, luminescent reagents involving irreversible binding of nitric oxide provide an important stopgap.

Nagano and coworkers at the University of Tokyo have made good progress in this matter [166] by constructing fluorescent PET reagents **89** [167] and **90**

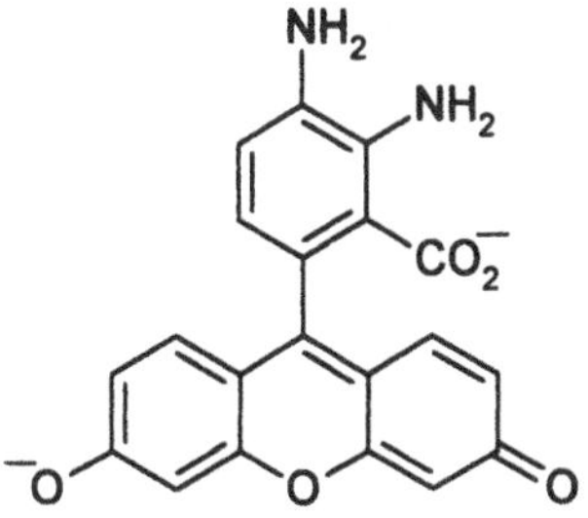

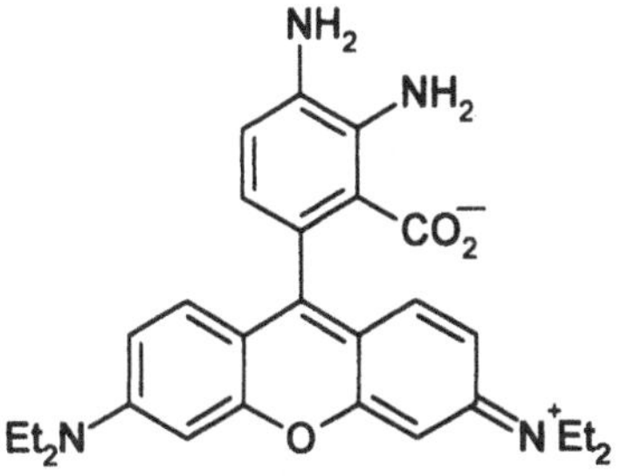

[168] with virtual C_0 spacers akin to the fluorescent PET sensor for Zn^{2+}, **61**, discussed previously. The o-phenylene diamine unit serves as the PET donor to the fluorescein unit in **89** and rhodamine unit in **90**. Weak fluorescence is the result in both cases. But nitric oxide transforms the o-phenylenediamine unit under aerated conditions into a benzotriazole moiety (**91**) which is much less

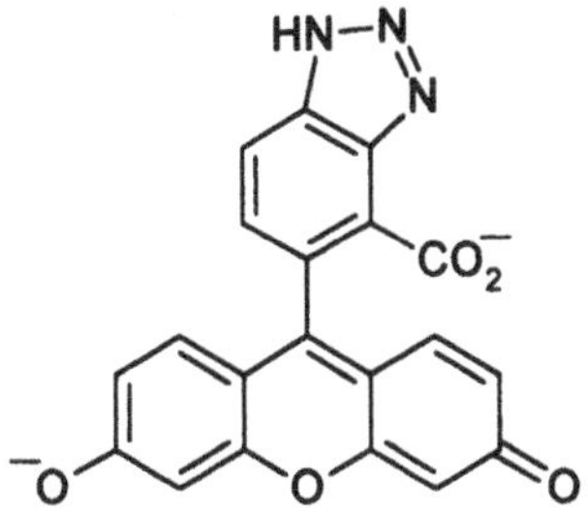

electron rich. So the interaction with nitric oxide produces a large enhancement of fluorescence (FE = 184 and 43 for **89** and **90**, respectively). **89** can be delivered intracellularly by exposing live cells to its diacetate derivative, in line with 30-year-old experiments with fluorescein [169]. **90** improves upon **89** in terms of photostability, pH-insensitive fluorescence, and longer wavelength excitation.

The PET genealogy of **89** has been traced back by these authors to Walt et al.'s
91a [170] from Tufts University, Medford, MA, whose fluorescence is strongly

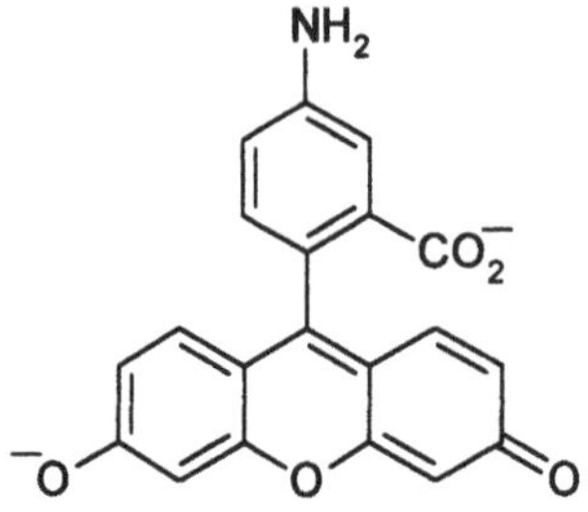

91a 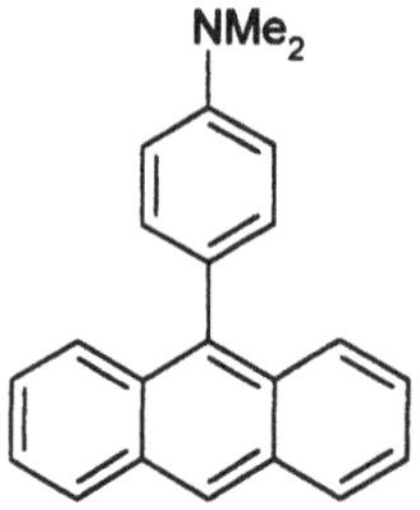   92

enhanced upon protonation. Shizuka et al.'s **92** and its proton-induced fluorescence recovery from Gunma University, Japan, shows that such genes go back
further in time [171].

XII. PET SIGNALING VIA REVERSIBLE COVALENT INTERACTIONS

One of the clearest pieces of evidence for the reversibility of some covalent interactions came from the work of Gunther Wulff at the University of Dusseldorf,
Germany [172]. He used imprinted polymers of o-aminomethyl phenylboronic
acids as chromatographic stationary phases for the separation of saccharides.
Older studies [173] also point to the reversible nature of the boronic acid–
saccharide interaction. The pioneering studies of fluorescent transduction of this
phenomenon by Czarnik and Yoon [174] (Ohio State University), Aoyama et al.
[175] (Kyushu University, Japan), and Shinkai et al. [176] (Kyushu University,
Japan) have been reviewed previously [9]. Our concern in this review is particularly with the systems that clearly involve PET. Czarnik and Yoon's **93** [177]
which interacts with catechol derivatives to produce **94** also belongs here. It

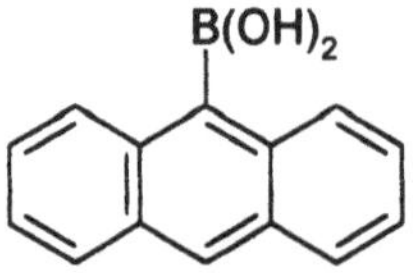

93 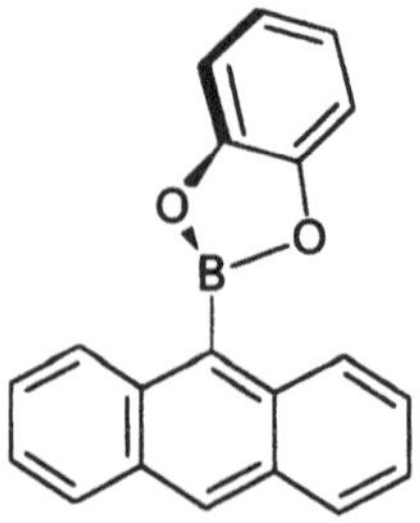   94

is probable that **94** fits the format of a twisted PET system with a virtual C_0 spacer. So it is understandable why the fluorescence of **93** is sharply quenched upon transformation to **94**.

But the bulk of results have emerged from Shinkai's laboratories [178] in Fukuoka and Kurume, Japan. His prototypical example, **24**, is essentially a classical aminomethylaromatic lumophore–spacer–receptor system with a twist that the weak B—N bond allows rather rapid PET from the amine unit to the anthracene resulting in weak fluorescence [176]. The binding of glucose to produce the boronate ester leads to a much stronger B—N bond that arrests the PET process and produces a large fluorescence enhancement.

The success of **24** and its cousins can be extended by attaching extra functionalities to improve selectivity towards chosen saccharide derivatives. For instance, glucosamine can be specified in its protonated form by modifying **24** with an azacrown ether. The latter unit is known to bind ammonium ions and 9-anthrylmethyl azacrown is a luminescent PET system of 15 years' standing [76]. **95**, due to Cooper and James at the University of Birmingham, England,

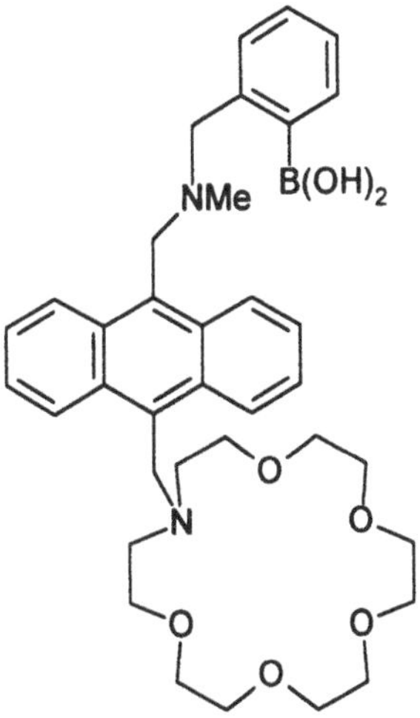

95

responds well with a substantial fluorescence enhancement to glucosamine at physiological pH and is largely insensitive to simple ammonium ions and to glucose [179]. This excellent selectivity can be assigned to the two PET processes available within **95**, both of which must be suppressed by appropriate guest components binding to the two receptors. **96** is an early case using the same concept in a homobireceptor context [180]. **97** [181] is the other "parent" since **95** is the hybrid of **96** and **97**. Nevertheless, **95** is an original and beautiful example of a heterobireceptor PET signaling system sensing a multifunctional

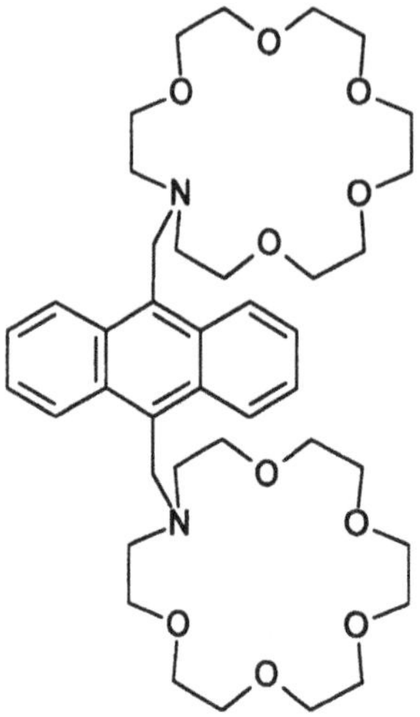

96 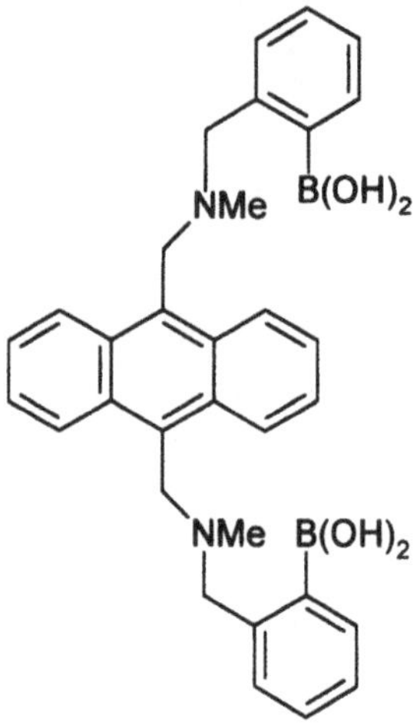   97

biomolecule. In contrast, the heterobireceptor system **98** [182] only has one PET process (originating from the azacrown nitrogen) which limits the selectivity of its fluorescence response to γ-aminobutyric acid. **95** can also be operated as a molecular AND logic gate [183], with saccharide and ammonium (or alkali) ion inputs, further increasing its appeal.

Shinkai's team also has an efficient heterobireceptor signaling system **99** which has as its target D-galactouronic acid [184–185]. The phenanthroline–Zn^{2+} complex serves as the fluorophore and also as the receptor for the car-

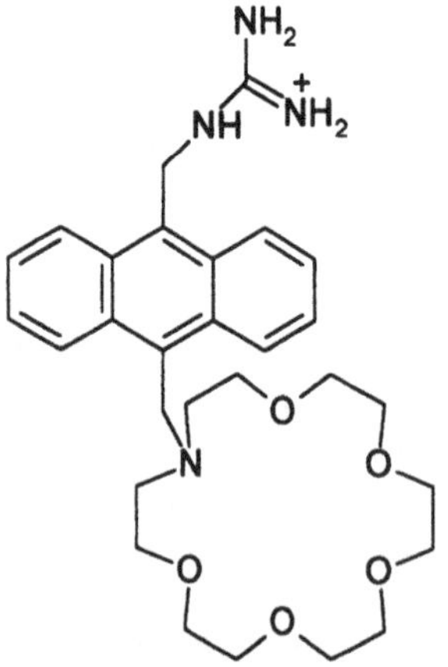

98 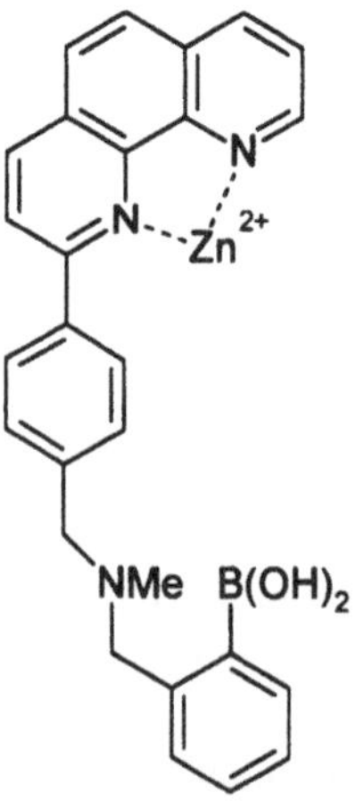   99

boxylate group of the uronic acid. Fluorescence enhancement factors of 3.5 are achievable. Saccharides lacking the acid group, such as fructose, produce enhancement factors of 1.6—small but not insignificant. This is because **99** has only one PET process centered on the aminomethylphenyl boronic acid moiety.

Lakowicz and his colleagues at the University of Maryland, Baltimore, have gone after a particularly difficult target for PET signaling, CO_2 [186]. The vitally important measurement of CO_2 is commonly conducted by exploiting its acidity. The acidification of a weakly basic solution upon CO_2 entry can be monitored by a pH sensor of PET or ICT design. A more direct, and hence more robust, measurement would require a direct interaction between the signaling agent and CO_2. Primary and secondary amines have long been known to reversibly absorb CO_2 to produce carbamic acids [187]. The similarity of these CO_2-signaling systems to those described earlier for specifying isocyanates is striking. Again, PET is feasible in **100** but not in **101**, owing to the delocaliza-

tion of the electron density on the nitrogen atom into the carbonyl group. This time, fluorescence enhancements of 2 are seen upon bubbling CO_2 through a dioxan solution of **100** previously flushed with Argon. In this instance, another path for fluorescence enhancement is available: acidity of **101** can protonate another molecule of **100**. Of course, secondary aminomethylanthracenes are rather efficient proton signaling systems. An attempt to control protonation effects has been made by adding triethylamine as a proton scavenger. Even the primary amine unit in **102** produces adequate PET activity because of the lower energy

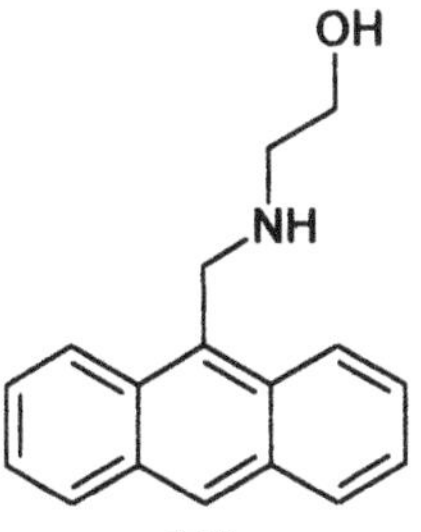

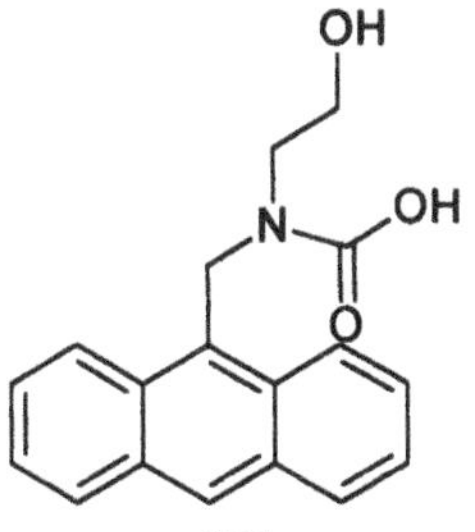

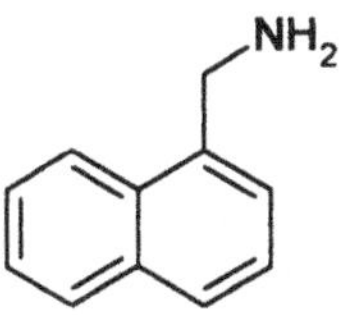

of the highest occupied molecular orbital of naphthalene compared to that of anthracene. In both **102** and **100**, the response time is about 10 minutes.

Although not based on PET designs, **103** [188] strengthens the signaling role played by reversible covalent interactions. Mohr and his colleagues at ETH,

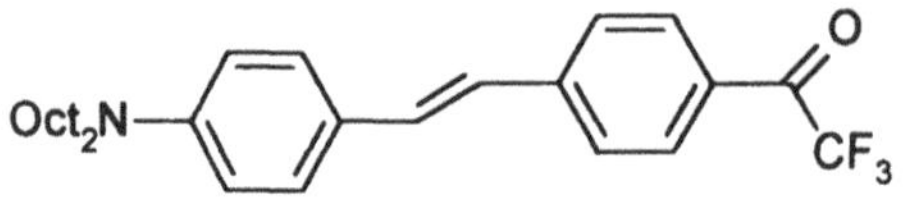

103

Zurich, exploit the susceptibility of trifluoromethyl ketones to nucleophilic addition by amines (primary or even secondary) producing hemiaminals **104**. Tertiary

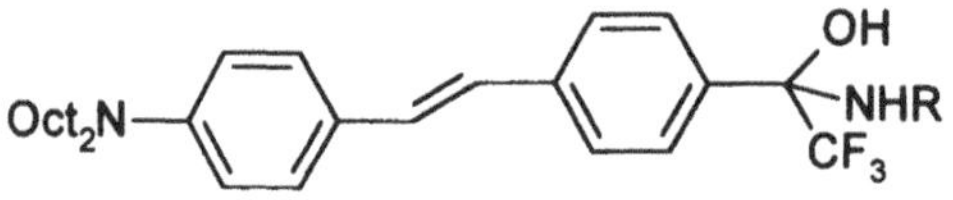

104

amines work in a related manner. This causes a significant reduction in electron delocalization within the lumophore since the electron-accepting component is taken out of play. The excitation and emission maxima shift accordingly from 468 nm and 576 nm (for **103**) to 384 nm and 426 nm (for **104**). Intramembrane operation allows an increase in selectivity. Of course, the amines need to be maintained in their nucleophilic form in solution with alkaline pH values.

It is remarkable that the set of analytes tackled in this and the previous section (thiols, NO, CO_2, CH_3NCO, and saccharides) are of undeniable biological significance for better or for worse. This must surely underscore the importance of research into luminescent PET signaling involving covalent interactions, whether irreversible or reversible.

XIII. LOGIC GATES AND ARITHMETIC SYSTEMS VIA PET SIGNALING

We have made the case on several occasions that luminescent signaling systems can be useful in information processing contexts, however small. Since several of these occasions have been rather recent [184–191] we only discuss the newest material here.

As far as we are aware, the first claims of molecular logic appeared from Birge's laboratory at Syracuse University, in the conference literature [192]. However, these reports concerning porphyrin molecules have never appeared in the primary literature in sufficient detail to allow corroboration via the usual scientific process. In addition, we have been unable to locate any work that followed up the original claims. It was in 1993 that we published the first general approach to molecular logic [183]. This approach has subsequently fueled work in several other laboratories besides our own. Luminescent PET signaling systems form a substantial fraction of the molecular logic gates currently known. However, those based on other mechanisms are included in the discussion for a rounded view.

To briefly illustrate the idea, consider **105** [183]. **105** is a lumophore–spacer$_1$–receptor$_1$–spacer$_2$–receptor$_2$ system where either receptor can launch a

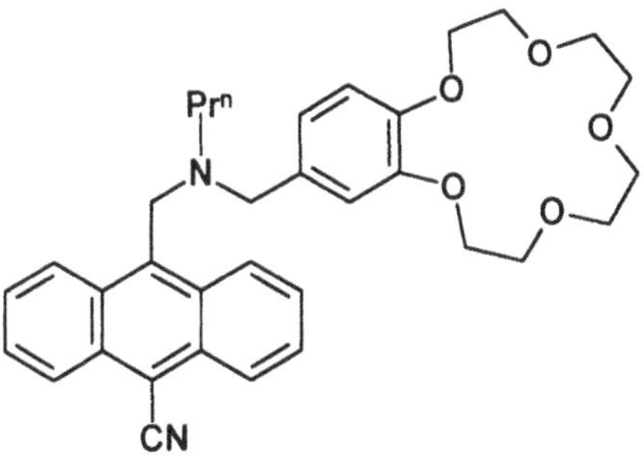

105

PET process to knock out the emission unless both receptors are blocked with their particular guest ions. These two guest ions are H^+ for the amine receptor$_1$ and Na^+ for the benzocrown receptor$_2$. The success of the simpler system **46** [99] was an inspiration behind **105**. If we consider that IN$_1$ is H^+ and IN$_2$ is Na^+ and the output is fluorescence, then we have the AND logic truth table (Fig. 2) being obeyed. Several more AND gates are available now [179,193–195].

The first XOR logic operation was produced by the combined effort of Balzani and Stoddart et al. [196]. Pseudorotaxane **106.107** is nonemissive owing

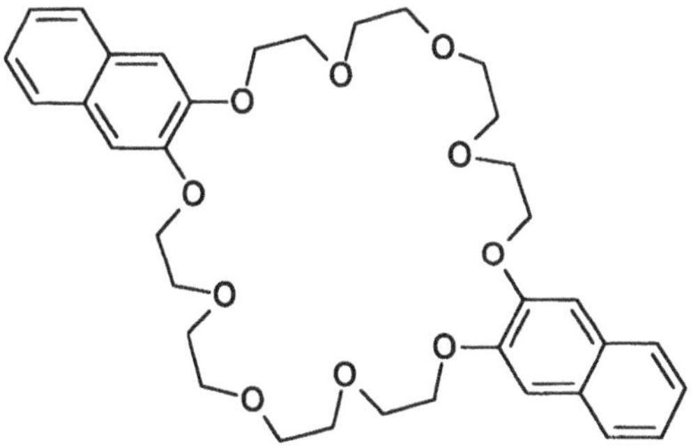

106 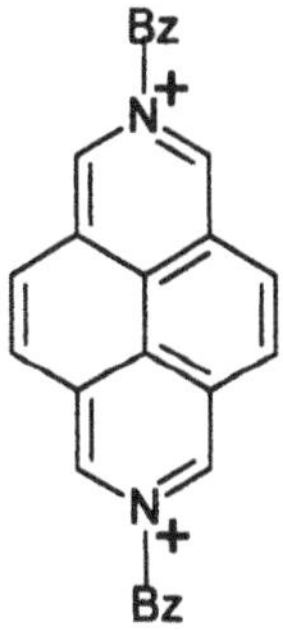   **107**

to the PET-type CT processes discussed above. Either H^+ or Bu_3N can dissociate the pseudorotaxane components by binding with **106** or **107**, respectively. Then

IN$_1$	0011		IN$_1$	00110011
IN$_2$	0101		IN$_2$	01010101
AND	0001		IN$_3$	00001111
XOR	0110		INHIBIT	00010000
NAND	1110		ENABLED OR	00000111
TRANSFER	0011			
INHIBIT	0010			
Half-adder (carry)	0001			
Half-adder (sum)	0110			

Figure 2 Truth tables for logic gates and arithmetic systems.

bright fluorescence from **106** is observed. Clearly, the addition of 1:1 H$^+$ and Bu$_3$N gives no change from the original nonemissive situation. Thus, the four rows of the XOR truth table (Fig. 2) are reproduced. This is a clear case of a cyclophane-based PET-type signaling system with multiple uses.

The "off-on-off" signaling system of **76** and [Co(CN)$_6$]$^{3-}$ at pH 3.6 from Pina and colleagues also behaves as an XOR logic gate [147]. While this system can be operated in several interesting formats, we consider only the simplest. The two inputs are two isoenergetic pulses (of 355 nm), each of which causes sufficient photoreaction to give maximal production of **77**. If both inputs are applied, the photoaquation of [CoCN)$_6$]$^{3-}$ also occurs to increase the pH and to return **77** back to **76**. Thus, the **77** concentration is minimal, just as if no input were applied at all. The irreversibility of the photoaquation is an aspect which it is hoped will be improved in later versions of this system.

The originators of the XOR gate involving **76** also have used a close cousin **108** to develop AND [197,198] and ENABLED OR logic systems [199]. Such compounds have the added interest of acting as photochromics capable of locking and erasing besides the usual write–read operations. The AND logic function of **108** is arranged with protons as one input and irradiation as the other. As discussed under "off-on-off" behavior involving **76**, both low pH values and irradiation are necessary to produce **109**. High proton densities can be arranged in the locality of **108** by either acidification or by adding an anionic micellar medium. Our **32** [79] clearly showed the highly perturbed H$^+$ concentrations near micelles. Thus, in the case of **108**, the three inputs are acidification

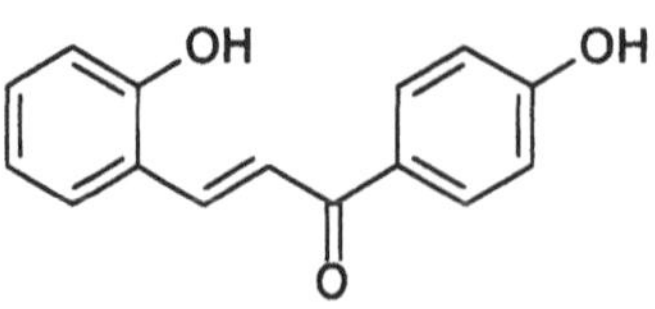

108 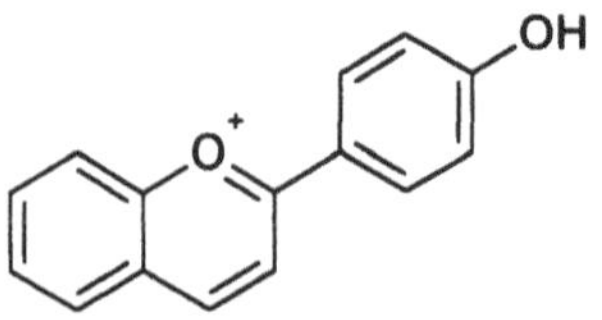   **109**

(to pH 1), sodium dodecylsulfate micelles, and irradiation. Irradiation is essential, of course, to form **109**. So it is the enabling input (IN$_3$ in Fig. 2). Either or both of the other two inputs will guarantee production of **109** upon irradiation of **108** [199].

Garcia-Espana and Pina's **73** [142], discussed earlier under "off-on-off" signaling, is also relevant in a logic context. At pH 6, **73** is brightly emissive. Addition of protons to take it to pH 2 causes no change in fluorescence. Addition of ATP at pH 6 also maintains the status quo. But the addition of ATP and protons to produce pH 2 causes strong quenching due to the complexation of the multiprotonated forms of **73** and ATP. This is clearly NAND logic behavior (Fig. 2).

A very clever system that uses three commercially available compounds **110** to **112** to demonstrate molecular NAND and TRANSFER logic has re-

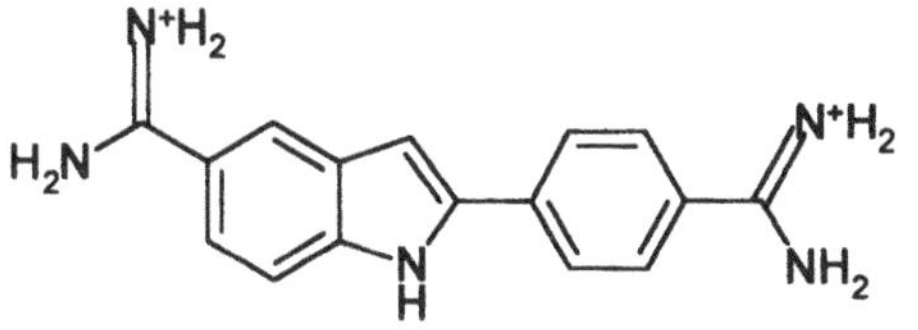

110

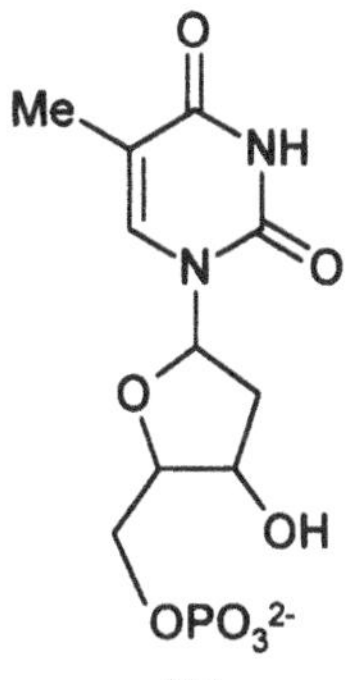

111 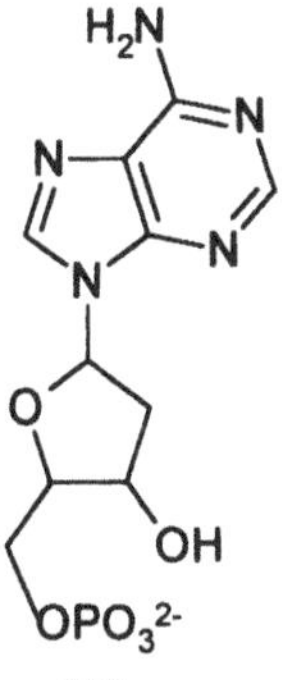   112

cently been invented by Akkaya and Baytekin from the Middle East Technical University, Ankara [200]. Another appealing feature is that Watson–Crick base pairing is used for the first time for molecular logic design. Therefore this work provides a nice bridge between molecular logic concepts discussed here and DNA computation [201,202]. **110**, being a popular DNA binder, interacts with the matching mononucleotide pair of **111** and **112** but not as much with either one in isolation. Observation at the chosen wavelength of 455 nm then produces a reduced emission intensity (50%) only if both inputs (**111** and **112**) are present. This is NAND logic (Fig. 2), which has several precedents [203,204]. The novelty in this work is the dual logic character exposed by observing the emission at two wavelengths. We note in passing that related dual logic is available in two PhD theses from Credi in Bologna, Italy [205], and McClenaghan in Belfast, Northern Ireland [206]. The spectral perturbations for **110** are such that the observation of the fluorescence at 411 nm now produces a truth table corresponding to TRANSFER logic (Fig. 2). It is hoped that the future will bring more understanding of the spectral perturbations so that they can be maximized at will.

Our **113** represents the 3-input INHIBIT logic operation [74], which possesses a substantial truth table by current standards in molecular logic (Fig. 2).

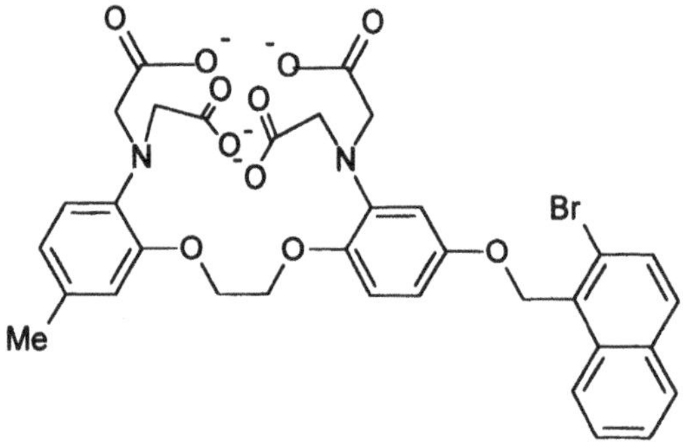

113

This uses Ca^{2+} as IN_1 to the amino acid receptor to block a PET process from this receptor to the bromonaphthalene lumophore. **113** also uses β-cyclodextrin as IN_2 which actually binds to the lumophore to offer it protection against triplet–triplet annihilation. The inhibitory IN_3 is O_2 which wrecks the phosphorescence emitted from the lumophore triplet excited state whether it is enveloped by β-cyclodextrin or not. The corresponding 2-input INHIBIT logic gate [207], is available from Gunnlaugsson and his colleagues at Trinity College, Dublin and the University of Durham in the form of **114**. IN_1 is H^+ which binds to the quinoline nitrogen to bring the absorption band of the latter into the region

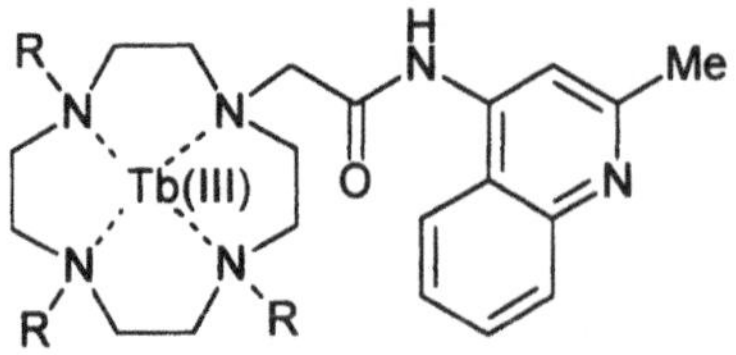

114; R = $CH_2P(Me)O_2^-$

of the exciting wavelength (330 nm). Strong atomic line emission is observed by sensitization of the terbium center by the quinolinium triplet. The inhibitory IN_2 is again O_2 which quenches the long-lived emission from the terbium ion, because the terbium excited state equilibriates with the pumping triplet of the quinolium group.

We reach the end of this review of luminescent PET signaling systems by noting their recent expression of numeracy. We believe this is special because people become (and remain) numerate via mysterious, but molecular, processes in their brains. **115** [208] is an AND gate very much in the mold of **105** [183].

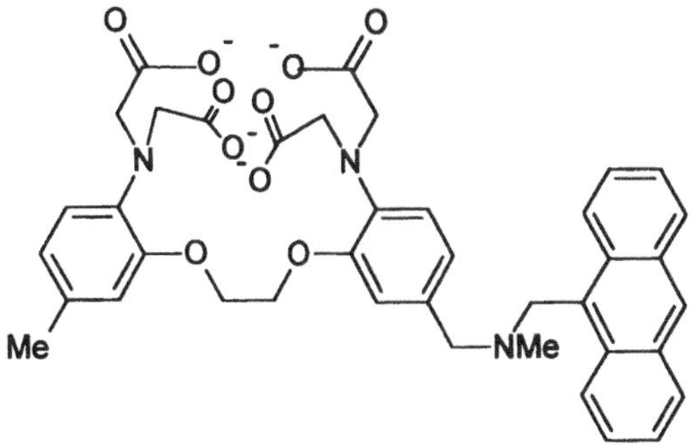

115

But when combined in parallel with the compatible XOR gate **116** [208], we have molecular-scale arithmetic for the first time. Molecular arithmetic was

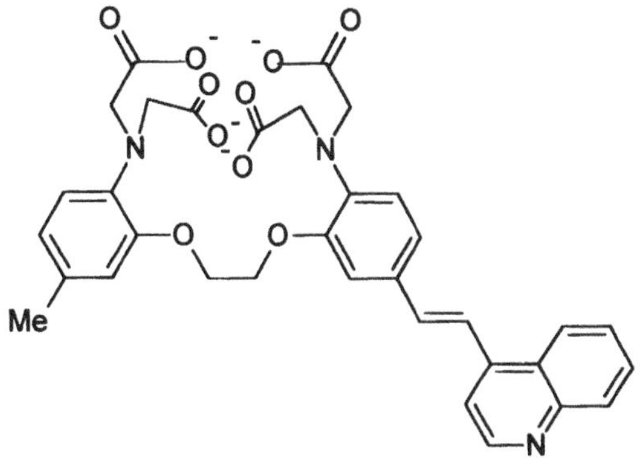

116

hampered until now because the available AND and XOR gate molecules were not compatible with each other to permit their parallel operation. For now, **116** is neither a PET system nor is it fluorescent, but a start has been made. The truth table for a half-adder is included in Fig. 2.

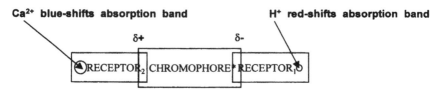

Figure 3 ICT-based XOR logic system design.

The mechanism of action of XOR gate **116** deserves a mention and is best shown with the schematic in Fig. 3. **116** is a push–pull system that has selective receptors at opposite terminals. The energy of the ICT excited state of **116** is perturbed in opposite directions when each receptor is blocked by its guest. So the absorption blue-shifts with Ca^{2+} and red-shifts with H^+. When both guests are present, the shifts cancel and the status quo is regained. Observation of the transmittance at 390 nm now gives the XOR truth table (Fig. 2).

A valuable analysis from Credi [205], from Balzani's Bologna laboratory shows how ideas underlying the XOR gate system **106.107** can be combined with EET-based AND-type systems [209] to give a half-adder capable of arithmetic functions. The practical realization of this analysis would be most welcome.

XIV. CONCLUSION

The foregoing pages showed how the competition between two cornerstones of photochemistry—luminescence and photoinduced electron transfer—can be managed. We hope the reader will be persuaded that such an exercise is worthwhile for its own sake and for practical applications concerned with information handling.

ACKNOWLEDGMENTS

We thank DENI and Avecia Ltd. for their support.

REFERENCES

1. AJ Bryan, AP de Silva, SA de Silva, RADD Rupasinghe, KRAS Sandanayake. Biosensors 4:169, 1989.
2. RA Bissell, AP de Silva, HQN Gunaratne, PLM Lynch, GEM Maguire, KRAS Sandanayake. Chem Soc Rev 21:187, 1992.

3. RA Bissell, AP de Silva, HQN Gunaratne, PLM Lynch, GEM Maguire, CP McCoy, KRAS Sandanayake. Top Curr Chem 168:223, 1993.
4. AP de Silva, CP McCoy. Chem Ind 992, 1994.
5. AP de Silva, HQN Gunaratne, T Gunnlaugsson, CP McCoy, PRS Maxwell, JT Rademacher, TE Rice. Pure Appl Chem 68:1443, 1996.
6. AP de Silva, T Gunnlaugsson, TE Rice. Analyst 121:1759, 1996.
7. AP de Silva, T Gunnlaugsson, CP McCoy. J Chem Educ 74:53, 1997.
8. AP de Silva, HQN Gunaratne, T Gunnlaugsson, AJM Huxley, CP McCoy, JT Rademacher, TE Rice. Adv Supramol Chem 4:1, 1997.
9. AP de Silva, HQN Gunaratne, T Gunnlaugsson, AJM Huxley, CP McCoy, JT Rademacher, TE Rice. Chem Rev 97:1515, 1997.
10. AP de Silva, AJM Huxley. In W Rettig, B Strehmel, S Schrader, H Seifert, eds. Applied Fluorescence in Chemistry, Biology and Medicine. Berlin: Springer Verlag, 1999, p 179.
11. AP de Silva, DB Fox, AJM Huxley, ND McClenaghan, J Roiron. Coord Chem Rev 185, 186:297, 1999.
12. AP de Silva, J Eilers, G Zlokarnik. Proc Natl Acad Sci USA 96:8336, 1999.
13. AP de Silva, DB Fox, TS Moody, SM Weir. Trends Biotechnol 19:27, 2001.
14. B Valeur. In JR Lakowicz, ed. Topics in Fluorescence Spectroscopy, vol. 4, Probe Design and Chemical Sensing. New York: Plenum, 1994, p 21.
15. B Valeur. In SG Schulman, ed. Molecular Luminescence Spectroscopy. Part 3. New York: Wiley, 1993, p 25.
16. B Valeur, E Bardez. Chem Brit 31:216, 1995.
17. AW Czarnik. Acc Chem Res 27:302, 1994.
18. AW Czarnik, ed. Fluorescent Chemosensors of Ion and Molecule Recognition. ACS Symposium Series 538, Washington DC: American Chemical Society, 1993.
19. AW Czarnik. Chem Biol 2:423, 1995.
20. AW Czarnik. Adv Supramol Chem 3:131, 1993.
21. AW Czarnik, J-P Desvergne, eds. Chemosensors of Ion and Molecule Recognition. Dordrecht: Kluwer, 1997.
22. J-P Desvergne, F Fages, H Bouas-Laurent, P Marsau. Pure Appl Chem 64:1231, 1992.
23. L Fabbrizzi, A Poggi. Chem Soc Rev 24:197, 1995.
24. L Fabbrizzi, M Licchelli, P Pallavicini. Acc Chem Res 32:846, 1999.
25. L Fabbrizzi, ed. Special issue on luminescent sensors. Coord Chem Rev 205:1–232, 2000.
26. R Bergonzi, L Fabbrizzi, M Licchelli, C Mangano. Coord Chem Rev 170:31, 1998.
27. L Fabbrizzi, M Licchelli, P Pallavicini, D Sacchi, A Taglietti. Analyst 121:1763, 1996.
28. TD James, P Linnane, S Shinkai. J Chem Soc Chem Commun 281, 1996.
29. KRAS Sandanayake, TD James, S Shinkai. Pure Appl Chem 68:1207, 1996.
30. K Kubo, T Sakurai. Heterocycles 52:945, 2000.
31. FL Carter, RE Siatkowski, H Wohltjen, eds. Molecular Electronic Devices. Amsterdam: Elsevier, 1988.
32. MA Fox, M Chanon, eds. Photoinduced Electron Transfer. Amsterdam: Elsevier, 1988.

33. Mattay J, ed. Photoinduced Electron Transfer. Parts I-V. Top Curr Chem 156:158, 1990; 159, 1991; 163, 1992; 168, 1993.
34. MR Wasielewski. Chem Rev 92:435, 1992.
35. GJ Kavarnos. Fundamentals of Photoinduced Electron Transfer. Weinheim: VCH, 1993.
36. J Mattay, ed. Electron Transfer. Part I. Top Curr Chem 169, 1994.
37. AP de Silva, HQN Gunaratne, J-L Habib-Jiwan, CP McCoy, TE Rice, J-P Soumillion. Angew Chem Int Ed Engl 34:1728, 1995.
38. A Weller. Pure Appl Chem 16:115, 1968.
39. H Siegerman. In NL Weinberg, ed. Techniques of Electroorganic Synthesis. Part II. New York: Wiley, 1975.
40. ZR Grabowski, J Dobkowski. Pure Appl Chem 55:245, 1983.
41. RA Marcus. Angew Chem Int Ed Engl 32:1111, 1993.
42. MS Alexiou, V Tychopoulos, S Ghorbanian, JHP Tyman, RG Brown, PI Brittain. J Chem Soc Perkin Trans 2:837, 1990.
43. A Pardo, JML Poyato, E Martin, JJ Camacho, D Reyman. J Lumin 46:381, 1990.
44. AP de Silva, HQN Gunaratne, J-L Habib-Jiwan, CP McCoy, J-P Soumillion. (Results to be published.)
45. MA Fox, E Galoppini. J Am Chem Soc 119:5277, 1997.
46. A Knorr, E Galoppini, MA Fox. J Phys Org Chem 10:484, 1997.
47. D Gosztola, MP Niemczyk, MR Wasielewski. J Am Chem Soc 120:5118, 1998.
48. MP O'Neil, MP Niemczyk, WA Svec, D Gosztola, GL Gaines III, MR Wasielewski. Science 257:63, 1992.
49. AP de Silva, Te Rice. Chem Commun 163, 1999.
50. J Deisenhofer, O Epp, K Miki, R Huber, H Michel. J Mol Biol 180:385, 1984.
51. H Tian, T Xu, Y Zhao, K Chen. J Chem Soc Perkin Trans 2:545, 1999.
52. MJP Leiner, H He, A Boila-Gockel. USP 5,952,491.
53. AP de Silva, RADD Rupasinghe. J Chem Soc Chem Commun 1669, 1985.
54. AP de Silva, HQN Gunaratne, PLM Lynch, AL Patty, GL Spence. J Chem Soc Perkin Trans 2:1611, 1993.
55. RP Haugland. Handbook of Fluorescent Probes and Research Chemicals. 7th ed. Eugene, OR: Molecular Probes, 1999.
56. G Copeland, S Miller. J Am Chem Soc 121:4306, 1999.
57. M Schuster, E Unterreitmaier. Fresenius J Anal Chem 346:630, 1993.
58. MY Chae, AW Czarnik. J Am Chem Soc 114:9704, 1992.
59. G Hennrich, H Sonnenschein, U Resch-Genge. J Am Chem Soc 121:5073, 1999.
60. M Schuster, M Sandor. Fresenius J Anal Chem 356:326, 1996.
61. M Sandor, F Geistmann, M Schuster. Anal Chim Acta 388:19, 1999.
62. U Resch, K Rurack, JL Bricks, YL Slominski. J Fluoresc 7:31S, 1997.
63. JL Bricks, K Rurack, R Radeglia, G Reck, B Schulz, H Sonneneschein, U Resch-Genger. J Chem Soc Perkin Trans 2:1209, 2000.
64. P Ghosh, PK Bharadwaj, S Mandal, S Ghosh. J Am Chem Soc 118:1553, 1996.
65. B Ramachandram, A Samanta. J Phys Chem A 102:10579, 1998.
66. B Ramachandram, A Samanta. Chem Phys Lett 290:9, 1998.
67. B Ramachandram, A Samanta. Chem Commun 1037, 1997.

68. K Mitchell, RG Brown, DW Yuan, SC Chang, RE Utecht, DE Lewis. J Photochem Photobiol A Chem 115:157, 1998.
69. W Wang, G Springsteen, S Gao, B Wang. Chem Commun 1283, 2000.
70. RA Bissell, E Calle, AP de Silva, SA de Silva, HQN Gunaratne, JL Habib-Jiwan, SLA Peiris, RADD Rupasinghe, TKSD Samarasinghe, KRAS Sandanayake, J-P Soumillion. J Chem Soc Perkin Trans 2:1559, 1992.
71. TD James, KRAS Sandanayake, S Shinkai. J Chem Soc Chem Commun 477, 1994.
72. J Ishikawa, H Sakamoto, S Nakao, H Wada. J Org Chem 64:1913, 1999.
73. J-E Sohna-Sohna, P Jaumier, F Fages. J Chem Res (S) 134, 1999.
74. AP de Silva, IM Dixon, HQN Gunaratne, T Gunnlaugsson, PRS Maxwell, TE Rice. J Am Chem Soc 121:1393, 1999.
75. AP de Silva, HQN Gunaratne, PLM Lynch. J Chem Soc Perkin Trans 2:685, 1995.
76. AP de Silva, SA de Silva. J Chem Soc Chem Commun 1709, 1986.
77. RE Gawley, Q Zhang, PI Higgs, S Wang, RM LeBlanc. Tetrahedron Lett 40:5461, 1999.
78. S Wang, Q Zhang, PK Datta, RE Gawley, RM LeBlanc. Langmuir 16:4607, 2000.
79. RA Bissell, AJ Bryan, AP de Silva, HQN Gunaratne, CP McCoy. J Chem Soc Chem Commun 405, 1994.
80. JH Chang, HJ Kim, JH Park, YK Shin, Y Chung. Bull Korean Chem Soc 2:796, 1999.
81. K Kubo, N Kato, T Sakurai. Bull Chem Soc Jpn 70:3041, 1997.
82. K Kubo, R Ishige, N Kato, E Yamamoto, T Sakurai. Heterocycles 45:2365, 1997.
83. K Kubo, R Ishige, T Sakurai. Heterocycles 48:347, 1998.
84. K Kubo, E Yamamoto, T Sakurai. Heterocycles 45:1457, 1997.
85. SL De Wall, ES Meadows, LJ Barbour, GW Gokel. Chem Commun 1553, 1999.
86. ES Meadows, SL De Wall, LJ Barbour, FR Fronczek, MS Kim, GW Gokel. J Am Chem Soc 122:3325, 2000.
87. H-F Ji, R Dabestani, RL Hettich, GM Brown. Photochem Photobiol 70:882, 1999.
88. S Quici, A Manfredi, M Maestri, I Manet, P Passaniti, V Balzani. Eur J Org Chem 2041, 2000.
89. H-F Ji, R Dabestani, GM Brown, RL Hettich. Photochem Photobiol 69:513, 1999.
90. H Shizuka, M Nakamura, T Morita. J Phys Chem 83:2019, 1979.
91. AP de Silva, HQN Gunaratne, PLM Lynch, AJ Patty, GL Spence. J Chem Soc Perkin Trans 2:1611, 1993.
92. JC Beeson, ME Huston, A Douglas, TK Pollard, AW Czarnik. J Fluoresc 3:65, 1992.
93. B Witulski, Y Zimmermann, V Darcos, J-P Desvergne, DM Bassani, H Bouas-Laurent. Tetrahedron Lett 39:4807, 1998.
94. MM Martin, P Plaza, YH Meyer, F Badaoui, J Bourson, JP Lefebvre, B Valeur. J Phys Chem 100:6879, 1996.
94a. R Mathevet, G Jonusauskas, C Rullière, J-F Létard, R Lapouyade. J Phys Chem 99:15709, 1995.
95. K Rurack, JL Bricks, A Kachkovski, U Resch. J Fluoresc 7:63S, 1997.
96. K Rurack, JL Bricks, B Schulz, M Maus, U Resch-Genger. J Phys Chem A 104:6171, 2000.

97. K Yoshida, T Mori, S Watanabe, H Kawai, T Nagamura. J Chem Soc Perkin Trans 2:393, 1999.
98. H Kawai, T Nagamura, T Mori, K Yoshida. J Phys Chem A 103:660, 1999.
99. AP de Silva, KRAS Sandanayake. J Chem Soc Chem Commun 1183, 1989.
100. H-F Ji, GM Brown, R Dabestani. Chem Commun 609, 1999.
101. Z Asfari, S Wenger, J Vicens. J Inclusion Phenom Mol Recognit Chem 19:137, 1994.
102. YA Ikeda, T Tsudera, S Shinkai. J Org Chem 62:3568, 1997.
103. L Prodi, F Bolletta, M Montalti, N Zaccheroni, A Casnati, F Sansone, R Ungaro. New J Chem 24:155, 2000.
104. RA Sachleben, A Urvoas, JC Bryan, TJ Haverlock, BP Hay, BA Moyer. Chem Commun 1751, 1999.
105. H-F Ji, R Dabestani, GM Brown, RA Sachleben. Chem Commun 833, 2000.
106. WS Xia, RH Schmehl, CJ Li. Chem Commun 695, 2000.
107. PD Beer. J Chem Soc Chem Commun 1678, 1986.
108. KRAS Sandanayake, K Nakashima, S Shinkai. J Chem Soc Chem Commun 1621, 1994.
109. M Takeuchi, T Mizuno, H Shinmori, M Nakashima, S Shinkai. Tetrahedron 52:1195, 1996.
110. F. Diederich. In JF Stoddart, ed. Cyclophanes. Cambridge: Royal Society of Chemistry, 1991.
111. V Balzani, M Gomez-Lopez, JF Stoddart. Acc Chem Res 31:405, 1998.
112. R Ballardini, V Balzani, A Credi, MT Gandolfi, D Marquis, L Perez-Garcia, JF Stoddart. Eur J Org Chem 81, 1998.
113. DG Hamilton, JE Davies, L Prodi, JKM Sanders. Chem Eur J 4:608, 1998.
114. PR Ashton, S Iqbal, JF Stoddart, ND Tinker. Chem Commun 479, 1996.
115. MA Bernardo, F Pina, E Garcia-Espana, J Latorre, SV Luis, JA Ramirez, C Soriano. Inorg Chem 37:3935, 1998.
116. W Rettig. Top Curr Chem 169:253, 1994.
117. K Rurack, M Kollmansberger, U Resch-Genger, J Daub. J Am Chem Soc 122:968, 2000.
118. SA Jonker, SI Van Dijk, K Goubitz, CA Reiss, W Schuddeboom, JW Verhoeven. Mol Cryst Liq Cryst 183:273, 1990.
119. M Kollmansberger, T Gareis, S Heinl, J Breu, J Daub. Angew Chem Int Ed Engl 36:1333, 1997.
120. T Gareis, C Huber, OS Wolfeis, J Daub. Chem Commun 1717, 1997.
121. R Grigg, WDJA Norbert. J Chem Soc Chem Commun 1300, 1992.
122. R Grigg, JM Holmes, SK Jones, WDJA Norbert. J Chem Soc Chem Commun 185, 1994.
123. T Hirano, K Kikuchi, Y Urano, T Nagano. Angew Chem Int Ed Engl 39:1052, 2000.
124. T Koike, T Watanabe, S Aoki, E Kimura, M Shiro. J Am Chem Soc 118:12696, 1996.
125. E Kimura, T Koike. Chem Soc Rev 27:179, 1998.
126. O Reany, T Gunnlaugsson, D Parker. Chem Commun 473, 2000.
127. F D'Souza, GR Deviprased, Y-Y Hsieh. Chem Commun 533, 1997.

128. F D'Souza. J Am Chem Soc 118:923, 1996.
129. GR Deviprasad, B Keshavan, F D'Souza. J Chem Soc Perkin Trans I 3133, 1998.
130. B Kratochvil, DA Zatko, Anal Chem 36:527, 1964.
131. V Goulle, A Harriman, J-M Lehn. J Chem Soc Chem Commun 1034, 1993.
132. RA Berthon, SB Colbran, GM Moran. Inorg Chim Acta 204:3, 1993.
133. G De Santis, L Fabrizzi, M Licchelli, C Mangano, D Sacchi. Inorg Chem 34:3581, 1995.
134. J Daub, M Beck, A Knorr, H Spreitzer. Pure Appl Chem 68:1399, 1996.
135. AP de Silva, HQN Gunaratne, CP McCoy. Chem Commun 2399, 1996.
136. V Balzani, F Scandola. Supramolecular Photochemistry. Chichester: Ellis-Horwood, 1991.
137. J-M Lehn, Supramolecular Chemistry. Weinheim: VCH, 1995.
138. SA de Silva, A Zavaleta, DE Baron, O Allam, EV Isidor, N Kashimura, JM Percarpio. Tetrahedron Lett 38:2237, 1997.
139. V Amendola, L Fabbrizzi, P Pallavicini, L Parodi, A Perotti. J Chem Soc Dalton Trans 2053, 1998.
140. L Fabbrizzi, M Licchelli, A Poggi, A Taglietti. Eur J Inorg Chem 35, 1999.
141. H Chao, B-H Ye, Q-L Zhang, L-N Ji. Inorg Chem Commun 2:338, 1999.
142. MT Albeda, MA Bernardo, E Garcia-Espana, ML Godino-Salido, SV Luis, MJ Melo, F Pina, C Soriano. J Chem Soc Perkin Trans 2:2545, 1999.
143. ME Huston, EU Akkaya, AW Czarnik. J Am Chem Soc 111:8735, 1989.
144. MW Hosseini, AJ Blacker, J-M Lehn. J Am Chem Soc 112:3896, 1990.
145. Y Kubo, S Obara, S Tokita. Chem Commun 2399, 1999.
146. Y Kubo, S Hamaguchi, A Numi, K Yoshida, S Tokita. J Chem Soc Chem Commun 305, 1993.
147. F Pina, M Maestri, V Balzani. J Am Chem Soc 122:4496, 2000.
148. F Pina, M Maestri, V Balzani. Chem Commun 107, 1999.
149. L Moggi, F Bolletta, V Balzani, F Scandola. J Inorg Nucl Chem 28:2589, 1966.
150. AP de Silva, HQN Gunaratne, T Gunnlaugsson, PLM Lynch. New J Chem 20:871, 1996.
151. S Encinas, KL Bushell, SM Couchman, JC Jeffery, MD Ward, L Flamigni, F Barigelletti. J Chem Soc Dalton Trans 1783, 2000.
152. P Wang, S Wu. J Photochem Photobiol A Chem 118:7, 1998.
153. L Fabbrizzi, M Licchelli, P Pallavicini, L Parodi. Angew Chem Int Ed Engl 37:800, 1998.
154. F Pina, MJ Melo, MA Bernardo, SV Luis, E Garcia-Espana. J Photochem Photobiol A Chem 126:65, 1999.
155. W Klopffer. Adv Photochem 312, 1977.
156. R Grigg, WDJA Norbert. J Chem Soc Chem Commun 1298, 1992.
157. AP de Silva, HQN Gunaratne, T Gunnlaugsson. Tetrahedron Lett 39:5077, 1998.
158. A Russo, EA Bump. Meth Biochem Anal 33:165, 1988.
159. Y Kanaoka. Angew Chem Int Ed Engl 16:137, 1977.
160. JK Weltman, RP Szaro, AR Frackelston, RM Dowben, JR Bunting, RE Cathou. J Biol Chem 218:3173, 1973.
161. HJ Verhey, CHW Bekker, JW Verhoeven, JW Hofstraat. New J Chem 20:809, 1996.

162. JW Verhoeven. Pure Appl Chem 62:1585, 1990.
163. SP Levine, JH Hoggatt, E Chladek, G Jungclaus, JL Gerlock. Anal Chem 51:1106, 1979.
164. C Sangoe, E Zimerson. J Liq Chromatogr 3:971, 1980.
165. S Moncada. J Roy Soc Med 92:164, 1999.
166. H Kojima, T Nagano. Adv Mater 12:763, 2000.
167. H Kojima, N Nakatsubo, K Kikuchi, S Kawahara, Y Kirino, H Nagoshi, Y Hirata, T Nagano. Anal Chem 70:2446, 1998.
168. H Kojima, M Hirotani, Y Urano, K Kikuchi, T Higuchi, T Nagano. Tetrahedron Lett 41:69, 2000.
169. B Rotman, BW Papermaster. Proc Natl Acad Sci USA 55:134, 1966.
170. C Munkholm, DR Parkinson, DR Walt. J Am Chem Soc 112:2608, 1990.
171. H Shizuka, T Ogiwara, E Kimura. J Phys Chem 89:4302, 1985.
172. G Wulff. Angew Chem Int Ed Engl 34:1812, 1995.
173. JP Lorand. J Org Chem 24:769, 1959.
174. J Yoon, AW Czarnik. J Am Chem Soc 114:5874, 1992.
175. Y Nagai, K Kabayashi, H Toi, Y Aoyama. Bull Chem Soc Jpn 66:2965, 1993.
176. TD James, KRAS Sandanayake, S Shinkai. J Chem Soc Chem Commun 477, 1994.
177. J Yoon, AW Czarnik. Bioorg Med Chem 1:267, 1993.
178. TD James, KRAS Sandanayake, S Shinkai. Angew Chem Int Ed Engl 35:1910, 1996.
179. CR Cooper, TD James. Chem Commun 1419, 1997.
180. AP de Silva, KRAS Sandanayake. Angew Chem Int Ed Engl 29:1173, 1990.
181. TD James, KRAS Sandanayake, S Shinkai. Angew Chem Int Ed Engl 33:2207, 1994.
182. AP de Silva, HQN Gunaratne, C McVeigh, GEM Maguire, PRS Maxwell, E O'Hanlon. Chem Commun 2191, 1996.
183. AP de Silva, HQN Gunaratne, CP McCoy. Nature 364:42, 1993.
184. M Takeuchi, M Yamamoto, S Shinkai. Chem Commun 1731, 1997.
185. M Yamamoto, M Takeuchi, S Shinkai. Tetrahedron 54:3125, 1998.
186. P Herman, Z Murtaza, JR Lakowicz. Anal Biochem 272:87, 1999.
187. G Gross, RE Forster, L Lin. J Biol Chem 251:4398, 1976.
188. GJ Mohr, C Demuth, UE Spichiger-Keller. Anal Chem 70:3868, 1998.
189. AP de Silva, ND McClenaghan, CP McCoy. In V Balzani, AP de Silva, eds. Handbook of Electron Transfer in Chemistry. Weinheim: Wiley-VCH, 2001 (in press).
190. AP de Silva, DB Fox, TS Moody. In M Shibasaki, JF Stoddart, F Vogtle, eds. Simulating Concepts in Chemistry. Weinheim: Wiley-VCH, 2000, p 307.
191. AP de Silva, ND McClenaghan, CP McCoy. In BL Feringa, ed. Molecular Switches. New York: Wiley-VCH, 2001, p 339.
192. RR Birge. In BC Crandall, BC Lewis, eds. Nanotechnology; Research and Perspectives. Cambridge, MA: MIT Press, 1992, p 156.
193. AP de Silva, HQN Gunaratne, CP McCoy. J Am Chem Soc 119:7891, 1997.
194. M Inouye, K Akamatsu, H Nakazumi. J Am Chem Soc 119:9160, 1997.
195. L Gobbi, P Seiler, F Diederich. Angew Chem Int Ed Engl 38:674, 1999.

196. A Credi, V Balzani, SJ Langford, JF Stoddart. J Am Chem Soc 119:2679, 1997.
197. F Pina, MJ Melo, M Maestri, R Ballardini, V Balzani. J Am Chem Soc 119:5556, 1997.
198. F Pina, A Roque, MJ Melo, M Maestri, L Belladelli, V Balzani. Chem Eur J 4:1184, 1998.
199. A Roque, F Pina, S Alves, R Ballardini, M Maestri, V Balzani. J Mater Chem 9:2265, 1999.
200. HT Baytekin, EU Akkaya. Org Lett 2:2000, 1725.
201. LM Adleman. Science 266:1021, 1994.
202. LM Adleman. Sci Am 279:54, 1998.
203. S Iwata, K Tanaka. J Chem Soc Chem Commun 1491, 1995.
204. D Parker, JAG Williams. Chem Commun 245, 1998.
205. A Credi. Molecular-Level-Machines and Logic Gates. PhD thesis, Universita di Bologna, Italy, 1998.
206. ND McClenaghan. Molecular Logic Systems. PhD thesis, Queen's University of Belfast, Northern Ireland, 1999.
207. T Gunnlaugsson, M MacDonail, D Parker. Chem Commun 93, 2000.
208. AP de Silva, ND McClenaghan. J Am Chem Soc 122:3965, 2000.
209. PR Ashton, R Ballardini, V Balzani, M Gomez-Lopez, SE Lawrence, MV Martinez-Diaz, M Montalti, A Piersanti, L Prodi, JF Stoddart, DJ Williams. J Am Chem Soc 119:2679, 1997.

3

Sensors Based on Electrogenerated Chemiluminescence

**Ann-Margret Andersson and
Russell H. Schmehl**
Tulane University, New Orleans, Louisiana

I. INTRODUCTION

Electrogenerated chemiluminescence, also referred to as electrochemilumines-
cence or ECL, is gaining increasing recognition as an analytical tool as reflected
in the growing number of publications presenting new applications of ECL for
selective and sensitive detection of numerous substrates, including several recent review articles [1–5]. The technique involves electrochemical generation of
strongly oxidizing and reducing reactants that, upon annihilation, exhibit luminescence from an energetically accessible excited state of one of the reactant
molecules. There are many reasons for the burgeoning interest in ECL sensor technology. Detection of analytes can be performed using relatively simple
instrumentation and inexpensive materials. In addition, the observation of luminescence, particularly in the absence of an excitation light source, affords
high sensitivity and a broad dynamic range. Since annihilation reactions, characteristic of ECL, often occur at near diffusion limited rates, signal onset and
equilibration are generally limited by factors other than reaction dynamics in
ECL systems. As the variety of reactions leading to ECL continues to grow,
the range of possibilities for development of ECL-based sensors is increasing

to envelop a wide variety of potential applications. The most fruitful employment of ECL in sensor development involves use of ruthenium diimine complex chromophores in the detection of biological molecules. Such complexes have been used in several commercially available sensors and examples of these ECL systems are presented in this chapter. The chapter begins with a discussion of fundamental aspects of electrogenerated chemiluminescence and is followed by a review of particular systems that have been used in sensor development.

II. AN OVERVIEW OF ELECTROGENERATED CHEMILUMINESCENCE

A. Chemical Processes Leading to ECL

The observation of molecular luminescence at electrode solution interfaces results from high-energy annihilation reactions between electrochemically generated radical ions that result in the formation of an electronically excited species [6–16]. The radical ions can be generated at two separate electrodes in close proximity to one another or at the same electrode by alternating between reductive and oxidative potentials. This is particularly useful when the radical ions are unstable since they can be produced *in situ* immediately prior to, or during, the reaction. The general mechanism of an ECL reaction is as follows.

$$A + e^- \rightarrow A^{-\cdot} \qquad \text{(electrochemical reduction)} \qquad (1a)$$

$$B - e^- \rightarrow B^{+\cdot} \qquad \text{(electrochemical oxidation)} \qquad (1b)$$

$$A^{-\cdot} + B^{+\cdot} \rightarrow {}^1A^* + B \qquad \text{(annihilation to yield singlet)} \qquad (2)$$

$${}^1A^* \rightarrow A + h\nu \qquad \text{(fluorescence)} \qquad (3)$$

or:

$$A^{-\cdot} + B^{+\cdot} \rightarrow {}^3A^* + B \qquad \text{(annihilation to yield triplet)} \qquad (4a)$$

$${}^3A^* + {}^3A^* \rightarrow {}^1A^* + A \qquad \text{(triplet–triplet annihilation)} \qquad (4b)$$

$${}^3A^* \rightarrow A + h\nu \qquad \text{(phosphorescence)} \qquad (4c)$$

$${}^1A^* \rightarrow A + h\nu \qquad \text{(fluorescence).} \qquad (3)$$

Following electrochemical reduction and oxidation of the reactants, the reaction can proceed in one of two ways. In *energy sufficient* systems, the free energy of the electron transfer reaction between the radical ions is high enough to populate the excited singlet state of one of the reactants [Eq. (2)]; the excited singlet then exhibits fluorescence [Eq. (3)]. This mechanism is also called the S-route. In an *energy deficient* system, or T-route, the electron transfer reaction between the two radical ions is only of sufficient free energy to generate the lower lying excited triplet state [Eq. (4a)] of one of the species. The fluorescent singlet state can then be populated via triplet–triplet annihilation [Eq. (4b)], effectively pooling the energy from two electron transfer reactions to create an excited singlet state. Alternatively, the triplet formed may relax to the ground state via phosphorescent emission [Eq. (4c)]. In both the energy sufficient and energy deficient systems, an important aspect of ECL reactions of this type is that they are potentially nondestructive and, ideally, both reagents are fully regenerated.

An alternate approach for ECL generation involves formation of the two radical ions through a chemical process following application of either an oxidizing or reducing potential. This has the practical advantage that both reagents necessary for the annihilation reaction are formed at the same electrode with the electrode potential fixed at a single potential rather than oscillating between oxidizing and reducing potentials. Both radical ions are formed at a single fixed potential and one of the species formed decomposes to make an intermediate that serves in the annihilation reaction. An example is given below for oxidation of a species that deprotonates following oxidation to yield a strongly reducing radical.

$$A - e^- \rightarrow A^{+\cdot} \qquad \text{(electrochemical oxidation)} \qquad (5a)$$

$$BH - e^- \rightarrow BH^{+\cdot} \qquad \text{(coreactant electrochemical oxidation)} \quad (5b)$$

$$A^{+\cdot} + BH \rightarrow A + BH^{+\cdot} \qquad \text{(coreactant chemical oxidation)} \qquad (6)$$

$$BH^{+\cdot} \rightarrow B^{\cdot} + H^+ \qquad \text{(coreactant deprotonation)} \qquad (7)$$

$$A^{+\cdot} + B^{\cdot} \rightarrow A^* + B^+ \qquad \text{(annihilation)} \qquad (8)$$

$$A^* \rightarrow A + h\nu \qquad \text{(emission).} \qquad (9)$$

In the process the species BH is oxidized by two electrons and the ultimate product B^+ may or may not be chemically stable. ECL reactions involving a coreactant can occur oxidatively as above, or in a reductive manner, in which case the first reaction step is an electrochemical reduction and the coreactant is concomitantly reduced either at the electrode or by the radical anion of the ECL chromophore. Common coreactants for oxidation processes are tertiary amines and oxalate. The oxalate ion loses CO_2 upon oxidation, producing $CO_2^{-\cdot}$, a powerful reductant (*vide infra*). Tertiary amines deprotonate to yield strongly reducing radical species. Reductive variations of the above reaction sequence also exist with coreactants such as peroxydisulfate, which, upon reduction, forms $SO_4^{-\cdot}$, a strong oxidant. It is important to note that only one of the reactants is regenerated when ECL occurs with a coreactant. The majority of ECL sensor technology is based on variations of the mechanism outlined in Eqs. (5) through (9).

B. Systems Having Organic Chromophores

In organic ECL reactions, the luminescent species are generally derivatives of polyaromatic hydrocarbons where A and B in Eqs. (1) through (4) can be either the same species (leading to self-annihilation) or two different PAHs with either being the analyte (mixed system). Some examples of both self-annihilation and mixed system ECL reactions of organic molecules are listed in Tables 1 and 2. One well-studied example is the self-annihilation reaction between the anion and cation radicals of 9,10-diphenylanthracene (DPA) via an S-route in acetonitrile resulting in blue fluorescence characteristic of DPA [17]:

$$DPA + e^- \rightarrow DPA^{-\cdot} \tag{10a}$$

$$DPA - e^- \rightarrow DPA^{+\cdot} \tag{10b}$$

$$DPA^{-\cdot} + DPA^{+\cdot} \rightarrow {}^1DPA^* + DPA \tag{10c}$$

$${}^1DPA^* \rightarrow DPA + h\nu. \tag{10d}$$

In contrast, the ECL reaction between DPA and N,N,N′,N′-tetramethyl-p-phenylene diamine (TMPD), a mixed system reaction, proceeds via the T-route resulting in triplet–triplet annihilation to yield singlet DPA:

$$DPA^{-\cdot} + TMPD^{+\cdot} \rightarrow {}^3DPA^* + TMPD \tag{11a}$$

$${}^3DPA^* + {}^3DPA^* \rightarrow {}^1DPA^* + DPA. \tag{11b}$$

In this system, radical ions are created by oscillation of the potential between positive and negative voltages; the reductive potential applied selectively reduces DPA (and the TMPD cation radical) and the oxidative potential used only results in oxidation of TMPD (and the DPA anion radical).

Another example of an organic self-annihilation ECL reaction is given in Fig. 1. Perylene diimide radical anions and cations are readily generated by pulsing the applied potential between the parent compound's first oxidation and reduction waves [18]. The proposed one-electron oxidation mechanism shows the formation of a radical anion that has the major portion of the electron density residing on the electron-withdrawing carbonyl oxygen, whereas the unpaired electron on the radical cation is believed to be delocalized over the conjugated system. ECL results from annihilation of the two ions to yield the triplets followed by triplet–triplet annihilation (T-route).

C. Systems Having Transition Metal
Complex Chromophores

ECL from inorganic chromophores has been observed from a variety of transition metal complexes of ruthenium, osmium, palladium, platinum, and a few other transition metal chromophores, some of which are listed in Tables 2 and 3. For example, ECL has been observed from tetrakis(pyrophosphito)diplatinate(II),

Table 1 Organic Compounds Exhibiting ECL Via Self-Annihilation

Compound	$E^o(ox)$ V	$E^o(red)$ V	λ^{em}_{max} nm (V)	ϕ_{em}	Ref.	Comment
9,10-Diphenylanthracene	+1.22	−1.92	395 (3.14)	0.95	17	
Luminol	+0.22	—	425 (2.92)	—	69	
Phenanthrene	+1.8	−2.47	495 (2.50)	—	70	Phosphorescence
Perylenetetracarboxylic diimide	+1.62[b]	−0.81[b]	540 (2.30)	—	18	vs. SCE
Rubrene	+0.95	−1.37	540 (2.30)	0.98	71	vs. SCE
1-Methyl-2,5-diphenylindene	+1.33	−2.39[a]	390 (3.18)	0.82	72	

[a]Peak potential for irreversible CV.
[b]Peak potential for reversible CV.

Table 2 Mixed Systems Exhibiting ECL

Compound	E_{pa}, V[a]	E_{pc}, V[a]	λ_{max}^{em}, nm (eV)	ϕ_{em}	Ref.	Comment
$[Ru(bpy)_3]^+ + S_2O_8^{2-}$	+1.30	—	610 (2.03)	0.07	24	vs. Ag/AgCl
$[Ru(bpy)_3]^{3+} + C_2O_4^{2-}$	+1.30	—	610 (2.03)	0.07	25	vs. Ag/AgCl
$(TMPD)^+ + (Rubrene)^{\cdot-}$	+0.24	-1.3	540 (2.30)	0.98	73	vs. SCE; ECL from rubrene singlet
$(10\text{-}MP)^{\cdot+} + (Fluoranthene)^{\cdot-}$	+0.77	-1.70	405 (3.06)	0.35	74	vs. Pt wire QRE; ECL from singlet FA
$(TTF)^{\cdot+} + (Anthracene)^{\cdot-}$	+0.33[b]	-2.09	400 (3.10)	0.30	75	vs. SCE; ECL from singlet AN
$(Thianthrene)^{\cdot+} + (PPD)^{\cdot-}$	+1.25	-2.17	430 (2.88)	0.04	76	vs. SCE; ECL from singlet TH
$[Cr(bpy)_3]^{3+} + Unknown$	—	-0.26	730 (1.70)	0.005	77,78	vs. Ag wire; oxidant not identified

[a] Potentials are reported as peak potentials and are reversible unless noted otherwise.
[b] E^0(ox), not peak potential.

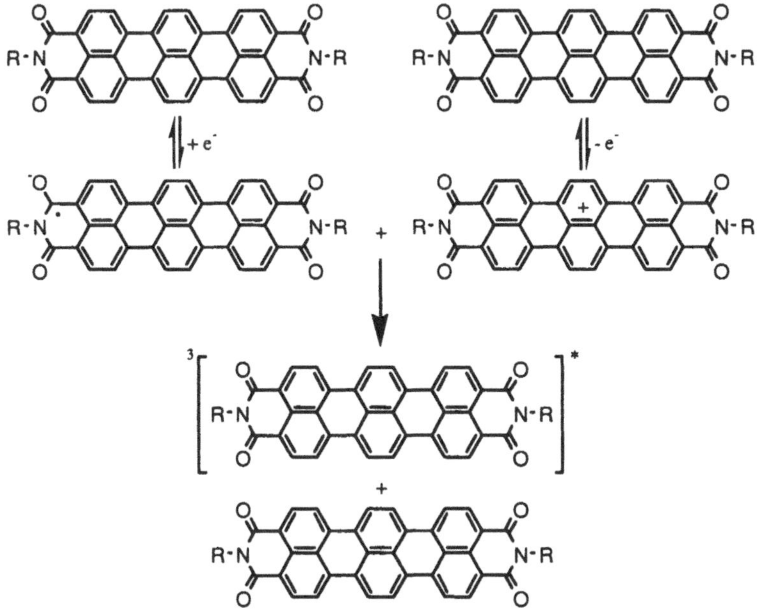

Figure 1 Proposed mechanism for the generation of the radical anion and radical cation of perylene diimide and self-annihilation of the two to yield the triplet excited state.

$Pt_2(P_2O_5H_2)_4^{4-}$, also known as $Pt_2(POP)_4^{4-}$, in the presence of tetrabutylammonium (TBA) salts as electrolyte [19,20]. ECL has also been observed from palladium and platinum $\alpha,\beta,\gamma,\delta$-tetraphenylporphyrin complexes, and also via self-annihilation of the electrogenerated anion and cation radicals [21].

The most widely studied and most exploited of inorganic ECL reactions, however, is that of tris(2,2'-bipyridine) ruthenium(II), $[(bpy)_3Ru]^{2+}$. The emissive excited state is a metal-to-ligand charge transfer state of triplet spin multiplicity; the emission yield is approximately 10% and oxygen does not significantly quench the luminescence [22]. The complex can exhibit ECL following sequential oxidation at the metal center and reduction of the coordinated 2,2'-bipyridine ligand [23]:

$$[(bpy)_3Ru]^{2+} + e^- \rightarrow [(bpy)_2Ru(bpy^{-\cdot})]^{1+} \tag{12a}$$

$$[(bpy)_3Ru]^{2+} - e^- \rightarrow [(bpy)_3Ru(III)]^{3+} \tag{12b}$$

$$[(bpy)_2Ru(bpy^{-\cdot})]^{1+} + [(bpy)_3Ru(III)]^{3+} \rightarrow [(bpy)_3Ru]^{2+*}$$
$$+ [(bpy)_3Ru]^{2+} \tag{12c}$$

$$[(bpy)_3Ru]^{2+*} \rightarrow [(bpy)_3Ru]^{2+} + h\nu. \tag{12d}$$

Table 3 Inorganic Complexes Exhibiting ECL Via Self-Annihilation

Complex	E^o (ox) V	E^o (red) V	λ_{max}^{em} nm (eV)	ϕ_{em}	Ref.	Comment
$[Ru(bpy)_3]^{2+}$	$+1.26^a$	-1.28	610 (2.03)	0.07	23,78	vs. SSCE; $\phi_{ecl} = 0.05$
$[Ru(dpphen)_3]^{2+}$	$+1.09$	-1.47	615 (2.01)	0.37	79,22	vs. NHE; $\phi_{ecl} = 0.24$
$[Os(bpy)_3]^{2+}$	$+0.82$	-1.21	725 (1.71)	0.005	80,81	vs. SSCE
$[Pt_2(P_2H_2O_5)_4]^{4-}$	$+0.45^a$	—	510 (2.43)	0.5	19,20,82	E^o(ox) is peak potential
$[Pd(TPP)]$	$+1.40^a$	-1.45^a	690 (1.80)	—	21	
$[Mo_6Cl_{14}]^{2-}$	$+1.36$	-1.70	765 (1.62)	0.18	83	vs. SCE; Cross rxn. w/$W_6Cl_8Br_6^-$ gives ECL
$[Cu(py)I]_4$	$+0.28$	—	690 (1.80)	—	84,85	vs. ferrocene

^{a}Potential vs. Ag wire quasi reference electrode.

This complex has been widely used in sensing applications since both radical ions of the complex are relatively stable to decomposition reactions. Many systems using this chromophore exist in which ECL is produced at a single electrode via coreactant oxidation or reduction schemes as discussed in the first segment of this section [Eqs. (5) through (9)]. For example, the reduction product of the peroxydisulfate dianion, $S_2O_8^{2-}$, can function as an oxidant in the ECL reaction by annihilation with the electrochemically generated Ru^{1+} to yield the MLCT excited state of the Ru(II) complex by the mechanism [24]:

$$[(bpy)_3Ru]^{2+} + e^- \rightarrow [(bpy)_3Ru]^{1+} \tag{13a}$$

$$S_2O_8^{2-} + e^- \rightarrow SO_4^{2-} + SO_4^{-\cdot} \tag{13b}$$

$$[(bpy)_3Ru]^{1+} + S_2O_8^- \rightarrow [(bpy)_3Ru]^{2+} + SO_4^{2-} + SO_4^{-\cdot} \tag{13c}$$

$$[(bpy)_3Ru]^{1+} + SO_4^{-\cdot} \rightarrow [(bpy)_3Ru]^{2+*} + SO_4^{2-} \tag{13d}$$

$$[(bpy)_3Ru]^{2+*} \rightarrow [(bpy)_3Ru]^{2+} + h\nu \tag{13e}$$

and

$$[(bpy)_3Ru]^{2+} + SO_4^{-\cdot} \rightarrow [(bpy)_3Ru]^{3+} + SO_4^{2-} \tag{13f}$$

$$[(bpy)_3Ru]^{3+} + [(bpy)_3Ru]^+ \rightarrow [(bpy)_3Ru]^{2+*} + [(bpy)_3Ru]^{2+}. \tag{13g}$$

In this system the strongly oxidizing $SO_4^{-\cdot}$ can also oxidize $[(bpy)_3Ru]^{2+}$ and subsequent annihilation of the $[(bpy)_3Ru]^{3+}$ with the electrogenerated $[(bpy)_3Ru]^{1+}$ also leads to excited state formation.

While persulfate is one of a small number of oxidants known to react with Ru^{1+} to form the electronically excited Ru^{2+} species, there are many molecules, such as oxalate, $C_2O_4^{2-}$, that act as reductants following direct electro-oxidation or oxidation by $[(bpy)_3Ru]^{3+}$ and subsequent reactions. Rubinstein and Bard proposed the mechanism for oxalate [25]:

$$[(bpy)_3Ru]^{2+} - e^- \rightarrow [(bpy)_3Ru]^{3+} \tag{14a}$$

$$[(bpy)_3Ru]^{3+} + C_2O_4^{2-} \rightarrow [(bpy)_3Ru]^{2+} + C_2O_4^{-\cdot} \tag{14b}$$

$$C_2O_4^{-\cdot} \rightarrow CO_2^{-\cdot} + CO_2 \tag{14c}$$

$$[(bpy)_3Ru]^{3+} + CO_2^{-\cdot} \rightarrow [(bpy)_3Ru]^{2+*} + CO_2 \tag{14d}$$

$$[(bpy)_3Ru]^{2+*} \rightarrow [(bpy)_3Ru]^{2+} + h\nu \tag{14e}$$

and

$$[(bpy)_3Ru]^{2+} + CO_2^{-\cdot} \rightarrow [(bpy)_3Ru]^{1+} + CO_2 \tag{14f}$$

$$[(bpy)_3Ru]^{1+} + [(bpy)_3Ru]^{3+} \rightarrow [(bpy)_3Ru]^{2+*} + [(bpy)_3Ru]^{2-} \tag{14g}$$

$$[(bpy)_3Ru]^{2+*} \rightarrow [(bpy)_3Ru]^{2+} + h\nu. \tag{14h}$$

The $CO_2^{-\cdot}$ radical anion formed is a very strong reducing agent and the MLCT state can be generated by direct reaction of the $CO_2^{-\cdot}$ with $[(bpy)_3Ru]^{3-}$ or reduction of $[(bpy)_3Ru]^{2+}$ followed by annihilation of the reduced and oxidized forms of the Ru complex.

Other molecules capable of serving as coreactants in this reaction type are aliphatic amines [26], amino acids [27–30], NAD(P)H [31], hydroxide ions [32], hydrazine [32], and $NaBH_4$ [32], although each reaction is different and there are variations in the mechanism of each analyte interacting with $[(bpy)_3Ru]^{2+}$.

D. ECL Systems Based on Conventional CL Reactions

A third class of ECL reactions is that of conventional chemiluminescent, or CL, reactions that are initiated electrochemically. The most widely studied example of these is that of luminol, a very versatile reaction illustrated in Fig. 2, that has been used in a variety of ECL determinations [1,2,33]. However, interest in this type of ECL reaction has diminished somewhat in recent years. This is because the emitting species is the product of an irreversible chemical reaction and this species therefore cannot be regenerated as in the ECL reaction types mentioned above. For example, the energy required for the creation of the emitting electronically excited state is derived from bond cleavage reactions in systems involving CL molecules such as luminol, lucigenin, peroxyoxalates, and dioxetanes. The reactant species also tend to be stable in air and many solvents so that there is no practical advantage to generating them by electrochemical means.

E. Energetics of ECL Reactions

In order to create an excited species via an annihilation reaction, the free energy of the reaction must be great enough to populate either singlet or triplet excited states of one of the reaction products. In the simplest case of self-annihilation between the one electron-oxidized and one-electron reduced forms of a species S, three possible processes are competing. The products of the annihilation are the excited singlet state via the S-route, the excited triplet state via the T-route (both potentially yielding luminescence), or the ground state of the reactant species:

$$S^{+\cdot} + S^{-\cdot} \rightarrow {}^1S^* + S \tag{15a}$$

$$S^{+\cdot} + S^{-\cdot} \rightarrow {}^3S^* + S \tag{15b}$$

$$S^{+\cdot} + S^{-\cdot} \rightarrow S + S. \tag{15c}$$

To determine whether ECL is possible, the free energy of the annihilation reaction can be determined from the one-electron oxidation and reduction potentials of the reactant species and can be expressed as

$$-\Delta G^\circ = E^\circ(S^{+\cdot}/S) - E^\circ(S/S^{-\cdot}). \tag{16}$$

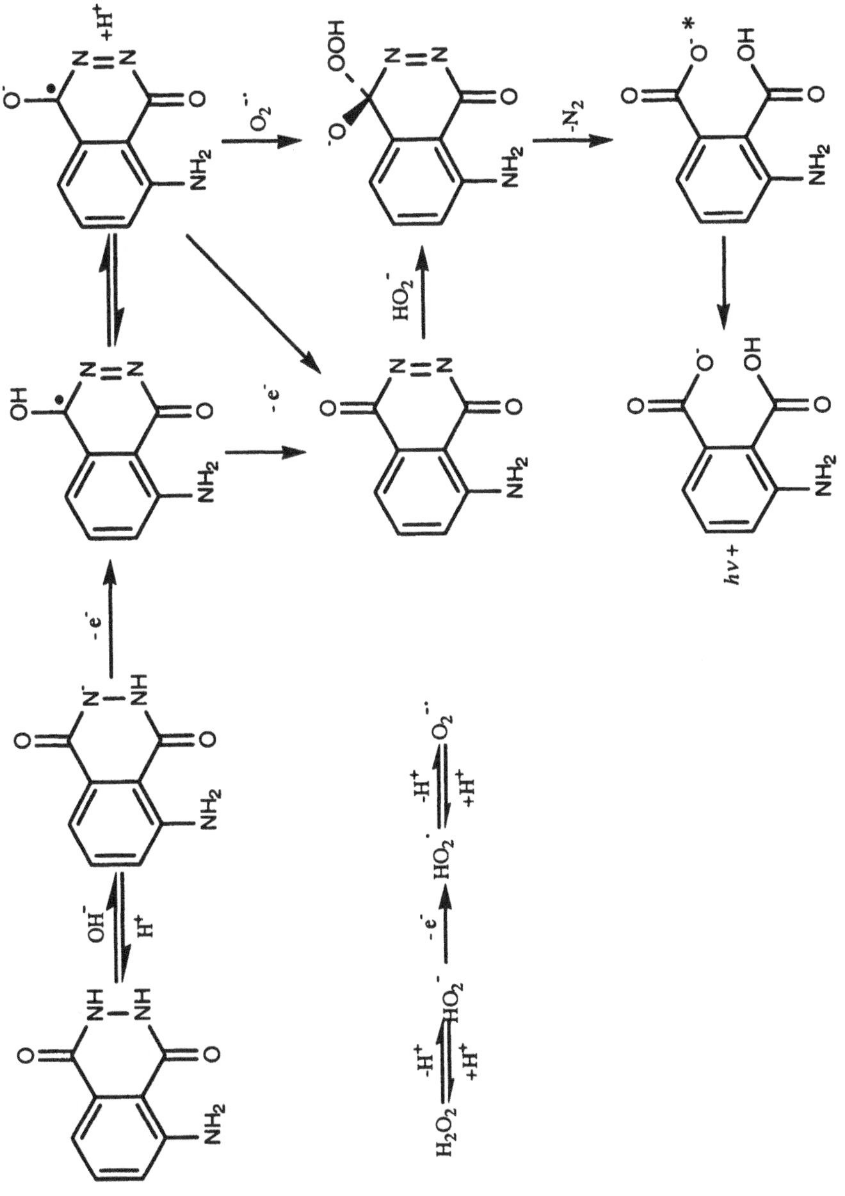

Figure 2 Proposed mechanism for the ECL reaction of luminol.

The values of the potentials can be determined from cyclic voltammetry or other electroanalytical measurements. It should be noted that, in general, systems exhibiting ECL via self-annihilation will exhibit reversible electrochemistry on the time scale of standard cyclic voltammograms (0.2–1 s). In energy sufficient (S-route) systems, the free energy of the annihilation reaction is greater than the energy required to populate the emitting excited singlet state. Alternatively, if the free energy is not sufficient to produce the singlet state directly, but large enough to produce a triplet excited state, luminescence may still be observed following annihilation of two triplets via the T-route discussed above. Figure 3 is an energy-level diagram for the reactions involving the ions of DPA and TMPD that illustrates the three outcomes of Eq. (10) as well as for the cross electron transfer reaction between DPA$^{-\cdot}$ and TMPD$^{+\cdot}$. All the reagents are produced electrochemically and the reactions occur between DPA$^{-\cdot}$ and DPA$^{+\cdot}$ or TMPD$^{+\cdot}$ in an organic solvent. Both processes produce emission from ^{1}DPA* which lies at 3.0 eV. Since the annihilation reaction between the radical anion and radical cation of DPA has a free energy of approximately 3.2 eV, the singlet

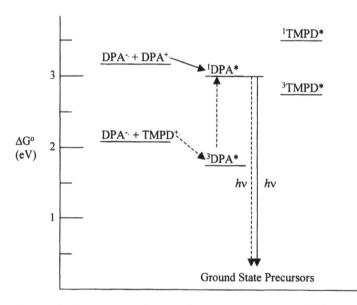

Figure 3 Energy state diagram of the reactions between DPA$^{-\cdot}$ and either DPA$^{+\cdot}$ or TMPD$^{+\cdot}$ yielding emission from ^{1}DPA*. The solid arrow shows the S route for the DPA$^{-\cdot}$/DPA$^{+\cdot}$ process and the dashed arrow shows the T route for the DPA$^{-\cdot}$/TMPD$^{+\cdot}$ process. Note that the T route is also possible for DPA$^{-\cdot}$/DPA$^{+\cdot}$ and that promotion from ^{3}DPA* to ^{1}DPA* requires a second triplet, whereas neither ^{1}TMPD* nor ^{3}TMPD* are accessible by DPA$^{-\cdot}$/TMPD$^{+\cdot}$.

excited state of DPA can be populated directly and ECL emission occurs. However, the annihilation reaction DPA^- and $TMPD^+$ is exergonic by only about 2.1 eV, resulting in the formation of the DPA triplet. ECL occurs indirectly by subsequent triplet–triplet annihilation. In comparison, neither the singlet nor excited triplet states of TMPD can be populated with the free energy available from the $DPA^{-\cdot}/TMPD^{+\cdot}$ annihilation reaction.

A question that arises in consideration of the annihilation pathways is why the reactions between radical ions lead preferentially to the formation of excited state species rather than directly forming products in the ground state. The phenomenon can be explained in the context of electron transfer theory [34-38]. Since electron transfer occurs on the Franck–Condon time scale, the reactants have to achieve a structural configuration that is along the path to product formation. The transition state of the electron transfer corresponds to the area of intersection of the reactant and product potential energy surfaces in a multidimensional configuration space. Electron transfer rates are then proportional to the nuclear frequency and probability that a pair of reactants reaches the energy in which they have a common conformation with the products and electron transfer can occur. The electron transfer rate constant can then be expressed as

$$k_{et} \propto \chi \exp\left(\frac{-\Delta G^*}{kT}\right), \tag{17}$$

where χ is related to the degree of adiabaticity and ΔG^* the activation free energy. Marcus and others developed expressions relating ΔG^* and the free energy of the reaction ΔG^o. An expression for the activation-free energy, derived classically, is given in Eq. (18):

$$\Delta G^* = \left[\left\{\frac{w^r + w^p}{2}\right\} + \left\{\frac{\lambda}{4}\right\} + \left\{\frac{\Delta G^o}{2}\right\} + \left\{\frac{(\Delta G^o + w^p + w^r)^2}{4\lambda}\right\}\right]. \tag{18}$$

In this relationship w^r and $-w^p$ describe the energy necessary for creating the reactant encounter complex and separating the product pair, and λ is the reorganization energy, which in ECL systems are estimated to be associated solely with the solvent reorganization.

A consequence of Eq. (18) is that the maximum electron transfer rate occurs when $\Delta G^o = -\lambda$. A plot of $\ln(k_{et})$ versus ΔG^o is shown in Fig. 4. Electron transfer reactions with ΔG^o more positive than $-\lambda$ define the normal region, where the rate of electron transfer increases with increasing exergonicity, whereas free energies negative relative to $-\lambda$ define the inverted region in which the rate decreases as ΔG^o becomes more negative. For most ECL reactions, the

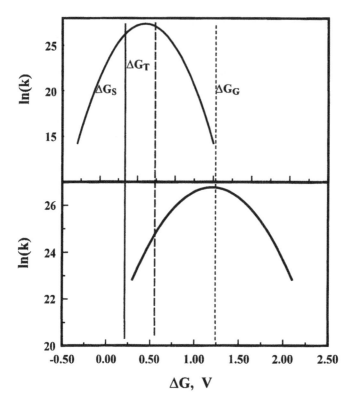

Figure 4 Free energy dependence of electron transfer reaction rate for case A where $-\lambda$ is approximately equal to ΔG_s (or ΔG_T) and case B where $-\lambda$ is closer in energy to ΔG_G.

free energy for generation of products containing an excited state lies in the normal region. Direct formation of ground state products upon annihilation, however, is necessarily much more exergonic than excited state formation and, in many cases, is in the inverted region (Fig. 4a). In this case, direct formation of the excited state can be kinetically favored over recombination to form ground state products. However, in systems having large reorganizational energies, recombination to the ground state will be favored over excited state formation (Fig. 4b). This is clearly illustrated in Fig. 3, where both the DPA$^-$/DPA$^+$ and DPA$^-$/TMPD$^+$ ion pairs are relatively close in energy to the ^{1}DPA and ^{3}DPA excited states, respectively. Since the reorganizational energy for recombination to the ground state is unlikely to be much greater than 1 eV and the process is exergonic by more than 2 eV in both cases, the recombination process likely lies well into the inverted region.

III. EXPERIMENTAL ASPECTS OF ECL DETECTION AND SENSOR DEVELOPMENT

Unlike other optically based techniques for analysis, chemiluminescence and ECL are simpler methods in some respects since they require no light source and associated focusing optics for sample excitation. Provided stray light is excluded from the system, light signals generated in electrolytic processes arise solely from specific reactions capable of generating luminescence. Most often ECL experiments are performed in three electrode electrochemical cells where the potential of the working electrode relative to a reference (such as the saturated calomel electrode) is known and the third electrode (auxiliary electrode) operates at whatever potential is necessary to maintain a current that is equal and opposite to that at the working electrode. With cells of this type there are several principal approaches for producing electrochemically generated luminescence. One involves production of a radical cation at the anode and a radical anion at the cathode followed by annihilation of the two radical ions. This approach is not of importance in sensing applications and requires the working and auxiliary electrodes to be in close proximity (such as in microelectrode interdigitated arrays). A second approach involves production of both a radical cation and radical anion at the working electrode via oscillation of the applied potential. Finally, a single fixed potential can be used to oxidize or reduce the ECL active chromophore and a coreactant. The oxidized/reduced coreactant in solution will react to produce a highly reactive radical ion that will react as described above (see Sec. II.A).

A key difference in the methods is that the profiles of the concentrations of radical ions from the electrode(s) differ and yield different annihilation regions for generation of luminescence. A qualitative representation of the concentration of species in solution as a function of distance from the electrode is shown in Fig. 5 for the case of the double potential step for sequential generation of the anion and cation radicals ($A^+ + A^-$) at a single electrode. Figure 5a shows the potential of the working electrode as a function of time. Initially, application of a positive potential leads to the formation of A^+ at the electrode surface. The concentration of A^+ decays away to zero at some point in solution and the concentration of A at the electrode surface drops to zero also. When the potential is switched to reduce A to A^-, ECL is observed. Immediately after switching the potential, the concentration of A^+ at the electrode surface drops to zero and A^- begins to accumulate. Figure 5b shows the concentrations of A, A^+, and A^- at a fixed time (t of 5A) following application of the potential to produce A^-. The profile includes consideration of the annihilation reaction between A^+ and A^-; this annihilation region is a plane parallel to the electrode surface as shown by the point in Fig. 5 where the concentrations of A^+ and A^- are zero. A significant ramification of the profile is that, in the vicinity of the

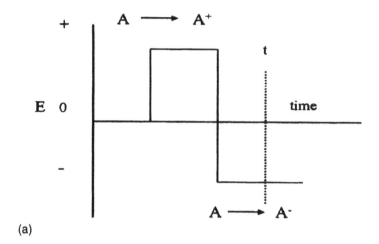

(a)

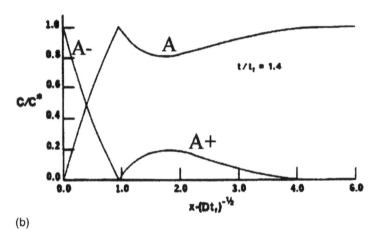

(b)

Figure 5 Concentration-distance profiles for reactants generated in double potential step experiments involving initial oxidation of reagent A followed by reduction of A. (From Ref. 36.)

plane where annihilation of radical ions occurs, the steady-state concentration of radical ion species is low and quenching of the luminescence by these reactive species generally does not occur to any significant extent. Since this approach to luminescence analysis relies on the formation of highly reactive radical ions, any analytical application of ECL must take into account excited state quenching possibilities.

In systems where ECL arises upon application of a single potential and reaction of a coreactant, as outlined in Eqs. (5) through (9), concentration distance profiles differ and depend on many more factors. In cases where a positive potential is applied and both the ECL chromophore and coreactant are oxidized at the electrode surface, concentrations of the two initial radical ion species will decay with increasing distance from the electrode, as will the concentration of the strong reducing agent formed upon decomposition of the coreactant. The zone where luminescence arises depends on relative rate constants for

1. reaction of the reducing radical with the oxidized chromophore,
2. reaction of the reducing radical with the chromophore (to generate reduced chromophore), and
3. the annihilation of the oxidized and reduced chromophores.

In general rate constants for reactions (1) and (3) will be diffusion limited since they are likely to be highly exergonic.

Bard and Faulkner discussed practical aspects of cell design for observation of ECL in an early review [12]. The article includes a description of concentration-distance profiles and the different approaches for generating and observing ECL used in the late 1970s. While significant changes have occurred in both optical and electrochemical instrumentation since the time of the review, factors such as solvent and electrolyte selection and purification, electrode materials, and electrode configuration are thoroughly discussed in the article and provide an excellent overview of practical considerations and limitations of the technique. In general, the electrochemical cell must contain working, reference and auxiliary electrodes as mentioned above. Assuming luminescence is generated from annihilation reactions at the working electrode, the optical system configuration will depend on the geometry of the working electrode and its relation to the detection optics. It is simple to use planar electrodes made from reflective materials and oriented parallel to an optically flat window, allowing collection of emitted light with a simple lens assembly. If the ECL intensity from a single emitting species is to be measured, optimum sensitivity can be obtained using an appropriately situated photomultiplier (PMT). If emission spectra are to be obtained, the emitted light can be focused onto the entrance slit of a monochromator/PMT or spectrograph/diode array (or CCD).

Since ECL is generated upon applying a potential (or oscillating potential), the intensity of emitted light depends on a variety of factors including potential sequencing and timing, solution convection, and the concentrations of the ECL chromophore (and coreactant) and electrolyte. A simple system for observation of ECL can be made by placing a spectrophotometric cuvette in the sample compartment of a fluorimeter and equipping the cuvette with a "flag" working electrode and appropriately placed reference and counter electrodes. However, such simple systems are far from optimal. Recently Preston and Nieman developed an immersible cell for measurement of ECL in solutions containing appropriate chromophores [39]. The probe, shown in Fig. 6, has a working electrode embedded in the center of an enclosed cylindrical compartment and a fiber optic bundle mounted directly opposite the working electrode. Channels in the compartment allow analyte to enter the probe and the cylinder is machined to accommodate counter and reference electrodes and a stir rod. Figure 7 shows the luminescence intensity measured for the probe immersed in a solution containing luminol and H_2O_2 as a function of time following a positive DC potential step in both stirred and unstirred solutions. In the absence of stirring, the ECL intensity decays rapidly due to depletion of reactants in the vicinity of the electrode. Earlier work by Nieman focused on development of cells for ECL analysis by flow injection methods [40]. One such cell, shown in Fig. 8a, has two working electrodes and a counter electrode at the outlet of the flow cell. The response of the cell as a function of coreactant concentration in solutions containing a fixed concentration of the injected ECL reagent ($[Ru(bpy)_3]^{2+}$) is shown in Fig. 8b. A recent modification of this cell eliminates the need for injection of the chemilu-

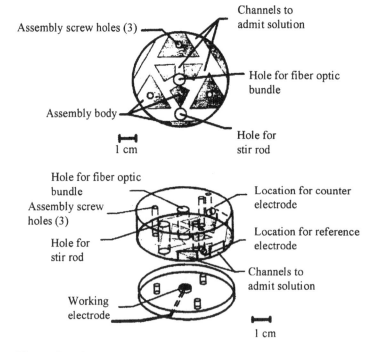

Figure 6 ECL probe devised by Nieman and coworkers. (From Ref. 39. With permission from the American Chemical Society.)

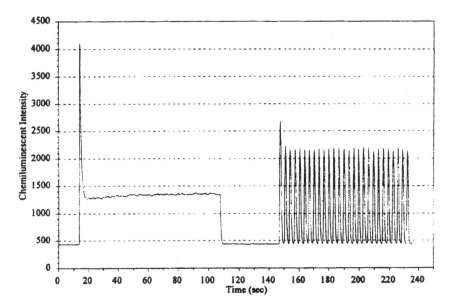

Figure 7 Intensity of measured ECL for the luminol/H_2O_2 system as a function of time using the Nieman probe in stirred and unstirred solutions. (From Ref. 39. With permission from the American Chemical Society.)

minescent reagent by incorporating the reagent in a nafion film adsorbed to the working electrode surfaces. Luminescence is only observed when the flow contains an appropriate coreactant that penetrates the film and reacts either with the radical ion of the chemiluminescent reagent or directly with the electrode. Use of physisorbed or chemisorbed chemiluminescent reagents in sol-gels and other supports is an active area of research related to the development of ECL-based sensors [41–43].

IV. ADVANTAGES AND LIMITATIONS OF SENSORS BASED ON ECL

Chemiluminescence has been recognized as a valuable tool for the analysis of a variety of molecules for some time and is firmly established as an analytical technique. ECL methods not only retain the benefits of CL analysis, but also have several additional advantages. In general, CL reactions are very specific, frequently have very low limits of detection, and offer a wide dynamic working range (several orders of magnitude). In addition, CL methods require simple and inexpensive reagents and instrumentation. Analysis using ECL has

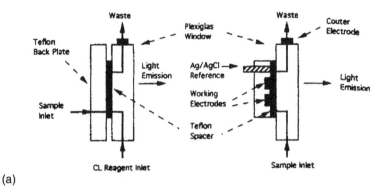

(a)

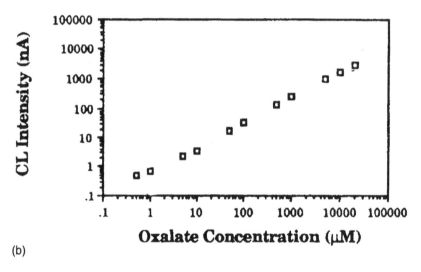

(b)

Figure 8 (a) A typical flow cell for flow injection ECL analysis with *in situ* generation of radical ion species. (b) Response of the flow cell for injection of varying concentrations of sodium oxalate into solutions containing $[Ru(bpy)_3]^{2+}$. (From Ref. 39. With permission from the American Chemical Society.)

the same advantages but since the reactants can be generated at an electrode as needed, the production of highly reactive species can be regulated through manipulation of the applied potential. Also, the emission occurs near the electrode surface, further increasing the sensitivity of ECL measurements. ECL can easily be turned on and off, facilitating in the screening of background light, which is generally negligible since there is no excitation source. In addition, the ECL chromophore is usually regenerated, thus smaller concentrations are required. If the luminescent reagent is immobilized on the electrode surface, the chemiluminescent species does not need to be added during analysis. In fact, this may lead to sensors that can be employed indefinitely in detecting analytes that serve as coreactants in flow systems. A few applications of immobilized ECL reagents used in immunoassays are discussed below.

Despite the many benefits of ECL as an analytical tool, there are still a number of limitations of ECL analysis often due to the complexity of a system that combines electrochemical and optical techniques. As previously mentioned, in cases that employ conventional CL reactions, the advantages of ECL methodology can be outweighed by the complication and cost of added electrochemical equipment, consisting minimally of an adapted electrochemical cell and a potentiostat that can be incorporated into the existing system. There is no appreciable advantage to the electrochemical generation of the reactant species immediately prior to the reaction since, in conventional chemiluminescent reactions, reactants are relatively stable and do not need to be created in a restricted volume. In some systems the emitting species is a chemically altered product of the CL reaction and cannot be regenerated at the electrode. As a result the ECL active chromophore has to be supplied continually.

In the case of organic energy deficient ECL processes, luminescence is observed only in organic solvents from which oxygen has been meticulously excluded to prevent quenching of the emitting excited state. Also, most organic ECL systems function only in nonaqueous solvents, although there have been some attempts to circumvent this limitation by using water soluble PAHs. Richards and Bard observed ECL from 9,10-diphenylanthracene-2-sulfonate by oxidation in the presence of tripropylamine and reduction in the presence of persulfate in 1:1 acetonitrile/water mixtures [44]. It has been suggested that these reactions could be helpful in the design of labels for ECL immunoassays. Nonetheless, organic ECL reactions have not yet found any significant applications in analytical systems.

It is also possible for species that are created in ECL reactions to interact with each other in ways that interfere with the generation of ECL and partially, if not completely, quench the emission. For example, one difficulty in direct sensing of coreactants is that the coreactant may also quench the luminescence of the excited state generated in the annihilation process. This difficulty was recognized several years ago by Bard and coworkers in the examination of the $[(bpy)_3Ru]^{2+}/S_2O_8^{2-}$ system [24]. Luminescence arises upon reduction of the Ru(II) complex and reduction of $S_2O_8^{2-}$ mediated by the Ru(I) complex formed. The intermediate $SO_4^{-\cdot}$ ion formed is a powerful oxidant and annihilation with the Ru(I) complex will yield the excited state of Ru (II) complex [Eq. (13d)]. However, the persulfate ion is an effective quencher of the MLCT excited state of the Ru(II) complex. Figure 9 shows the observed ECL intensity for this system

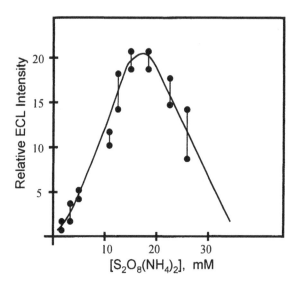

Figure 9   Quenching of $[Ru(bpy)_3]^{2+}$ by the added coreactant $S_2O_8^{2-}$. (From Ref. 24. With permission from the American Chemical Society.)

as a function of $S_2O_8^{2-}$ concentration, clearly illustrating both the increase in intensity as more excited states are formed and the quenching of the excited states by the persulfate coreactant.

V.   EXAMPLES OF SENSORS AND DEVICES EMPLOYING ECL

While there is a wide variety of molecules that exhibit ECL, the overwhelming majority of publications concerned with analytical applications of ECL are based on chemistry involving $[(bpy)_3Ru]^{2+}$ or closely related analogues as the emitting species. There are several reasons for this. Both the electrochemistry and photochemistry of this chromophore and numerous derivatives have been extensively investigated [45]. Luminescence from $[(bpy)_3Ru]^{2+}$ is from a metal-to-ligand charge transfer state of triplet spin multiplicity (^{3}MLCT). The emission has a maximum at approximately 620 nm ($\Phi_{em} \sim 0.08$) and a lifetime in the 0.5 to 1 microsecond range. The quantum yield of ECL, Φ_{ECL}, defined as the quantitative probability of emission per redox event, for the reaction between $[(bpy)_3Ru]^{2+}$ and oxalate in aqueous buffer is around 2%, which is relatively high in comparison to other systems that exhibit ECL [46]. Unlike many other ECL active compounds, ECL reactions involving $[(bpy)_3Ru]^{2+}$ as the emitting species can generally be performed in aqueous buffers *in the presence of dissolved oxygen* and at room temperature since the chromophore is quite stable under these conditions and the excited state is only partially quenched by dissolved oxygen. The reduced and oxidized species, $[(bpy)_3Ru]^+$ and $[(bpy)_3Ru]^{3+}$, can be rapidly generated at electrode surfaces and potentials for both oxidation and reduction are within the useful window of several solvents ($+1.3$ V and -1.3 V vs. Ag/AgCl). Finally, both radical ions are stable on the second to minute time scale in aqueous solutions lacking a coreactant.

A. ECL Sensors in Which the Coreactant is the Analyte

In 1987, Noffsinger and Danielson reported the observation of ECL from the reaction of $[(bpy)_3Ru]^{3+}$ with aliphatic amines, proposing the mechanism below, in which the alkyl amine is oxidized to form an intermediate radical capable of reducing Ru^{2+} to Ru^{1+} [26]. The two Ru containing radical ions then react to yield an excited state.

$$[(bpy)_3Ru]^{2+} \rightarrow [(bpy)_3Ru]^{3+} + e^- \tag{19a}$$

$$R_2NCH_2R' + [(bpy)_3Ru]^{3+} \rightarrow R_2N^{+\cdot}CH_2R' + [(bpy)_3Ru]^{2+} \tag{19b}$$

$$R_2N^{+\cdot}CH_2R' + H_2O + [(bpy)_3Ru]^{2+} \rightarrow 2H^+ + R_2NH + COHR'$$
$$+ [(bpy)_3Ru]^{1+} \tag{19c}$$

$$[(bpy)_3Ru]^{1+} + [(bpy)_3Ru]^{3+} \rightarrow [(bpy)_3Ru]^{2+*} + [(bpy)_3Ru]^{2+} \tag{19d}$$

$$[(bpy)_3Ru]^{2+*} \rightarrow [(bpy)_3Ru]^{2+} + h\nu. \tag{19e}$$

Leland and Powell also studied ECL obtained from reaction of $[(bpy)_3Ru]^{3+}$ with trialkylamines [47]. Since the mechanism involves an electron transfer from the amine to Ru^{3+}, there exists an inverse relationship between the first ionization potential of the amine and ECL intensity. The relative intensity of $[(bpy)_3Ru]^{2+*}$ ECL was found to be ordered tertiary $>$ secondary $>$ primary. Quaternary ammonium ions and aromatic amines do not produce ECL with Ru(II) diimine complexes. Brune and Bobbitt subsequently reported the detection of amino acids by $[(bpy)_3Ru]^{2+*}$ ECL [28,29]. Employing capillary electrophoresis for separation, the presence of various amino acids can be detected directly by reaction with $[(bpy)_3Ru]^{3+}$ generated *in situ* with up to femtomolar sensitivity and with a selectivity for proline and leucine over other amino acids. The formation of an amine radical cation intermediate is characteristic of proposed mechanisms of both aliphatic amines and amino acids.

Jameison et al. [48] demonstrated the quantitative ECL detection of amine-containing β-nicotinamide adenine dinucleotide (NADH) and the phosphate

NADPH. They proposed the following mechanism involving direct oxidation of the NAD(P)H and the Ru complex at the electrode.

$$[(bpy)_3Ru]^{2+} - e^- \rightarrow [(bpy)_3Ru]^{3+} \tag{20a}$$

$$NAD(P)H \rightarrow NAD(P)H^{+\cdot} + e^- \tag{20b}$$

$$NAD(P)H^{+\cdot} \rightarrow NAD(P)^{\cdot} + H^+ \tag{20c}$$

$$[(bpy)_3Ru]^{3+} + NAD(P)^{\cdot} \rightarrow [(bpy)_3Ru]^{2+*} + NAD(P)^+ \tag{20d}$$

$$[(bpy)_3Ru]^{2+*} \rightarrow [(bpy)_3Ru]^{2+} + h\nu. \tag{20e}$$

The intermediate, NAD· or NADP·, is a radical on the nicotinamide that can react with $[(bpy)_3Ru]^{3+}$. Any enzyme that produces or consumes either NADH or NADPH can be directly monitored by ECL since only the reduced forms NAD(P)H but not the oxidized forms NAD(P)$^+$ can function as a coreactant [31,49]. This difference has been exploited in the clinical chemistry assays of ethanol, glucose, bicarbonate, cholesterol, and glucose-6-phosphate dehydrogenase.

Another interesting application involves use of ECL in the detection of pathogenic bacteria that produce β-lactamases, enzymes produced by the bacteria in order to resist the toxic effects of clinically administered β-lactam antibiotics such as penicillins. These enzymes function by hydrolytically opening the cyclic amide of the antibiotic, thereby rendering it harmless to the bacteria (Fig. 10). Liang et al. have developed a method for indirectly screening bacteria for the presence of β-lactamases by utilizing the ability of β-lactam antibiotics to act as the amine-containing reductant in the $[(bpy)_3Ru]^{2+}$ ECL reaction [50,51]. As described earlier in Eqs. (20a) to (20e), amines, particularly tertiary amines, can promote $[(bpy)_3Ru]^{2+}$ ECL by acting as a coreactant and β-lactam antibiotics, such as benzylpenicillin in Fig. 10, can also function this way. Hydrolysis of the β-lactam rings by the bacterial enzyme causes dramatic changes in the ability of the antibiotic to promote $[(bpy)_3Ru]^{2+}$ ECL emission, which affords sensitive and quantitative detection of β-lactams and β-lactamases. Increases in ECL signal were observed with three antibiotics when they were hydrolyzed by enzyme; systems were optimized to have sub ppm detection limits. ECL-based antibiotic detection was also accomplished in *untreated* whole milk, and β-lactamases were

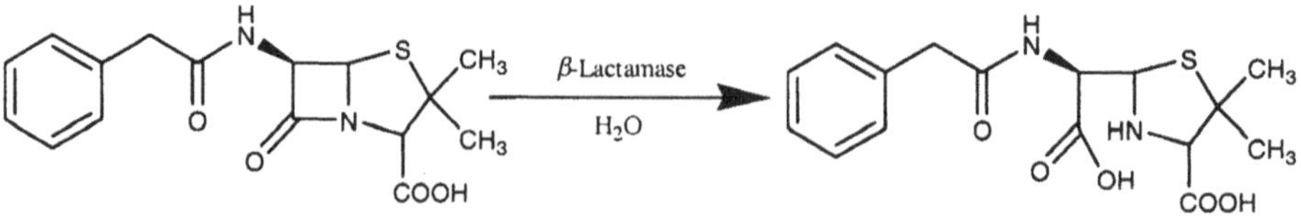

Figure 10   The hydrolysis of penicillin by β-lactamase.

detected in crude bacterial broth cultures. Light emission was further increased by covalently linking β-lactam antibiotics to $[(bpy)_3Ru]^{2+}$ by a flexible alkyl chain in a separate experiment. This technique could potentially be valuable in a variety of practical applications, such as screening foods (dairy and meat, e.g.) for unacceptable levels of residual antibiotics.

Dong and Martin developed a sensitive assay that detects the catalytic activity of the enzymes pig liver esterase (PLE) and porcine kidney leucine aminopeptidase (LAP) by using substrates modified with metal-binding ligands [52]. The enzymes catalyze changes in the substrates that affect their ability to bind to nonluminescent $[(bpy)_2Ru(L)_2]^{2+}$ complexes to form mixed-ligand complexes capable of ECL. For example, in the presence of PLE the substrate picolinic acid ethyl ester, which will not bind to the metal complex, is hydrolyzed to picolinic acid, which forms an ECL active complex following reaction with $[(bpy)_2Ru(L)_2]^{2+}$. It is possible to detect PLE in the low picomolar range and the hydrolysis of micromolar quantities of the substrate. In a second assay, LAP hydrolyzed 8-(L)-leucylamino-quinoline to leucine and aminoquinoline. In this case the Ru complex precursor forms an ECL active mixed ligand compound with the unhydrolyzed substrate but not with the hydrolyzed products, also allowing for detection of the picomolar concentrations of enzyme and micromolar concentrations of substrate.

Pyruvate analysis can also be accomplished by ECL using a Ru(II) complex. Since $[(bpy)_3Ru]^{2+}$ is not a strong enough oxidant to directly oxidize pyruvate, cerium(III) nitrate is added. By applying a higher electrode potential (+1.55 V vs. Ag wire), Ce^{4-} is generated, which can mediate the oxidation of pyruvate which reacts to form CO_2 and the strongly reducing intermediate $CH_3CO\cdot$. This radical species reacts with $[(bpy)_3Ru]^{3+}$ in a manner analogous to $CO_2^{-}\cdot$, resulting in luminescence from the Ru(II) complex [53].

B. ECL Sensors in Which the Coreactant is Not the Analyte

The use of ECL processes for the detection of biological compounds is a rapidly growing area of interest, both for quantitation of analytes and to measure biomolecular interactions. By using ECL active chromophores as labels for biological compounds, a variety of applications is possible, including assays for enzymatic activity, binding assays, immunoassays, and nucleic acid probe assays.

Since Blackburn et al. reported the first use of ECL detection for the development of immunoassays and DNA probe assays [54], several automated systems have become commercially available. An activated species that readily labels proteins and nucleic acids is obtained by chemically modifying one of the bipyridyl ligands of $[(bpy)_3Ru]^{2+}$ with N-hydroxysuccinimide (Fig. 11a). The

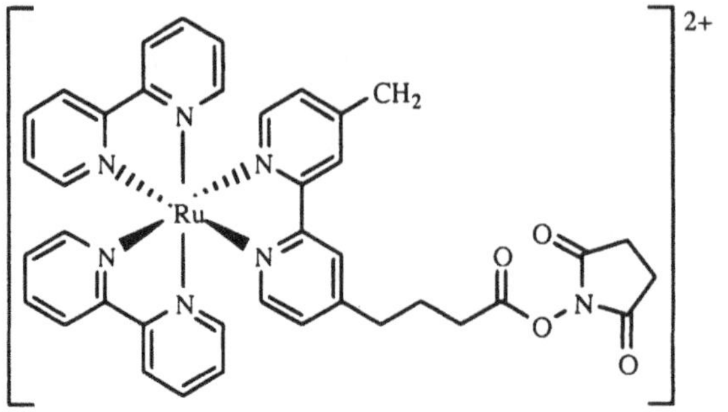

(a)

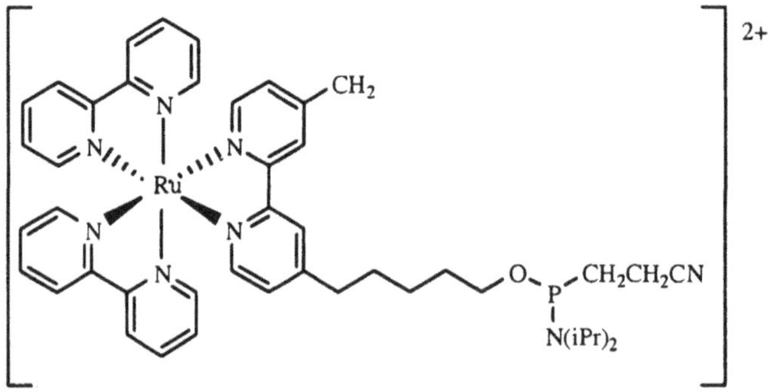

(b)

Figure 11   Structures of $[(bpy)_3Ru]^{2+}$ labels. (a) Origen label. (b) Perkin-Elmer label.

Origen system was subsequently developed by IGEN, Inc.* The instrumental principle of operation is based on the employment of magnetic beads as a support phase. A selected antibody or analyte-specific antigen will bind to beads having a streptavidin-biotin coating. The beads are then exposed to the analyte, which will in turn bind to the immobilized group on the bead. Antibody labeled with the ECL active Ru(II) complex binds to the analyte and the beads are then captured on the working electrode in a flow system by means of an applied magnetic field while residual labels are swept away. ECL is produced by the application of an oxidizing potential in the presence of tripropylamine. Various immunoassays are possible with this technology as outlined in Fig. 12. ECL methods exhibit promise for widespread use in immunoassays since the

---

*See the home page of IGEN for more information: http://www.igen.com.

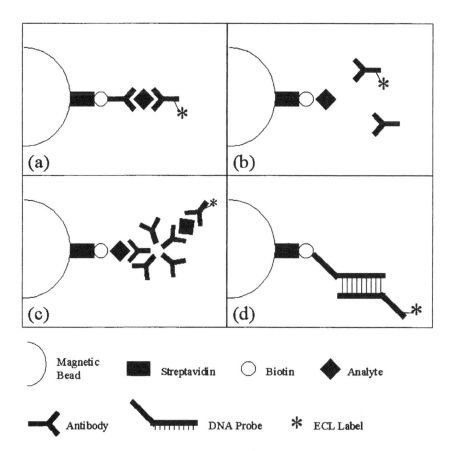

Figure 12 Examples of immunoassay systems and DNA assays performed with magnetic bead technology: (a) sandwich assay; (b) competitive assay; (c) bridge assay to determine proteins; (d) DNA and RNA assays.

technique offers very low detection limits (subpicomolar concentrations) and eliminates the use of radioactive isotopes employed in many current methods. The Ru complex labels are versatile and long-lived, and even multiple labels do not interfere with antibody immunoreactivity.

ECL has also been applied to the analysis of polymerase chain reaction (PCR) products [53,55]. PCR is first used to amplify the specific genes by use of two primers, one of which is biotinylated. The double-stranded DNA is then captured on streptavidin-coated magnetic beads and washed with an alkaline solution to denature and separate the strands. The particle-bound, single-strand DNA is used to capture products hybridized with Ru-labeled complementary strands. The beads are then captured at an electrode and ECL analysis is performed in the presence of TPA. DiCesare et al. developed a Ru(II) complex label (Fig. 11b) similar to that of Blackburn et al. [56] which was used by Perkin-Elmer in the production of an automated system (QPCR 5000 system) for post-PCR determinations; the device, however, is no longer commercially available. Boehringer Mannheim also markets automated instruments for ECL immunoassays and DNA probe studies: the ELECSYS 1010 and ELECSYS 2010.

Another novel approach to identification of particular DNA fragments using ECL was presented recently by Xu and Bard [57]. The approach was to modify an Au electrode with difunctional hydrocarbon having a thiol to bind to the gold surface and a phosphonic acid moiety to ion pair with Al^{3+} in solution. The resulting modified electrode efficiently binds the phosphate group of DNA single strands (i.e., poly-dC). Exposing this electrode to $[Ru(bpy)_3]^{2+}$ labeled single strands having the complementary sequence (i.e., poly-dG) results in ECL from the surface bound Ru(II) complex in the presence of tripropylamine [Eq. (11)].

There has been significant effort to develop analytical systems involving ECL active Ru(II) diimine complexes immobilized on electrode surfaces [58,59]. Electrode bound Ru(II) complexes immobilized in nafion films by electrostatic binding have been used in flow injection analysis of oxalate, alkyl amines, and NADH [31]. Martin and Nieman have also employed electrode bound Ru(II) bipyridyl complexes in the analysis of glucose by detecting NADH produced in oxidation of glucose by glucose dehydrogenase [49].

It is also possible to exploit quenching of ECL in the detection of various substances. Recently Richter and coworkers have shown that ECL from $[(bpy)_3Ru]^{2+}$, generated following oxidation in the presence of trialkylamines, is quenched by quinones and other aromatic hydrocarbons in nonaqueous solvents [60].

Several other chromophores have been used in the development of sensors based upon ECL. For example, the luminol reaction is a conventional chemiluminence reaction that has been studied in detail and it is believed that the mechanism of the ECL reaction is similar, if not identical, to that of the chemiluminescence. As shown in Fig. 2, the luminol ion undergoes a one-electron oxidation to yield a diazaquinone, which then reacts with peroxide or superoxide (^-OOH) to give the excited 3-aminophthalate which has an emission maximum of 425 nm. This reaction is particularly versatile and has been utilized in a variety of ECL assays, many of which have been previously summarized by Knight [1]. The luminol ECL reaction can be used for the determination of any species labeled with luminol derivatives, hydrogen peroxide, and other peroxides or enzymatic reactions that produce peroxides. A couple of examples are described later.

Sekura and Terao reported a luminol-based ECL sensor for lipid hydroper-oxides, which are of great importance since they are involved in the regulation of prostaglandin biosynthesis, which affects cancer development, aging, and other pathological conditions [61]. In this method, there are two approaches for generation of ECL. By applying potentials between 0.5 and 1.0 V, luminol is oxidized to a diazaquinone, which reacts with free lipid hydroperoxide to yield luminescence as in Fig. 2. When the applied potential is greater than 1.0 V, both luminol and the lipid hydroperoxide are oxidized electrochemically and a more complicated emission occurs by the reaction of the diazaquinone with the oxidized product of lipid hydroperoxide.

Marquette and Blum have reported a luminol-based fiber optic biosensor for glucose and lactate [62]. In this method glucose oxidase or lactate oxidase is immobilized on preactivated membranes. Substrates present in solution, either glucose or lactate, are oxidized by the respective enzyme, producing hydro-gen peroxide. Luminol, which has been previously oxidized at a glassy carbon electrode, reacts with hydrogen peroxide producing an ECL signal.

A variety of analytical applications exists that involve electrochemical generation of luminescence in systems for which the chemistry is not clearly understood. A few recent examples are mentioned here. A novel ECL assay for the determination of 2,4- and 3,4-diaminotoluene (DAT) isomers is based on reaction of these molecules with Au^+ and C^{2+}, respectively, in aqueous solution under oxidizing conditions in a buffer containing tripropylamine [63]. Luminescence is observed upon potential ramping from 0 to $+2.8$ V. The nature of the emitting species was not specified, but could involve a charge transfer excited state of the metal complex with DAT or an oxidized form of DAT. DAT isomers were screened for ECL enhancement against 32 metals; the apparent specificity of Au^+ for 2,4-DAT and Cu^{2+} for 3,4-DAT is believed to be linked to the radii of the metal ions. This ECL approach could lead to applications in the determination of some aminoaromatics from degradation of explosives (e.g., TNT) as well as detection and quantitation of various transition metals in water supplies.

Chen et al. characterized ECL from indole and tryptophan in the presence of hydrogen peroxide under oxidizing conditions [64]. In both cases, there is only tryptophan or indole, electrolyte, and H_2O_2 in the cell. A triangular pulse voltage was applied (0.0–1.0–0.0 V) to oxidize tryptophan or indole. Tryptophan and indole are both very weakly electrochemiluminescent on their own, but the ECL signal is enhanced in the presence of H_2O_2. The process is believed to involve a dioxetane intermediate obtained by a one-electron oxidation of the substrate. The presence of 17 other amino acids (in excess) did not interfere in the determination of tryptophan, making the method remarkably selective.

Egashira et al. have observed ECL from various hydroxyl compounds, which has led to the development of alcohol and sugar sensors [65]. ECL is observed upon voltage cycling basic aqueous solutions of various sugars and simple vicinal alcohols between $+1.4$ and -1.2 V (vs. Ag/AgCl). While the mechanism is not clear, the authors speculate that aldehyde excited states may be formed. No spectral data are provided, as the emission is very weak.

A completely new source of electrochemiluminescent compounds may have been recently discovered as well. Weak luminescence was observed from fluids extracted from tunicates, a type of marine invertebrate, and synthetic tunichromes (the chromophore in the tunicates) upon oxidation [66]. The structure of the synthetic tunicate chromophore is shown in Fig. 13. A 10-fold enhancement of ECL is observed when the synthetic analogues of the tunichromes are complexed to Hg^{2+}. The mechanism of the luminescent enhancement has not yet been elucidated.

C. Device Applications

The observation of luminescence from laser dyes by ECL methods offers the possibility of using this approach to create dye lasers. A laser operating by ECL would not require an additional pump laser, and enhanced power, tunability, and wavelength selection are additional factors. While the pumping rate achieved by ECL previously has been two orders of magnitude lower than the optimal, Horiuchi et al. have reported a device structure designed to enhance the ECL efficiency and realize laser action driven by ECL [67]. This experiment is illustrated in Fig. 14. A pair of sputter-deposited platinum film electrodes were positioned facing each other 2 to 7 microns apart. One electrode functioned

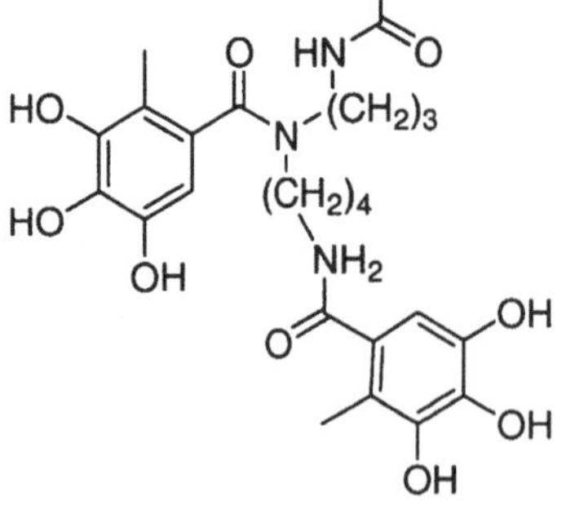

3,4,5-Tunichrome (synthetic)

Figure 13 Tunichrome structure.

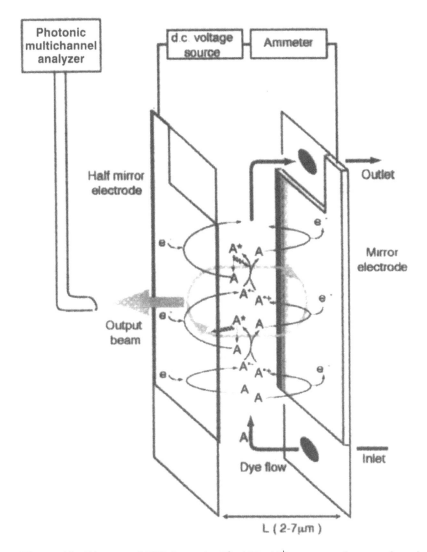

Figure 14 Diagram of ECL laser. A, A*, A·⁻, A·⁺ represent the ground, excited, anion radical, and cation radical states of 9,10-diphenylanthracene, respectively. (From Ref. 66. With permission from McMillin Publishers, Ltd.)

as a mirror while the other was a half-mirror with a transmittance of 0.8%. A solution of DPA dissolved in DMF was continuously introduced into the cavity between the electrodes at which the ground state DPA molecules were oxidized and reduced. This highly efficient ECL reaction emits blue light with which one can realize lasing upon DC electrolysis. The report of Horiuchi and coworkers will, it is hoped, serve as an entrée to a potentially significant application.

Light emission by ECL at scanning electrochemical microscope (SECM) tips is also under current development [68]. Bard and coworkers have demonstrated that ECL can be generated at SECM tips when $[(bpy)_3Ru]^{2+}$ is used as

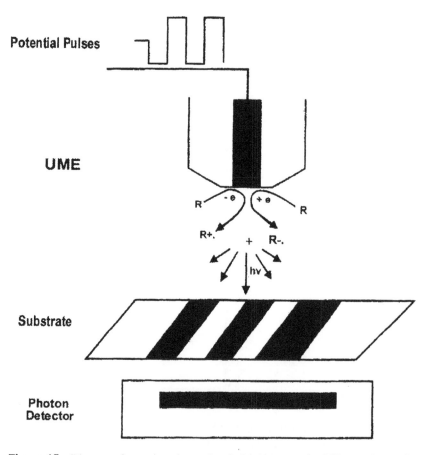

Figure 15 Diagram of scanning electrochemical microscope for ECL-based detection. ECL generated at an ultramicroelectrode via annihilation of R^- and R^+ and detected with a photomultiplier. (From Ref. 85. With permission from the American Chemical Society.)

the chromophore either with potential cycling to generate ions that annihilate or cycling to generate only oxidized complex in the presence of tripropylamine. Figure 15 illustrates the general experimental arrangement for luminescence generation and detection. The luminescence intensity is sensitive to the substrate-tip distance as the tip is moved in the vicinity of both insulating and conductive substrates. The ECL intensity decreases with both insulating and conductive substrates as the tip-substrate distance decreases. Although the resolution is still poor (at best 1 μm), the technique shows potential for applications in the studies of the kinetics and mechanisms of coreactant ECL reactions.

ACKNOWLEDGMENTS

The authors wish to thank Professor Alan J. Bard for a helpful discussion. A portion of the work reported in this chapter was supported by the US Department of Energy, Office of Basic Energy Sciences (Grant DE-FG-02-96ER14617).

REFERENCES

1. AW Knight. Trends in Anal Chem 18:47, 1999.
2. AW Knight, GM Greenway. Analyst 119:879, 1994.
3. WY Lee. Mikrochim Acta 127:19, 1997.
4. NN Rozhitskii. J Anal Chem USSR (Engl Transl) 47:1288, 1992.
5. NN Rozhitskii, EM Belash, AI Bykh. J Anal Chem USSR (Engl Transl) 49:829, 1994.
6. AJ Bard, LR Faulkner. In Electrochemical Methods. New York: Wiley, 1980, Ch 14.
7. T Kuwana. In AJ Bard, ed. Electroanalytical Chemistry, vol 1. New York: Marcel Dekker, 1966, Ch 3.
8. EA Chandross. Trans NY Acad Sci Ser 2 32:571, 1969.
9. DM Hercules. In A Weissberger, B Rossiter, eds. Methods of Organic Chemistry, Part II. New York: Academic, 1971.
10. AJ Bard, CP Keszthelyi, H Tachikawa, NE Tokel. In DM Hercules, J Lee, M Cormier, eds. Chemiluminescence and Bioluminescence. New York: Plenum, 1973.
11. LR Faulkner. Int Rev Sci Phys Chem Ser 29:213, 1975.
12. LR Faulkner, AJ Bard. Electroanal Chem 10:1, 1977.
13. F Pragst. Z Chem 18:41, 1978.
14. LR Faulkner. Methods Enzymol 57:494, 1978.
15. RS Glass, LR Faulkner. In W Adam, G Cilento, eds. Chemical and Biological Generation of Excited States. New York: Academic, 1982, Ch 6.
16. JG Velasco. Bull Electrochem 10:29, 1994.
17. J Kim, LR Faulkner. J Electroanal Chem 242:107, 123, 1988.
18. KS Lee, Y Zu, A Herrmann, Y Geerts, K Mullen, AJ Bard. J Am Chem Soc 121:3513, 1999.

19. A Vogler, H Kunkely. Angew Chem Int Ed. 23:316, 1984.
20. J Kim, FF Fan, AJ Bard, C-M Che, HB Gray. Chem Phys Lett 121:543, 1985.
21. NE Tokel-Takvoryan, AJ Bard. Chem Phys Lett 25:235, 1974.
22. A Juris, B Balzani, F Barigelletti, S Campagna, P Belzer, A vonZelewsky. Coord Chem Rev 84:85, 1988.
23. NE Tokel, AJ Bard. J Am Chem Soc 94:2862, 1972.
24. HS White, AJ Bard. J Am Chem Soc 104:6891, 1982.
25. I Rubinstein, AJ Bard. J Am Chem Soc 103:512, 1981.
26. JB Noffsinger, ND Danielson. Anal Chem 59:865, 1987.
27. L He, KA Cox, ND Danielson. Anal Lett 23:195, 1990.
28. SN Brune, DR Bobbitt. Talanta 38:419, 1991.
29. SN Brune, DR Bobbitt. Anal Chem 64:166, 1992.
30. K Uchikura, M Kirisawa. Chem Lett 1373, 1991.
31. TM Downey, TA Nieman. Anal Chem 64:261, 1992.
32. DM Hercules, FE Lytle. Photochem Photobiol 13:123, 1971.
33. G Merenyi, J Lind, TE Eriksen. J Biolumin Chemilumin 5:53, 1990.
34. RA Marcus. J Chem Phys 43:679, 1965.
35. RA Marcus. J Chem Phys 43:2654, 1965.
36. AJ Bard, ed. Electroanalytical Chemistry, vol 10, New York: Marcel Dekker, 1977, p 19.
37. RA Marcus. J Chem Phys 52:2083, 1970.
38. RA Marcus. Ann Rev Phys Chem 15:155, 1984.
39. JP Preston, TA Nieman. Anal Chem 68:966, 1996.
40. W-Y Lee, TA Nieman. Anal Chem 67:1789, 1995.
41. (a) MA Collinson, J Taussig, SA Martin. Chem Mater 11:2594, 1999; (b) AN Khramov, MA Collinson. Anal Chem 72:2943, 2000.
42. A-M Andersson, RH Schmehl. J Chem Soc Chem Commun 6:505, 2000.
43. M Sykora, TJ Meyer. Chem Mater 11:1186, 1999.
44. TC Richards, AJ Bard. Anal Chem 67:3140, 1995.
45. (a) TJ Meyer. Pure Appl Chem 58:1193–1206, 1986; (b) GA Crosby. Acc Chem Res 8:231–238, 1975; (c) RJ Watts. J Chem Educ 60:834–842, 1983.
46. NE Tokel-Tokvoryan, RE Hemingway, AJ Bard. J Am Chem Soc 95:6582, 1973.
47. JK Leland, MJ Powell. J Electrochem Soc 137:3127, 1990.
48. F Jameison, RI Sanchez, L Dong, JK Leland, D Yost, MT Martin. Anal Chem 68:1298, 1996.
49. AF Martin, TA Nieman. Anal Chim Acta 281:475, 1993.
50. P Liang, RI Sanchez, MT Martin. Anal Chem 68:2426, 1996.
51. P Liang, L Dong, MT Martin. J Am Chem Soc 118:9198, 1996.
52. L Dong, MT Martin. Anal Biochem 236:344, 1996.
53. AW Knight, GM Greenway. Analyst 120:2543, 1995.
54. GF Blackburn, HP Shah, JH Kenten, J Leland, RA Kamin, J Link, J Peterman, MJ Powell, A Shah, DB Talley, SK Tyagi, E Wilkins, TG Wu, RJ Massey. Clin Chem 37:1534, 1991.
55. JH Kenten, J Casadei, J Link, S Lupold, J Willey, M Powell, A Rees, R Massey. Clin Chem 37:1626, 1991.

56. J DiCesare, B Grossman, E Katz, E Picozza, R Regusa, T Woudenberg. BioTechniques 15:152, 1993.
57. X-H Xu, AJ Bard. J Am Chem Soc 117:2627, 1995.
58. HD Abruna, AJ Bard. J Am Chem Soc 104:2641, 1982.
59. I Rubinstein, AJ Bard. J Am Chem Soc 102:6641, 1980.
60. C Alexander, J McCall, MM Richter. Anal Chem 71:2523, 1999.
61. S Sakura, J Terao. Anal Chim Acta 262:59, 1992.
62. CA Marquette, LJ Blum. Anal Chim Acta 381:1, 1999.
63. JG Bruno, JC Cornette. Microchem J 56:305, 1997.
64. GN Chen, RE Lin, ZF Zhao, JP Duan, L Zhang. Anal Chim Acta 341:251, 1997.
65. N Egashira, Y Nabeyama, Y Kurauchi, K Ohga. Anal Sci 12:793, 1996.
66. JG Bruno, SB Collard, RJ Andrews. J Biolumin Chemilumin 12:155, 1997.
67. T Horiuchi, O Niwa, N Hatakenaka. Nature 394:659, 1998.
68. F-RF Fan, D Cliffel, AJ Bard. J Anal Chem 70:2941, 1998.
69. B Epstein, T Kuwana. Photochem Photobiol 4:1157, 1965.
70. A Zweig, DL Maricle, JS Brinen, AH Maurer. J Am Chem Soc 89:473, 1967.
71. DL Maricle, AH Maurer. J Am Chem Soc 89:188, 1967.
72. JR Wilson, S-M Park, GH Daub. J Electrochem Soc 128:2085, 1981.
73. J Kim, LR Faulkner. J Am Chem Soc 110:112, 1988.
74. R Bezman, LR Faulkner. J Am Chem Soc 94:6331, 1972.
75. WL Wallace, AJ Bard. J Electrochem Soc 125:1430, 1978.
76. CP Keszthelyui, H Tachikawa, AJ Bard. J Am Chem Soc 94:1522, 1972.
77. F Bolletta, M Ciano, V Balzani, N Serpone. Inorg Chim Acta 62:207, 1982.
78. N Kane-Maguire, JA Guckert, PJ O'Neill. Inorg Chem 26:2340, 1987.
79. P McCord, AJ Bard. J Electroanal Chem 318:91, 1991.
80. HD Abruna. J Electroanal Chem 175:321, 1984.
81. EM Kober, JV Caspar, RS Lumpkin, TJ Meyer. J Phys Chem 90:3722, 1986.
82. WA Fordyce, JG Brummer, GA Crosby. J Am Chem Soc 103:7061, 1981.
83. RD Mussell, DG Nocera. Inorg Chem 29:3711, 1990.
84. PC Ford. Coord Chem Rev 132:129, 1994.
85. A Vogler, H Kunkely. Am Chem Soc Symp Ser 333:155, 1987.
86. F-RF Fan, D Cliffel, AJ Bard. Anal Chem 70:2941, 1998.

4

From Superquenching to Biodetection: Building Sensors Based on Fluorescent Polyelectrolytes

David Whitten, Robert Jones, Troy Bergstedt, and Duncan McBranch
QTL Biosystems, LLC, Santa Fe, New Mexico

Liaohai Chen
Los Alamos National Laboratory, Los Alamos, New Mexico

Peter Heeger
Case Western Reserve University, Cleveland, Ohio

I. BACKGROUND

Excited states of conjugated molecules such as aromatic hydrocarbons, dyes, and heteroaromatics are quenchable by a variety of small molecules. Of the many types of quenching processes that have been studied, the most extensive include energy and electron transfer processes [1–8]. Energy transfer quenching, especially singlet–singlet energy transfer can occur over both long and short ranges and is not restricted to collisional encounter or precomplexing between the donor and acceptor. In contrast, electron transfer quenching of excited states typically involves contact or near contact between the substrate and quencher and the rate falls off sharply with substrate–quencher separation [9–11]. Most commonly, the effectiveness of a quenching process is assessed by the empirical Stern–Volmer (SV) Eq. (1),

$$\frac{I_0}{I} = 1 + K_{SV}[Q], \tag{1}$$

where I_0 and I denote the fluorescence intensity in the absence and presence of quencher, respectively. K_{sv} is the quenching constant.

A linear Stern–Volmer relationship is usually observed when the quenching is a simple process involving a single mode of interaction between substrate and quencher. Two limiting cases are commonly observed. When the quenching occurs by a dynamic (often diffusional) process, the SV plot is linear and K_{sv} is the product of the rate constant for quenching k, and the excited state lifetime τ. In cases where the quenching is diffusion-limited, the quenching constant cannot exceed k_{diff}. In cases where the excited state is a singlet having a lifetime in the range of nanoseconds or less, K_{sv} is typically on the order of 10 or less. The second limiting case involves precomplexation of the substrate and quencher and thus the K_{sv} measured in this situation generally reflects the equilibrium constant for the complex formation. A typical situation where the second limiting case is encountered involves the quenching of excited singlet states of molecules with very short excited state lifetimes. An example involves the olefin, trans-stilbene (**1**), whose singlet lifetime in fluid solution is very short (ca. 80 ps) due to com-

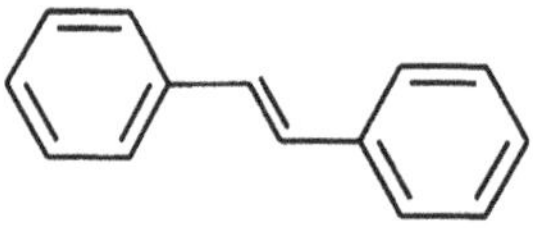

1

petition between strongly allowed fluorescence and irreversible conversion of a near-planar transoid excited state to a twisted or near-perpendicular excited state by rotation about the central double bond [12,13]. The solution fluorescence of trans-stilbene and related molecules can be quenched by addition of powerful electron acceptors such as the organic dication methyl viologen (**2**) with a

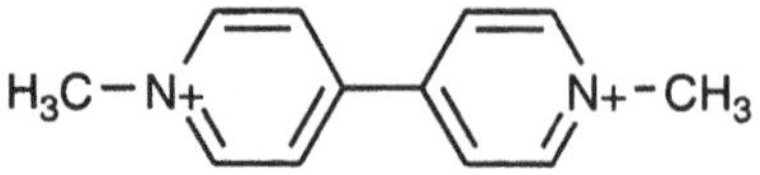

2

Stern–Volmer constant $K_{sv} \sim 15$ (in acetonitrile at room temperature) that is clearly not compatible with dynamic quenching [14–16]. Close spectroscopic examination of solutions having high quencher concentrations shows that there is a weak complex formed between the stilbene and viologen that is characterized by small changes in the stilbene transitions and a weak new transition in the visible where there is no absorption for either stilbene or viologen alone [14–16]. Such weak ground state charge transfer complexes have been observed for a variety of systems and it is nearly always the case that the complex is nonemissive even where both substrate and quencher are individually luminescent [17–20].

Much of this chapter focuses on quenching processes that have been observed with substrates that can be thought of as polymers of simple chromophores such as trans-stilbene and quenchers such as methyl viologen or related small molecules. Herein, we show that similar ground state complexation between the quencher and one of the repeat units of the polymer can result in a remarkable amplification or superquenching. Thus, Stern–Volmer constants more than a millionfold higher than those observed between **1** and **2** can be observed for polymeric substrates [21]. From a slightly different perspective it can be seen that addition of a single quencher to one repeat unit of the polymer results in quenching of the fluorescence from the entire polymer. We show that this "superquenching" can be explained in terms of a combination of enhanced complexation under certain conditions with efficient energy migration or exciton delocalization. The ability of such small amounts of a specific small molecule quencher to extinct polymer fluorescence by formation of a relatively weak ground state complex provides an opportunity for a number of sensing applications. Thus, the chapter first examines the origins of the superquenching and follows with a discussion of several specific applications.

A. Excited States of Conjugated Polymers

Understanding the excited states associated with conjugated polymers remains an active area of investigation. For conjugated polymers it is clear that extending the conjugation from that present in simple monomers or oligomers to a very large number of repeat units does not produce the anticipated effect of monotonic decrease in excitation and fluorescence onsets that would be associated with a monotonic increase in conjugation. Indeed, for most conjugated polymers a maximum in both absorption and fluorescence is quickly reached for oligomers containing a relatively small number of repeat units. Thus, it is clear that, from a spectroscopic point of view, long polymer chains consist of broken segments that for the most part are no longer than several repeat units. However, the possibility for both intrachain and interchain excited states and the presence of a rich array of low-lying excited states resulting or coexisting due to the different conformations present on solubilized conjugated polymers provides for a rich array of radiative and nonradiative decay paths. The question of how these segments "communicate" with one another is interesting and is most important from the perspective of utilizing conjugated polymers in device applications.

The spectroscopic signatures of excited states of derivatives of the conjugated polymer poly (p-phenylene vinylene) have been widely investigated both on ultrafast [22–24] and steady-state timescales [25,26]. It is known that the dominant excited states of pristine separated chains are singlet excitons with a strongly allowed radiative transition and a fluorescence lifetime of approximately one ns [22]. However, chain folding in poor solvents, or the proximity of adjacent chains in concentrated solutions or the condensed phase, leads to a variety of interchain excited states, including excimers [25,27] and charged excitations (polarons) formed by interchain electron transfer [22]. These interchain states result in strong self-quenching, dramatically reducing the quantum efficiency of fluorescence [27,28]. The self-quenching most likely results from a combination of efficient through-space Förster energy transfer among the many conjugated segments, together with quenching at a nonradiative or weakly radiative interchain excited state that provides the lowest energy available state [22,29].

Controlling the degree of "parasitic" self-quenching has proved possible by adding bulky side chains to prevent aggregation [30], and by tuning the solvent from which films are cast [28]. However, a systematic control over and understanding of the relationships between polymer morphology and the resulting optical and electronic properties have still not been achieved. In particular, these issues are just beginning to be explored for conjugated polyelectrolytes, for which the morphology can be a sensitive function of complex Coulombic and hydrophobic interactions. These issues are discussed in more detail below.

B. Polyelectrolytes, Detergents, and Micelles

The self-assembly of detergents to form micelles has been well studied as have the notable properties of micelles in terms of solubilizing or concentrating in aqueous environments a wide variety of reagents ranging from nonpolar organics to macromolecules and organic and inorganic ions. The ability of micelles to collect diverse molecules near the detergent–water interface is often reflected in terms of novel or rapid reactions not readily observable in homogeneous environments [31]. Several of the characteristic features of detergent micelles as solubilizing reagents are mimicked by water-soluble polyelectrolytes but with the potential advantage that the properties of the polyelectrolyte are preserved intact even at very low dilution whereas many of the properties of detergent micelles vanish at or below the critical micelle concentration [32,33]. In addition to solubilizing small molecules and macromolecules, polyelectrolytes exhibit the ability to associate with oppositely charged detergents to form "polymer micelles" [34–38]. Various structures have been demonstrated or proposed for these assemblies [34–38].

Although the structural and micellar properties of "conventional" polyelectrolytes have been extensively studied, only recently have conjugated polyelectrolytes become widely available. As we discuss below, the combination of polymer micelles with extended fluorescent excited states leads to a rich variety of new phenomena and applications.

II. SUPERQUENCHING OF FLUORESCENT POLYELECTROLYTES BY SMALL MOLECULE QUENCHERS

A. Quenching by Oppositely Charged Organic Ions

Several years ago it was observed that the quenching of substrates such as trans-stilbene by positively charged electron acceptors such as Cu^{2+} or methyl viologen (2) could be significantly enhanced by the addition of anionic detergents such as sodium dodecyl sulfate (SDS) at concentration above its critical micelle concentration [14–16]. For example, for a series of trans-stilbene derivatives, ranging from 1 itself to an array of amphiphilic molecules including the trans-stilbene chromophore in the backbone of a fatty acid chain, it was found that the Stern–Volmer quenching constant was enhanced by roughly two orders of magnitude. Interestingly, it was found that there were only minor differences between different stilbenes in the series [14,15]. When the quenching constant was "corrected" for the small volume (i.e., the micelle) in which the stilbene and viologen were confined, it was found that the corrected constant was virtually the same as that for stilbene and 2 in homogeneous solution. These results are consistent with a structure in which the micelle provides an extensive interface between water and detergent and wherein both the stilbene and viologen reside in interfacial regions. Similar results for the "micelle binding sites" have been obtained in a number of other studies [17]. Inorganic ions such as Cu^{2+} also exhibit enhanced quenching of stilbene and other aromatics in the presence of anionic detergent which again indicates that the stilbene is in an interfacial site rather than "buried" within a highly hydrophobic region [16].

The findings with stilbene and viologen as "guests" in anionic micelles suggested that similar effects might be obtained when a polyelectrolyte was used to "concentrate" the two reagents. More important, these findings suggested that when a conjugated polyelectrolyte was used as both the concentrating reagent and excited substrate even more dramatic quenching behavior might be observed. For example, it has been shown that incorporation of a potential quencher such as a fullerene or viologen into an uncharged conjugated polymer can result in efficient quenching [39]. Studies by Swager et al. [40–44] have shown that efficient migration to synthetically engineered internal traps or "defect sites" takes place within mixed conjugated polymer systems [40,41]. These studies also show modest quenching enhancements when viologen derivatives are added to nonionic conjugated polymers [42,44]. Thus, it was reasonable to expect that two complementary amplification mechanisms might be operative if a conjugated polyelectrolyte of the appropriate structure were available. The availability of the water-soluble ionic derivative of poly(phenylene)vinylene, MPS-PPV (3)

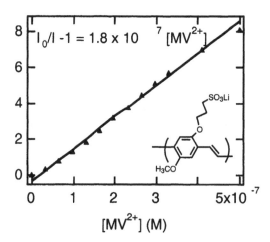

3

provided a candidate to test this hypothesis. This compound, which can be thought of as a "poly-trans-stilbene" shows absorption and fluorescence in the visible that has weak structure similar to that of **1**. Not surprisingly, we found that the strong fluorescence of **3** could be quenched by addition of **2** or Cu^{2+} [21]. What was remarkable was the finding of quenching at extremely low quencher concentrations (Fig. 1). At these low quencher concentrations it is clear that insignificant "dynamic" quenching can occur.

Figure 1 Stern–Volmer plot of fluorescence intensity of the polymer MPS-PPV (inset) in the presence of methyl viologen (MV^{2+}). Slope of line gives $K_{sv} = 1.8 \times 10^7$ M^{-1}.

Evidence that the quenching results from ground state complexation comes from a minor but clear change in the absorption spectrum of **3** as **2** is added. The amplified quenching sensitivity, or "superquenching," shown in Fig. 1 leads to K_{sv} values of 10^7 or higher. This superquenching (amplification by $>10^6$ compared to homogeneous solution) can be analyzed from several perspectives. For the present polymer (**3**) the number of repeat units is ca. 1000. The concentration of polymer used in the experiment leading to the quenching shown in Fig. 1 is 5×10^{-5} M in polymer repeat units and significant quenching is observed with concentrations of **2** as low as 100-fold lower. Thus, on the basis of quencher per polymer repeat unit it appears that between 1 and 10 quenchers per polymer chain is sufficient to quench the fluorescence resulting from excitation of the polymer at any site. In recent experiments using a different type of polymer with a smaller number of repeat units we have observed several cases where a single quencher per polymer chain is sufficient to entirely quench polymer fluorescence [45].

Thus, it is clear that a major portion of the amplified quenching may be attributable to the fact that the presence of a quencher anywhere on the polymer chain has a high probability of quenching any excitation delivered to the polymer. The second source of the amplification is the anticipated enhanced association of the quencher with the polymer due to a combination of Coulombic and hydrophobic interactions similar to those that account for the association of reagents such as **2** with anionic micelles [14–16]. As pointed out above, the enhanced binding of **2** to SDS micelles is only anticipated at SDS concentrations above the critical micelle concentration. The average "size" of an SDS micelle is about 80 detergent molecules. Given the much larger number of repeat units (and hence charges) on **3**, we can anticipate that the binding of **2** to **3** may be significantly enhanced. Thus, it is not unreasonable to presume the total quenching amplification may be the product of the enhanced binding constant between **2** and the polymer ($K \sim 10^5–10^6$) and the ratio of polymer repeat units to quencher that result in effective quenching ($10^2–10^3$).

This leads to a prediction that K_{sv} values in the range 10^7 to 10^9 may be anticipated for polymer systems of moderate molecular weight. It also predicts that the quenching sensitivity may be tuned by varying the charge on the polymer as well as the number of polymer repeat units. Although this concept has not yet been tested quantitatively, we have found that lower molecular weight polymers related to **3** show reduced quenching sensitivity. In recent studies, we have demonstrated that positively charged electrolytes bearing chromophores on each repeat unit exhibit an analogous superquenching in the presence of small molecule anionic electron acceptors [45]. Comparable Stern–Volmer quenching constants are observed ($10^7–10^8$) even though the number of repeat units is lower. Thus, it seems quite reasonable that superquenching is a general phenomenon between polyelectrolytes and countercharged electron transfer quenchers.

B. Quenching by Inorganic Ions

Previous studies have shown that paramagnetic ions such as Cu^{2+} quench excited states of **1** and related compounds and that the quenching can be enhanced in a similar way as for **2** by the addition of anionic detergent [16,46–49]. A similar superquenching might therefore be anticipated when solutions of **3** are exposed to Cu^{2+}. Consequently, we were not surprised to find that addition of Cu^{2+} does result in quenching of the fluorescence of **3**; however, the measured K_{sv} of 2×10^4 M^{-1} is considerably lower than that for **2** [21]. Other diamagnetic ions that might not be expected to specifically quench the fluorescence of **3** via redox or electronic interactions have also been found to quench the fluorescence of **3** [21]. These include Mg^{2+} and Ca^{2+}, both of which are closed-shell metal ions that would not be anticipated to quench simple molecular excited states (e.g., neither ion quenches stilbene fluorescence).

The K_{sv} values for these ions are similar to those for Cu^{2+} and it may be anticipated that binding to the polymer by all three ions occurs with comparable strength due to reasonably similar Coulombic interactions between the divalent ions and the polymer. Since the polymer fluorescence is known to be strongly attenuated when interchain or intermolecular interactions occur, it may be that all or part of the quenching observed for these ions may be attributed to "aggregation" effects induced by association of the divalent cations with the polymer.

C. Quenching by Neutral Organics

Since the excited states of **3** and related polymers can be quenched by electron acceptors such as viologens, it appeared possible that some degree of superquenching might occur with neutral organics that possess the ability to quench excited states by electron transfer and that have low solubility in water. It was anticipated that these molecules might associate strongly with the polymer in aqueous solution and perhaps be solubilized to a limited extent in the same way detergents solubilize organic reagents [21]. Three neutral organic electron acceptors, 2,4,6-trinitrotoluene (TNT), 2,4-dinitrotoluene (DNT), and 9,10-dicyanoanthracene (DCA), were all found to exhibit "enhanced" quenching (as measured by Stern–Volmer constants on the order of 10^4 M^{-1}) compared to small molecule substrate–quencher interactions [21]. Since these quenching constants are much too large to be associated with any dynamic quenching process, it is concluded that a ground state complex may be formed in each case, probably due to hydrophobic interactions.

D. Effects of Detergents on Polyelectrolyte Photophysics and Quenching

As mentioned previously, polyelectrolytes have been shown to interact with oppositely charged detergents or other surfactants to form micelle-like assemblies [34–38]. Several different models and types of structures have been proposed for these assembly processes [34]. In the present case we employ a cationic detergent, decyltrimethylammonium bromide (DTAB) which is expected to have a moderately high critical micelle concentration and thus associate individually with the polymer when added at low concentrations as opposed to absorbing the polymer into a preformed micellar structure. Addition of DTAB at very low concentrations results in a sharp increase of the polymer fluorescence (and a small change in the absorption spectrum) that saturates at a level of one detergent per three polymer repeat units. The time-resolved photoluminescence also shows major changes upon addition of the detergent; in water alone the early fluorescence decay is dominated by two components with lifetimes of 0.2 and 2.4 ps. Addition of detergent results in a fluorescence decay largely dominated by a single component with a lifetime of 1.2 ps.

The more than twentyfold increase in the polymer fluorescence quantum efficiency (accompanied by increased fluorescence lifetime) is clearly due to elimination of nonradiative decay pathways upon formation of a complex with the detergent; a tentative explanation is that addition of the detergent to the polymer results in an expansion of the polymer, resulting in reduced chain folding and fewer "kink" defects [50]. As indicated above, interchain interactions result in an increase in nonradiative decay pathways and a consequent decrease in fluorescence. A reduction of these should be manifested in higher fluorescence yields, and also in a more homogeneous distribution of polymer segments. That this occurs is suggested by the red-shifted and narrowed absorption and fluorescence, the reduced Stokes shift, and the more pronounced vibronic structure in the detergent-complexed polymer.

The complexing of the polymer with detergent also has a striking effect on its quenching by the various neutral and charged organics discussed above. For quenching of **3** by viologen (**2**) there is a sharp attenuation upon addition of the detergent. Thus, the K_{sv} for **2** decreases from 2×10^7 M^{-1} to 6.6×10^4 M^{-1} upon addition of detergent at the level of one detergent per three repeat units [51]. In contrast, the Stern–Volmer constants for neutral organic molecules such as DCA, DNT, and TNT all increase by severalfold upon addition of the same level of detergent. The origin of the more than two orders of magnitude attenuation of the quenching of polymer fluorescence by 2 is somewhat difficult to understand. One possibility would be that the binding of the detergent to the polymer suppresses the binding of viologen to the polymer. However, on first consideration this seems not likely to be the dominant source of the attenuation since at the levels where significant suppression of quenching occurs there is relatively little "neutralization" of the net charge on the polymer. For anionic detergent micelles it has been found that binding of viologen to the micelles is not appreciably attenuated by the addition of previous viologens as might be anticipated on the basis of charge "screening" [16,46]. This has been interpreted as due to the fact that the binding of viologen to micellar SDS is largely entropically controlled and that additional viologens are added to the micelle with relatively little attenuation until the surface area of the micelle is exhausted [46]. In contrast inorganic ions such as Cu^{2+} do exhibit attenuation when the same type of study is carried out, suggesting that screening plays a major role for these ions where few, if any, hydrophobic interactions occur [46]. Since the Stern–Volmer constant for superquenching has been suggested to be the product of the binding constant between the quencher and polymer and the number of repeat units quenched by a single quencher on the polymer chain, a possibility is that the addition of detergent can produce a significant attenuation in both of these parameters. Thus, it seems possible that addition of the detergent may provide a modest attenuation of the binding of the viologen to the polymer but a more significant effect may be the separation of the bound viologen from the π electron system of the polymer and hence a reduction in the efficiency of quenching of bound viologen.

The opposite effect on quenching by neutral molecules when small amounts of detergent are added to aqueous solutions of 3 could be attributed to several possible factors. First, the addition of detergent to the polymer may help in affording a more hydrophobic environment in its vicinity that may enhance the polymer–small molecule association. Also, since surfactants such as DTAB are well known to solubilize organic molecules in water, it appears reasonable that clusters of the detergent may combine with the neutral quenchers and function as "chaperones" to increase their effective solubility and thus enhance the complex formation with the polymer.

E. Formation of Films and Assemblies from Fluorescent Polyelectrolytes

Several recent studies have been focused on the transfer of polyelectrolytes from aqueous solutions to surfaces having moderate to high countercharge densities. In many examples it has been shown that the sequential transfer of layers of oppositely charged polyelectrolytes can occur in a stepwise and regular fashion to build relatively homogeneous and robust layered assemblies [53,54]. Our studies with moderately concentrated solutions of **3** have shown that this polymer can be transferred and that the layered polymer retains relatively strong fluorescence. The polymer has been layered directly on derivatized glass, over layers of cationic polyelectrolyte, and on various charged bead supports. It has been found possible to layer the polymer that has been treated with the detergent DTAB as described above. It is also possible to layer polymer **3** on a positive support and to treat this with detergent by dipping the polymer-coated sample into an aqueous solution of DTAB. It has also been shown that polymer complexed with certain quenchers, such that solution fluorescence is largely quenched, can be transferred from aqueous solution to positively charged support such that the layered polymer remains quenched to a similar extent. As discussed below, the polymer in these layered formats can be used in sensing applications similar in several cases to those for solutions of the polymer.

F. Chemical Sensing by Fluorescent Polyelectrolyte Quenching

Since polymer **3** is electron-rich and its fluorescence can be quenched by electron-deficient ions and molecules, it seemed attractive to determine whether films of the polymer, prepared as described above, could be useful in sensing electron-deficient organics of low-moderate volatility at very low concentration levels. Several recent studies have focused on the sensing of nitroaromatics as potential signatures of explosives, especially in the detection of landmines [54–56]. It was found that exposure of layered polyelectrolyte assemblies containing an outermost layer of **3** exhibited strong fluorescence that is rapidly quenched on exposure to nitrobenzene, dinitrotoluene, or trinitrotoluene vapor. For example, nearly complete quenching of the fluorescence of **3** was obtained within one minute on exposure to dinitrotoluene vapor at a concentration (from its vapor pressure) of $<8 \times 10^{-9}$ M; significant quenching can be detected with concentrations an order of magnitude lower.

As has been observed for other fluorescent polymers whose fluorescence is quenched by similar levels of nitroaromatics [54,55], the fluorescence quenching observed under these conditions was irreversible and even pumping on the sample under vacuum for prolonged periods did not result in significant ($>10\%$) recovery of the fluorescence from **3**. It was also observed that the absorption spectrum of the polymer changed upon exposure and quenching by the nitroaromatics. Interestingly, we found that layers of **3** prepared either from aqueous solutions with DTAB added (as described above) or prepared by layering **3** from a pure aqueous solution followed by addition of an "overlayer" of DTAB exhibited similar degrees of quenching on exposure to nitroaromatic vapor. However, in these cases there was very little change in the absorption spectrum of the polymer and upon pumping of the sample (10^{-3} torr) for 10 min about 90% of the fluorescence is recovered. Samples so treated can be used again in "sensing" the nitroaromatics and "recycled" with a similar recovery after evacuating the film to remove the nitroaromatic.

Perhaps the most interesting format for use of the polymer for sensing nitroaromatics is the isolation of a 1:1 complex between DTAB and **3** and the use of this complex in spin-cast films. Thus, when the ratio of DTAB to polymer repeat units in aqueous solutions is raised to 1:1, the polymer:DTAB complex precipitates as a bright red solid. This solid is insoluble in water but soluble in several organic solvents. The organic solutions of the DTAB:**3** complex show enhanced fluorescence and greater stability than the corresponding solutions of polymer alone [50,51]. Spin-coating this complex from methanol leads to a solid film whose fluorescence is quenched by similar levels of nitroaromatics but whose fluorescence can be ~98% recovered by vacuum treatment.

III. THE QTL APPROACH TO SENSING OF BIOMOLECULES

From the previous discussion we have developed a picture of the chemical quenching for both organic cations and neutral molecules wherein amplified quenching is observed as a consequence of relatively weak association between the quencher and polymer, coupled with a highly efficient fluorescence quenching due to a combination of energy transfer and/or exciton delocalization. As shown above, this quenching can be the basis of a sensitive but not very selective fluorescence "turn-off" chemical sensing.

We suspected that the quencher–polymer association might be sufficiently weak that it could be easily reversed under conditions where stronger forces intervened. Thus, we found that addition of other dense polyelectrolytes could reverse the quenching by competitive removal of the quencher in the case of methyl viologen. For example, addition of small amounts of Laponite clay (anionic) evidently can compete for the positively charged viologen and completely reverse the quenching of fluorescence from **3**. This suggested the possibility of constructing a conjugate in which a quencher would be linked via a short tether to a ligand, capable of forming a strong and specific complex with a given bioagent. In the general case it would be anticipated that the bioagent–ligand complex should be much stronger than the quencher–polymer association, and that steric effects between the high molecular weight polymer and bioagent should prevent simultaneous associations of the small conjugate (QTL) with both macromolecules. Thus, it would be anticipated that addition of the QTL conjugate to solutions or films of the polymer should result in strong fluorescence quenching in the absence of a bioagent recognizing the ligand of the QTL.

Addition of the bioagent should result in removal of the QTL from the polymer and a turning on of its fluorescence. Thus, the biosensing using this approach would result in a "turn-on" fluorescence detection.

A. Biotin-Avidin

As a first test of the QTL approach the viologen–biotin conjugate **4** was synthesized. Since the small molecule biotin is known to bind strongly and specifically

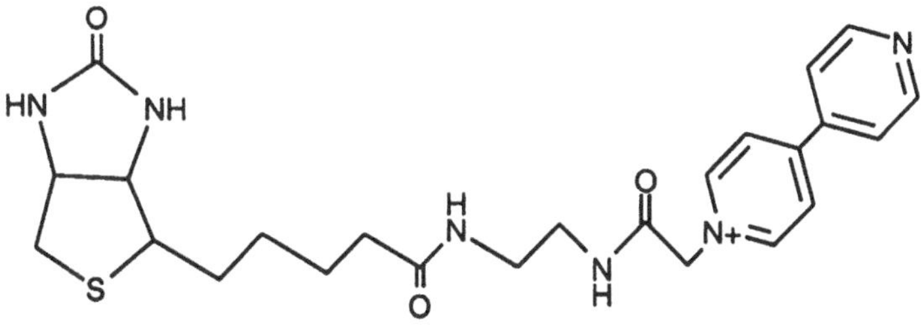

4

to the proteins avidin and streptavidin [57–60], it was felt that these systems would provide an ideal initial case for testing the QTL approach. The crystal structure of the biotin–avidin complex indicates that the bound biotin is in a relatively deep pocket of the avidin and confirms the expectation that the conjugate **4** can not likely associate simultaneously with both the protein and the polymer [59]. Quenching experiments established that **4** exhibited superquenching towards polymer **3**, but with a slightly lower K_{sv} than that obtained with **2** [21]. Thus, it was easy to observe partial quenching of the fluorescence of **3** with concentrations of **4** in the range 10^{-7} M or less. As anticipated from the above discussion, it was found that addition of very small amounts of avidin ($\sim 1 \times 10^{-8}$ M) or streptavidin reversed the quenching induced by **4** [21]. A quantitative examination of the quenching–quenching reversal showed that there was close to the anticipated fourfold removal of **4** by addition of avidin, consistent with the four known biotin binding sites per molecule of the protein [21]. As controls, it was found that proteins such as albumin or choleratoxin (see below), which lack a biotin binding site, do not induce "unquenching" of the polymer quenched by **4**. Similarly, it was found that monomethylviologen (**5**), which does not contain the biotin ligand, can quench the fluorescence of **3** about

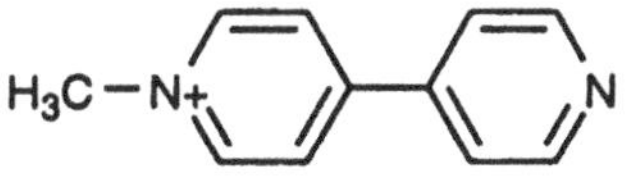

5

as efficiently as **4**; however, these quenched solutions exhibit no fluorescence recovery when solutions of avidin are added.

B. GM1–Choleratoxin

To further test the QTL hypothesis we investigated a structurally quite different bioagent–ligand combination. Choleratoxin protein, similar to several other toxins such as Shigella or botulinum consists of two subunits, one containing the enzymic component while the other (pentameric B subunit) is the recognition portion which attaches to a cell membrane by binding with a specific glycolipid ganglioside that protrudes from a lipid bilayer membrane into the outer aqueous solution. For choleratoxin binding, the specific ligand is the ganglioside GM1. The complex polysaccharide is the portion of the glycolipid that binds to the protein. A crystal structure of the Cholera toxin B subunit pentamer with GM1 shows that the polysaccharide resides in a relatively deep pocket of the protein [61]. We were able to construct a QTL molecule containing the polysaccharide portion of GM1 by removing one of the chains of the parent molecule and replacing it with a chain containing a viologen monocation. The GM1–viologen QTL (**6**) quenches the polymer fluorescence; addition of choleratoxin protein containing only the recognition portion results in a reversal of the quenching at very low levels. Other proteins such as avidin and albumin that lack a GM1 binding site produce little or no change in the polymer fluorescence and thus indicate the specificity for GM1 binding to remove the QTL from the polymer.

C. F_V–Hepatitis C Core Protein

One of the most crucial areas for detection involves protein–protein interactions. In some respects this becomes a very important test for the QTL approach. In the two foregoing examples the ligand incorporated into the QTL bioconjugate is a relatively small organic molecule. As pointed out in Sec. III.A, the QTL approach should work best where there is strong association between the ligand and bioagent to be sensed and a relatively large difference between the size of the ligand and the ligand:bioagent complex such that the increased steric size of the latter forces a removal of the quencher from the polymer. It would be anticipated that as either the tether length is greatly increased, or the difference between size of the ligand and ligand:bioagent is decreased, the likelihood of removal of the QTL conjugate by the bioagent should decrease. An important test therefore is the use of a protein or peptide chain as both the ligand and bioagent. Since antibody–antigen interactions are widely used in various biosensing applications it was important to see if the QTL approach could be adapted for this analysis.

In our initial studies in this area we have approached antibody–antigen interactions through the use of an antibody fragment (F_V) of relatively low

6

molecular weight (25 kD) that has been expressed with a tag containing a sequence of several histidines. We felt that the relatively small antibody fragment might offer sufficient size distinction compared to the F_v:protein complex to elicit a QTL response such as observed above. In our initial experiments, we felt that an expedient way of constructing a QTL from the F_v fragment was to take advantage of the fact that the histidine tag forms a complex with Cu^{2+} and generate the Cu^{2+} complex of the F_v as the QTL. The F_v fragment–Cu^{2+} complex was formed *in situ* by adding the F_v to a solution containing an excess of Cu^{2+} (excess Cu^{2+} was necessary to maintain a concentration of the complex with the histidines). Since Cu^{2+} is a quencher for the polymer (Sec. II.B), this implies that adding a solution of F_v containing some free Cu^{2+} to the polymer results in quenching from both complexed and free Cu^{2+}.

We were able to demonstrate that sequential addition of sufficient Cu^{2+} to form the histidine complex to a solution of polymer **3** does result in significant but not complete quenching of the polymer fluorescence. Addition of the F_v fragment results in additional quenching of the polymer fluorescence. Addition of F_v to the polymer alone results in no fluorescence quenching; therefore it is reasonable that the "additional" quenching produced by addition of the F_v fragment to the solutions containing polymer and Cu^{2+} is due to complexing of the *in situ* generated QTL to the polymer.

The antibody fragment F_v that we have used in these studies binds specifically to a Hepatitis C core protein. Thus, we find that addition of the Hepatitis C core protein antigen to a quenched (by the sequential addition of solutions of Cu^{2+} and F_v fragment) polymer solution results in nearly full recovery of the polymer fluorescence. In contrast, addition of an unrelated Hepatitis B core protein (which does not bind to the Hepatitis C antibody) results in no change in the quenched polymer fluorescence. Thus, while these results are preliminary, the initial indications are that at least for antibody fragments, the QTL approach may be applicable to protein–protein fragment sensing. We are currently working on an approach where the antibody–QTL employs a quencher that can be added without providing an extra quenching pathway, and that presumably will provide a general approach to QTL synthesis from different protein derivatives.

IV. ADVANCED SENSING APPLICATIONS AND PROSPECTS

The fundamental "fluorescence turn-on" sensor described above for biomolecule detection can be used in many different formats for homogeneous assays. For example, one attractive formulation is a "reverse assay" (Fig. 2) wherein a molecule similar in structure and function to the ligand is sensed by having the fluorescent polymer and the QTL:bioagent complex together. In this case,

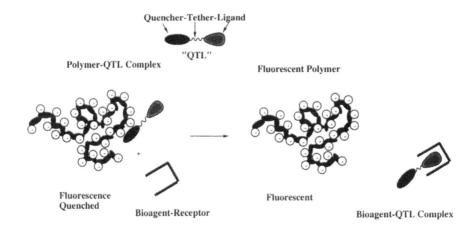

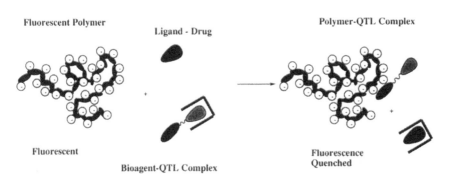

Figure 2 Two-stage use of the QTL approach for competitive assay and/or drug discovery. In the upper part the polymer QTL complex (nonfluorescent) is treated with a bioagent-receptor and formation of the bioagent:QTL complex results in freeing the polymer of the quencher and turning on the polymer fluorescence. If a mixture of polymer and bioagent:QTL complex (fluorescent) is exposed to a new molecule capable of binding with the bioagent, release of the QTL from the complex leads to its association with the polymer and a turning off of the polymer fluorescence.

the sensing is a "fluorescence turn-off" assay or competition analysis since the polymer is unquenched in the absence of a ligand-competitor for the bioagent and the addition of this reagent releases the QTL which in turn quenches the polymer fluorescence. Other modes of operation include various anchored formats where either the polymer or a connected polymer–QTL ensemble is attached to a support and contacted with a solution of analyte. We are currently exploring the QTL approach with an array of polymers, various custom synthesized QTL molecules, and a range of bioagents.

V. SUMMARY—CONCLUSIONS

The QTL approach provides a sensitive means of sensing a potentially vast array of biomolecules in an assay that offers several advantages. The sensing is homogeneous, rapid, sensitive, and selective. The key to the high sensitivity is the remarkable amplification of quenching sensitivity provided by using a conjugated polyelectrolyte as the fluorescent substrate.

ACKNOWLEDGMENT

The work at QTL Biosystems has been carried out with support from the Defense Advanced Research Projects Agency under Contract #MDA972-00-C-006.

REFERENCES

1. SM Hubig, JK Kochi. J Am Chem Soc 121:8, 1688–1694, 1999.
2. S Fukuzumi, T Suenobu, M Patz, T Hirasaka, S Itoh, M Fujitsuka, O Ito. J Am Chem Soc 120:32, 8060–8068, 1998.
3. TP Le, JE Rogers, LA Kelly. J Phys Chem A 104:29, 6778–6785, 2000.
4. L Brancaleon, D Brousmiche, V Rao, LJ Johnston, VJ Ramamurthy. Am Chem Soc 120:20, 4926–4933, 1998.
5. D Graff, JP Claude, TJ Meyer. Adv Chem Ser 253(Electron Transfer Reactions): 183–198, 1997.
6. IR Gould, D Ege, JE Moser, S Farid. J Am Chem Soc 112:11, 4290–301, 1990.
7. X Ci, R Silveira da Silva, D Nicodem, DG Whitten. J Am Chem Soc 111:4, 1337–1343, 1989.
8. KS Schanze, LYC Lee, C Giannotti, DG Whitten. J Am Chem Soc 108:10, 2646–55, 1986.
9. A Ponce, HB Gray, JR Winkler. J Am Chem Soc 122:8187–8191, 2000.
10. K Weidemaier, HL Tavernier, SF Swallen, MD Fayer. J Phys Chem A 101:1887–1902, 1997.
11. LT Calcaterra, GL Closs, JR Miller. J Am Chem Soc 105:670–671, 1983.
12. J Saltiel, AS Waller, DF Sears Jr, EA Hoburg, DM Zeglinski, DH Waldeck. J Phys Chem 98:10689–98, 1994.
13. J Saltiel, AS Waller, DF Sears Jr. J Am Chem Soc 115:2453–2465, 1993.
14. JC Russell, SB Costa, RP Seiders, DG Whitten. J Am Chem Soc 102:5678–5679, 1980.

15. JC Russell, DG Whitten, AM Braun. J Am Chem Soc 103:3219–3220, 1981.
16. JBS Bonilha, TK Foreman, DG Whitten. J Am Chem Soc 104:4215–4220, 1982.
17. FM Martens, JW Verhoeven. J Phys Chem 85:1773–1777, 1981.
18. E Bosch, SM Hubig, SV Lindeman, JK Kochi. J Og Chem 63:3, 592–601, 1998.
19. JK Kochi. Chimia 45:10, 277–281, 1991.
20. JM Masnovi, JK Kochi, EF Hilinski, PM Rentzepis. J Am Chem Soc 108:6, 1126–1135, 1986.
21. L Chen, DW McBranch, H-L Wang, R Helgeson, F Wudl, DG Whitten. Proc Natl Acad Sci 96:12287–12292, 1999.
22. B Kraabel, VI Klimov, R Kohlman, S Xu, H-L Wang, DW McBranch. Phys Rev B 61:8501–8515, 2000.
23. D McBranch, MB Sinclair. Ultrafast photoinduced absorption in nondegenerate ground-state conjugated polymers: Signatures of excited states. In NS Sariciftci, ed. Nature of the Photoexcitations in Conjugated Polymers: Semiconductor Band vs. Exciton Model, New York: World Scientific, 1997.
24. LJ Rothberg, M Yan, S Son, ME Galvin, EW Kwock, TM Miller, HE Katz, RC Haddon, F Papadimitrakopoulos. Synth Met 78:231, 1996.
25. IDW Samuel, G Rumbles, CJ Collison, SC Moratti, AB Holmes. Intra- and intermolecular photoexcitations in a cyano-substituted poly(p-phenylenevinylene). Chem Phys 227:1,2, 75–82, 1998.
26. NS Sariciftci, ed. Nature of the Photoexcitations in Conjugated Polymers: Semiconductor Band vs. Exciton Model, New York: World Scientific, 1997.
27. R Jakubiak, CJ Collison, WC Wan, LJ Rothberg, BR Hsieh. Aggregation quenching of luminescence in electroluminescent conjugated polymers. J Phys Chem A 103:14, 2394–2398, 1999.
28. T-Q Nguyen, IB Martini, J Liu, BJ Schwartz. Controlling interchain interactions in conjugated polymers: The effects of chain morphology on exciton-exciton annihilation and aggregation in MEH-PPV films. J Phys Chem B 104:2, 237–255, 2000.
29. T-Q Nguyen, J Wu, V Doan, BJ Schwartz, SH Tolbert. Control of energy transfer in oriented conjugated polymer-mesoporous silica composites. Science (Washington, DC) 288:5466, 652–656, 2000.
30. F Wudl, P-M Allemand, G Srdanov, Z Ni, D McBranch. Polymers and an unusual molecular crystal with nonlinear optical properties. In SR Marder, J Sohn, GD Stucky, eds. Materials for Nonlinear Optics: Chemical Perspectives. ACS Symposium Series 455: 683, 1991.
31. Micelles, Microemulsions, and Monolayers: Science and Technology. D Shah, ed. New York: Marcel Dekker, 1998.
32. W Guo, H Uchiyama, EE Tucker, SD Christian, JF Scamehorn. Colloids Surf A 123–124:695–703, 1997.
33. A Shioi, M Harada, M Adachi. Recent Res Dev Phys Chem 3:149–182, 1999.
34. WJ MacKnight, Ponomarenko EA, DA Tirrell. Acc Chem Res 31:781–788, 1998.
35. AV Kabanov, TK Bronich, VA Kabanov, K Yu, A Eiseneberg. 120:9941–9942, 1998.
36. PS Kuhn, Y Levin, MC Barbosa. Chem Phys Lett 298:51–56, 1998.
37. AF Thuenemann. Adv Mater (Weinheim) 11:127–130, 1999.

38. K Faid, M Leclerc. J Am Chem Soc 120:5274–5278, 1998.
39. NS Sariciftci, L Smilowitz, AJ Heeger, F Wudl. Science (Washington, DC, 258: 5087, 1992, 1474–1476. B Kraabel, CH Lee, D McBranch, D Moses, NS Sariciftci, A Heeger. J Chem Phys Lett 213:3–4, 389–394, 1993.
40. TM Swager, CJ Gil, MS Wrighton. J Phys Chem 99:4886–4893, 1995.
41. MB Goldfinger, TM Swager. J Am Chem Soc 116:7895–7896, 1994.
42. Q Zhou, TM Swager. J Am Chem Soc 117:7017–7018, 1995.
43. PM Cotts, TM Swager, Q Zhou. Macromolecules 29:7323, 1996.
44. TM Swager. Acc Chem Res 31:201–207, 1998.
45. R Jones, D McBranch, D Whitten. (unpublished results).
46. TK Foreman, WM Sobol, DG Whitten. J Am Chem Soc 103:5333, 1981.
47. DG Whitten, JC Russell, RH Schmehl. Tetrahedron 38:2455, 1982.
48. CA Backer, DG Whitten. J Phys Chem 91:865, 1987.
49. CA Backer, JR Corvan-Cowdery, SP Spooner, B Armitage, GL McLendon, DG Whitten. J Surface Sci Technol 6:59, 1990.
50. L Chen, S Xu, D McBranch, D Whitten. J Am Chem Soc 122:9302–9303, 2000.
51. L Chen, D McBranch, R Wang, D Whitten. Chem Phys Lett 330:27–33, 2000.
52. G Decher. Science 277:1232–1237, 1997.
53. I Place, J Perlstein, TL Penner, DG Whitten. Langmuir 16:9042–9048, 2000.
54. J-Y Yang, TM Swager. J Am Chem Soc 120:11864–11873, 1998.
55. DT McQuade, AE Pullen, TM Swager. Chem Rev 100:2537–2574, 2000.
56. X Yang, J Shi, S Johnson, B Swanson. Langmuir 14:1505–1507, 1998.
57. S Ghafouri, M Thompson. Langmuir 15:564–572, 1999.
58. LM Torres-Rodriguez, A Toget, M Billon, G Bidan. Chem Commun 18:1993–1994, 1998.
59. VT Moy, E Florin, HE Gaub. Science 266:257–259, 1994.
60. O Livnah, EA Bayer, M Wilcher, JL Sussman. Proc Natl Acad Sci USA 90:5076–5080, 1993.
61. R-G Zhang, ML Westbrook, EM Westbrook, DL Scott, Z Otwinowski, PR Maulik, RA Reed, GG Shipley. J Mol Biol 550–559, 1995.

5

Luminescent Metal Complexes as Spectroscopic Probes of Monomer/Polymer Environments

Alistair J. Lees

State University of New York at Binghamton, Binghamton,
New York

I. INTRODUCTION

Industry utilizes a wide variety of technologies based on both photochemically and thermally initiated polymerization mechanisms. For instance, ultraviolet (UV)-curable coatings are used extensively in industries as diverse as microelectronics, packaging, and graphic arts, and they are also becoming increasingly employed in medicine and dentistry. Indeed, thin films comprised of acrylates are some of the most important photoresist materials used in the manufacture of printed circuit boards [1–3]. In addition, thermosetting materials such as epoxy resins offer an excellent combination of mechanical and electrical properties and were widely employed as adhesives, protective coatings, electrical mouldings, encapsulants, and as matrices for dielectrics in various electronic applications [4–7]. With any of these applications, though, it is highly desirable to know about the chemical and physical changes taking place in the polymer material throughout various stages of the curing reaction. Unfortunately, industrial procedures typically lack the practical methods necessary for obtaining this information and so the processing usually involves selecting the batch conditions for polymerization in an empirical manner.

Finding new ways to study the kinetics of the polymerization mechanism is, therefore, of considerable importance. Recently, several experimental techniques that are able to monitor polymerization have been used; these include UV-visible (UV-vis) fluorescence and Fourier-transform infrared (FTIR) spectroscopy, and differential scanning calorimetry (DSC) [8–11]. Whereas FTIR spectroscopy is commonly used to determine the extent of polymerization during online processing [9,12], its application, and that of UV-visible spectroscopy, is usually limited to thin films. Furthermore, while DSC is undoubtedly suitable for monitoring the overall extent of the cure reaction, it is a destructive technique and cannot be used to perform analysis *in situ*. In contrast, fluorescence spectroscopy offers some decided advantages over these other methods. A major one is that fluorescence measurements are nondestructive and so analysis can be performed on a variety of samples *in situ*. Also, fluorescence spectroscopy is a very sensitive technique and it is usually rather selective, enabling the emission characteristics of a particular chromophore to be distinguished within a mixture of absorbing and emitting species. The key to further advancing fluorescence spectroscopy as a method to monitor polymerization is the development of appropriate probe molecules that are sensitive to the physical and chemical changes occurring during the polymer cure. Currently, several fluorescence probes have been employed to monitor the curing kinetics of thermal polymerization [13–16], but there are only a few cases where spectroscopic probes have been developed that are able to monitor kinetic changes during photochemically initiated processes [17,18]. Unfortunately, the absorption and emission features of most molecules are in the UV and/or near-UV regions and they typically interfere with the spectral bands of photoinitiators used for light-induced polymerizations.

A number of directions have been taken to circumvent this problem. One of these utilizes a probe molecule that exhibits multiple fluorescence [19–24]. Here, both the fluorescence energy maximum and efficiency are influenced by the formation of the polymer network as a result of the nonradiative relaxation rates from the excited states being dependent on the rotational ability of the probe molecule. A second direction involves the application of a fluorescence probe combined with a suitable quencher molecule [25–27]. In these cases, the fluorescence is intensified as the microviscosity increases in the polymer network and the rate of energy transfer from the probe to the acceptor species is reduced. Third, the stabilization of excimers (generally napthalene, anthracene, or pyrene and related derivatives) during the polymerization mechanism has been used to generate a shift in the emission spectrum of the probe species [28–34]. Other instances of fluorescent probe molecules monitoring changes in the monomer/polymer environment have recently been reviewed [35].

However, in another approach, it is just becoming apparent that some types of luminescent metal complexes display spectroscopic properties that facilitate the monitoring of kinetic events during a variety of industrially important polymerization processes. These probes are derivatives of metal carbonyl complexes which exhibit low-energy phosphorescence bands that are sensitive to the viscosity changes occurring during polymer formation. In this chapter, we discuss these recent findings and focus on two types of organometallic complexes that act well as spectroscopic probes in monomer/polymer environments. These complexes are $W(CO)_4L$ and *fac*-ClRe$(CO)_3L$, where X is Cl, Br, or I, and L is an α,α'-diimine ligand such as 2,2'-bipyridine or 1,10-phenanthroline (and related derivatives). It is demonstrated how both these types of complexes (**1**) exhibit

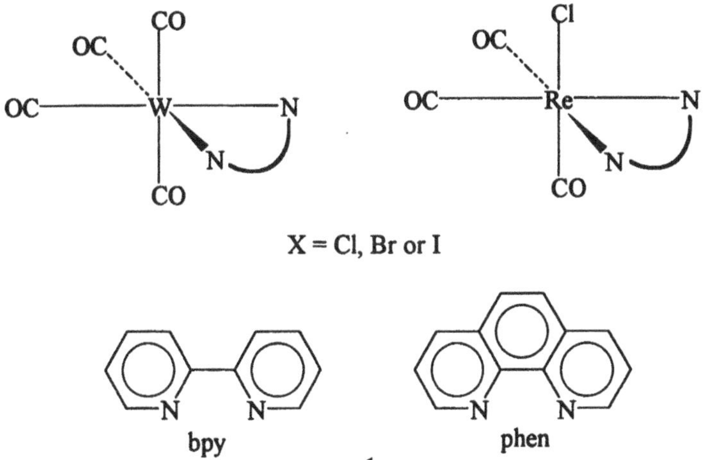

X = Cl, Br or I

bpy   phen

1

key spectroscopic features that enable them to be used as a probe in monitoring polymerization and why they offer some advantages over more conventional fluorescent organic probes.

II. ELECTRONIC AND RIGIDOCHROMIC PROPERTIES

A. *fac*-XRe(CO)₃(α,α'-diimine) Complexes

The photophysics of the *fac*-XRe$(CO)_3(\alpha,\alpha'$-diimine) system have received considerable attention in recent years with the observation that their lowest energy excited states exhibit readily detectable luminescence in the visible region in both frozen glassy and fluid solution environments. Indeed, the phenomenon of luminescence rigidochromism from organometallic complexes was first determined for a series of *fac*-ClRe$(CO)_3L$ (L = bpy, phen, 5-Me-phen, 4,7-Ph₂-phen, 5-Cl-phen, 5-Br-phen, 5-NO₂-phen, phen-5,6-dione, and biquin) derivatives [36]. Each of these ligands exhibits a vacant low-lying π^*-acceptor orbital and the complexes themselves display intense metal-to-ligand charge transfer (MLCT) electronic absorption bands. The luminescence from these complexes is strong and has been associated with the triplet manifold from these lowest energy MLCT excited states. Figure 1 illustrates typical electronic absorption and emission spectra for *fac*-ClRe(CO)$_3$(phen) in ether/isopentane/ethanol (EPA, 5:5:2 by volume) at both 298 and 77 K [36]. Clearly, the MLCT emission band is moved substantially to higher energy upon formation of the rigid glass, whereas the MLCT absorption band is shifted considerably less. Table 1 summarizes the emission characteristics for the above series of *fac*-ClRe(CO)$_3$L complexes [36]. Noticeably, both emission quantum yields (ϕ_e) and emission lifetimes (τ_e) are significantly increased when the solution is cooled to 77 K and forms a glassy matrix, implying that the radiative decay routes are substantially favored in the rigid environment. For each complex that luminescences in fluid solution, there is a distinct hypsochromic shift in the emission maximum (on the order of 940 to 1820 cm^{-1}) when the frozen glass is formed.

Although the MLCT emitting levels in organometallic complexes such as *fac*-XRe(CO)$_3$L are considered to be mainly of triplet character [37,38], it

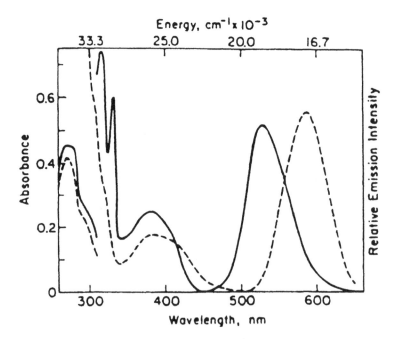

Figure 1 Electronic absorption (left) and emission (right) spectra of *fac*-ClRe(CO)$_3$(phen) in EPA at (---) 298 K and (——) 77 K. Room temperature absorption maxima are 26,100 cm^{-1} ($\varepsilon = 4000$ M^{-1} cm^{-1}) and 37,030 cm^{-1} ($\varepsilon = 30,600$ M^{-1} cm^{-1}); the low-temperature spectrum is not corrected for solvent contraction. Emission spectra at 298 and 77 K were not recorded at the same sensitivity. (From Ref. 36.)

Table 1 Emission Properties of *fac*-ClRe(CO)₃L Complexes[a]

L	Emission max. 10^{-3} cm^{-1}		τ_e, μs		ϕ_e ($\pm15\%$)[c]	ϕ_e ($\pm15\%$)
	298 K	77 K	298 K	77 K	298 K	77 K
phen	17.33	18.94	0.3	9.6	0.036	0.33
bpy		18.87	0.6	3.8	—	—
5-Me-phen	17.01	18.83	<0.65	5.0	0.030	0.30
4,7-Ph₂-phen	17.24	18.18	0.4	11.25	—	—
5-Cl-phen	17.12	18.69	<0.65	6.25	—	—
5-Br-phen	17.12	18.69	<0.65	7.6	0.020	0.20
5-NO₂-phen	[b]	18.28	—	11.8	—	0.033
phen-5,6-dione	[b]	18.45	—	2.5	—	—
biquin	[b]	14.58	—	—	—	—

[a]Measurements in EPA at 77 K or in CH₂Cl₂ at 298 K. (Data taken from Ref. 36.)
[b]Luminescence was not detectable from these complexes in solution at 298 K.
[c]Quantum yields determined in benzene at 298 K.

is recognized that the heavy metal in these compounds precludes a pure multiplicity designation [39]. The primary evidence for the triplet assignment in *fac*-ClRe(CO)₃(phen) arises from energy transfer studies, as the emission from the metal complex is quenched effectively via collisional energy transfer [36] to the triplet levels of anthracene [40] and *trans*-stilbene [41]. In the case of quenching by the stilbene, a *trans* → *cis* isomerization also occurs, further implicating that it is the triplet excited state of *trans*-stilbene which participates in the quenching mechanism [42]. Furthermore, the observed reaction quantum yields for *trans* → *cis* isomerization of the stilbene are essentially the same as those obtained for the benzophenone sensitization of *trans*-stilbene [41], revealing that the efficiency of intersystem crossing in these Re complexes is approximately unity [36,43].

A variety of spectroscopic methods has been used to determine the nature of the MLCT excited state in the *fac*-XRe(CO)₃L system. Time-resolved resonance Raman measurements of *fac*-XRe(CO)₃(bpy) (X = Cl or Br) have provided clear support for the Re → π* (bpy) assignment of the lowest energy excited state [44]. Intense excited-state Raman lines have been observed that are associated with the radical anion of bpy, and the amount of charge transferred from Re to bpy in the lowest energy excited state has been estimated to be 0.84 [45]. Fast time-resolved infrared spectroscopy has been used to obtain the vibrational spectrum of the electronically excited states of *fac*-ClRe(CO)₃(bpy) and the closely related *fac*-XRe(CO)₃(4,4′-bpy)₂ (X = Cl or Br) complexes. In each case, the spectra of the ^{3}MLCT excited state reveal a shift to higher frequency for the carbonyl-stretching bands compared to the ground state; this is consistent with oxidation of the metal center and electron transfer to the bipyridyl ligands [46–49].

Comparing all the energy shifts observed in the absorption and emission bands of fac-ClRe(CO)$_3$(α,α'-diimine) (see Fig. 1 and Table 1) leads one to conclude that it is mainly the ^{3}MLCT excited states which exhibit the rigidochromism and there is a considerably smaller effect on the corresponding ^{1}MLCT levels. Table 2 illustrates environmental effects on the MLCT absorption and emission maxima for several fac-ClRe(CO)$_3$L complexes in various media [36]. The data clearly reveal that the luminescence rigidochromism is associated with the rigidity change of the solution and it is not simply brought about by a temperature change. Each complex displays a substantial hypsochromic shift in its emission band (of up to 1830 cm^{-1}) as its environment becomes rigid. This is irrespective of the temperature as significant rigidochromic shifts are also observed in rigid polyester resins at room temperature.

Table 2 Environmental Effects on MLCT Absorption and Emission Maxima of fac-ClRe(CO)$_3$L Complexes[a]

L	Environment	Absorption max. 10^{-3} cm^{-1}	Emission max. 10^{-3} cm^{-1} (τ_e, μs)
phen	CH$_2$Cl$_2$, 298 K	26.53	17.33 (0.3)
	polyester resin, 298 K	—	18.52 (3.67)
	EPA, 77 K	—	18.94 (9.6)
5-Me-phen	benzene, 298 K	25.65	17.00 ($\leq$0.65)
	CH$_2$Cl$_2$, 298 K	26.32	17.01
	CH$_3$OH, 298 K	27.05	17.00
	pure solid, 298 K	—	18.42
	polyester resin, 298 K	—	18.48 (3.5)
	EPA, 77 K	—	18.83 (5.0)
5-Br-phen	benzene, 298 K	25.32	17.15 ($\leq$0.65)
	CH$_2$Cl$_2$, 298 K	25.84	17.12
	CH$_3$OH, 298 K	26.88	17.04
	pure solid, 298 K	—	17.83
	polyester resin, 298 K	—	18.32 (2.2)
	EPA, 77 K	—	18.69 (7.6)
5-Cl-phen	CH$_2$Cl$_2$, 298 K	25.91	17.12
	pure solid, 298 K	—	17.99
	EPA, 77 K	—	18.69 (6.25)

[a]Data taken from Ref. 36.

B. W(CO)$_4$(α,α'-diimine) Complexes

The electronic and photophysical properties of the M(CO)$_4$(α,α'-diimine) (M = Cr, Mo, or W) system have been studied extensively [50–63]. Electronic absorption spectra obtained from W(CO)$_4$(4-Me-phen) (4-Me-phen = 4-methyl-1,10-phenanthroline) in deoxygenerated benzene at 293 K and in glassy EPA solution at 80 K are depicted in Fig. 2. These spectra are typical of those observed from M(CO)$_4$(α,α'-diimine) complexes. The room-temperature absorption spectrum of W(CO)$_4$(4-Me-phen) is dominated by an intense MLCT band that is centered at 500 nm (ε_{max} = 9250 M^{-1} cm^{-1}), and weaker ligand field (LF) transitions are also seen at 391 nm (sh) and 340 nm (ε_{max} = 3550 M^{-1} cm^{-1}) [59]. Noticeably, it is the lowest energy MLCT absorption band of W(CO)$_4$(4-Me-phen) and those of a number of closely related derivatives that are especially solvent dependent [58–60,64–76]. It is important to note that the MLCT absorption band envelope substantially blue shifts, undergoes band sharpening, and reveals structure depicting two MLCT features when the solution is cooled and forms a frozen glass (see Fig. 2).

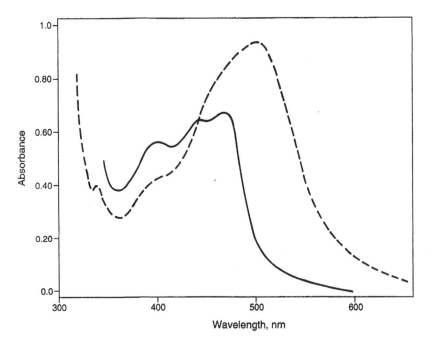

Figure 2   Electronic absorption spectra of W(CO)$_4$(4-Me-phen) in (—–) deoxygenated benzene at 293 K and (——) an EPA glass at 80 K. (From Ref. 60.)

Extensive resonance Raman (RR) and magnetic circular dichroism (MCD) measurements have been carried out on the $M(CO)_4(\alpha,\alpha'$-diimine) system and these indicate that the lowest energy absorption band actually comprises three MLCT transitions. The most intense one has been assigned to a z-polarized (d_{yz}) $b_2 \rightarrow b_2$ (π^*) transition that is directed along the dipole vector of the complex [61,63]. Moreover, the solvatochromism exhibited by these complexes has been attributed to the $b_2 \rightarrow b_2^*$ transition and, specifically, with the degree of mixing that occurs between the metal d_{yz} and ligand π^* orbitals. In the case of $W(CO)_4$(4-Me-phen), the extent of mixing between these metal and ligand orbitals is relatively small and the MLCT band is extremely solvent dependent [56,58–60,63,68]. By way of contrast, the degree of mixing of the corresponding orbitals is considerably lower in other types of α,α'-diimine species, such as in $W(CO)_4$(R-dab) and $W(CO)_4$(R-pyca) (R-dab = 1,4-diaza-1,3-butadiene and R-pyca = pyridine-2-carbaldehyde imine), and their solvent sensitivity is much less (**2**).

R-dab   R-pyca

2

RR excitation profiles from a series of $M(CO)_4(\alpha,\alpha'$-diimine) complexes have identified additional y-polarized $(d_{x^2-z^2})$ $a_1 \rightarrow b_2$ (π^*) and x-polarized (d_{xy}) $a_2 \rightarrow b_2$ (π^*) MLCT transitions in the absorption band envelope, although the latter component is relatively weak for substituted phen derivatives [63]. These absorption features are observable in the 80 K spectrum of $W(CO)_4$(4-Me-phen) in EPA (see Fig. 2). The most intense MLCT components in the low-temperature spectrum appear at 444 and 468 nm, and these are associated with the $a_1 \rightarrow b_2^*$ and $b_2 \rightarrow b_2^*$ transitions, respectively [60].

Some key photophysical observations have been made following analysis of emission spectra from the $M(CO)_4(\alpha,\alpha'$-diimine) complexes. Figure 3 illustrates spectra obtained from $W(CO)_4$(4-Me-phen) in deoxygenated benzene at 293 K and an EPA glass at 80 K [60]. Dual ^{3}MLCT emission bands are observed at either temperature; the spectra reveal a weak higher energy feature and a more intense lower energy band. Significantly, when the EPA solution is cooled from 293 to 150 K, the intensity ratio of these two bands remains virtually unaltered, but once the solution passes through the glass transition temperature (120 to 140 K) the overall emission intensity increases approximately 100-fold, the lower band becomes even more dominant, and both bands undergo a discernible blue shift. The luminescence rigidochromism phenomenon is particularly apparent for the lower energy band. Emission has also been detected

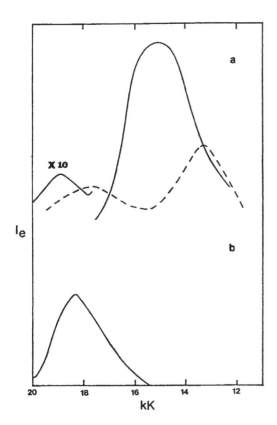

Figure 3 Emission spectra of (a) W(CO)$_4$(4-Me-phen) and (b) W(CO)$_4$(en) in (———)
deoxygenated benzene at 293 K and (———) EPA glasses at 80 K. Excitation wavelength
is 400 nm. The spectrum at 293 K is uncorrected whereas the spectra at 80 K are fully
corrected for wavelength variations in detector response (the emission intensity at low
temperature is approximately 100 times more intense than that at room temperature). No
luminescence was observed from W(CO)$_4$(en) at room temperature. (From Ref. 60.)

from W(CO)$_4$(en) (en = ethylenediamine) at 80 K and is included in Fig. 3
for comparison purposes; this spectrum reveals the close proximity of the ^{3}LF
excited state to the ^{3}MLCT levels. Table 3 summarizes the emission maxima
observed for a range of M(CO)$_4$(α,α'-diimine) (M = Mo or W) complexes at
both room temperature and low temperature.

The dual luminescence bands exhibited by the M(CO)$_4$(α,α'-diimine) complexes in room-temperature solution are unusual and have been further explored
by excitation wavelength studies [60]. The higher energy emission band is seen
to increase substantially in intensity as shorter exciting wavelengths are used

Table 3 Emission Maxima of $M(CO)_4(\alpha,\alpha'$-diimine) Complexes in Various Solutions[a]

| | λ_{em} nm | | |
| | Benzene | EPA | EPA |
Complex	293 K	293 K	80 K
$Mo(CO)_4$(4-Me-phen)	546, 752	553, 765	533, 647
$Mo(CO)_4$(5-Me-phen)	568, 764	572, 772	530, 658
$Mo(CO)_4$(4,7-Ph_2-phen)	582, 770	585, 785	545 (sh), 660
$Mo(CO)_4$(en)	b	b	b
$W(CO)_4$(4-Me-phen)	585, 782	584, >780	527, 677
$W(CO)_4$(4,7-Ph_2-phen)	595, 780	595, >800	550 (sh), 675
$W(CO)_4$(en)	b	b	542

[a]Excitation wavelength is 400 nm, and the spectra are fully corrected for wavelength variations in detector response; sh = shoulder. (Data taken from Ref. 60.)
[b]No emission observed.

(see Fig. 4). Following detailed emission and excitation studies, it has been rationalized that in fluid solution the ^{3}MLCT emitting levels undergo thermal equilibration. However, upon forming the frozen glass, the thermal equilibrium between the excited states is lost and the ^{3}MLCT states emit independently. As noted above, the ^{3}LF state also emits at 80 K and the overlap of emission from this level probably contributes to the reduced rigidochromic shift shown by the higher ^{3}MLCT energy band.

Additional photophysical properties of $W(CO)_4$(4-Me-phen) are revealed by a comparison of its luminescence parameters with those of other $M(CO)_4(\alpha,\alpha'$-diimine) complexes [63]. Figure 5 illustrates luminescence spectra obtained from $W(CO)_4L$ (L = 4,7-Ph_2-phen, i-Pr-pyca, and i-Pr-dab) in benzene at 293 K and 2-Me-THF (2-methyl-tetrahydrofuran) at 80 K. The $W(CO)_4$(4,7-Ph_2-phen) complex exhibits emission features similar to the $W(CO)_4$(4-Me-phen) derivative yet, in contrast, the room temperature spectra of the corresponding i-Pr-pyca and i-Pr-dab complexes are highly dependent on the excitation wavelength. Although single emission bands are observed for these latter complexes, they apparently comprise two emitting levels. The $W(CO)_4$(i-Pr-pyca) complex exhibits a low energy feature when the solution is cooled to 80 K; this resembles those from the bpy and phen complexes. In contrast, the emission spectrum of $W(CO)_4$(i-Pr-dab) undergoes hardly any change upon cooling to 80 K and still only displays a high energy feature.

The spectral variations in the $W(CO)_4L$ system have been attributed to changes in the MLCT character of the z-polarized ($b_2 \rightarrow b_2^*$) transition. In the

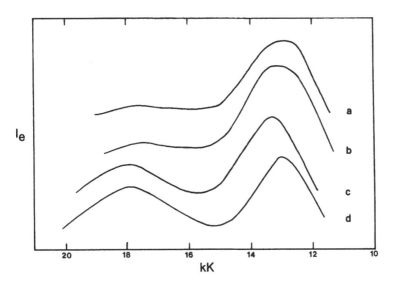

Figure 4   Emission spectra of W(CO)$_4$(4-Me-phen) in deoxygenated benzene at 293 K. Excitation wavelengths are (a) 475, (b) 450, (c) 400, and (d) 350 nm. The spectra are uncorrected for detector response; the lowest energy band is slightly blue-shifted on 400-nm excitation due to interference from solvent scatter. (From Ref. 60.)

dab complexes, the mixing of the metal d_{yz} and ligand π^* orbitals is much larger than in the bpy or phen derivatives, but in the pyca complexes the mixing appears to lie somewhat in between [63,77]. RR excitation profiles have clearly indicated that MLCT $b_2 \rightarrow b_2^*$ excitation predominantly influences the metal-skeletal vibrations of the R-dab complexes, whereas the internal vibrations of the α,α'-diimine ligands are the most affected in the bpy and phen derivatives. It is concluded that the R-dab complexes undergo efficient nonradiative decay and do not exhibit a lowest energy emission band because the metal-ligand skeletal modes most strongly affect the electronic integral and, subsequently, the matrix element connecting the ground and excited states [78]. In the case of the W(CO)$_4$(i-Pr-pyca), the RR excitation profiles indicate that the $b_2 \rightarrow b_2^*$ transition has an intermediate MLCT character. Indeed, the photophysical properties of W(CO)$_4$(i-Pr-pyca) are similar to the R-dab derivatives in room temperature solution. In rigid conditions at low temperature, though, the nonradiative deactivation processes are again reduced and the emission spectra are similar to the bpy and phen complexes [63]. Furthermore, solvatochromic studies of luminescence spectra have associated the solvent sensitivity of the M(CO)$_4$(α,α'-diimine) system with the z-polarized $b_2 \rightarrow b_2^*$ transition [63,79].

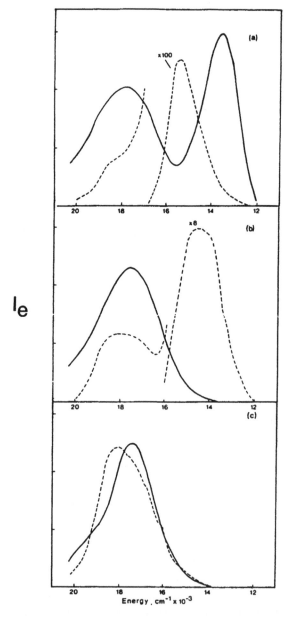

Figure 5 Emission spectra of (a) W(CO)$_4$(4,7-Ph$_2$-phen), (b) W(CO)$_4$(i-Pr-pyca), and (c) W(CO)$_4$(i-Pr-dab) in (——) benzene at 293 K and (–––) a 2-Me-THF glass at 80 K. Excitation wavelength is 488 nm. (From Ref. 63.)

III. THERMOSETTING MATERIALS

Metal complexes have been used as spectroscopic probes in several thermosetting resins, including various epoxy/anhydride mixtures. The systems investigated (**3**) have consisted of the cycloaliphatic resin, (3,4-epoxycyclohexyl)methyl-3,4-

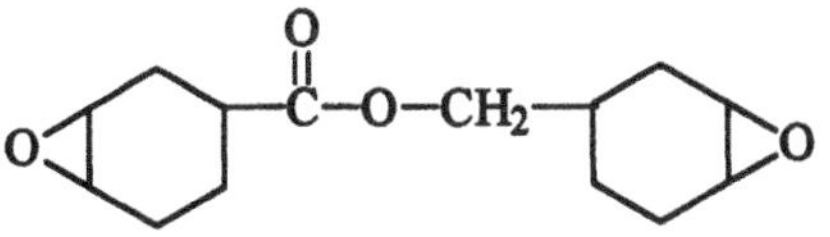

cycloaliphatic epoxy

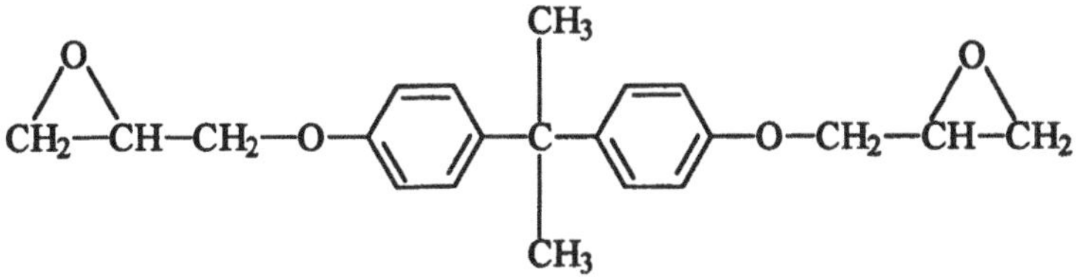

3

diglycidyl ether of bisphenol-A (DGEBA)

epoxycyclohexylcarboxylate (Union Carbide, ERL-4221), or the diglycidyl ether of bisphenol-A (DGEBA) (Ciba-Geigy, Aratronics® 5001), combined with *cis*-cyclohexanedicarboxylic anhydride and N-benzyldimethylamine as the curing agent [80,81]. Spectral measurements from these epoxy mixtures were typically obtained from samples of 1 mm thickness following a curing process where the resins were heated at 393 K.

Spectroscopic data observed from *fac*-ClRe(CO)$_3$L complexes in the thermosetting cycloaliphatic epoxy (ERL-4221)/anhydride resin reveal the probe applications of this system [80,81]. Significantly, these organometallic complexes are sufficiently soluble to facilitate their use as spectroscopic probes in the resin. In contrast, most transition-metal complexes are charged species and not readily soluble in organic media. In addition, it is important that these metal complexes are thermally stable, and show no significant extent of thermal degradation during heating and effecting the polymer cure. Typically, the metal carbonyl complexes only need to be added in small amounts (0.008 to 0.01%, by weight) to the resin materials.

Luminescence spectra recorded from the probe complex, *fac*-ClRe(CO)$_3$ (4,7-Ph$_2$-phen) (4,7-Ph$_2$-phen = 4,7-diphenyl-1,10-phenanthroline) in the cycloaliphatic epoxy (ERL-4221)/anhydride resin are shown in Fig. 6. Three stages of the polymerization reaction are depicted: before curing, at an intermediate stage of the cure, and at the completion of the cure. Two emission bands are clearly seen prior to heating at 393 K; these are centered at 480 and 610 nm. Once the curing reaction proceeds, the relative intensity of the higher energy feature diminishes compared to the lower energy band. A combination of both fluorescence and scattered light has been attributed to the higher energy band as it apparently arises from the resin itself; the energy position of the emission band depends on the excitation wavelength and the emission is observed even in the absence of any added organometallic probe. On the other hand, the lower energy band is only detected when the rhenium complex is present and it is assigned to the well-characterized MLCT emission of this organometallic species [36–38]. Significantly, this emission feature increases approximately 10-fold in intensity and it shifts to 556 nm following polymerization, representing a hypsochromic shift (ΔE_{em}) of 1592 cm^{-1}.

Electronic absorption and luminescence spectra have been recorded for several other closely related fac-XRe(CO)$_3$L complexes in both methylene chlo-

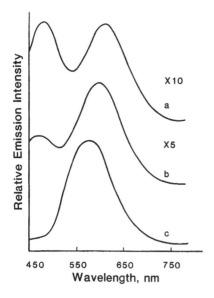

Figure 6 Emission spectra of fac-ClRe(CO)$_3$(4,7-Ph$_2$-phen) in the cycloaliphatic epoxy (ERL-4221)/anhydride resin at 293 K: (a) before curing, (b) partially cured by heating at 393 K for 30 min, and (c) cured by heating at 393 K for 60 min. Excitation wavelength is 400 nm. Intensities of spectra (a) and (b) have been scaled ×10 and ×5, respectively. (From Ref. 80.)

ride solution and the cycloaliphatic epoxy (ERL-4221)/anhydride resin [80,81]. In each case, a hypsochromic shift is observable as the polymer network is formed; these shifts range from 764 to 1633 cm^{-1}, depending on which of the probe complexes are incorporated (see Table 4). These hypsochromic energy shifts are brought about by the formation of a rigid network around the organometallic molecule and are attributed to the rigidochromic effect [36,76]. It is noticeable that the iodo derivative displays substantially lower luminescence rigidochromism ($\Delta E_{em} = 764$ cm^{-1}) than the other complexes. Recently, it has been demonstrated that in the case of the iodo species the lowest energy excited state is mainly of X $\rightarrow$ L charge transfer character, rather than a MLCT transition [82] and, consequently, the change in dipole moment from the ground to the excited state will be quite different from the other complexes.

Luminescence excitation spectra have been determined from the rhenium complexes in the epoxy resin while monitoring emission at the ^{3}MLCT maxima and a typical spectrum is shown in Fig. 7. Analogous excitation spectra have been recorded for *fac*-ClRe(CO)$_3$(4,7-Ph$_2$-phen) in deoxygenated methylene chloride solution, confirming the ^{3}MLCT assignment of the lowest energy emission band. Emission quantum yields and emission lifetimes have also been obtained for a series of *fac*-XRe(CO)$_3$L complexes and these results are shown in Table 5. Photophysical deactivation rate constants have been derived accord-

Table 4 Electronic Absorption and Emission Maxima of the MLCT Bands Observed from *fac*-XRe(CO)$_3$L Complexes in Deoxygenated Methylene Chloride and the Cycloaliphatic Epoxy (ERL-4221)/Anhydride System at 293 K[a]

		λ_{em} nm			
Complex	λ_{abs} nm CH$_2$Cl$_2$	CH$_2$Cl$_2$	Uncured epoxy	Cured[b] epoxy	ΔE_{em}[c] cm^{-1}
fac-ClRe(CO)$_3$(phen)	378	591	592	543	1,524
fac-ClRe(CO)$_3$(4-Me-phen)	372	587	575	543	1,025
fac-ClRe(CO)$_3$(4,7-Ph$_2$-phen)	381	598	610	556	1,592
fac-BrRe(CO)$_3$(4,4'-Me$_2$-bpy)	382	587	585	534	1,633
fac-IRe(CO)$_3$(4,7-Ph$_2$-phen)	410	602	608	581	764

[a]Emission spectra are uncorrected for wavelength variations in photomultiplier response; excitation wavelength is 400 nm. (Data taken from Ref. 81.)
[b]Cured by heating at 393 K for 180 min; observed emission intensity is approximately 10-fold greater than uncured sample.
[c]Energy difference between observed emission bands of the organometallic probe complexes in uncured and cured epoxy samples.

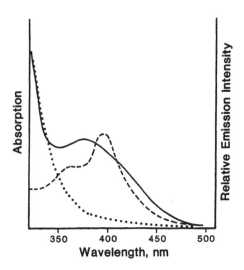

Figure 7 Electronic absorption (——) and excitation (– – –) spectra of *fac*-ClRe(CO)₃
(4,7-Ph₂-phen) in the cycloaliphatic epoxy (ERL-4221)/anhydride resin at 293 K. The
electronic absorption spectrum (●●●●) of the uncured epoxy/anhydride resin without the
incorporation of the organometallic complex is also shown. The excitation spectrum was
recorded with emission monitored at 610 nm. (From Ref. 80.)

Table 5 Photophysical Parameters for *fac*-XRe(CO)₃L Complexes in the Uncured
and Cured Cycloaliphatic (ERL-4221)/Anhydride Epoxy System at 293 K[a]

Complex	Cure state[b]	ϕ_e[c]	τ_e ns	k_r s^{-1}	k_{nr} s^{-1}
fac-ClRe(CO)₃(4,7-Ph₂-phen)	uncured	0.042	280	1.5×10^5	3.4×10^6
	cured	0.44	4100	1.1×10^5	1.3×10^5
fac-ClRe(CO)₃(4-Me-phen)	uncured	0.039	306	1.3×10^5	3.1×10^6
	cured	0.56	2360	2.4×10^5	1.8×10^5
fac-IRe(CO)₃(4,7-Ph₂-phen)	uncured	0.048	1049	4.6×10^4	9.1×10^5
	cured	0.56	8100	6.9×10^4	5.4×10^4
fac-BrRe(CO)₃(4,4′-Me₂-bpy)	uncured	0.035	77	4.5×10^5	1.3×10^7
	cured	0.37	600	6.2×10^5	1.0×10^6

[a]Excitation wavelength is 400 nm. (Data taken from Ref. 81.)

[b]Cured samples were heated at 393 K for 180 min.

[c]Obtained from corrected emission spectra using emission quantum yield of *fac*-ClRe(CO)₃(phen)
in deoxygenated CH₂Cl₂ as a calibrant [36].

ing to Eqs. (1) and (2), again with the assumption that the emitting state is formed with unity efficiency [36–38]. It is important to note that in each probe complex the nonradiative decay constant (k_{nr}) is decreased by at least an order of magnitude, whereas the radiative rate constant (k_r) is relatively unchanged. It is concluded that k_{nr} of the probe molecule decreases as polymer viscosity increases during the polymerization and the free volume is reduced [83,84].

$$k_r = \phi_e / \tau_e. \tag{1}$$

$$k_{nr} = (1/\tau_e) - k_r. \tag{2}$$

Figure 8 illustrates the energy position of the MLCT emission maximum from fac-ClRe(CO)$_3$(4,7-Ph$_2$-phen) at various stages of the cure reaction in the cycloaliphatic epoxy/anhydride resin. The spectra indicate that the energy position of the emission maximum slowly changes at the beginning of the sequence, but it then rises sharply after 20 min of heating to reach a plateau after about 60 to 80 min of heating. In separate measurements, dynamic mechanical analyses of the resin have been carried out in a parallel plate geometry and the resulting stress amplitude and phase angle have been obtained, leading to a determination of the dynamic moduli (G' and G''). These are measures of the stress/strain ratio during deformation [85]. The complex shear modulus and viscosity have also been derived, according to Eqs. (3) and (4), where G* is the complex shear

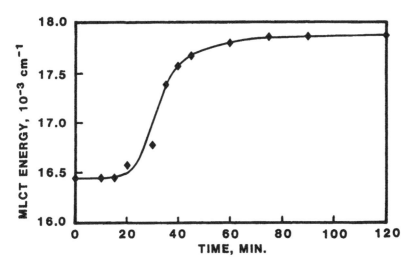

Figure 8 Energy position of the ^{3}MLCT emission band maximum of fac-ClRe(CO)$_3$(4,7-Ph$_2$-phen) in the cycloaliphatic epoxy (ERL-4221)/anhydride resin as a function of time at the isothermal temperature of 393 K. The spectra were recorded at 293 K following excitation at 400 nm. (From Ref. 81.)

modulus, G' is the storage or elastic modulus, G'' is the loss modulus, η^* is the complex viscosity, and ω represents the angular frequency in rad/s.

$$G^* = [(G')^2 + (G'')^2]^{1/2}. \tag{3}$$

$$\eta^* = G^*/\omega. \tag{4}$$

Figure 9 illustrates the dynamic moduli and complex viscosity results for the cycloaliphatic epoxy/anhydride resin. The t_{gel} value denotes the time to gelation based on the dynamic moduli crossover ($G' = G''$) and is estimated at

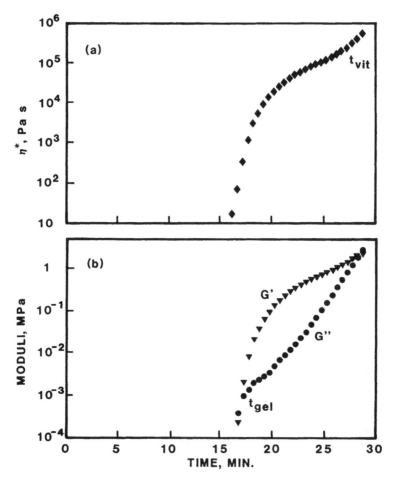

Figure 9 (a) Complex viscosity (η^*) and (b) dynamic moduli (G' and G'') data in the cycloaliphatic epoxy (ERL-4221)/anhydride resin as a function of cure time at 393 K. Here t_{gel} and t_{vit} represent the times to gelation and vitrification, respectively. (From Ref. 81.)

18 min [86–88]. The t_{vit} value represents the time to vitrification based on the second inflection point in the η^* data and this occurs at approximately 28 min.

The emission band maximum of fac-ClRe(CO)$_3$(4,7-Ph$_2$-phen) is clearly sensitive to the changing physical properties of the cycloaliphatic epoxy/anhydride material. During the curing process, the epoxy resin forms a three-dimensional crosslinked network of increasing molecular weight and the resin's viscosity increases by approximately five orders of magnitude from about 10 Pa s to almost 10^6 Pa s (see Fig. 9a). Concomitantly, when the epoxy resin reaches its final cure point the energy position of the ^{3}MLCT emission band of the probe complex undergoes a substantial blue shift (up to 1633 cm^{-1}; see Table 4). The emission data reveal an S-shaped dependence (see Fig. 8) that occurs in three stages. On heating there is initially no change in the ^{3}MLCT band position, this is subsequently followed by a sharp increase in the energy of the emission band accompanying the onset of the gelation, and then the changes ultimately become more gradual and a plateau is reached after vitrification has taken place. From the results in Fig. 8 it can be estimated that t_{gel} is in the range 18 to 25 min (indicated by the sharp change in slope of emission energy against cure time) and t_{vit} is at about 40 to 45 min (from the second slope change on the emission curve). These values are rather longer than the data obtained from the dynamic mechanical analyses but one needs to realize that there is a lag time in the emission readings compared to the rheological measurements.

Analogous results have been obtained for the fac-ClRe(CO)$_3$(4,7-Ph$_2$-phen) probe in related room temperature curing epoxy systems involving diglycidyl ether of bisphenol A (DGEBA) [81]. Measurements of ^{3}MLCT emission intensity and η^* from DGEBA resins containing the probe complex with either polymercaptan or amine curing agents are shown in Fig. 10. In these experiments the cure reaction was monitored *in situ* by determining the emission intensity at a constant wavelength. It is found that the ^{3}MLCT emission intensity and complex viscosity data display similar behaviors with the curing time in both the DGEBA/polymercaptan and DGEBA/amine resins.

A series of temperature-dependence measurements have been performed to explore the close relationship between the ^{3}MLCT energy of fac-ClRe(CO)$_3$(4,7-Ph$_2$-phen) and η^* of the cycloaliphatic epoxy/anhydride resin [81]. Figure 11 depicts data recorded for both the ^{3}MLCT emission energy position of the probe complex and η^* of the neat cycloaliphatic epoxy resin (without anhydride or added amine curing agent) at temperatures between 210 and 300 K. The viscosity of the resin material rises by over five orders of magnitude and the probe emission blue shifts substantially as the epoxy molecules freeze; the results again exhibit a reversed S-shaped function. Clearly, there is an excellent correlation between the energy maximum of the MLCT emission and the η^* data; indeed, this correlation exists for over five orders of magnitude variation in η^* (see Fig. 12). This is a significant finding because it shows that the rigidochromic shift of the probe complex can be applied in monitoring kinetic

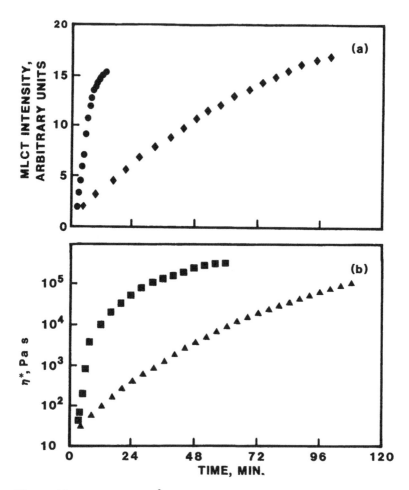

Figure 10 (a) Intensity of ^{3}MLCT emission at 600 nm of *fac*-ClRe(CO)$_3$(4,7-Ph$_2$-phen) in the (●) DGEBA/polymercaptan and (◆) DGEBA/amine epoxy resins, and (b) complex viscosity (η^*) data from (■) DGEBA/polymercaptan and (▲) DGEBA/amine epoxy resins as a function of time at the isothermal cure temperature of 293 K. Emission data recorded at 293 K following excitation at 400 nm. (From Ref. 81.)

changes within the monomer/polymer environment for a large fraction of reaction conversion.

Another type of film investigated recently has been aromatic cyanate esters prepared from the Ciba-Geigy products AroCy® L-10 (1,1-bis(4-cyanatophenyl) ethane) and AroCy® B-30 (the prepolymer form of 2,2′-bis(4-cyanatophenyl) isopropylidene, AroCy® B-10) [89]. The chemical structures of these molecules are shown in **4**. Spectral measurements from these aromatic cyanate ester ma-

NCO—⟨benzene⟩—C(CH₃)(H)—⟨benzene⟩—OCN

AroCy® L-10

NCO—⟨benzene⟩—C(CH₃)(CH₃)—⟨benzene⟩—OCN

AroCy® B-10

4

terials were obtained from samples of approximately 1 mm thickness. When *fac*-ClRe(CO)₃(4,7-Ph₂-phen) is incorporated in these cyanate ester resins, its emission band appears weakly at 580 nm (the band is also at 580 nm in a methylene chloride solution). Following thermal cure of the cyanate ester film, the emission band becomes greatly intensified and shifts to 565 nm. The complex apparently acts well as a spectroscopic probe for cyanate ester polymerization. On curing, cyanate esters crosslink to form triazine-based networks that display excellent thermal stability and good adhesion to metals [90]. They also show good resistance to solvents, low dielectric constants, and high glass transition temperatures, with somewhat lower thermal expansion properties than epoxide materials [90–96]. These materials are, consequently, being used increasingly as resists in photolithographic applications for the microelectronics industry and as composites in the aerospace industry [97].

An alternative procedure has been taken by other authors in that the curing agent itself has been monitored spectroscopically [98]. Fluorescence emission and excitation spectra of the commonly employed aromatic diamine during agent 4,4′-diaminodiphenyl sulfone (DDS) have been obtained in various epoxies, including diglycidyl ether of bisphenol A (DGEBA, **3**) and diglycidyl ether of butanediol (DGEB, **5**). Red shifts of approximately 25 nm are observed in both

H_2C—CH—CH_2—O—$(CH_2)_4$—O—CH_2—CH—CH_2

5

diglycidyl ether of butanediol (DGEB)

the fluorescence and excitation spectra as the primary amine groups in DDS are converted to tertiary amine groups. It is noticeable that the excitation spectra yield somewhat sharper bands and this helps the monitoring process, especially during the later stages of the cure reaction. Figure 13 illustrates representative fluorescence excitation spectra for the DGEB/DDS epoxy system as a function of cure time. Plots of the excitation spectral peak position of DDS as a function of the cure time in the DGEB/DDS and DGEBA/DDS epoxy systems are shown for three different curing temperatures (see Fig. 14).

FTIR spectroscopy has also been used to examine the extent of the curing reaction in the DGEB/DDS and DGEBA/DDS epoxy systems [98]. Here, the decline in the infrared band at 910 cm^{-1}, representing the epoxide ring, has been

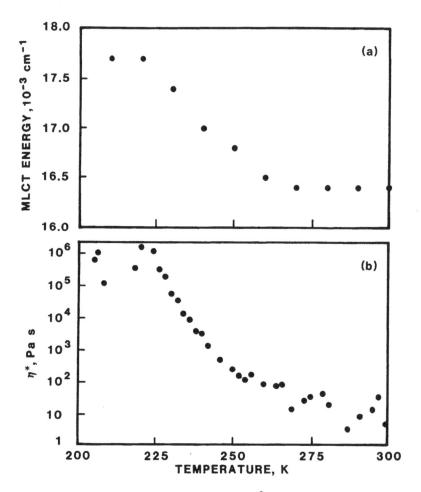

Figure 11 (a) Energy position of the uncorrected ^{3}MLCT emission band maximum of *fac*-ClRe(CO)$_3$(4,7-Ph$_2$-phen) and (b) complex viscosity (η^*) changes as a function of temperature in the neat cycloaliphatic epoxy (ERL-4221) resin. The excitation wavelength is 400 nm for the data in (a). (From Ref. 81.)

followed as a function of the cure time. These results are depicted in Fig. 15 and it can be observed that the changes exhibit similar behavior to the corresponding fluorescence excitation spectra (see Fig. 14). Recorded fluorescence spectra also exhibit analogous changes, but these spectra are not as clearly resolved and, subsequently, the correlations with the infrared data are not as good [98].

In addition, the luminescence features of DDS have been demonstrated to be useful for spectroscopically monitoring epoxide curing reactions [99].

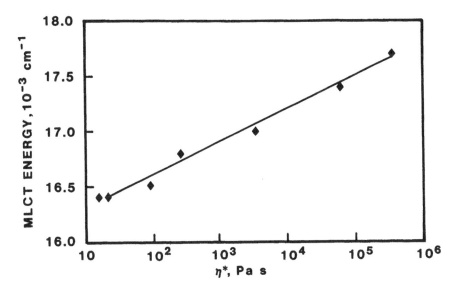

Figure 12 Plot of the energy position of the ^{3}MLCT emission band maximum of *fac*-ClRe(CO)$_3$(4,7-Ph$_2$-phen) against determined complex viscosity (η^*) values for the neat cycloaliphatic epoxy (ERL-4221)/anhydride resin. (From Ref. 81.)

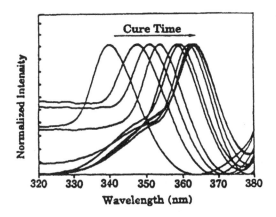

Figure 13 Fluorescence excitation spectra of the DGEB/DDS epoxy system as a function of cure time at 433 K. Cure times are 0, 10, 20, 30, 46, 60, 90, 120, 150, 180, and 240 min (from left to right). Emission was monitored at 390 nm. (From Ref. 98.)

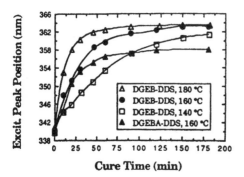

Figure 14 Plots of the excitation spectral peak position for the DGEB/DDS and DGEBA/DDS epoxy systems as a function of cure time at various cure temperatures. (From Ref. 98.)

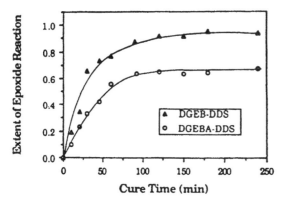

Figure 15 A comparison of the change in the extent of the epoxide reaction as a function of cure time for the DGEB/DDS and DGEBA/DDS epoxy systems. (From Ref. 98.)

Figure 16 depicts emission spectra obtained from the DGEBA/DDS epoxy system as a function of the cure time. Before curing, there is very little phosphorescence observed from DDS because the long-lived triplet state is effectively quenched by oxygen as diffusion occurs readily in the low-viscosity matrix. During curing, though, the intensity of this phosphorescence band increases dramatically and red shifts slightly, as radiative decay from the triplet state of DDS in the tertiary amine form becomes more favorable. Figure 17 illustrates a plot of the intensity of the observed phosphorescence at 500 nm as a function of the cure time in the DGEBA/DDS epoxy system. This graph reveals that the

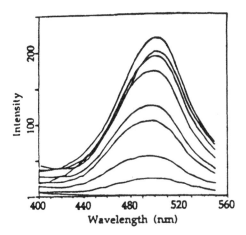

Figure 16 Phosphorescence spectra of the DGEBA/DDS epoxy system as a function of cure time at 433 K. Cure times are 0, 10, 20, 31, 46, 60, 120, 150, and 180 min (from bottom to top). Excitation wavelength is 330 nm. (From Ref. 99.)

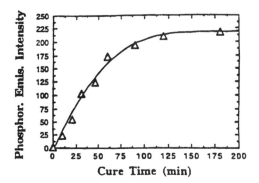

Figure 17 Plot of the phosphorescence intensity at 500 nm for the DGEBA/DDS epoxy system as a function of cure time at 433 K. (From Ref. 99.)

phosphorescence behavior closely parallels that shown in the above fluorescence excitation spectra (see Fig. 14).

IV. PHOTOSENSITIVE MATERIALS

The fac-ClRe(CO)$_3$(α,α'-diimine) system is also an excellent spectroscopic probe for acrylate polymerization. Photosensitive acrylate resins investigated were constituted by mixing medium weight poly(methyl methacrylate) (PMMA) with trimethylolpropane triacrylate (TMPTA) and 2,2'-dimethoxy-2-phenylaceto-phenone photoinitiator (Ciba-Geigy, Irgacure® 651) [100,101]. The spectral measurements from photosensitive samples were obtained from films of 0.254 mm thickness that were coated on a polyester sheet. A 350 W mercury lamp with an intensity of 3.5 mW/cm^2 was used for light exposure in these thin films. Emission spectra obtained from a 0.25-mm thin film of 1:1 (by weight) TMPTA/PMMA containing the metal complex (0.01%, by weight) and the 2,2'-dimethoxy-2-phenylacetophenone photoinitiator (1%, by weight) are shown in Fig. 18, revealing substantial changes upon UV light exposure. Throughout acrylate crosslinking the intensity of the MLCT probe emission increases substantially and the band shifts from 576 nm (prior to irradiation) to 562 nm (after 120 s of UV irradiation), corresponding to a hypsochromic shift of 432 cm^{-1}. The emission lifetime of the complex has also been measured in these acrylate thin

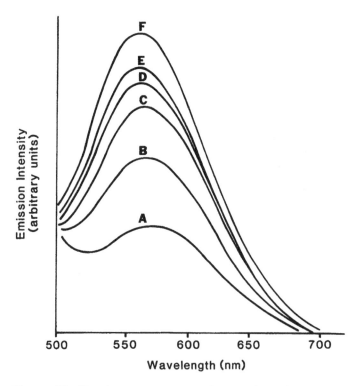

Figure 18 Phosphorescence spectra of a 1:1 (by weight) acrylate-based TMPTA/PMMA 0.25-mm thin film containing an acetophenone photoinitiator and *fac*-ClRe(CO)$_3$(4,7-Ph$_2$-phen) as a function of UV exposure: (A) 0 s, (B) 5 s, (C) 10 s, (D) 20 s, (E) 30 s, and (F) 60 s. Excitation wavelength is 420 nm. (From Ref. 100.)

films; before UV excitation the emission lifetime is 0.85 μs and this length-ens to 2.5 μs after light irradiation for 120 s [100]. These spectral changes are entirely in accordance with a decrease in the nonradiative decay rate (k_{nr}) of the MLCT excited state during the acrylate polymerization. This photophysical effect is understood to relate to the changes in rotational and vibrational relaxation pathways as the polymer becomes more rigid and the matrix free volume is reduced [100–102].

Previous studies on paraffins, rhodamine dyes, and 1,3-bis(N-carbozoyl) propane excimers have concluded that there is a relationship between k_{nr} and polymer viscosity and free volume [103–105]. Indeed, this dependence has been investigated in the context of decreasing free volume during methyl methacrylate polymerization [83,84]. It has been shown that the nonradiative decay processes follow an exponential relationship with polymer free volume (v_f), in which k_{nr} reduces as free volume is decreased [see Eq. (5)]. Here, k_{nr}° represents the intrinsic rate of molecular nonradiative relaxation, v_0 is the van der Waals volume of the probe molecule, and b is a constant that is particular to the probe species. Clearly, the experimentally observed changes in both emission intensity and lifetime for fac-ClRe(CO)$_3$(4,7-Ph$_2$-phen) in the TMPTA/PMMA thin film are entirely consistent with this rationale.

$$k_{nr} = k_{nr}^\circ \exp(-v_0/bv_f). \tag{5}$$

A comparison of the changes in the emission intensity with infrared attenuated total reflectance (ATR) spectra obtained from the TMPTA/PMMA thin film is informative (see Fig. 19) [100]. There is a substantial increase in the MLCT emission band intensity immediately upon UV light excitation of the TMPTA/PMMA photosensitive film. After 60 s of irradiation, no further enhancement of the probe luminescence occurs. In the absence of added photoinitiator, however, the emission intensity of the probe molecule in the film was observed to remain constant, confirming that the metal complex does not itself initiate polymerization of the film or undergo significant photodegradation. Analogous behavior is seen in the infrared ATR results; immediately after UV light exposure there is a substantial decrease in the integrated area of the infrared band at 808 cm^{-1}, representing the CH$_2$ wagging vibration of the residual acrylate monomer. Taken together, the emission and infrared ATR spectra reveal that there is a simultaneous increase in the ^{3}MLCT emission intensity from the probe complex as the acrylate monomer is consumed.

Table 6 summarizes the spectral data that have been recorded with various compositions of TMPTA/PMMA. It is apparent that the magnitude of the rigidochromic shift (ΔE_{em}) is significantly reduced when the starting mixture has a higher polymer content. Similarly, the ratio (I_f/I_0) of the final (I_f) and initial (I_0) intensities of the MLCT phosphorescence is considerably reduced with a higher PMMA content in the unirradiated resin. Both these effects appear to relate to

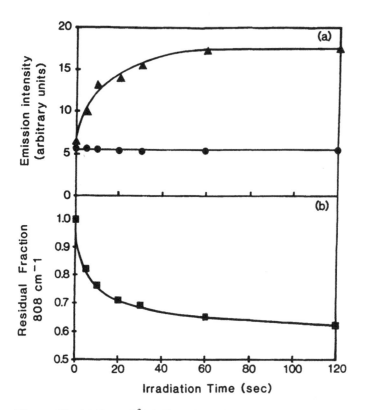

Figure 19   (a) Plots of ^{3}MLCT emission intensity (recorded at the maximum) of *fac*-ClRe(CO)$_3$(4,7-Ph$_2$-phen) in the 1:1 (by weight) acrylate-based TMPTA/PMMA 0.25-mm thin film as a function of UV-irradiation time: (▲) with and (●) without added acetophenone photoinitiator. Excitation wavelength is 420 nm. (b) Plot of normalized area of acrylate monomer infrared ATR band at 808 cm^{-1} as a function of UV-irradiation time in the TMPTA/PMMA thin film. (From Ref. 100.)

the viscosity of the material. When the ratio of PMMA to TMPTA is increased, the viscosity of the unirradiated resin mixture is also raised and, consequently, the overall viscosity change upon curing will be less.

The W(CO)$_4$(4-Me-phen) complex also displays a number of spectroscopic features that lend themselves to utility as a probe [106]. Figure 20 depicts electronic absorption spectra recorded from a 0.25-mm thickness film of a photosensitive acrylate system comprising a 1:1 (by weight) ratio of TMPTA and medium weight PMMA. Incorporated are the photoinitiators benzophenone (4% by weight) and 4,4′-bis(dimethylamino)benzophenone (0.5% by weight), and the probe complex W(CO)$_4$(4-Me-phen) (0.3% by weight). The major absorption feature in this acrylate film is observed at 353 nm and is due to the photoini-

Table 6 Emission Data of *fac*-ClRe(CO)$_3$(4,7-Ph$_2$-phen) in the Photosensitive Acrylate (TMPTA/PMMA) System as a Function of Resin Composition[a]

| TMPTA:PMMA ratio (wt%) | λ_{em} nm | | $\Delta E_{em}{}^c$ cm^{-1} | $I_f/I_o{}^d$ |
	Uncured resin	Cured[b] resin		
60:40	567	562	160	2.1
50:50	576	562	432	2.5
40:60	580	561	584	4.6
30:70	583	561	672	6.9

[a]Emission maxima are uncorrected for photomultiplier response; excitation wavelength is 420 nm. (Data taken from Ref. 100.)
[b]Following UV-light exposure for 120 s.
[c]Energy difference of the MLCT band maxima in the uncured and cured materials.
[d]Ratio of the initial (I_o) and final (I_f) emission intensities recorded at the MLCT band maxima.

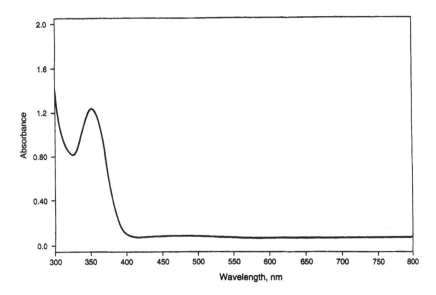

Figure 20 Electronic absorption spectrum of a 1:1 (by weight) TMPTA/PMMA 0.25-mm thin film containing benzophenone photoinitiators and W(CO)$_4$(4-Me-phen) at 293 K. (From Ref. 106.)

tiators present. However, the organometallic complex is incorporated in such a low concentration that its normally intense MLCT absorption transitions are barely detectable (A < 0.1) in the visible region between 400 and 550 nm. Significantly, this spectrum was found to not change following UV-light irradiation of the acrylate thin film, indicating that the metal complex does not appreciably photodegrade during the polymerization. Although $W(CO)_4$(4-Me-phen) exhibits photoreactive LF states in the UV region, these absorptions are weak and completely masked by the intense photoinitiator absorption bands.

Figure 21 depicts emission spectra recorded from the same composition of TMPTA/PMMA thin film at 293 K following excitation at 400 nm. Prior to UV irradiation, the characteristic dual ^{3}MLCT emission features of $W(CO)_4$-(4-Me-phen) are observable at 520 and 750 nm, and the spectrum is similar to that determined in fluid solution (see Fig. 3). However, after UV irradiation for 60 s the two ^{3}MLCT emission bands become almost equivalent in intensity and their maxima have moved to 525 and 715 nm, respectively. Without addition of the $W(CO)_4$(4-Me-phen) complex, the unirradiated thin film gives rise to no detectable emission in the 450 to 800-nm region. Following UV irradiation, however, substantial scattered light is observed in the 420 to 620-nm region.

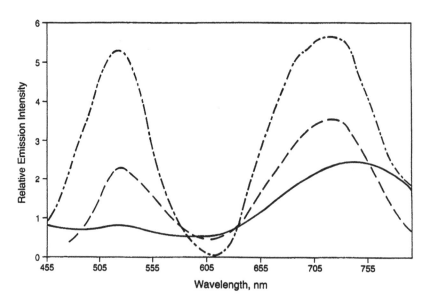

Figure 21 Emission spectra of a 1:1 (by weight) TMPTA/PMMA 0.25-mm thin film containing benzophenone photoinitiators and $W(CO)_4$(4-Me-phen) following (———) 0 s, (–––) 10 s, and (- - -) 60 s UV exposure at 293 K. Excitation wavelength is 400 nm in each case. (From Ref. 106.)

Consequently, the increase in intensity of the short-wavelength emission component is attributed to both the ^{3}MLCT emission and increased scattered light from the polymer film surface. On the other hand, the long-wavelength emission band is associated solely with the ^{3}MLCT emission from the W(CO)$_4$(4-Me-phen) complex.

Figure 22 illustrates observed changes in intensity of the long-wavelength emission band of W(CO)$_4$(4-Me-phen) in the TMPTA/PMMA thin film after various periods of light exposure. There is a steep increase in the emission intensity immediately following irradiation, followed by a more gradual change,

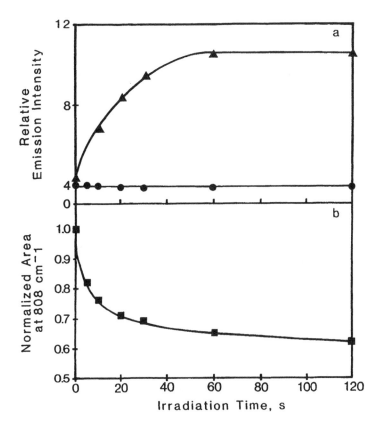

Figure 22 (a) Plots of emission intensity at 715 nm of W(CO)$_4$(4-Me-phen) in a 1:1 (by weight) TMPTA/PMMA 0.25-mm thin film as a function of UV-irradiation time (▲) with and (●) without the benzophenone photoinitiators. Excitation wavelength is 400 nm in each case. (b) Plot of the normalized area of acrylate monomer vibration at 808 cm^{-1} in a 1:1 (by weight) TMPTA/PMMA 0.25-mm thin film as a function of UV irradiation time. (From Ref. 106.)

and then a plateau is reached after approximately 60 s of UV irradiation. A similar, but inverse, dependence is observed in the intensity of the infrared band at 808 cm^{-1}, corresponding to the CH$_2$ wagging mode of the acrylate monomer. Clearly, the emission changes displayed by the metal complex are able to reflect the extent to which the acrylate monomer is consumed during the photopolymerization. It is important to note that the emission intensity of the long-wavelength ^{3}MLCT emission band remains constant in the absence of added photoinitiator, providing further evidence that the organometallic complex does not undergo degradation during the polymerization or itself act as a photoinitiator in these acrylate films.

Several organic molecules have been employed as spectroscopic probes of free-radical polymerization. For example, julolidine malanonitrile has been employed as a fluorescence probe for high conversion bulk polymerization of methyl methacrylate, ethyl methacrylate, n-butyl methacrylate, ethyl acrylate, styrene, and the copolymerization of styrene/n-butyl methacrylate [107]. In each of these cases, the fluorescence intensity from the probe species increases substantially upon polymer formation. This effect has again been related to the changes in the viscosity and mean free volume of the medium [83,84,108]. Furthermore, a series of intramolecular charge transfer (ICT) fluorescence probes, such as 5-dimethylaminonaphthalene-1-sulfonyl-n-butylamide (DASB) and 6-propionyl-2-(dimethylamino)napthalene (PRODAN), have been used to monitor the degree of cure and coating thicknesses in photocurable acrylate, acrylic, and polyester resins [35,109]. These ICT probe molecules display hypsochromic shifts in their emission bands during their polymerization reactions. It is rationalized that the excited molecule is less able to relax to its twisted charge transfer state as the matrix viscosity increases. Dansylamides have also been found to be useful as ICT fluorescence probes of acrylate polymerization and the molecules 4-dimethylamino-4'-nitrobiphenyl and 4-dimethylamino-4'-nitrostilbene (**6**) are

6

4-dimethylamino-4'-nitrobiphenyl   *4-dimethylamino-4'*-nitrostilbene

particularly sensitive to changes in solvent polarity and medium microviscosity [110,111].

A series of organic salts, comprising an alkylated pyridinium ion linked to a dimethylamino group by a π-system and an inorganic anion, have recently been used as spectroscopic probes of dimethylacrylate photopolymerization [112]. An example is 2-[4-(4-(dimethylamino)phenyl)-1,3-butadienyl]-1-

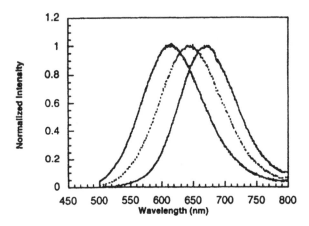

7

*2-[4-(4-(dimethylamino)phenyl)-1,3-butadienyl]-1-
ethylpyridinium tetrafluoroborate ("pyridine1")*

ethylpyridinium tetrafluoroborate ("pyridine 1", **7**) and emission spectra obtained from this probe molecule during the curing of ethylene glycol dimethylacrylate (EGDMA) are illustrated in Fig. 23. Clearly, the fluorescence band of the probe substantially blue shifts upon polymerization; a plot of the emission maximum of the probe molecule versus the extent of double-bond conversion is shown in Fig. 24. A mechanism involving the stabilization of the excited state by the position of the anion has been implicated, as this type of probe does not exhibit significant solvatochromism.

The *fac*-ClRe(CO)$_3$(4,7-Ph$_2$-phen) complex has been demonstrated to be a useful spectroscopic probe in the curing of photosensitive epoxy-based materials [100]. Emission spectra obtained from a 0.05-mm thickness film of a mixed epoxy system of bisphenol-A/novalac resin (Interz, SU8) and diglycidyl ether of bisphenol-A (DGEBA) containing cation-generating triarylsulfonium hexafluoroantimonate salts as photoinitiators (Union Carbide, Cryacure® UVI-6974) and

Figure 23 Normalized emission of "pyridine 1" in EGDMA before irradiation (right spectrum), after 2 min irradiation (middle spectrum), and after 20 min irradiation (left spectrum). Double-bond conversions, determined by FTIR spectroscopy, are 0, 38, and 76%, respectively. (From Ref. 112.)

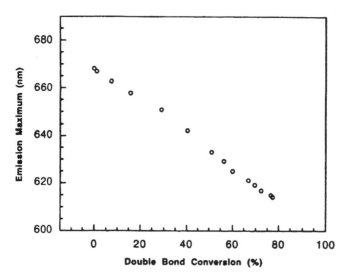

Figure 24 A plot of emission maximum of "pyridine 1" in EGDMA as a function of the double-bond conversion. (From Ref. 112.)

0.005% (by weight) of the organometallic probe complex *fac*-ClRe(CO)$_3$(4,7-Ph$_2$-phen) are shown in Fig. 25. The photosensitive epoxy films were heating for 10 min at 348 K both before and after irradiation.

Prior to light exposure, the spectra reveal an emission band that is centered at 460 nm. This is observed even in the absence of added organometallic probe complex and has been attributed primarily to Raman scattering arising from the resin itself [100,101]. A long-wavelength emission band is initially observed as a shoulder at 582 nm and this feature intensifies greatly upon UV exposure and appears as a maximum at 522 nm. This emission band is associated with the ^{3}MLCT emission from the *fac*-Re(CO)$_3$(4,7-Ph$_2$-phen) complex. Notably, both the increase in emission intensity and the hypsochromic shift (1945 cm^{-1}) of the long-wavelength band are significantly larger than those observed for the same metal complex in the acrylate mixtures (160 to 672 cm^{-1}; see Table 6). In addition, the emission lifetime of *fac*-ClRe(CO)$_3$(4,7-Ph$_2$-phen) in this cured epoxy resin is 6.4 μs, which is appreciably longer than the 2.5 μs recorded in the 1:1 TMPTA/PMMA acrylate system.

The increases in both emission intensity and emission lifetime of *fac*-ClRe(CO)$_3$(4,7-Ph$_2$-phen) are again related to a reduction in the nonradiative decay pathways as the epoxy material polymerizes and forms a rigid network. Significantly, though, the observed rigidochromic shifts in the epoxy-based net-

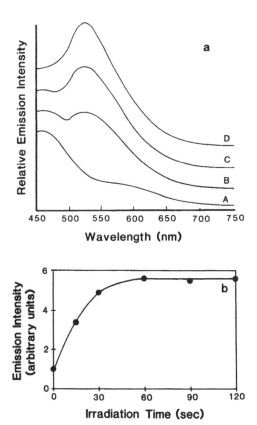

Figure 25 (a) Emission spectra at 293 K of *fac*-ClRe(CO)$_3$(4,7-Ph$_2$-phen) in the mixed epoxy system of bisphenol-A/novalac and DGEBA resin (0.05-mm thin film) containing a triarylsulfonium hexafluoroantimonate photoinitiator as a function of UV-irradiation time: (A) 0 s, (B) 15 s, (C) 30 s, and (D) 60 s. The emission spectra are uncorrected for photomultiplier response and vertically displaced for clarity. Excitation wavelength is 420 nm in each case. (b) Plot of emission intensity at the MLCT band maximum of *fac*-ClRe(CO)$_3$(4,7-Ph$_2$-phen) as a function of UV-irradiation time. (From Ref. 100.)

works are very much greater than those observed from the acrylate-based systems. However, we recognize that the cured epoxy matrix is highly polar [$\delta = 9.7$ to 10.9 (cal/cm^3)$^{1/2}$] [113] and this environment will substantially influence the excited state of the probe complex. In comparison, the dipolar interactions in acrylates, such as the TMPTA/PMMA mixtures described above, will be considerably less because their polymerized networks are not as polar [$\delta \sim 9.4$ (cal/cm^3)$^{1/2}$ [113].

V. LUMINESCENCE RIGIDOCHROMISM

The luminescence rigidochromism behavior exhibited by metal complexes is clearly one that can be very useful in designing spectroscopic probes to monitor environmental changes. In order to provide more insight into this phenomenon, a summary of electronic absorption and emission results obtained for the organometallic complexes in the range of photosensitive and thermosetting polymers studied is presented in Table 7. These results are revealing as it is immediately apparent that the absorption maxima of the metal complexes exhibit relatively small hypsochromic shifts when the environment rigidity increases, whereas the MLCT emission bands are subject to much more substantial en-

Table 7 Energy Shifts Observed at the MLCT Absorption (ΔE_{abs}) and Emission (ΔE_{em}) Maxima for Organometallic Complexes Upon Curing in Thermosetting and Photosensitive Polymers[a]

Complex	Resin material	ΔE_{abs} cm^{-1}	ΔE_{em} cm^{-1}
fac-ClRe(CO)$_3$(4,7-Ph$_2$-phen)	thermosetting cycloaliphatic epoxy, ERL-4221	211	1,592
	thermosetting epoxy, DGEBA	—	711
	photosensitive epoxy, DGEBA/bisphenol-A	—	1,945
	photosensitive acrylate, TMPTA/PMMA[b]	—	432
	aromatic cyanate ester, AroCy® L-10	—	460
	aromatic cyanate ester, AroCy® B-30	—	460
fac-ClRe(CO)$_3$(phen)	thermosetting cycloaliphatic epoxy, ERL-4221	—	1,524
fac-ClRe(CO)$_3$(4-Me-phen)	thermosetting cycloaliphatic epoxy, ERL-4221	73	1,025
fac-BrRe(CO)$_3$(4,4'-Me$_2$-bpy)	thermosetting cycloaliphatic epoxy, ERL-4221	292	1,633
fac-IRe(CO)$_3$(4,7-Ph$_2$-phen)	thermosetting cycloaliphatic epoxy, ERL-4221	369	764
W(CO)$_4$(4-Me-phen)	photosensitive acrylate, TMPTA/PMMA[b]	—	653

[a]Data compiled from Refs. 80, 81, 89, 100–102, 106.
[b]Composition is 1:1 TMPTA/PMMA (by weight).

ergy shifts. The cause of this apparent decoupling between the absorption and emission behavior may arise from the fact that the initially formed ^{1}MLCT state is extremely short-lived and decays rapidly via efficient nonradiative relaxation to the ground state and effective intersystem crossings, whereas the emissive ^{3}MLCT state is relatively long-lived [37,38]. The relaxed ^{3}MLCT level will be much more susceptible to changes in the rigidity of the matrix environment because it is this state which is the most affected by changes in dipole–dipole interactions with the surrounding solvent molecules.

A diagram representation of the varying effects of dipolar interactions that can occur both between the ground state (GS) or ^{3}MLCT excited state and the local solvent dipoles in the medium is shown in Fig. 26 [100]. The surrounding solvent molecules are readily able to orient about the ground-state complex to best accommodate its dipole moment. However, the dipole moment of the ^{3}MLCT excited state of a substituted metal carbonyl complex is considered to be reversed from that of the ground state [73,114], so its dipole moment will be immediately destabilized by the environment upon excitation, until the complex and surrounding solvent molecules are able to reorient themselves to facilitate a more favorable interaction [71]. In fluid solution this relaxation process will take place quickly, but in a more rigid environment it will be restrained. The net result will be a destabilization of the ^{3}MLCT excited state under rigid conditions compared to the nonrigid situation and, subsequently, a hypsochromic shift in the emission from that level (see Fig. 27). Continuing the same logic, therefore, it is expected that these variations in solvation will be much less influential on the short-lived ^{1}MLCT excited states and that these absorption bands will experience considerably smaller energy perturbations.

The luminescence rigidochromic effect may also be related to differing amounts of excited-state distortion in the fluid and rigid environments. Clearly, the ^{3}MLCT excited states of these organometallic complexes are greatly distorted compared to their corresponding ground states, as evidenced by the sizable energy gaps between the emission and excitation spectra (e.g., see Figs. 1 and 7).

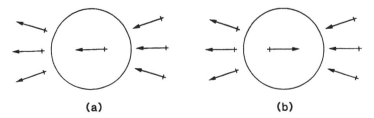

(a)   (b)

Figure 26 Diagram showing solvation of (a) a molecule in the ground state and (b) a molecule in the ^{3}MLCT excited state. The dipole moment of the molecule is depicted to change direction upon excitation. (From Ref. 100.)

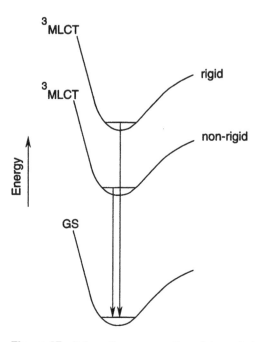

Figure 27 Schematic representation of the emission from the ^{3}MLCT excited state in rigid and nonrigid environments. (From Ref. 106.)

A hypsochromic shift in the emission band would be observed if the amount of excited-state distortion were lower in a rigid matrix than in fluid solution. Such an interpretation has been recently suggested for $Cu_4I_4py_4$ and related copper clusters, and this constitutes an unusual case of molecules that exhibit rigidochromism but not solvatochromism in its emission from a cluster-centered excited state [115]. It is likely that both excited-state destabilization and distortion arise during the polymerization processes in the tungsten and rhenium organometallic systems.

Further insight on the nature of the rigidochromism has arisen from recent fast time-resolved infrared (TRIR) studies of *fac*-ClRe(CO)$_3$(bpy) in fluid and glassy solutions [116]. Figure 28 depicts TRIR spectra taken 100 ns after laser excitation from this complex in a butyronitrile/propionitrile (PrCN/EtCN; 5:4 by volume) solution at 135 and 77 K. For comparison purposes, the FTIR spectrum at room temperature is also included. Upon excitation, the three ν(CO) bands of the ground state decline in absorbance while three new ν(CO) bands representing the MLCT excited state appear at higher frequency. Similar behavior has been observed from *fac*-ClRe(CO)$_3$(bpy) in CH_2Cl_2 solution [117] and for other metal carbonyl systems in fluid environments [118–123]. However, it

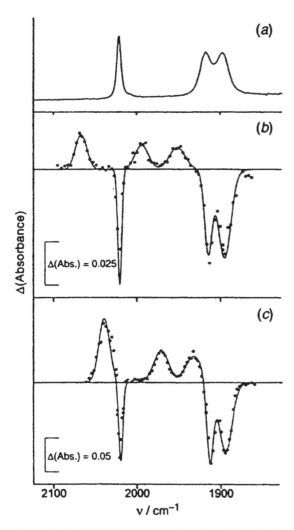

Figure 28 (a) FTIR spectrum of *fac*-ClRe(CO)$_3$(bpy) in PrCN/EtCN (5:4 by volume) at room temperature; (b) TRIR of similar solution at 135 K, 100 ns after laser excitation; (c) TRIR of similar solution at 77 K, 100 ns after laser excitation. The spectral points are shown such that decreasing absorbance values represent parent loss and increasing values represent product gain. The lines represent multiple Gaussian fits to the spectra. (From Ref. 116. Reproduced by permission of The Royal Society of Chemistry.)

is interesting to see in this recent work that the infrared bands of the MLCT excited state are shifted from the parent bands by a lower amount (mean value of shift approximately 20 cm^{-1}) in the frozen glass compared to fluid solution, despite the parent bands being almost unaffected by the change from fluid to glassy solution (see Table 8).

This phenomenon has been termed "infrared rigidochromism" and its origin is not yet obvious. One consideration is that the vibrational potential functions of the excited state are steeper in the fluid than in the glass, but this should also pertain to the ground state and is not reflected in the results which are virtually constant over the temperature range (see Table 8). In addition, one would really expect a greater shift in the ν(CO) bands as the glass is formed, as any change in the potential function ought to be more pronounced for the excited state. Another consideration is the charge transfer nature of the excited state, and one would again forecast a greater shift for the ν(CO) bands on going from the fluid solution to the glass because the charge separation between ground and excited states is increased [124]. Similarly, insight arising from spectral band

Table 8 Energy of ν(CO) Bands of *fac*-ClRe(CO)$_3$(bpy) in PrCN/EtCN (5:4 by Volume) at Various Temperatures and in CH$_2$Cl$_2$ at Room Temperature[a]

	Ground state	Excited state	Difference	Mean difference
Room temp. (FTIR)	2021.5	—	—	—
	1917.0	—	—	—
	1986.0	—	—	—
Room temp. (TRIR)	2022	2065	43	—
	1915	1991	76	58
	1895	1951	56	—
135 K	2020	2066	46	—
	1914	1993	79	60
	1895	1950	55	—
110 K	2018	2039	21	—
	1913	1976	63	41
	1896	1936	40	—
77 K	2020	2040	20	—
	1913	1971	58	38
	1894	1932	36	—
Room temp. in CH$_2$Cl$_2$	2024	2064	40	—
	1921	1987	66	55
	1899	1957	58	—

[a]Data taken from Refs. 116, 119. Reproduced by permission of The Royal Society of Chemistry.

fitting of UV-vis bands also indicated that in the glass there is a larger distortion of the normal coordinates and predicted an increase in $\nu(CO)$ band shifts in the glassy medium [125–127]. To rationalize the unexpected shift in direction of the MLCT excited state $\nu(CO)$ bands in the glass compared to fluid solution, it has been postulated that the nearby π,π^* intraligand (IL) state may mix with the MLCT state in this particular system. Consequently, the MLCT excited state may involve less electron transfer in the glass relative to the fluid solution [116].

VI. CONCLUDING REMARKS

Organometallic complexes that luminesce from low-lying MLCT excited states offer a new type of spectroscopic probe for monitoring kinetic changes during polymerization, including those occurring in the thin films. The metal complexes exhibit a number of key features that, taken together, provide advantages over conventional probe molecules. These properties include their solubility and thermal stability in monomer/polymer media, their intense visible absorptions, their strong and long-lived phosphorescence which appears in the yellow-red region of the visible spectrum, and also the sensitivity of their emission features to environmental rigidity changes during polymerization. Significantly, the luminescence parameters from these organometallic complexes are sensitive over a wide range of viscosities within the monomer/polymer solution.

Future work can be expected to involve more detailed design of ligands and metal centers with the intention of providing metal complexes that are especially sensitive to changes in viscosity and polarity within different types of monomer/polymer environments. In taking this type of approach, it will be possible to more specifically tailor the emission properties of the probe molecules to applications in a variety of polymerization processes.

ACKNOWLEDGMENTS

My work on spectroscopic probes in polymers has been supported by the Division of Chemical Sciences, Office of Basic Energy Sciences, Office of Science, US Department of Energy (Grant DE-FG02-89ER14039), the Petroleum Research Fund, administered by the American Chemical Society, and IBM Corporation. I thank Ms. Martha Gahring for invaluable assistance in the preparation of this manuscript. Gratitude is also extended to students in my group and collaborators who have participated in the development of this research, as listed in the referenced publications.

ABBREVIATIONS

η^*	complex viscosity
G^*	complex modulus
G'	storage modulus
G''	loss modulus
k_{nr}	nonradiative decay rate constant
k_r	radiative decay rate constant
λ_{abs}	absorption maximum
λ_{em}	emission maximum
ν_f	polymer-free volume
ν_0	van der Waals volume
ω	angular frequency
ϕ_e	emission quantum yield
t_e	emission lifetime
t_{gel}	time to gelation
t_{vit}	time to vitrification
AroCy® B-10	2,2'-bis(4-cyanatophenyl)isopropylidene
AroCy® B-30	2,2'-bis(4-cyanatophenyl)isopropylidene (in prepolymer form)
AroCy® L-10	1,1-bis(4-cyanatophenyl)ethane
ATR	attenuated total reflectance
biquin	2,2'-biquinoline
bpy	2,2'-bipyridine
4,4'-bpy	4,4'-bipyridine
5-Br-phen	5-bromo-1,10-phenanthroline
5-Cl-phen	5-chloro-1,10-phenanthroline
Cyracure® UVI-6974	propylene carbonate/triarylsulfonium hexafluoroantimonate
DGEBA	diglycidyl ether of bisphenol A
DASB	5-dimethylaminonaphthalene-1-sulfonyl-n-butylamide
DDS	4,4'-diaminodiphenyl sulfone
DSC	differential scanning calorimetry
en	ethylenediamine
EGDMA	ethylene glycol dimethylacrylate
EPA	ether/isopentane/ethanol (5:5:2 by volume)
ERL-4221	(3,4-epoxycyclohexyl)methyl-3,4-epoxycyclohexylcarboxylate
FTIR	Fourier-transform infrared
ICT	intramolecular change transfer
IL	intraligand

Irgacure® 651	2,2'-dimethoxy-2-phenylacetophenone
LF	ligand field
MCD	magnetic circular dichroism
MLCT	metal-to-ligand charge transfer
4,4'-Me$_2$-bpy	4,4'-dimethyl-2,2'-bipyridine
4-Me-phen	4-methyl-1,10-phenanthroline
5-Me-phen	5-methyl-1,10-phenanthroline
2-Me-THF	2-methyl-tetrahydrofuran
5-NO$_2$-phen	5-nitro-1,10-phenanthroline
phen	1,10-phenanthroline
phen-5,6-dione	1,10-phenanthroline-5,6-dione
4,7-Ph$_2$-phen	4,7-diphenyl-1,10-phenanthroline
PMMA	poly(methyl methacrylate)
PrCN/EtCN	butyronitrile/propionitrile (5:4 by volume)
PRODAN	6-propionyl-2-(dimethylamino)naphthalene
pyridine 1	2-[4-(4-(dimethylamino)phenyl)-1,3-butadienyl]-1-ethylpyridinium tetrafluoroborate
R-dab	1,4-diaza-1,3-butadiene
R-pyca	pyridine-2-carbaldehyde imine
RR	resonance Raman
SU8	bisphenol-A/novalac resin
TMPTA	trimethylolpropane triacrylate
TRIR	time-resolved infrared
UV	ultraviolet
vis	visible

REFERENCES

1. AA Gamble. In DR Randell, ed. Radiation Curing of Polymers. London: Royal Society of Chemistry, 1987, p 48.
2. B Klingert, M Riedliker, A Roloff. Comments Inorg Chem 7:109, 1988.
3. E Reichmanis, FL Thompson. Chem Rev 89:1273, 1989.
4. CA May, Y Tanaska, eds. Epoxy Resins Chemistry and Technology. New York: Marcel Dekker, 1973.
5. LH Lee. Adhesive Chemistry. New York: Plenum, 1983.
6. WD Callister. Materials Science and Engineering: An Introduction. New York: J Wiley, 1991.
7. DR Randell, ed. Radiation Curing of Polymers. London: Royal Society of Chemistry, 1987.
8. NS Allen, SJ Hardy, AF Jacobine, DM Glaser, B Yang, D Wolf. Eur Polym J 26:1041, 1990.
9. CJ Decker. Polym Sci A Polym Chem 30:913, 1992.

10. WD Cook. Polymer 33:2152, 1992.
11. D Wang, L Carrera, MJM Aradie. Eur Polym J 29:1379, 1993.
12. C Kutal, PA Grutsch, DB Yang. Macromolecules 24:6872, 1991.
13. FW Wang, RE Lowry, WH Grant. Polymer 25:690, 1984.
14. CSP Sung. In CE Hoyle, JM Torkelson, eds. Photophysics of Polymers. ACS Symposium Series 358, Washington, DC: American Chemical Society, 1987, p 463.
15. A Stroeks, M Shmorhun, AM Jamieson, R Simha. Polymer 29:467, 1988.
16. P Dousa, C Konak, V Fidler, K Dusek. Polym Bull (Berlin) 22:585, 1989.
17. SF Scarlatta, JA Ors. Polym Commun 27:41, 1986.
18. EW Meijer, RJM Zwiers. Macromolecules 20:332, 1987.
19. R Hayashi, S Tazuke, CW Frank. Macromolecules 20:983, 1987.
20. R Hayashi, S Tazuke, CW Frank. Chem Phys Lett 135:123, 1987.
21. S Tazuke, RK Guo, R Hayashi. Macromolecules 21:1046, 1988.
22. J Paczkowski, DC Neckers. Macromolecules 24:3013, 1991.
23. LW Jenneskens, HJ Verhey, HJ van Ramesdonk, AJ Witteveen, JW Verhoeven. Macromolecules 24:4038, 1991.
24. J Paczkowski, DC Neckers. J Polym Sci A Polym Chem 31:841, 1993.
25. JAJ Burrows, GW Haggquist, RD Burkhart. Macromolecules 23:988, 1990.
26. FM Winnik. Macromolecules 23:1647, 1990.
27. G Wang, L Chen, MA Winnik. Macromolecules 23:1650, 1990.
28. I Yamazaki, FM Winnik, MA Winnik, S Tazuke. J Phys Chem 91:4213, 1987.
29. P Chandar, P Somasundaran, NJ Turro. Macromolecules 21:950, 1988.
30. FM Winnik. Macromolecules 22:734, 1989.
31. J Naciri, RG Weiss. Macromolecules 22:3928, 1989.
32. O Valdes-Aguilera, CP Pathak, DC Neckers. Macromolecules 23:689, 1990.
33. A Pattikottu, WL Mattice. Macromolecules 23:867, 1990.
34. M Wilhelm, C-L Zhao, Y Wang, R Xu, MA Winnik, J-L Mura, G Riess, MD Croucher. Macromolecules 24:1033, 1991.
35. JC Song, DC Neckers. In MW Urban, T Provder, eds. Multidimensional Spectroscopy of Polymers. ACS Symposium Series 598. Washington, DC: American Chemical Society, 1995, p 472.
36. M Wrighton, DL Morse. J Am Chem Soc 96:998, 1974.
37. GL Geoffroy, MS Wrighton. Organometallic Photochemistry. New York: Academic, 1979.
38. AJ Lees. Chem Rev 87:711, 1987.
39. GA Crosby, KW Hipps, WH Elfring. J Am Chem Soc 96:629, 1974.
40. JB Birks. Photophysics of Aromatic Molecules. London: Wiley, 1970, p 182.
41. J Saltiel, J D'Agostino, ED Megarity, L Metts, KR Neuberger, MS Wrighton, OC Zafiriou. Org Photochem 3:1, 1973.
42. G Fischer, KA Muszkat, E Fischer. J Chem Soc B 1156, 1968.
43. NJ Turro. Modern Molecular Photochemistry. Menlo Park, CA: Benjamin-Cummings, 1978, p 179.
44. WK Smothers, MS Wrighton. J Am Chem Soc 105:1067, 1983.
45. JV Caspar, TD Westmoreland, GH Allen, PG Bradley, TJ Meyer, WH Woodruff. J Am Chem Soc 106:3492, 1984.

46. P Glyn, MW George, PM Hodges, JJ Turner. J Chem Soc Chem Commun 1655, 1989.
47. DR Gamelin, MW George, P Glyn, F-W Grevels, FPA Johnson, W Klotzbücher, SL Morrison, G Russell, K Schaffner, JJ Turner. Inorg Chem 33:3246, 1994.
48. MW George, FPA Johnson, JR Westwell, PM Hodges, JJ Turner. J Chem Soc Dalton Trans 2977, 1993.
49. S-S Sun, AJ Lees. J Am Chem Soc 122:8956, 2000.
50. H Bock, H tom Dieck. Angew Chem Int Ed Engl 5:520, 1966.
51. H Saito, J Fujita, K Saito. Bull Chem Soc Jpn 41:863, 1968.
52. J Burgess. J Organomet Chem 19:218, 1969.
53. H tom Dieck, IW Renk. Angew Chem Int Ed Engl 9:793, 1970.
54. D Walther. Z Anorg Allg Chem 396:46, 1973.
55. D Walther. J Prakt Chem 316:604, 1974.
56. MS Wrighton, DL Morse. J Organomet Chem 97:405, 1975.
57. J Burgess, JG Chambers, RI Haines. Transition Met Chem (Weinheim, Ger.) 6:145, 1981.
58. DM Manuta, AJ Lees. Inorg Chem 22:3825, 1983.
59. DM Manuta, AJ Lees. Inorg Chem 25:1354, 1986.
60. KA Rawlins, AJ Lees. Inorg Chem 28:2154, 1989.
61. RW Balk, DJ Stufkens, A Oskam. Inorg Chim Acta 28:133, 1978.
62. RW Balk, DJ Stufkens, A Oskam. Inorg Chem 19:3015, 1980.
63. PC Servaas, HK van Dijk, TL Snoeck, DJ Stufkens, A Oskam. Inorg Chem 24:4494, 1985.
64. AJ Lees, AW Adamson. J Am Chem Soc 104:3804, 1982.
65. AJ Lees, JM Fobare, EF Mattimore. Inorg Chem 23:2709, 1984.
66. JA Connor, C Overton, N El Murr. J Organomet Chem 277:277, 1984.
67. ES Dodsworth, ABP Lever. Chem Phys Lett 112:567, 1984.
68. DW Manuta, AJ Lees. Inorg Chem 25:3212, 1986.
69. W Kaim, S Kohlmann. Inorg Chem 25:3306, 1986.
70. W Kaim, S Kohlmann, S Ernst, B Olbrich-Deussner, C Bessenbacher, A Schulz. J Organomet Chem 321:215, 1987.
71. ES Dodsworth, ABP Lever. Inorg Chem 29:499, 1990.
72. MM Zulu, AJ Lees. Inorg Chem 27:3325, 1988.
73. MM Glezen, AJ Lees. J Am Chem Soc 111:6602, 1989.
74. KA Rawlins, AJ Lees, AW Adamson. Inorg Chem 29:3866, 1990.
75. PNW Baxter, JA Connor. J Organomet Chem 486:115, 1995.
76. AJ Lees. Comments Inorg Chem 17:319, 1995.
77. HK van Dijk, PC Servaas, DJ Stufkens, A Oskam. Inorg Chim Acta 104:179, 1985.
78. DJ Robbins, A Thompson. J Mol Phys 25:1103, 1973.
79. AJ Lees. Coord Chem Rev 177:3, 1998.
80. TG Kotch, AJ Lees, SJ Fuerniss, KI Papathomas. Chem Mater 3:25, 1991.
81. TG Kotch, AJ Lees, SJ Fuerniss, KI Papathomas. Chem Mater 4:675, 1992.
82. BD Rossennar, DJ Stufkens, A Vlček Jr. Inorg Chem 35:2902, 1996.
83. RO Loutfy. Macromolecules 14:270, 1981.

84. RO Loutfy. In MA Winnik, ed. Photophysical and Photochemical Tools in Polymer Science: Conformation, Dynamics, Morphology. NATO ASI Series, Series C 182. Dordrecht, The Netherlands: D. Reidel, 1986, p 429.

85. P Sherman. Industrial Rheology. London: Academic, 1970, p 24.

86. C-YM Tung, PJ Dynes. J Appl Polym Sci 27:569, 1982.

87. HH Winter, FJ Chambon. Rheology 30:367, 1986.

88. MS Heise, GC Martin, JT Gotro. Polym Eng Sci 30:83, 1990.

89. JT Grant, N Dunwoody, V Jakúbek, AJ Lees. Polym Polym Compos 6:47, 1998.

90. I Hamerton. In I Hamerton, ed. Chemistry and Technology of Cyanate Ester Resins. Glasgow: Blackie Academic and Professional, 1994, p 1.

91. AW Snow. In I Hamerton, ed. Chemistry and Technology of Cyanate Ester Resins. Glasgow: Blackie Academic and Professional, 1994, p 7.

92. S Das, F DeAntonis. Polym Mater Sci Eng 71:627, 1994.

93. T Fang. Polym Mater Sci Eng 71:682, 1994.

94. DA Shimp, B Chin. In I Hamerton, ed. Chemistry and Technology of Cyanate Ester Resins. Glasgow: Blackie Academic and Professional, 1994, p 230.

95. DA Shimp. Polym Mater Sci Eng 71, 1994.

96. SL Simon, JK Gillham. In I Hamerton, ed. Chemistry and Technology of Cyanate Ester Resins. Glasgow: Blackie Academic and Professional, 1994, p 87.

97. DA Shimp. In I Hamerton, ed. Chemistry and Technology of Cyanate Ester Resins. Glasgow: Blackie Academic and Professional, 1994, p 282.

98. JC Song, CSP Sung. Macromolecules 26:4818, 1993.

99. JC Song, CSP Sung. Macromolecules 28:5581, 1995.

100. TG Kotch, AJ Lees, SJ Fuerniss, KI Papathomas, R Snyder. Inorg Chem 32:2570, 1993.

101. TG Kotch, AJ Lees, SJ Fuerniss, KI Papathomas, R Snyder. Polymer 33:657, 1992.

102. TG Kotch, AJ Lees, SJ Fuerniss, KI Papathomas, R Snyder. Inorg Chem 30:4871, 1991.

103. AK Doolittle. J Appl Phys 23:236, 1952.

104. GE Johnson. J Chem Phys 63:4047, 1975.

105. T Karstens, K Koh. J Phys Chem 84:1871, 1980.

106. KA Rawlins, AJ Lees, SJ Fuerniss, KI Papathomas. Chem Mater 8:1540, 1996.

107. RO Loutfy. J Polym Sci B Polym Phys 20:825, 1982.

108. RO Loutfy, BA Arnold. J Phys Chem 86:4205, 1982.

109. JC Song, DC Neckers. Polym Eng Sci 36:394, 1996.

110. ZJ Wang, JC Song, R Bao, DC Neckers. J Polym Sci B Polym Phys 34:325, 1996.

111. WF Jager, A Lungu, DY Chen, DC Neckers. Macromolecules 30:780, 1997.

112. WF Jager, D Kudasheva, DC Neckers. Macromolecules 29:7351, 1996.

113. G Salomon. In R Houwink, G Salomon, eds. Adhesion and Adhesives. Amsterdam: Elsevier, 1962, p 17.

114. PJ Giordano, MS Wrighton. J Am Chem Soc 101:2888, 1979.

115. D Tran, JL Bourassa, PC Ford. Inorg Chem 36:439, 1997.

116. IP Clark, MW George, FPA Johnson, JJ Turner. Chem Commun 1587, 1996.

117. P Glyn, PA Johnson, MW George, AJ Lees, JJ Turner. Inorg Chem 30:3543, 1991.

118. JJ Turner, MW George, FPA Johnson, JR Westwell, Coord Chem Rev 125:101, 1993.

119. MW George, FPA Johnson, JR Westwell, PM Hodges, JJ Turner. J Chem Soc Dalton Trans 2977, 1993.

120. JR Schoonover, KC Gordon, R Argazzi, WH Woodruff, KA Peterson, CA Bignozzi, RB Dyer, TJ Meyer. J Am Chem Soc 115:10996, 1993.

121. DR Gamelin, MW George, P Glyn, F-W Grevels, FPA Johnson, W Klotzbücher, SL Morrison, G Russell, K Schaffner, JJ Turner. Inorg Chem 33:3246, 1994.

122. FPA Johnson, MW George, SL Morrison, JJ Turner. J Chem Soc Chem Commun 391, 1995.

123. JR Schoonover, GF Strouse, RB Dyer, WD Bates, TJ Meyer. Inorg Chem 35:273, 1996.

124. TJ Meyer. Pure Appl Chem 9:1193, 1986.

125. EM Kober, JV Caspar, RS Lumpkin, TJ Meyer. J Phys Chem 90:3722, 1986.

126. RS Lumpkin, TJ Meyer. J Phys Chem 90:5307, 1986.

127. E Danielson, RS Lumpkin, TJ Meyer. J Phys Chem 91:1305, 1987.

6

Photorefractive Effect in Polymeric and Molecular Materials

Liming Wang, Man-Kit Ng, Qing Wang, and Luping Yu
University of Chicago, Chicago, Illinois

I. INTRODUCTION

The index of refraction of a material possessing both photoconductivity and electro-optic (EO) response can be modulated when the material is subjected to nonuniform optical illumination. This is the so-called photorefractive (PR) effect. Its origin is the Pockels effect that is in response to the photoinduced space-charge field. Several features of the PR effect are attractive and interesting, including the characteristic that the refractive index change is reversible. A second feature is that the magnitude of the index modulation is independent of the illumination intensity and a saturated refractive index modulation is achievable under the illumination of a low-power laser source. In contrast, light intensity affects the speed of the PR response. In addition, it is a nonlocal effect; the largest index change does not occur at the brightest region. There is a nonzero phase-shift of the resultant index grating with respect to the light intensity distribution.

The PR effect was first observed in 1966 as an undesirable effect in an EO crystal, $LiNbO_3$ [1]. It was soon realized that this effect is reversible and its PR mechanism was then identified [2,3]. These PR materials can be applied to high-capacity optical memories, dynamic hologram formations, massive interconnections, high-speed tunable filters, and true-time relay lines for phase array

antenna processing [4–6]. Therefore, extensive experimental, theoretical studies
were carried out all over the world and numerous applications were suggested
and demonstrated.

However, before 1990, inorganic PR crystals such as ferroelectric crystals
($BaTiO_3$, $LiNbO_3$, $LiTaO_3$, etc.), semiconductor materials (GaAs, InP, CdTe,
etc.) as well as sillenites ($Bi_{12}SiO_{20}$) had been the dominant materials. In 1990,
the PR effect was observed in an organic material, a carefully grown and doped
molecular crystal [7]. In a short time, a PR polymeric composite was disclosed
[8]. These works marked the beginning of a new research direction for PR materials, one that was more cost-effective, processable, and versatile in structural
modification. The particular advantages of amorphous organic PR materials include a low dielectric constant and a high EO coefficient. Since then, organic
PR materials have become an exciting research topic presenting challenges in
rationally synthesizing new and better materials, and opportunities for revealing
new physical properties.

Recently, the advances in molecular engineering have led to the development of numerous polymeric and molecular PR materials. Two fundamentally
different strategies were developed for the preparation of PR polymers: composite materials [9,10] and fully functionalized polymers [11–13]. Composite
materials are composed of polymer hosts (EO polymers, photoconductive polymers, or inert polymers) doped with different functional species necessary for
the PR effect. This approach has been successful in developing many PR systems. In particular, composites based on photoconductive poly(N-vinylcarbazole)
(PVK) (Scheme 1a) showed excellent properties. A diffraction efficiency close
to 100% in 105-μm thick films and a high net optical gain coefficient or larger
than 200 cm^{-1} have been achieved, which well exceeded the performance of
inorganic counterparts such as $BaTiO_3$ (gain coefficient: 40 to 50 cm^{-1}). However, composite systems exhibit thermodynamic instability associated with phase
separation. The dopants, which are usually small molecules, tend to crystallize

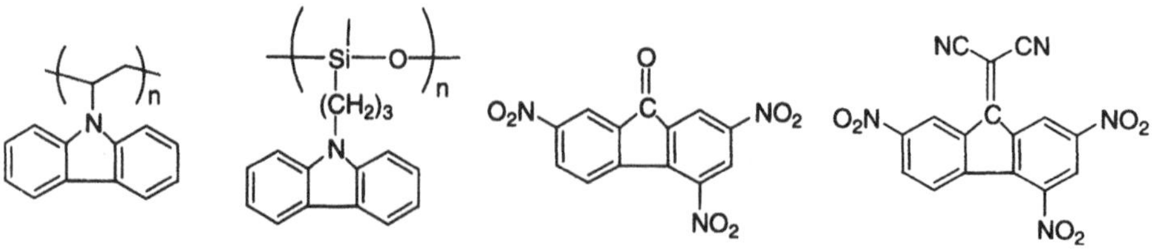

(a) **PVK**   (b) **PSX**   (c) **TNF**   (d) **TNFDM**

Scheme 1 Some hole transport polymers (a) and (b), and sensitizers (c) and (d) that
are used in composite PR materials.

or aggregate because of their incompatibility with macromolecules. These aggregates may cause severe optical scattering and may ruin the optical properties of the materials completely. A logical solution to the problems associated with composite systems is to incorporate all necessary functional species covalently within a single polymer so that the phase separation can be minimized. The challenge is the requirement of precise molecular engineering to design and optimize the structure and the ratio of each polymer component. Due to the difficulties in optimization and synthesis, the development of the fully functionalized PR polymer or molecule is relatively slower than the polymeric composites. Recently, it has been discovered that a simple class of amorphous PR molecular materials exhibits high PR performances. These glassy molecular materials are based upon methine dyes and have shown a net PR gain of 215 cm^{-1} and a diffraction efficiency of 87.6% in a 130-μm thick film at a semiconductor laser wavelength of 780 nm [14]. Some interesting photophysical and optical phenomena such as complementary grating competition, self-defocusing, and optical instability have been observed in these glassy molecular materials.

The PR effect is a complex phenomenon that relates to fields such as optics, materials science and chemistry. Significant development of organic PR materials has been achieved in the last 10 years, the progress of which has greatly benefited from the physics and materials science of the corresponding inorganic materials. Principles of the PR effect based on PR crystals have laid a solid foundation. Fundamental knowledge developed in the areas of organic photoconductive materials and nonlinear optic polymers has played a very important role. However, due to the limitation of space, we are unable to cover the entire topic related to the PR effect, but we do make an effort to summarize the physical principles of organic PR materials. A significant amount of discussion is devoted to typical polymer composites and fully functionalized PR materials.

In Sect. II, a brief review of the fundamentals of the PR effect is provided. The energy transfer and light diffraction of the wave mixings in a PR medium is introduced, and the optical gain coefficient and diffraction efficiency are defined. The process of light-induced refractive index modulation is considered, and the main results of Kukhtarev's PR model (commonly used in inorganic and organic materials) are presented.

In Sect. III, the photoinduced refractive index variation in organic materials is discussed. The four elementary processes necessary for a PR effect are considered: photogeneration, transport, and trapping of charged carriers, as well as the EO effect.

Section IV reviews the development of the three categories of the PR materials: polymer composites, fully functionalized polymers, and glassy molecular materials.

The detailed physical properties of glassy molecular materials based on methine dyes is given in Sect. V. Some special photophysical, and optical phenomena such as complementary grating competition, self-defocusing, and optical instability are discussed.

The final section summarizes a possible trend for the future development of organic PR materials.

II. FUNDAMENTAL PRINCIPLES OF PR EFFECT

A. Brief Summary of Physical Principles of PR Effect

1. Degenerate Two-Wave Coupling in a PR Medium

We first consider the interaction of two laser beams with the same frequency inside a PR medium. When two coherent laser beams intersect inside a PR medium (Fig. 1), a periodic variation of the intensity is formed due to interference, which will induce a periodic charge distribution and lead to the formation of volume index grating through a PR mechanism. The grating wave vector is given by

$$\vec{K} = \pm(\vec{k}_2 - \vec{k}_1), \tag{1}$$

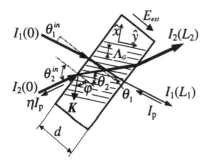

Figure 1   The setup geometry for grating recording experiments. The incident angles θ_i^{in} ($i = 1, 2$) relate to the angle in the medium θ_i by Snell's law: $\theta_i = Arcsin[(\sin \theta_i^{in})/n_0]$. The two beams form an intensity grating with spacing Λ_G in the medium and the grating vector $\vec{K}$, which forms an angle φ with respect to the film normal. For a two-wave mixing experiment, two laser beams are usually p-polarized to write the grating. Beam 1 refers to the attenuated beam; Beam 2 is the amplified one. Two writing laser beams for a degenerate four-wave mixing experiment are s-polarized and the probe beam is p-polarized.

where $\vec{k}_1$ and $\vec{k}_2$ are wave vectors of the beams. The grating spacing is therefore

$$\Lambda_G = \frac{2\pi}{K} - \frac{\lambda}{2\sin[(\theta_2 - \theta_1)/2]}, \tag{2}$$

where K is the magnitude of the grating vector ($K = |\vec{K}|$), and λ is the wavelength of the light in the medium, which is related to the free-space wavelength λ_0 and the refractive index of the medium n_0 by $\lambda = \lambda_0/n_0$. Because the Bragg scattering is perfectly phase-matched, the two beams are strongly diffracted by the index grating; the diffracted portion of Beam 1 is propagating along the direction of Beam 2 and, similarly, that of Beam 2 is propagating along the direction of Beam 1. This leads to the energy coupling and self-diffraction.

The unique feature of the PR gratings, different from refractive index modulation such as $\chi^{(3)}$ and those with other thermal, chemical, and electronic origins, is its nonlocal nature. The index gratings are phase-shifted with respect to the light intensity grating (from 0 to 90 degrees). This phase shift is due to the macroscopic movement of photoinduced charge carriers. The presence of such a phase shift allows nonreciprocal steady-state transfer of energy between beams. In the case of codirectional two-wave mixing as shown in Fig. 1, the following coupled equations can be derived from the Maxwell wave equations under the slow varying amplitude approximation [15].

$$\frac{d}{dz}I_1 = -\Gamma\frac{I_1 I_2}{I_1 + I_2} - \alpha I_1$$

$$\frac{d}{dz}I_2 = \Gamma\frac{I_1 I_2}{I_1 + I_2} - \alpha I_2 \tag{3}$$

with two-beam coupling gain coefficient defined as [16]:

$$\Gamma = \frac{4\pi}{\lambda}\Delta n(\hat{e}_1 \cdot \hat{e}_2^*)\sin\Phi, \tag{4}$$

where Δn is the modulation magnitude of the photorefractive grating, $\hat{e}_1$ and $\hat{e}_2$ are wave vectors of the two beams, and Φ is the phase-shift of the index grating with respect to the intensity grating.

By examining the coupled equations, Eq. (3), it is clear that the maximum energy exchange occurs when $\Phi = 90°$. In the case of $\Phi = 0$, Beam 1 and Beam 2 will be uncoupled and Eqs. (3) become Beer–Lambert equations. Moreover, in the absence of material absorption, I_2 is an increasing function of z if Γ is positive, while I_1 is a decreasing function of z. This indicates that the energy can be transferred from Beam 1 to Beam 2. The direction of energy transfer is determined by the sign of Γ, which depends on the sign of the charge carrier and the direction of the applied field.

The solution of Eq. (3) is

$$I_1(z) = I_1(0) \frac{1 + \beta^{-1}}{1 + \beta^{-1}\exp(\Gamma z)}\exp(-\alpha z)$$

$$I_2(z) = I_2(0) \frac{1 + \beta}{1 + \beta\exp(-\Gamma z)}\exp(-\alpha z),$$

(5)

where $\beta = I_1(0)/I_2(0)$ is the intensity ratio of Beam 1 over Beam 2 at $z = 0$.

The two-beam coupling gain coefficient can be derived from Eqs. (5) for Beam 2 (the gain beam):

$$\Gamma = \frac{1}{L_2}\ln\left(\frac{\gamma_2\beta}{1 + \beta - \gamma_2}\right),$$

(6)

where $L_2(= d/\cos\theta_2)$ is the optical path of Beam 2 and $\gamma_2\{= I_2(L)/[I_2(0)\exp(-\alpha L_2)]\}$ is the beam-coupling ratio, which is defined as the intensity ratio (after sample) of Beam 2 in the presence of Beam 1, versus Beam 2 in the absence of Beam 1. If we refer to Beam 2 as the signal beam and Beam 1 as the pump beam, the signal beam is amplified. Note that in the undepleted condition (i.e., $\beta \gg 1$) the second equation of Eq. (5) becomes $I_2(z) = I_2(0)\exp[(\Gamma - \alpha)L_2)]$, and Eq. (6) becomes $\gamma_2 = \exp(\Gamma L_2)$. To give a general view about how the energy is exchanged between the two beams, we plotted beam-coupling ratios γ_1 and γ_2 as a function of normalized length Γz at different beam ratios β (Fig. 2). For the entire range of beam ratios, Beam 1 is gradually depleted and Beam 2 gains energy correspondingly with the increase of interaction length. At a large beam ratio, for example, $\beta = 400$, the intensity of the pump beam (Beam 1) can be regarded as unchanged at beginning and the signal beam (Beam 2) increases exponentially. Knowing the β value and the sample thickness d, the optical gain coefficient Γ can be experimentally determined by measuring the beam-coupling ratio γ_2. The phase-shift Φ can be experimentally determined by the grating translation [7,17] and phase-modulation [18] techniques.

2.   Degenerate Four-Wave Mixing in a PR Medium

With the index grating recorded by the two strong writing beams (Beams 1 and 2), we then consider a probe beam with the same wavelength as the writing beams. The ideal situation is that the probe beam is weak enough to neglect the erasure on the grating and the incident angle θ_{in} (angles between the wave-vector of probe beam and the grating vector) obey the Bragg diffraction condition:

$$2\Lambda_G \sin\theta_{in} = N\lambda,$$

(7)

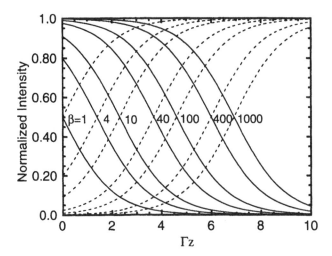

Figure 2 The normalized intensities of the two coupling beams; $(1 + \beta^{-1})/[1 + \beta^{-1} \exp(\Gamma z)]$ for pump beam (Beam 1, the solid lines) and $(1 + \beta)/[1 + \beta \exp(-\Gamma z)]$ for signal beam (Beam 2, the dashed lines) as the function of the normalized interaction length Γz for the different beam ratios β.

where N is an integer. In practice, for a PR characterization of polymeric films, the experimental geometry is usually arranged in such a way that the probe beam counter-propagates in the direction of one writing beam. Under a thick grating assumption, the diffraction efficiency defined as the intensity ratio of the beam diffracted by the grating with the probing beam, is derived by Kogelink as [19]:

$$\eta = \exp(-\zeta) \frac{\sin^2 (\nu^2 - \xi^2)^{1/2}}{1 - \xi^2 / \nu^2} \tag{8}$$

with

$$\nu = \frac{\pi \Delta n d (\hat{e}_p \cdot \hat{e}_d)}{\lambda (c_p c_d)^{1/2}}, \xi = \frac{\alpha d}{2} \left(\frac{1}{c_p} - \frac{1}{c_d} \right) \text{ and } \zeta = \frac{\alpha d}{2} \left(\frac{1}{c_p} + \frac{1}{c_d} \right),$$

where $\hat{e}_p$ and $\hat{e}_d$ are the polarization vectors of the probe beam and the diffracted beam, respectively. If the probe beam counterpropagates along Beam 1, $c_p = \cos\theta_1$, $c_d = \cos\theta_2$, and $\hat{e}_p \cdot \hat{e}_d = \cos(\theta_2 - \theta_1)$ when the probe beam is p-polarized; $\hat{e}_p \cdot \hat{e}_d = 1$ when the probe beam is s-polarized. If the probe beam counterpropagates along Beam 2, $c_p = \cos\theta_2$ and $c_d = 2\cos\theta_2 - \cos\theta_1$.

If $v \gg \xi$, which means contribution from the absorption modulation is negligible, and the p-polarized probe beam counterpropagates along Beam 1, Eq. (8) can be simplified as

$$\eta = \exp(-\alpha L) \sin^2 v = \exp(-\alpha L) \sin^2 \left(\frac{\pi \Delta n d \cos(\theta_2 - \theta_1)}{\lambda (\cos \theta_2 \cos \theta_1)^{1/2}} \right), \qquad (9)$$

where $L = d(\csc \theta_2 + \csc \theta_1)/2$.

According to Eq. (9), when $v = \pi/2$, the diffraction efficiency reaches its maximum value $\eta = \eta_{max} = \exp(-\alpha L)$ and begins to decrease for larger Δn value. The magnitude of the index modulation at maximum diffraction efficiency is

$$\Delta n_{\pi/2} = \frac{\lambda_0 (\cos \theta_2 \cos \theta_1)^{1/2}}{2d \cos(\theta_2 - \theta_1)}. \qquad (10)$$

Since the index modulation in organic PR materials is strongly dependent on the external field, the field $E_{\pi/2}$ at which the overmodulation occurs is now regarded as a new figure of merit for characterizing a PR polymeric and molecular material. A small $E_{\pi/2}$ value is desirable for practical applications.

Unlike the two-beam coupling gain coefficient, diffraction efficiency bears no relationship to the phase-shift of the index grating with respect to the intensity grating. Any kind of local or nonlocal index grating can diffract light. Therefore, light diffraction is not a unique property of the PR medium, but it is a very useful characterization of the order of magnitude of index change and property for practical applications.

B. PR Mechanism

1. Formation of Dephased Index Grating

Now let us take a closer look into the process of formation of the dephased PR grating. When two coherent laser beams intersect inside a PR medium, a periodic variation of light intensity due to interference is formed (Fig. 3a). The spatially modulated intensity distribution is

$$I(x) = I_0 \left(1 + m \cos \frac{2\pi x}{\Lambda_G} \right), \qquad (11)$$

where $I_0 = I_1 + I_2$ is the sum of the intensity of the two beams and $m = 2\beta^{1/2}/(1 + \beta)$ is the fringe visibility. The bright regions are located near $x = N\Lambda_G$, where N is an integer, and the dark regions are located near $x = (N + 1/2)\Lambda_G$. If $I_1 = I_2$, $I(N\Lambda_G) = 2I_0$, and $I[(N + 1/2)\Lambda_G] = 0$ (Fig. 3a).

As indicated by Eq. (2), the smaller the intersect angle $\theta_2 - \theta_1$, the larger the grating spacing Λ_G. The grating spacing value can vary from sub-micron to several tens of micrometers. In a photoconductive medium, upon steady irradiation

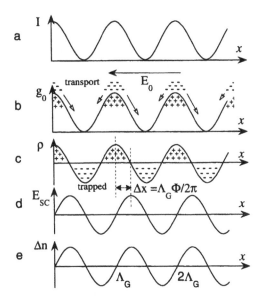

Figure 3 The response of a photorefractive medium to the two conference light beams with the same intensities. The coordination is indicated in Fig. 1. (a) The intensity distribution is formed in the medium. (b) The steady-state photogeneration rate g_0 is proportional to the light intensity. The electrons are defused and drifted to the dark region by the external field E_0 ($E_0 = E$ ext $\cos \zeta$). (c) The steady-state density of space-charge ρ is in phase with the intensity grating. (d) The space-charge generates an electric field E_{SC}, which is dephased with respect to the intensity grating by distance $\Lambda_G \Phi/2\pi$. (e) The refractive index is then modified by the space-charge field.

with light of intensity I, free carriers are produced at a rate g_0 (carriers/m^3/sec) and their lifetime before recombination is τ_R. Eventually, a steady-state concentration of carriers, n_c, is achieved where the rate of carrier generation equals the rate of recombination

$$n_c = g_0 \tau_R \tag{12}$$

with

$$g_0 = I\phi/d \tag{13}$$

where ϕ is the quantum efficiency of charge generation (number of carrier pairs produced per absorbed photon) and d is the sample thickness. Therefore, the carrier concentration is proportional to the local light intensity. As a result of the illumination with periodic intensity, charge carriers are generated with the density pattern in phase with the intensity pattern (Fig. 3b). In the simplest model based on single-polar transport, the major carriers, say electrons, migrate into the dark region through diffusion due to carrier concentration gradients or drift under the applied field. The immobile carriers (e.g. holes) are left in the bright region (Fig. 3b). Localized sites with lower potential, known as traps, then capture the major carriers in the dark region. These traps can be chemical impurities and/or structural defects. The buildup of space-charge density continues until the limitation of the number density of the traps presented in the materials is reached. Therefore, the dark region is negatively charged and the bright region is positively charged. This space-charge density is inhomogeneously distributed, and its fundamental component can be written as

$$\rho(x) = \rho_0 \cos\left(\frac{2\pi x}{\Lambda_G}\right), \tag{14}$$

where ρ_0 is a constant. The space-charge density grating is in phase with the intensity grating (Fig. 3c). The electric field generated by these space charges obeys the Poisson equation $\nabla \cdot \varepsilon \vec{E} = \rho(\vec{r})$ (in SI units), which takes the form of $dE_{SC}(x)/dx = \rho(x)/\varepsilon$ in this case, where ε is the static dielectric constant of the material. An integration of the Poisson equation leads to a space-charge field as

$$E_{SC}(x) = \frac{\rho_0}{\varepsilon} \sin\left(\frac{2\pi x}{\Lambda_G}\right). \tag{15}$$

We note from Eq. (15) that the space-charge field is shifted in space by $\pi/2$ with respect to the intensity pattern [Eq. (11)], which corresponds to a distant shift of $\Lambda_G/4$ in the x direction (Fig. 3d). This space-charge field induces an index volume grating via the Pockels effect (Fig. 3e). The refractive index including the fundamental component of refractive index modulation with magnitude of Δn can be written as

$$n(x) = n_0 + \Delta n \sin\left(\frac{2\pi x}{\Lambda_G}\right) \tag{16}$$

and Δn is related to the modulation magnitude of the space-charge field E_{SC} as

$$\Delta n = -\frac{1}{2}n_0^3 r_{\text{eff}} E_{SC}, \tag{17}$$

where r_{eff} is the effective EO coefficient.

In summary, a PR effect consists of the following fundamental processes:

1. photoionization of photo-sensitizer molecule and generation of charged carriers;
2. transport of the carrier from the bright region to the dark region. The transport length is usually in the order of micrometers;

3. trapping of the carriers in the dark region and the formation of the distributed space-charge density; and
4. modulation of the refractive index through the Pockels effects or any other field response mechanisms.

2. Kukhtarev's Model

Although Eqs. (11 to 17) give an easily understandable picture of the PR grating formation, the real processes involved in the PR effect are much more complicated. Mathematically, a complete set of nonlinear equations that consist of the rate equations of photogeneration and recombination, the transport equations, the electric continuity, and Poisson's equation have to be solved. Kukhtarev and coworkers [20,21] developed a PR model based on inorganic photorefractive single crystals such as ferroelectric crystals ($BaTiO_3$, $LiNbO_3$, $LiTaO_3$, etc.) and sillenites ($Bi_{12}SiO_{20}$). This model is a very useful foundation for the understanding of and explanations in organic PR materials.

Kukhtarev's model predicts the fundamental component of the steady-state space-charge field created by intensity distribution in Eq. (11) as

$$E_{SC}(x) = -m E_{SC} \cos\left(\frac{2\pi x}{\Lambda_G} + \Phi\right) \tag{18}$$

with the amplitude E_{SC} given by

$$E_{SC} = E_q \left[\frac{E_0^2 + E_D^2}{E_0^2 + (E_D + E_q)^2}\right]^{1/2} \tag{19}$$

and Φ, the phase shift between $E_{SC}(x)$ and $I(x)$, by

$$\Phi = Arc\tan\left[\frac{E_D}{E_0}\left(1 + \frac{E_D}{E_q} + \frac{E_0^2}{E_D E_q}\right)\right], \tag{20}$$

where

$$E_D = \frac{k_B T}{e}\frac{2\pi}{\Lambda_G} \quad \text{and} \quad E_q = \frac{2\pi e}{\varepsilon\Lambda_G}\frac{(N - N_0^+)N_0^+}{N},$$

are the diffusion field and the trap-limited space-charge field, respectively, k_B is the Boltzmann's constant, T is the absolute temperature, e is the charge of an electron, and N and N_0^+ are the total concentration of impurities and the concentration of ionized impurities, respectively. Note that if $E_0 = 0$ (diffusion only) or $E_0 \gg E_q$ (high field), the phase-shift $\Phi = \pi/2$. The model also predicts a single exponential function for the grating formation and erasure. For erasure by uniform illumination, the space charge field is given by [22]:

$$E_{SC}(x, t) = m E_{SC} \cos\left(\frac{2\pi x}{\Lambda_G} + \Phi\right) \exp(-t/\tau_r), \tag{21}$$

where τ_r is the response time.

III. MOLECULAR DESIGN AND PHYSICS OF POLYMERIC AND MOLECULAR PR MATERIALS

The necessary elements for a PR polymer, therefore, are

1. photogeneration centers of the mobile charge carrier;
2. transport channels for the mobile charge carriers under the driving force of diffusion and drift;
3. trapping sites that restrain the mobile charge carrier;
4. mechanism by which the refractive index can be changed in response to the electric field.

When the space-charge field can be constructed, the index modulation can also be built correspondingly. This mechanism can be a linear EO effect or an index modulation via alignment of the NLO chromophore in a low glass transition temperature (T_g) PR material.

A. Photogeneration of Charge Carriers

The study of photoconductivity in organic materials began in the late 1950s and continues today due to their important applications such as xerographic photoreceptors. Several excellent reviews deal with photoconduction in polymers. These studies, in fact, provided the solid foundation and knowledge for understanding the PR effect and designing new PR materials. Since polymers are usually disordered materials, the band model developed for semiconductors cannot be applied to explain polymer photoconductive behavior. Organic materials usually possess small dielectric constants so that the germinate recombination (due to Coulombic attraction) dominates, leading to very low quantum yields for the photogeneration of charge carriers.

The models that are most widely used to describe the photogeneration of charge carriers in the presence of a dominating germinate recombination process are based on the theories developed by Onsager [23]. In these models, the generation of free carriers is treated as a two-step process. In the first step, a bound electron–hole pair with the localized hole and a free hot electron is formed upon absorbing a photon. The hot electron rapidly loses its excess kinetic energy by scattering and becomes thermalized at a mean distance r_0 from the immobile hole. The formation of this intermediate charge-transfer state has an initial quantum efficiency ϕ_0 which is independent of the applied field. In the second step, the intermediate charge-transfer state can dissociate into a free electron or free hole, or recombine. Therefore, photogeneration efficiency is the product of the probability of the formation of the charge-transfer state and the probability of dissociation. The Onsager expression for the dissociation probability is developed by Mozumder [24], and quantum efficiency can be written as (SI system)

$$\phi(r_0, E) = \phi_0 \left[1 - \frac{k_B T}{e E r_0} \sum_{m=0}^{\infty} I_m \left(\frac{e^2}{4\pi\varepsilon_0\varepsilon k_B T r_0} \right) I_m \left(\frac{e E r_0}{k_B T} \right) \right], \quad (22)$$

where the recurrence formula $I_m(x)$ is given by

$$I_m(x) = I_{m-1}(x) - \frac{\exp(-x)x^m}{m!} \quad \text{and} \quad I_0(x) = 1 - \exp(-x). \quad (23)$$

The theory predicts a strong dependence of photogeneration efficiencies on the field and it approaches unity at high field. The temperature sensitivity decreases with the increase in field. The theory has found satisfactory explanations in the photogeneration process in many organic disordered systems, such as PVK (Scheme 1a) [25], and triphenylamine doped in polycarbonate [26]. Figure 4 shows an example of the field dependence of ϕ calculated from Eq. (22) (the solid lines) to fit the quantum efficiency data at room temperature for hole and electron generation in an amorphous material. The material consists of a sexithiophene covalently linked with a methine dye molecule (compound **1**) (Scheme 2).

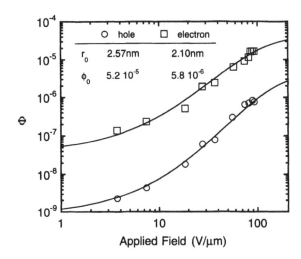

Figure 4 Photogeneration quantum efficiencies for hole and electron generation in compound **1**. The solid lines are Onsager theory fittings with parameters listed in the figure. (Reprinted from Ref. 27. Copyright 2000 The American Physical Society.)

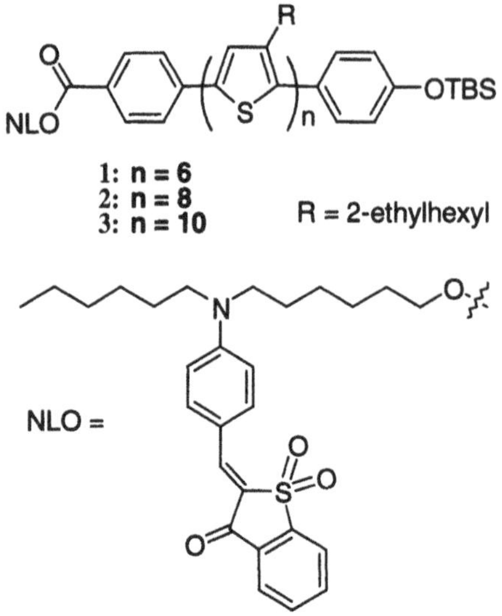

Scheme 2

The strong dependence of photogeneration of the charge carrier on an electric field implies that a strong external field is required for good performance on organic PR materials. It is also understandable that the PR materials must have enough absorption at the operating wavelength to generate electron–hole pairs. A paradigm for designing new PR materials is to ensure that photons are mostly absorbed by the photogenerators and the absorption by the other components is minimized. However, our recent results show that the PR effect can be remarkably large when the EO chromophores also play the role of a photoconductive component. The situation can occur if the bound electron–hole pairs generated by the NLO chromophores can efficiently dissociate into a free charge carrier.

B. Transport of Charge Carriers

1. Relationship of Response Time and Sensitivity of PR Effect to Carrier Mobility

Charge transport is one of the important processes that control the speed of the PR index grating formation and the PR sensitivity. According to the standard theory of photorefraction [21], the response time for the formation and erasure of the space-charge field [τ_r in Eq. (21)] is proportional to the dielectric relaxation time τ_{di} with a coefficient $c(\Lambda_G, E_0)$ that is a function of the grating spacing and the applied field; that is, $\tau_r = c(\Lambda_G, E_0)\tau_{di}$. Also, the dielectric relaxation time is inversely proportional to the conductivity $\sigma = e\mu n_c$ as

$$\tau_{di} = \frac{\varepsilon}{e\mu n_c}, \tag{24}$$

where μ is the carrier mobility and n_c is the number density of the carriers, which is proportional to the light intensity I, the quantum efficiency ϕ [Eq. (22)], and the recombination time τ_R. Since the initial quantum efficiency in Eq. (22) is proportional to the absorption coefficient of the photosensitizer α, we can write $\phi = \alpha\gamma_0$. The PR sensitivity S_{n1} is defined as the index change per absorption energy per unit volume at the initial formation of the PR grating $S_{n1} = \Delta n/\alpha I t$. Combining Eqs. (17), (21), and (24), the PR sensitivity can be expressed as

$$S_{n1} = \left.\frac{\partial n}{\alpha I \partial t}\right|_{t=0} \approx \frac{n_0^3 r_{eff}}{2} \frac{e}{\varepsilon} \frac{\mu E_{SC} \tau_R}{d} \frac{\gamma_0}{c(\Lambda_G, E_0)}. \tag{25}$$

Therefore, the response time is inversely proportional to the mobility of the carriers and the PR sensitivity is just the opposite.

It is worth pointing out that in low-T_g PR materials with molecular birefringence as the dominant contribution to the index modulation, the real response time and the PR sensitivity can also be limited by the diffusion time of angular rotation of the chromophores under the drive of the space-charge field, and even the time constant τ_e for the formation of the space-charge field may be smaller [27].

2. Charge Transport Mechanism in Organic Polymeric and Molecular Materials

Charge transport in organic polymeric photoconductors is believed to undergo hopping or tunneling. The excited π-electron may tunnel through a potential barrier, which can be considered equal to the molecule ionization potential. The probability of the tunneling transition must be greater than the probability of returning to the initial state in order to realize an actual transporting. The nature of the barrier depends on the Coulombic potential between the electron and positive ion, as well as the affinity of the neutral molecules. The process can be regarded as an oxidation-reduction process. The hopping model is well established during studies of hole-transporting polymers: PVK and their charge transfer complexes, especially with electron deficient molecule, trinitrofluorenone (TNF) (Scheme 1c), and molecularly doped polymers such as polycarbonate and triphenylamine or N-isopropylcarbazole. This group of polymers and compounds still plays a very important role in preparing PR materials, especially in composite polymers.

The typical hopping process occurs among point-like localized sites with their geometric size much smaller than their average separations. Upon photogeneration, some of the sites are positively charged under the action of the external field. Neutral molecules will repetitively transfer electrons to their neighboring cationic sites. The net result of this process is the motion of the positive charges across the bulk materials.

It is worth mentioning that conjugated polymers are another class of photoconductive materials. Conjugated polymers possess delocalized π-electrons along the polymer chain. The bands in these types of polymers are relatively broader than the saturated polymers, and they are analogues to the inorganic semiconductors. The excellent electronic conductivity along the conjugation chains has been well explained by the soliton, polaron, and bipolaron models. On the other hand, the similarity of the absorption spectra between the solid and solution states implies that the intermolecular interaction is insignificant. Therefore, the charge transfer mechanism can be considered to occur in a network of highly conductive parts and dielectric regions. The charge transport can be understood in the frame of a band model intraconjugated chain and as a hopping or tunneling model for the inter-chain transition. The decorrelation of the photoconductivity as a fast and a slow component supports the view that the slow component of the photoconductivity is determined by the hopping process. The mobility in disordered organic films has been discussed based upon the Poole–Frenkel effect, kinetic rate models, the Marcus theory, the dipole trap argument, and by disordered formalism [28]. The transport models are mostly distinguished by the form of charge transfer between transport molecules, by the distributions of site energy and position, and by the formalism for combining hopping events to calculate bulk transport. Among these models, the Poole–Frenkel effect is the most frequently used to describe charge transport. The Poole–Frenkel effect describes the reduction in ionization energy of a carrier in a Coulombic potential by an applied field. For the one-dimension case, the field dependence of mobility has the relationship:

$$\mu = \mu_0 \exp \frac{\beta_{PF} E^{1/2}}{k_B T}, \tag{26}$$

where $\beta_{PF} = (e^3/\pi\varepsilon\varepsilon_0)^{1/2}$, ε is the dielectric constant at high frequency, and ε_0 is the permittivity of free space. The field dependency of mobility ($\log \mu \propto \beta_{PF} E^{1/2}$) derived from the Poole–Frenkel effect is almost always observed in polymers. The disorder formalism and the dipole trap argument also predict a similar relationship.

Other models such as kinetic rate theory predict stronger field dependence, contrary to the observations in most organic materials. Kinetic rate theory assumes that the electron and hole transports occur by hopping among the localized states, and the mobility can be adopted as

$$\mu = 2\frac{\rho}{E}\nu\exp\left(-\frac{\Delta(E)}{k_B T}\right)\sinh\left(\frac{\rho e E}{2k_B T}\right), \tag{27}$$

where ρ is the average separation of the transporting agents, ν is the attempt frequency of electron exchange between charged and uncharged localized discrete chemical species, $\Delta(E) = \Delta_0 - \beta_{kR}E$ is the activation energy for the hopping process, and Δ_0 is the zero field intercept, and β_{kR} is a constant independent of E. The main assumptions for this model are: (1) the electronic carriers are localized on discrete chemical species, and (2) the transport is an activated process of charge motion from one site to another with an average energy barrier height Δ [29]. Therefore, more energy is necessary for each hop than ionization energy of a carrier in a Coulombic potential in the Poole–Frenkel effect, and the mobility exhibits a stronger dependence on the applied field.

Experimentally, the charged carrier mobility can be determined by a time of flight (TOF) technique (for details see [28]).

3. Effect of Polar Molecules Effect on Carrier Mobility in PR Materials

A special case in charge transport is polymeric materials with dipole chromophores. The dipole molecules, a necessary element in PR polymers, can strongly affect the charged carrier transport properties. Studies on the transport properties have been carried out in several systems consisting of combinations of two polymer binders, six charge transport agents (four for holes and two for electrons), and varying concentrations of two highly polar EO chromophores. It is revealed that the dipoles will generate a random electric field, which interferes with charge transport by increasing the width of the hopping site energy distribution. Thus, the carrier mobility and PR speed will be reduced greatly [30,31].

4. Special Considerations in Low Glass Transition Temperature Materials

The dipoles in a low-T_g material can be aligned *in situ* by an external field during the mobility measurement. The material can become more ordered at a high applied field than at a low one and can undergo a quasimorphological change with the increase of the applied field. The carrier mobility can behave differently in the field ranges studied. For example, Fig. 5 shows the field dependence of the mobility of holes and electrons in materials that consist of compound **1** [27]

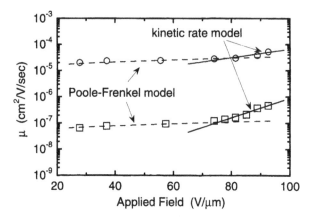

Figure 5 Hole and electron mobility (circles and squares) in compound **1** as functions of the applied field. The dashed and solid lines are best fittings using Eqs. (26) and (27), respectively. In the fitting, the dielectric constant ε is assumed to be 3.0, and the average distance value ρ for both hole and electron transport species is 13.9 Å. The value is calculated from the formula $\rho = [M/(Ad_0)]^{1/3}$, where M is the molecular weight of compound **1** (1946.16), d_0 is the density (1.2), and A is Avogadro's number (6.02×10^{23}). (Reprinted from Ref. 27. Copyright 2000 The American Physical Society.)

(Scheme 2). It can be seen that, in the low field range of $E < 70$ V/μm, the mobility shows much weaker dependence on the field and follows the relationship predicted by the Poole–Frenkel effect. In the high field range ($70 < E < 100$ V/μm), the data can be well fitted into the kinetic rate theory. This inconsistency in the mechanisms throughout the measurement range can be attributed to a quasi-morphological difference at different applied fields as a result of the low-T_g nature of the material. The chromophores are more randomized at a low field than at a high field. The dashed and solid lines of Fig. 5 are the best fitting results using Eqs. (26) and (27), respectively [27].

C. Traps and Photoinduced Space-Charge Fields

The presence of a trapping center is very important since Eq. (19) indicates that the steady-state strength of the photoinduced space-charge field depends on the number density of the deep traps. Nevertheless, the nature of the traps in organic PR materials is the least studied of all the elements for the PR effect. The main reason is the lack of structural information of the trapping centers. The amorphous nature of these materials warrants the existence of a variety of trapping centers, such as energy levels localized at impurities or structural defects. However, one can differentiate between deep traps, which are localized near the center of the band gap, and shallow traps, which are located near the highest occupied molecular orbital (HOMO) or lowest unoccupied molecular orbital (LUMO) of the transport molecule. The charges trapped in shallow traps can usually be released thermally. The trap density in polymeric and molecular materials can be determined from the PR grating formation experiments, such as four-wave mixing and two-beam coupling, with subsequent parameter fittings into the standard PR model [by using Eqs. (4), or (8), (17), and (19)] [8,26]. To achieve a high space-charge field, a large amount of *deep* traps is required. The importance of the deep traps for PR behavior in polymeric material has been demonstrated by an optical trap activation study [32]. Three-fold enhancement in the steady-state diffraction efficiency is observed in a composite consisting of poly(methyl methacrylate):(1,3-dimethyl-2,2-tetramethylene-5-nitrobenz-imidazoline):C_{60} by activation of the deep traps using uniform illumination prior to the measurement. However, the deep trap density should not be too high or they will provide an effective channel for charge migration.

A spectroscopic method has been recently developed to determine the trap density in C_{60}-sensitized PR polymers [33]. It monitored the extinction coefficient increase due to the increase in number density of C_{60} anion traps with irradiation time, and it was found that the measured concentration of C_{60} anion and that of the traps determined by the spectroscopic method and PR grating formation experiments, respectively, agree with each other to within a factor of three. The study also revealed that the trap density is not constant but history-dependent and varies with both light intensity and electric field, which is in contrast to the assumption of constant trap density in the standard model.

D. Nonlinear Optical Chromophores and Refractive Index Modulation

The last requirement for a PR effect is the mechanism for index modulation in the response of the space-charge field. Two mechanisms have been found to lead to index change: EO response and birefringence.

1. Origins for Index Modulations

In an organic molecule, the nonlinear optical effect originates from nonlinear polarization of the molecules. The polarizability of a molecule is the ability of a charge in the molecule to be displaced under the driving of the electric field. Under an intense optical field, induced polarizability p can be expressed as a polynomial function of local field strength e_i [34]:

$$p_i = \alpha_{ij}e_j + \beta_{ijk}e_je_k + \gamma_{ijkl}e_je_ke_l, \ldots, \tag{28}$$

where α_{ij} is linear polarizability and β_{ijk} and γ_{ijkl} are the first and second hyperpolarizabilities that are responsible for the second and third nonlinear optical effects, respectively where i, j, k, l represent molecule coordinates x, y, z. The value of β_{ijk} is nonzero only when the charge distribution of the molecule is asymmetric. In many nonlinear optical chromophores, the linear polarizability properties can be approximately treated as rod-shaped so that $\alpha_\perp(-\omega; \omega) = \alpha_{xx}(-\omega; \omega) = \alpha_{yy}(-\omega; \omega)$ and $\alpha_\parallel(-\omega; \omega) = \alpha_{zz}(-\omega; \omega)$. Therefore, the anisotropy of the optical polarizability is $\Delta\alpha(-\omega; \omega) = \alpha_\parallel(-\omega; \omega) - \alpha_\perp(-\omega; \omega)$. The first hyperpolarizability is assumed to be quasi-one-dimensional; only the dominant $\beta_{zzz}(-\omega; 0, \omega)$ component is taken into account, and others are neglected.

Macroscopically, the induced polarization P in an optically nonlinear material by an electric field $\vec{E}$ can be expressed as

$$P_I = \chi_{IJ}^{(1)} E_j + \chi_{IJK}^{(2)} E_J E_K + \chi_{IJKL}^{(3)} E_J E_K E_L, \ldots \tag{29}$$

The linear susceptibility $\chi^{(1)}$ is related to optical refraction and absorption. The most common effects due to second-order susceptibility $\chi^{(2)}$ are frequency doubling $\chi^{(2)}(-2\omega; \omega, \omega)$ and the EO (Pockels) effect $\chi^{(2)}(-\omega; 0, \omega)$. The third-order susceptibility $\chi^{(3)}$ is responsible for such phenomena as frequency tripling and the Kerr effect.

Since the dipoles of chromophore molecules are randomly distributed in an inert organic matrix in amorphous PR materials, the material is centrosymmetric and no second-order optical nonlinearity can be observed. However, in the presence of a dc external field, the dipole molecules tend to be aligned along the direction of the field and the bulk properties become asymmetric. Under the assumption that the interaction between the molecular dipoles is negligible compared to the interaction between the dipoles and the external poling field (oriented gas model), the linear anisotropy induced by the external field along Z axis at weak poling field limit ($\mu E / k_B T \ll 1$) is [35,36]:

$$\Delta\chi_{ZZ}^{(1)} = \frac{2}{45} N f^\omega \Delta\alpha(-\omega; \omega) \left(\frac{\mu E}{k_B T}\right)^2$$

$$\Delta\chi_{XX}^{(1)} = \Delta\chi_{YY}^{(1)} = -\frac{1}{45} N f^\omega \Delta\alpha(-\omega; \omega) \left(\frac{\mu E}{k_B T}\right)^2, \tag{30}$$

where f^ω are the local field factors at optical frequency ω, and E is the poling field. The opposite signs of these two equations mean that the polarizability along the direction of the external field is increased, while that perpendicular to the direction of the field is decreased. Since the material poled under an electric field shows a $C_{\infty v}$ symmetry, the nonzero components for the second-order susceptibility tensor are:

$$\chi_{ZZZ}^{(2)} = \chi_{33}^{(2)} = Nf^\omega f^\omega f^0 \beta_{zzz}(-\omega; 0, \omega)\frac{\mu E}{5k_B T}$$

$$\chi_{XXZ}^{(2)} = \chi_{YYZ}^{(2)} = \chi_{XZX}^{(2)} = \chi_{YZY}^{(2)} = \chi_{ZXX}^{(2)} = \chi_{ZYY}^{(2)} = \chi_{13}^{(2)} \tag{31}$$

$$= Nf^\omega f^\omega f^0 \beta_{zzz}(-\omega; 0, \omega)\frac{\mu E}{15k_B T},$$

where N is number density, f^0 are the static local field factors, μ is the dipole moment of the NLO chromophore along its dominant axis. Note that all of the nonlinear polarizabilities have the same sign.

The refractive index modulations due to the anisotropy of the optical polarizability and the EO effects are related to the $\Delta\chi^{(1)}$ and $\chi^{(2)}$:

$$\Delta n_I^{BR} = \frac{1}{2n_I}\Delta\chi_{II}^{(1)}, \tag{32}$$

and

$$\Delta n_I^{EO} = \frac{1}{n_I}\chi_{IIK}^{(2)}E_K, \tag{33}$$

where I, K represent laboratory coordinates, X, Y, Z. The $\chi^{(2)}$ term is related to the EO coefficient r by the equation

$$\chi_{IIK}^{(2)} = -\frac{1}{2}n_I^3 r_{IIK}. \tag{34}$$

In practical PR experiments, the $\Delta\chi^{(1)}$ and $\chi^{(2)}$ in Eqs. (32) and (33) must be regarded as effective values, which are functions of the polarization of the probe beam, the symmetry of anisotropic materials, and the geometry of the experiments.

2. PR Gratings Due to Pure EO Effect

EO response can be introduced into polymers containing the donor–conjugation–acceptor moieties by applying a very strong electric field while the polymer is heated to its T_g. The field-induced order is subsequently "frozen in" by cooling the polymer films well below T_g. The achievable second-order susceptibility is determined by Eq. (31), where the temperature is the poling temperature. Equation (31) reveals that the $\mu\beta$ value of a NLO chromophore is crucial for the achievable second-order susceptibility. For high T_g EO polymers, a commonly used figure of merit (*FOM*) of chromophores is

$$FOM^{EO} = \frac{\mu\beta}{M}, \tag{35}$$

where M is the molecular weight of the chromophore. The index modulation by the space-charge field follows Eq. (17), where the r_{eff} values at the experimental geometry shown in Fig. 1 for s-polarized and p-polarized probe beams are [37,38]:

$$
\begin{aligned}
r_s^{eff} &= r_{13} \cos\varphi \\
r_p^{eff} &= r_{33} \sin\theta_1 \sin\theta_2 \cos\varphi + r_{13} \cos\theta_1 \cos\theta_2 \cos\varphi \\
&\quad + r_{13} \sin(\theta_1 + \theta_2) \sin\varphi,
\end{aligned} \tag{36}
$$

where the first index represents the combination IJ and has the values $1 = XX$, $2 = YY$, $3 = ZZ$, $4 = YZ = ZY$, $5 = ZX = XZ$, and $6 = XY = YX$. The second index is $1 = X$, $2 = Y$, and $3 = Z$.

3. PR Gratings Enhanced by Birefringence

The origin of the PR effect can be quite different in amorphous organic materials with low-T_g. The materials are isotropic in the absence of an electric field but are poled easily at room temperature by an external field. The combined effect of external field and the periodic space-charge field results in the *in situ* poling of the PR chromophores. Under the experimental geometry of Fig. 1, the total poling field is the vector sum of the external field E_{ext} and space-charge field ΔE_{SC} [Eq. (18)]:

$$\vec{E}_T(x) = \left[E_{ext} \cos\varphi - m E_{SC} \cos\left(\frac{2\pi x}{\Lambda_G} + \Phi\right) \right] \hat{x} + (E_{ext} \sin\varphi)\hat{y}. \tag{37}$$

Both the amplitude and the direction of the total poling field are functions of position x. The dipole chromophore will align in response to the field $\vec{E}_T(x)$, and a periodic refractive index modulation will be formed due to both molecular anisotropy and the Pockels effect [37]. The contribution from molecular anisotropy is an important, sometimes even dominant, mechanism.

With the experimental geometry shown in Fig. 1 for an s-polarized or p-polarized probe beam, the first-order approximations of index modulation due to birefringence contribution are [38]:

$$\Delta n_s^{BR} = -\frac{1}{n_0}C^{BR}E_{ext}E_{SC}\cos\varphi$$

$$\Delta n_p^{BR} = \frac{1}{n_0}C^{BR}E_{ext}E_{SC}\left[2\sin\theta_1\sin\theta_2\cos\varphi - \cos\theta_1\cos\theta_2\cos\varphi \quad (38)\right.$$

$$\left. + \frac{3}{2}\sin(\theta_1+\theta_2)\sin\varphi\right]$$

and those due to EO contribution are [37]:

$$\Delta n_s^{EO} = \frac{2}{n_0}C^{EO}E_{ext}E_{SC}\cos\varphi$$

$$\Delta n_p^{EO} = \frac{2}{n_0}C^{EO}E_{ext}E_{SC}[3\sin\theta_1\sin\theta_2\cos\varphi + \cos\theta_1\cos\theta_2\cos\varphi \quad (39)$$

$$+ \sin(\theta_1+\theta_2)\sin\varphi],$$

where

$$C^{BR} = \frac{2}{45}Nf^\omega\Delta\alpha(-\omega;\omega)\left(\frac{\mu}{k_BT}\right)^2, \quad \text{and}$$

$$C^{EO} = Nf^\omega f^\omega f^0\beta_{zzz}(-\omega;0,\omega)\frac{\mu}{15k_BT}.$$

Note that the index modulation from the EO effect in a low-T_g material [Eq. 39] is about twice as large as that in a high-T_g material [Eqs. (17) and (36)]. The total index modulation in a low-T_g material is given by the sum of those from both birefringent and EO effects as in Eqs. (38) and (39). On a microscopic level, a new figure of merit for a PR chromophore is defined [38–40]:

$$FOM^{PR} = \frac{9k_BT\mu\beta + 2\mu^2\Delta\alpha}{k_BTM}. \quad (40)$$

Experimentally, the r_{eff} value originating from the EO effect and birefringence can be determined by using ellipsometric [41–43] or interferometric [44] methods.

Before we close this section, we make some conjectures. We assume a high-T_g polymer with achievable $r_{33} = 30$ pm/V, and therefore, $r_{13} \cong r_{33}/3 = 10$ pm/V. Assuming $E_{SC} = 50$ V/μm and $n_0 = 1.6$, an index modulation

$\Delta n \sim 0.004$ should be achieved, which is about the index modulation needed for the overmodulation. Although an EO coefficient of 30 pm/V can be easily achieved in high-T_g pure EO polymers, such a high EO coefficient and the large index modulation has never been obtained in high-T_g PR polymers.

IV. CHEMISTRY AND PHYSICAL PROPERTIES OF POLYMERIC AND MOLECULAR PR MATERIALS

Polymeric and glassy molecular PR materials can be classified into two categories: the polymer composites and fully functionalized polymers or molecules. Depending on the function of the matrix polymer, there are three classes of polymer composites that can be based on inert polymers, NLO polymers, or photoconducting polymers. Other functional components needed for a PR effect are small molecules as dopants. The third type of PR composites has been the most popular material. For example, PVK-based composite PR polymers have been extensively studied and have shown promising results. They are frequently sensitized by doping with electron–deficient TNF (Scheme 1c), which readily forms a charge-transfer complex with carbazole units, or with C_{60}. A plasticizer is usually added to the polymer matrix in order to lower the T_g. On the other hand, fully functionalized polymers or molecules contain all the functional components in a single polymer chain or molecule. These types of materials possess minimized phase separation and, therefore, long-term stability and high optical quality. Fully functionalized molecular materials are relatively less studied but have recently exhibited promising properties which are discussed in the next section.

Based on their thermal properties, polymeric and glassy molecular PR materials can be classified as high-T_g and low-T_g materials. The T_g values of polymer composites can be lowered by doping small molecules as plasticizer. Introduction of a long alkyl side chain can effectively reduce the T_g of fully functionalized polymers. The glassy molecular materials usually possess low T_g if they are amorphous. For the high-T_g and low-T_g materials, the figures of merit of the NLO chromophores are defined by Eqs. (35) and (40), respectively. The latter are obviously enhanced by contributions from the polarizability anisotropy of the NLO chromophore.

In this section, we review in chronological order the development of the polymeric composite, fully functionalized polymeric, and the glassy molecular PR materials, respectively. For the latter two categories, we mainly focus on the progress made in our laboratory. More detailed accounts of the materials developed in other groups can be found in the references [9,10,37]. Another promising class of PR materials is PR liquid crystals, which have been reviewed by G. Wiederrecht, in Chapter 7.

A. Polymeric Composite PR Materials

The first polymer composite that showed a PR effect was composed of a crosslinkable, NLO epoxy polymer (bisphenol-A-diglycidylether-4-nitro-1,2-phenylenediamine, bisA-NPDA) and a hole transporting agent, diethylamino-benzaldehydediphenylhydrazone [8]. Two-beam coupling (2BC) and four-wave mixing (FWM) experiments performed at 647 nm using a Kr$^+$ ion laser revealed the diffraction efficiency η of 5×10^{-5}, and an optical gain Γ of 0.33 cm^{-1}. The refractive index grating with magnitude 4.5×10^{-6} was found to have a phase shift of 90 degrees with respect to the intensity grating. After this report, numerous different types of polymeric and glassy molecular materials have appeared.

A net optical gain in organic PR materials was first obtained from a PVK-based composite material. The NLO chromophore, 3-fluoro-4-N,N-diethylamino-β-nitrostyrene (F-DEANST) (Scheme 3a) (33 wt%) and the sensitizer, TNF were doped into a PVK matrix [45]. The T_g of the resulting composite was about 40°C, a dramatic decrease from that of the pure PVK ($T_g = 212$°C). The photoconductivity is measured to be 2.0×10^{-12} cm/(ΩW). The EO measurements using the Mach–Zehnder interferometric technique revealed an $n^3 r_{33}$ value of 2.4 pm/V at $E = 40$ V/μm and λ = 830 nm. A net optical gain of 7.2 cm^{-1} was observed at λ = 753 nm at an external field of 40 V/μm (the Γ value was 8.6 cm^{-1}, while the α value was 1.4 cm^{-1}). The diffraction efficiency η at 647 nm was 1.2%, which is higher than that at longer wavelengths due to higher sensitivity. It was also pointed out that the NLO chromophore accounted for 10% of absorption at this wavelength. Therefore, the absorption grating may also have contributed to the diffraction of the reading beam in the FWM experiment. Grating growth rate on the order of 100 ms was observed. More importantly, the studies on the polymer lead to the discovery of the so-called *orientational enhancement* of PR performance [38], which has afterward been fully utilized in the preparation of PR composite materials. The discovery has also triggered searching of the new PR chromophores based on a new figure of merit for the PR chromophore as defined by Eq. (40).

Significant improvement in PR performance was obtained in a PVK:TNF composite doped with 2,5-dimethyl-4-(p-nitrophenylazo)anisole (DMNPAA) (Scheme 3b) as the NLO chromophore [46]. This polymer composite showed a gain coefficient of 30 cm^{-1}, and diffraction efficiency of 5% in a 105-μm thick film at a field of 40 V/μm and a wavelength of 647 nm. A net optical gain of 6 cm^{-1} is obtained. The PR performances of this polymer composite were further dramatically improved by adding N-ethylcarbazole (ECZ) (Scheme 4a) to decrease the T_g [47]. The composition of the polymer was PVK:DMNPAA:ECZ:TNF = 33:50:16:1 wt%. Over-modulation of the diffraction efficiency and an extremely large net optical gain of 207 cm^{-1} were achieved

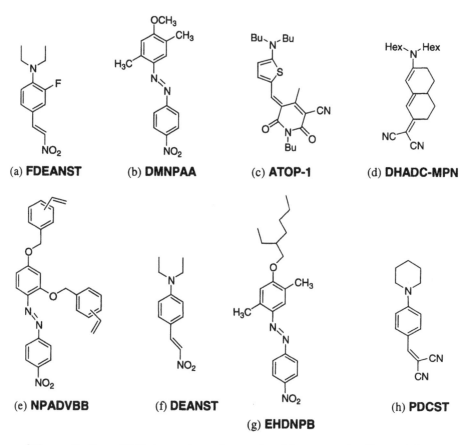

(a) FDEANST

(b) DMNPAA

(c) ATOP-1

(d) DHADC-MPN

(e) NPADVBB

(f) DEANST

(g) EHDNPB

(h) PDCST

Scheme 3 Some NLO chromophores that are doped into composite PR materials.

(a) ECZ

(b) DBP

(c) TCP

(d) BBP

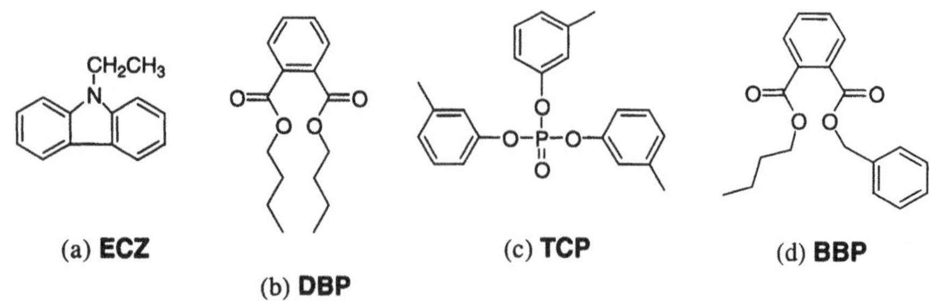

Scheme 4 Some compounds that are doped into composite PR materials as plasticizers to lower the T_g of the materials.

($\Gamma = 220$ cm^{-1} at a field of 90 V/μm). The authors used an ellipsometric technique to deduce the relative contributions to the index modulation from the poling birefringence, the Pockels effect, and the Kerr effect, respectively [48]. It was confirmed that large photorefractivity is mainly due to the periodic orientation of the NLO chromophore by the internal space-charge field. Based on the orientational enhancement effect and the new figure of merit of NLO chromophores, numerous examples of well-performing PVK-based systems were reported.

Wurthner et al. described a new series of merocyanine dyes with the main structural feature of aminothienyldioxopyridine (ATOP) [49]. These dyes possessed a highly improved PR figure of merit. A PR composite consisting of ATOP-1 (Scheme 3c) (20 wt%), PVK, the IR sensitizer (2,4,7-trinitro-9-fluorenyliden)malononitrile (TNFDM) (Scheme 1d), and the plasticizer ECZ was prepared and examined at 790 nm. The field-dependent diffraction efficiency reached a maximum at $E = 68$ V/μm on the 105-μm thick sample, which corresponded to complete internal diffraction of the readout beam ($\eta_{int} = 1$). Externally, the achievable diffraction was limited to $\eta_{ext} = 0.3$ due to the absorption of the material at 790 nm. The maximum index-modulation amplitude, Δn was 5.2×10^{-3} at 95 V/μm. The gain coefficient reached -280 cm^{-1} for p-polarized and $+70$ cm^{-1} for s-polarized beams. The response time for the grating buildup was about 2800 ms at $E = 68$ V/μm. However, at higher loading levels, the PR response was diminished due to the aggregation of dyes [50].

Kippelen et al. demonstrate a large Δn (0.0085) at 830 nm in PVK composites comprised of 2-N,N-dihexylamino-7-dicyanomethylideny-3,4,5,6,10-pentahydronaphthalene (DHADC-MPN) (Schcme 3d) and 4-(4$'$-nitrophenylazo)-1,3-di[3$''$ or 4$''$-vinyl) benzyloxy]-benzene (NPADVBB) (Scheme 3e) as NLO chromophores, respectively [51,52]. Imaging through scattering media was also demonstrated by using a holographic time-grating technique.

It is found that numerous factors affect the PR performance of the PVK-based materials. A peculiar observation is that it is critical to maintain the composition of the composite system. A slight change in the composition results in a dramatic decrease in the diffraction efficiency. The PR properties change tremendously just by altering the chromophore density or lowering the T_g. The exact reasons for these changes in properties are not fully understood [53]. Orientation enhanced effect may explain some of the changes. Silence et al. have investigated different NLO chromophores (different aminonitrobenzene derivatives) as a function of dopants' structures in PVK:TNF composites. Optical gain values varied in a wide range, from 1.2 to 8.0 cm^{-1} [54].

The plasticizer plays an important role in PR performance. Zhang et al. have prepared the composite using PVK doped with diethylamino-β-nitrostyrene (DEANST) (Scheme 3f) as the NLO chromophore and C$_{60}$ as the sensitizer (PVK:DEANST:C$_{60}$ = 56:26.5:1.6) [55]. The achievable EO coefficient after contact poling was 4 pm/V. A maximum diffraction efficiency of 2×10^{-5} was observed. After adding 20 wt% dibutyl phthalate (DBP) (Scheme 4b) as the plasticizer, the new composite (DEANST: 25 wt%, C_{60}: 0.56 wt%) showed a dramatic increase in PR responses [56]. At $E = 40$ V/μm, the r_{33} value was determined to be 10 pm/V. The maximum optical gain of 4 cm^{-1} ($E = 50$ V/μm) was deduced. The FWM experiments indicated a fast index writing and erasure process. Further depression in T_g was achieved by using a liquid plasticizer, tricresylphosphate (TCP) [57] (Scheme 4c). The polymer composite, which consisted of PVK, DEANST (3.75 wt%), TCP (36 wt%), and C_{60} (0.22 wt%), exhibited a T_g below 14°C. Although the chromophore content was lower than that of the previous system, a higher EO coefficient was observed ($r_{33} = 37.6$ pm/V at 140 V/μm), and the maximum values of η and Γ were found to be 33% and 133.6 cm^{-1} at $E = 110$ V/μm, respectively. This corresponds to a net gain of 116.6 cm^{-1}.

A systematic study on the effect of plasticization on the PR properties is reported by Bolink et al. [58]. It is shown that ECZ can be used as an efficient plasticizer, leading to a large increase in the gain coefficient and the diffraction efficiency, which arises solely due to an improvement in the orientational mobility of the dispersed NLO molecules.

By changing the ratio of PVK:ECZ, Bittner et al. have investigated the influence of T_g and chromophore content on the steady-state PR performance of the PVK:DMNPAA:ECZ:TNF system [59]. A performance optimum is found for highly doped materials with T_g around room temperature. This is a result of two counteracting effects: the poling of the chromophores becomes more efficient with decreasing T_g; and the relative strength of the space-charge field decreases for T_g below room temperature.

Based on the PVK:FDEANST system, the PR performance as a function of the photosensitizer has been studied [54]. It is apparent that varying the sensitizer changes the photogeneration rate and greatly influences both optical gain and diffraction efficiency. By replacing TNF with TNFDM, the spectral sensitivity of composites can be extended to the near infrared (IR) region.

A remarkable result based on polysiloxanes (PSX) (Scheme 1b) with pendant carbazole groups is obtained by Zobel et al. [60]. The T_g of this system can be varied in a wide range of −45 to 51°C by changing the length of the alkyl spacer linking the carbazole fo the polymer backbone. Doped with DM-NPAA (43 wt%) and TNF (1 wt%), this composite exhibits a net optical gain of 220 cm^{-1} and diffraction efficiency of 60%. The utility of the composite of PSX:FDEANST:TNF for storage applications is demonstrated by recording digital data at a density of 0.52 Mbit/cm^2 and reading it back without error up to 5 min after recording. This material also shows excellent optical quality with light scattering reduced by a factor of 400 compared with a similar PVK-based system [49].

One major problem associated with composite PR materials is the inherent metastability. In such guest/host systems, a gradual crystallization of the small molecule dopants will cause phase separation and lead to the loss of optical transparency, which severely limits the shelf lifetime of the samples. For instance, the lifetime of DMNPAA:PVK:ECZ:TNF samples with 50 wt% of DMNPAA is found to vary between a few hours and a few weeks depending on the starting materials and the processing conditions [61]. Several approaches have been proposed for improving the shelf lifetime, such as by reducing the chromophore content, increasing T_g, and using liquid chromophore [62]. Unfortunately, longer lifetimes are obtained in most attempts at the expense of reduced PR efficiency. Meerholz et al. have used a eutectic mixture of two isomeric EO chromophores, 2,5-DMNPAA and 3,5-DMNPAA, to improve the sample lifetime by a factor of five [63]. The steady-state PR performance is found to be independent of the chromophore's isomer ratio. A net optical gain of 20 cm^{-1} ($\Gamma = 80$ cm^{-1}) and index modulation amplitude $\Delta n = 5.0$–5.4×10^{-3} are obtained in a sample with 39 wt% chromophore at 690 nm and 102 V/μm.

A similar approach is utilized to prepare a PR composite based on isomeric mixtures of the chromophore NPADVBB [64]. The composition is NPADVBB: PVK:ECZ:TNF = 39:39:20:2 wt%, whereas NPADVBB is a mixture of four regioisomers, with vinyl groups in positions (3,3′), (3,4′), (4,3′) and (4,4′). Although it is claimed that this composite exhibited an improved shelf-life by two orders of magnitude, the gain coefficient achieved is significantly smaller than that obtained for a similar DMNPAA-based sample.

Another approach is to use better engineered chromophores with added side groups that improve compatibility. Cox et al. have incorporated a racemic ethylhexyl group into the azo chromophore (1-(2′-ethylhexyloxy)-2,5-dimethyl-4-(4′-nitrophenylazo)-benzene, EHDNPB) (Scheme 3g) to increase the solubility of the dye in PVK, inhibiting crystallization and, at the same time, acting as a plasticizer [65]. A 60% device diffraction efficiency and gain coefficient of 120 cm^{-1} are achieved with 55 wt% chromophore.

Grunnet-Jepsen et al. have reported a PVK composite material consisting of 35 wt% chromophore, 4-piperidinobenzylidene-malononitrile (PDCST) (Scheme 3h), 15 wt% liquid plasticizer, butyl benzyl phthalate (BBP) (Scheme 4d), and 1 wt% of C$_{60}$ [66,67]. These crystallization-resistant samples exhibit a gain coefficient of 200 cm^{-1}, a rise time of 50 ms, and a sensitivity of 3 cm^3/kJ at $E = 100$ V/μm. It is suggested that the improvement in stability is due to the use of a liquid plasticizer BBP instead of the crystalline ECZ.

B. Fully Functionalized PR Polymers

The first demonstration of fully functionalized PR polymer is a multifunctional polyurethane, on which all the necessary functional species (charge sensitizer, charge transporter and NLO chromophore) are incorporated as side chains [68–70]. The PR effect is small and the largest gain obtained is only 2.3 cm^{-1}. The poor performance is attributed to the low molecular weight of the polymers, the inert polyurethane backbone, and the difficulty in finding the optimum ratio in a multiple component system.

1. Conjugated PR Polymers

The lessons learned from these initial works led to the design of a new class of polymers—conjugated PR polymers. Conjugated polymer backbones have several advantages for fully functionalized PR polymers. First of all, conjugated polymers are well known to be good photoconductors. Numerous conjugated polymers were studied and demonstrated to exhibit photoconductive effects. Second, conjugated polymers have a wide range of structural variation, permitting chemical modifications and functionalization. Because of these versatile features, incorporations of NLO chromophore and photosensitizer become feasible and the T_g of the resulting polymers can be fine tuned. Furthermore, when the conjugated polymer plays the role of a backbone, one can have more room to manipulate the density of the NLO chromophore and photosensitizer. Based on these considerations, a new polymer system (polymers **I** to **IV**) was synthesized by utilizing the Stille coupling reaction [71–73]. The structures of the synthesized conjugated PR polymers are shown as polymers **I** to **IV** (Scheme 5), where the dihydropyrrolopyrroldione (DPPD) moiety is introduced as the photosensitizer to extend the photosensitive region into a longer wavelength. The conjugated backbone plays the triple role of charge generator, charge transporting species, and backbone. The index modulation is accomplished by using NLO chromophore as a side chain.

The photoconductivities for polymers **I**, **II**, and **III** are found to be *ca.* 8×10^{-11}, 1.8×10^{-11}, and 4×10^{-11} cm/(ΩW), respectively, under a field strength of 150 V/μm and a laser intensity of 311 mW/cm^2. These values are comparable to those of the well-known conjugated polymers such as poly(*p*-phenylene vinylene) (PPV). Photoconductivity studies of these polymers show a strong dependence on the external field strength. A typical photocurrent response time of *ca.* 100 ms is estimated. It is found that the spectral dependence of the photocurrent has a similar shape to the absorption spectrum of the conjugated PR polymer. This seems to indicate that the optical excitation of the conjugated backbone is the origin of the photocharge generation. The quantum yield of the photogeneration of charge carriers and the charge carrier mobility (μ) is determined by using the TOF technique. The polymers were corona-discharge poled at their glass transition temperatures. The maximum optical gain for polymer **II** is 45 cm^{-1}. Similar results are obtained for polymers **I** and **III**. The

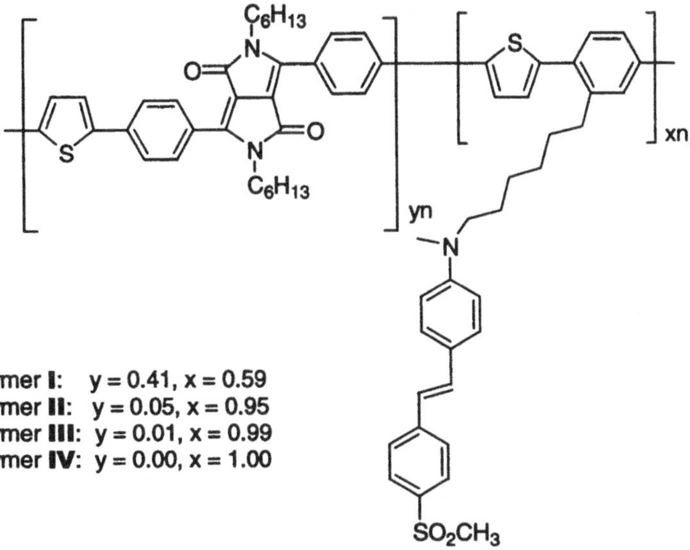

Polymer **I**:   y = 0.41, x = 0.59
Polymer **II**: y = 0.05, x = 0.95
Polymer **III**: y = 0.01, x = 0.99
Polymer **IV**: y = 0.00, x = 1.00

Scheme 5

effective density of the empty trap centers (N_e) and the maximum refractive index change (Δn) for polymer **II** are *ca.* 2×10^{14} cm^{-3} and 4×10^{-5}, respectively. For polymer **I**, an N_e of ca 1.9×10^{15} cm^{-3} and a Δn of 3×10^{-5} are deduced.

An unusual observation is that a large index change ($\Delta n \sim 7.1 \times 10^{-5}$) occurs under zero external field conditions. This result seems to imply that an internal field which assists the charge separation exists and enhances the PR effect under a zero field. It is also found that the optical gain decreases upon increasing the field in the region of 0 to 10 V/μm, and there exists a valley near the field of 10 V/μm. The optical gain increases as the field further increases. The results clearly indicate that the internal field is about 10 V/μm opposite to the direction of the poling field, which could be due to the orientation of the dipoles of the NLO chromophores after electric poling, or due to trapped charges from the corona poling process. However, due to its large absorption loss, the optical gain is not a net gain. The major issue is the low quantum efficiency of the photogeneration of charge carriers. Later on, the high T_g is also found to be responsible for the low optical gain coefficient. To circumvent this problem, the conjugated polymers containing a transition metal complex are designed and synthesized.

2. High-T_g Conjugated Polymers Containing Transition Metal Complexes

Ru(bpy)$_3$(II) (bpy = bipyridine) and its derivatives are known to exhibit an interesting metal-to-ligand charge transfer (MLCT) process, and their photochemistry, photophysics, electrochemistry, and electron/energy transfer properties have been extensively studied [74,75]. They have also been widely investigated as light-harvesting materials [76]. To utilize their interesting charge transfer properties in the synthesis of PR polymers, Ru(II) or Os(II) complexes are designed to chelate with the conjugated polymer backbone (macroligand) to accomplish a higher charge separation efficiency [77,78].

The new polymer system is designed based upon the idea illustrated in Fig. 6. In this system, a tris(bipyridyl) ruthenium complex is chosen as the photocharge generator and introduced into the backbone. The conjugated polymer backbone is chosen to play a dual role as the transporting channel for the charge carriers and as the macroligand to chelate with the transition metal complex. Upon excitation in the region of the MLCT transition of the Ru complex, electrons are injected into the polymer backbone (equivalent to n-doping of PPV), transported away through either intrachain migration or interchain hopping, and eventually trapped by the trapping centers.

The polymers are synthesized by utilizing the Heck coupling reaction. Their structures are shown as polymers **V** to **VII** (Scheme 6). The metal-to-ligand charge transfer of the ruthenium complexes of polymer **VI** results in

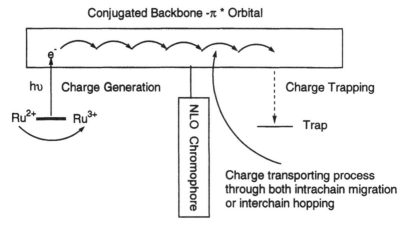

Figure 6   Schematic model showing the design concept of conjugated PR polymer containing transition metal complexes. (Reprinted from Ref. 77. Copyright 1996 American Institute of Physics.)

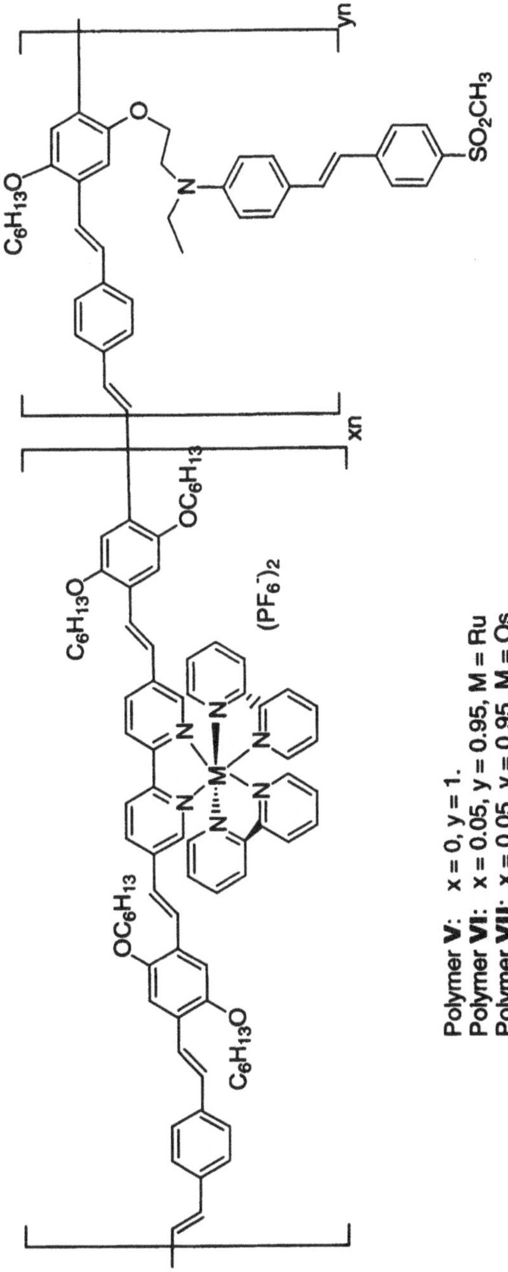

Polymer **V**:   x = 0, y = 1.
Polymer **VI**: x = 0.05, y = 0.95, M = Ru
Polymer **VII**: x = 0.05, y = 0.95, M = Os

Scheme 6

an absorption tail extending beyond 600 nm. Polymer **VI** exhibits a photoconductivity of 3×10^{-10} cm/(ΩW) and a photocharge generation efficiency of 0.2% under an external electric field of 95 kV/μm. This value is a significant improvement over polymers **I**, **II**, and **III**. After corona poling, an EO coefficient of 7 pm/V is obtained at 690 nm. An optical gain higher than 300 cm^{-1} at 690 nm is obtained. Since polymer **VI** exhibits an absorption coefficient of 102 cm^{-1}, a net optical gain of 198 cm^{-1} is obtained in this polymer at a zero external electric field. A diffraction efficiency of 0.5% on a 10 μm thin film is obtained. A phase shift close to 90° of the index grating with respect to the intensity distribution is determined. Polymer **V**, without the Ru-complexes, does not show clear photorefractivity.

Polymer **VII** containing 5 mol% Os-complex is synthesized by using the same methodology. Its UV/vis spectrum shows an absorption extending to 600 nm due to a spin allowed MLCT transition of the Os-complex. This polymer also has an absorption tail extending close to 800 nm, the wavelength region of commonly used diode lasers. The corresponding optical gain at 780 nm is 80.4 cm^{-1} which is not a net optical gain (α was measured to be 186 cm^{-1}). The data are taken under a zero external field and an asymmetric energy exchange is clearly observed.

3. Multifunctional Polyimides

The polyimide-base PR system [79,80] was designed on the premise that porphyrin-electron acceptor (quinones or imide moieties) systems are well-known model compounds for photosynthetic processes and exhibit very interesting charge transfer properties [81]. A high quantum yield of charge separation can be achieved in these systems. Polyimides are found to be photoconductive and allow charge transport [82]. Furthermore, polyimides possess high T_g and therefore, the electric field-induced dipole orientation can be fixed after imidization [83].

Polyimides carrying NLO chromophores are synthesized. The materials exhibit high T_g ($\sim$220°C). The photoconductivity of polyimide **VIII** (Scheme 7) is determined to be 1.1×10^{-12} cm/(ΩW) under an external field of 150 V/μm at $\lambda = 690$ nm and $I = 5.9$ mW/cm^2. After the sample is corona poled, an optical gain coefficient of 22.2 cm^{-1} under zero field conditions is detected. Polyimide **IX** exhibits an optical gain of 15 cm^{-1}, which is very close to the optical loss (17.6 cm^{-1}).

4. Low-T_g PR Polymers Containing Tris(bipyridyl) Ruthenium(II) Complexes as Photosensitizer

By introducing long alkyl side chains into the poly(p-phenylenevinylene) (PPV) backbones and to the amino groups on the NLO chromophores, a low-T_g polymer **X** (Scheme 8) containing ruthenium complexes and a conjugated system is

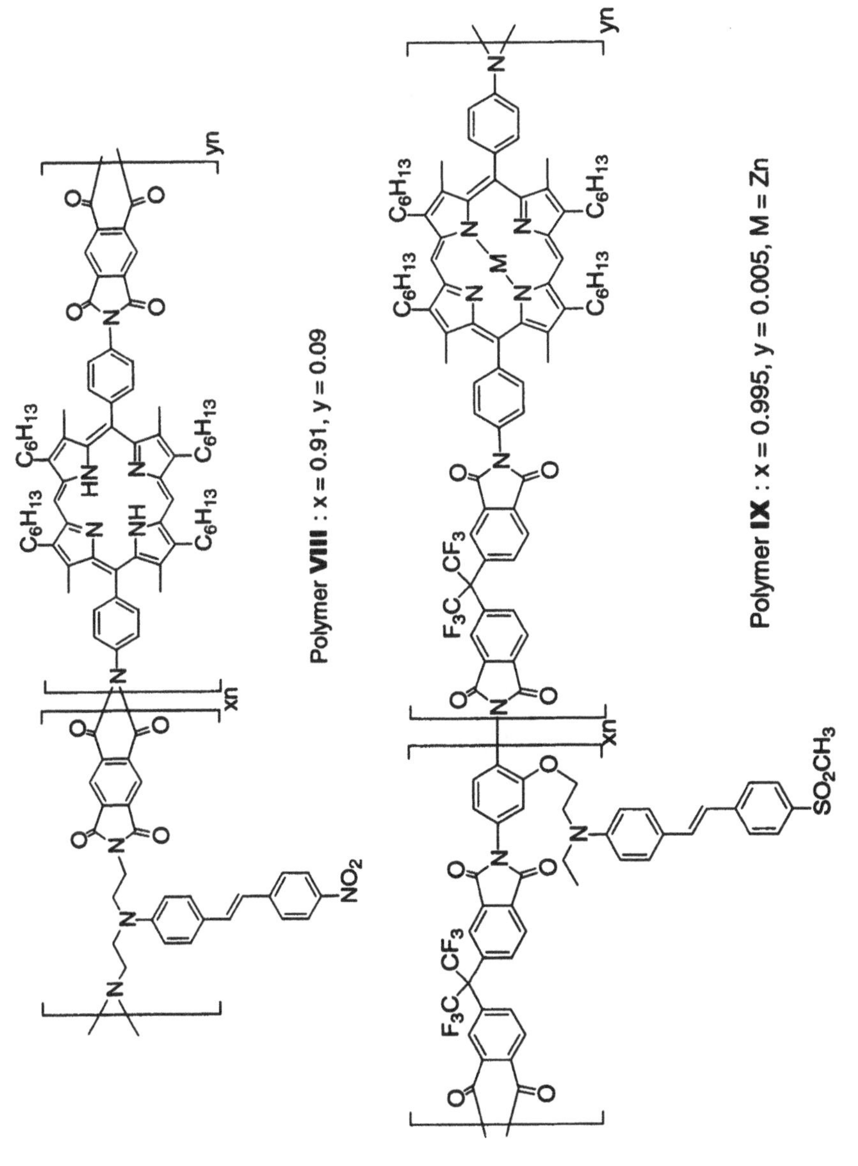

Polymer **VIII** : x = 0.91, y = 0.09

Polymer **IX** : x = 0.995, y = 0.005, M = Zn

Scheme 7

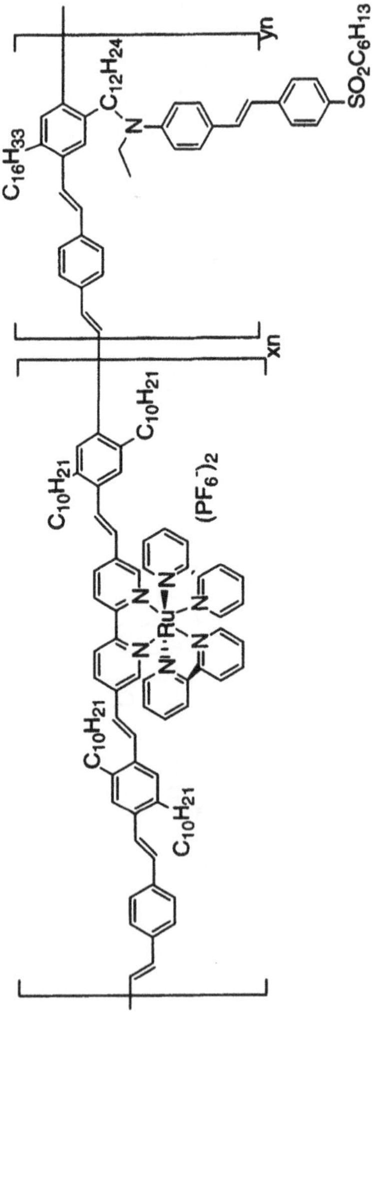

Polymer **X:** x = 0.01, y = 0.99

Scheme 8

synthesized [84]. The low T_g value (11°C) and the good solubility in organic solvents allow this polymer to be easily processed from its solution into high quality optical films thicker than 100 μm. Another advantage of introducing alkyl chains is that the absorption of the PPV backbone of the resulting polymer is blue-shifted compared to those with alkoxy substituents. This blue-shift minimizes the absorption overlap between the $\pi–\pi^*$ transition of the PPV backbone and the MLCT transition of the Ru complex. Thus, charge carriers can be selectively generated from the Ru complex center by using a longer wavelength laser (He–Ne, 632.8 nm).

At a field of 80 V/μm, an optical gain of 26.6 cm^{-1} is detected, while the absorption coefficient α is 28 cm^{-1}. It is observed, however, that the optical gain coefficient and diffraction efficiency initially increase with the external field and then level off in high applied fields (50 to 100 V/μm). This behavior is quite different from that of low-T_g composite PR materials, in which the gain coefficient increases nonlinearly with the applied field. Since this polymer system possesses a low T_g value, the dipole moments pointing from Ru(II) to $(PF_6^-)_2$ are readily aligned and generate an ionic dipole field which screens photogeneration sites, the Ru(II)-tris(bipyridyl) complex, from the applied field. This local field lowers the photogeneration efficiency and results in the saturation of photocurrent and PR gain at a high applied field [85]. This local field's effect on an ionic polymer is investigated by analysis of the dependence of photocurrent, PR gain, diffraction efficiency, and birefringence on the applied field. This conclusion is consistent with theoretical analysis under the assumptions that the detrapping rate is large enough that the condition of space-charge fields being limited by trap density is not reached. The space-charge field will be a function of photoconductivity [86] and has been confirmed experimentally by comparing the photoconduction of a similar nonionic polymer. For organic materials, especially for low-T_g materials, the trapped charge can be more easily detrapped than that in inorganic PR materials because of their amorphous nature [84,85].

The above results indicate that to synthesize a high-performance, low-T_g PR polymer, a neutral metal complex that has an efficient photoinduced charge generation mechanism can be utilized as a photosensitizer.

5. Low-T_g PR Polymers Containing Metalloporphyrin and PPV Backbones

Porphyrin and its derivatives are well-known model compounds for photosynthetic processes that involve charge separation [87]. A zinc porphyrin, a copper porphyrin, and a zinc phthalocyanine (Pc) are chosen and incorporated into PPV backbones. Four polymers are synthesized via the Heck polycondensation [88]. The incorporation of these metal complex moieties in polymers **XI** to **XIII** is manifested by the appearance of Q bands from metalloporphyrin and zinc Pc in the UV/vis spectra of the polymers. The Q bands in both polymers **XI** and **XII** (Scheme 9) are consistent with their corresponding porphyrin monomer, appearing at around 535, 571 nm and 540, 574 nm, respectively. On the other hand, Q bands in polymer **XIII** (Scheme 10) shows a significant red-shift of about 19 nm with respect to those in its Pc monomer (669, 684 nm). Presumably this is caused by the more extensive electron delocalization between the Pc and PPV backbone. These Q bands are the most interesting features because they allow us to selectively photoexcite polymers through metalloporphyrin and metallophthalocyanine complex sites using a He–Ne laser (i.e., 632.8 nm), far from the absorption maximum of the chromophore and PPV backbone around 380 nm.

Although these polymers have very similar structures except for the small amount of photosensitizers, their physical properties exhibit interesting differences. Polymers **XI**, **XII**, and **XIII** exhibit much higher photoconductivity than polymer **XIV** (Scheme 10), indicating that the incorporation of metalloporphyrin and zinc Pc into a conjugated backbone significantly enhances the photosensitivity of the resulting polymers in the visible region. Polymer **XIII** shows the highest photoconductivity and quantum yield of photocharge generation, due to the fact that Pc is easier to oxidize than the corresponding porphyrin. The quantum yield of photogeneration of charge carriers in polymer **XII** containing Zn porphyrin units is much higher than that of polymer **XI** containing Cu porphyrin. It seems that the unpaired spin in the Cu^{2+} center plays a crucial role. One possible reason is that the spin may facilitate the relaxation of the excited states through spin-orbital interaction. This will certainly reduce the quantum yield for the charge generation. The effect of metal ions is also observed on PR performances of polymers. The gain coefficients are found to strongly depend on the applied field and exhibit trends well correlated with the photoconductivity results. At the field strength of 60 V/μm, gain coefficients for polymers **XI**, **XII**, and **XIII** are 12, 65.7, and 103.5 cm^{-1}, respectively, while the absorption coefficients (α) are determined to be 20, 12.4, and 37 cm^{-1}. Thus, net optical gain coefficients ($\Gamma - \alpha$) of 53.3 and 66.5 cm^{-1} were obtained in polymers **XII** and **XIII**. For polymer **XIV**, there is no photorefractivity observed. Thus, both Zn phthalocyanine and porphyrin complexes are good photosensitizers for the PR process in these polymers. At $E = 60$ V/μm, the diffraction efficiencies for polymers **XII** and **XIII** of 13.0 and 17.8%, respectively, are observed. These are among the best results obtained for fully functionalized polymers.

C. Fully Functionalized Molecular PR Materials

Polymeric PR materials show advantages in mechanical strength and in forming amorphous films. However, if one considers the fact that most of the PR measurements and applications will be in the form of films sandwiched between two

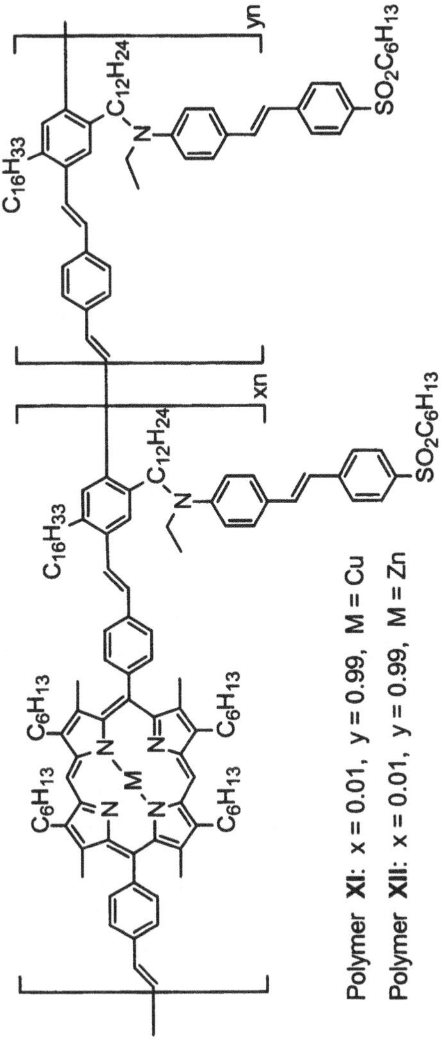

Polymer **XI**: x = 0.01, y = 0.99, M = Cu

Polymer **XII**: x = 0.01, y = 0.99, M = Zn

Scheme 9

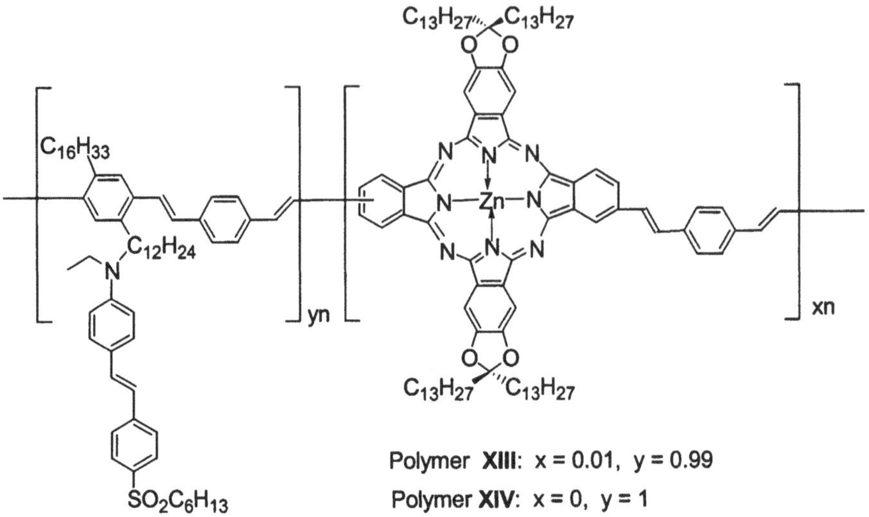

Scheme 10

electrodes, mechanical strength becomes an unnecessary requirement. As long as the materials possess all of the required functions and can form transparent films, they are suitable for PR studies. Organic glassy molecules constitute such a class of materials. Apart from their excellent processability and transparency, long-term stability and durability, small molecular systems are easy to purify and can benefit significantly from the orientational enhancement associated with low-T_g materials.

1. Amorphous Molecular PR Materials

Recently, a novel series of amorphous molecular materials based on carbazole and methine dyes has been synthesized [89]. These molecular materials exhibit a very interesting charge-transfer complex formation and large PR responses.

The molecule of general structure **4** (Scheme 11) possesses dual functions: photoconductivity and second-order NLO activity. The T_g of these materials are low (10 to 15°C). The amorphous films made from these compounds are very stable in an ambient atmosphere for over two years, and no crystallization phenomenon was observed even upon further heating. It is interesting to note that in the solid-state UV/vis spectra, new long-wavelength absorption bands are observed in addition to the peaks arising from the chromophore and carbazole moieties. For compound **4b**, at a wavelength of 632.8 nm and a field of 46 V/μm, an optical gain value of 65.4 cm^{-1} is achieved. The absorption coefficient of the material is only 1.4 cm^{-1}, thus giving us a net optical gain of 64 cm^{-1}. For

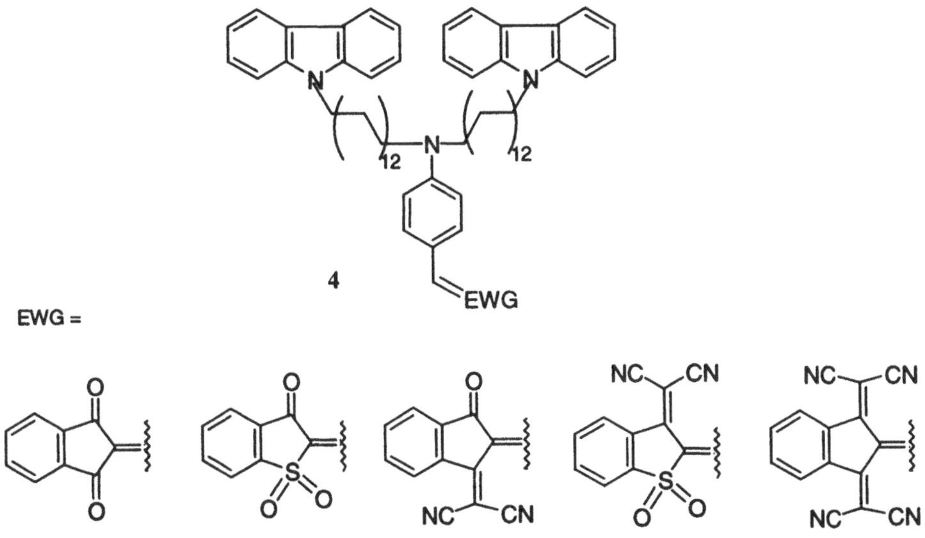

Scheme 11

compound **4c**, we obtain a PR gain of 53 cm^{-1} at a wavelength of 780 nm by applying a field of 62 V/μm, whereas the absorption coefficient is 4.3 cm^{-1}. In the FWM experiments, it is observed that the diffraction efficiency also has a strong dependence on the applied field and reaches 17% for compound **4b** and 9% for compound **4c** at $E = 42$ V/μm.

2. Oligothiophene Based PR Materials

Using the design idea from previous work on conjugated PR polymers, a molecular system that contains a 3-alkyl-substituted oligothiophene molecule covalently connected to a NLO chromophore was synthesized (compound **5**) (Scheme 12) [90,91]. Oligothiophene played a dual role as a photogenerator of charge carriers and charge transporting agent.

A photoconductivity of *ca.* 1.59×10^{-9} cm/(ΩW) is determined at 632 nm under an external electric field of 46.2 V/μm. Since the materials have a low T_g value, the dipoles of the NLO chromophore of the film can be aligned by an external electric field at room temperature. A Γ value of 102 cm^{-1} is obtained at $E = 70.6$ V/μm, which exceeds the absorption coefficient (19 cm^{-1} at 632.8 nm) in this sample, giving a net optical gain of 83 cm^{-1}. At the highest applied field of 70.6 V/μm, an energy exchange of almost 62% between the two beams is observed. A diffraction efficiency of almost 40% at $E = 70.6$ V/μm is obtained.

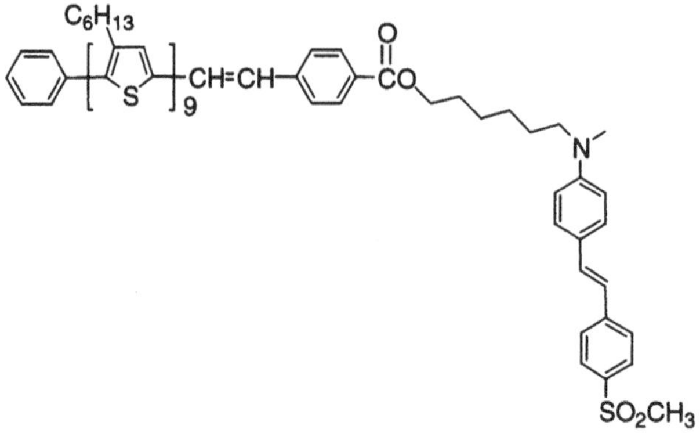

5

Scheme 12

This value corresponds to a refractive index modulation of $\Delta n = 2.55 \times 10^{-3}$. A response time of 42 ms is obtained at an applied field of 61.6 V/μm. The fast response time constant can be attributed to the relatively higher purity and fewer defects in this material, compared to those in polymeric counterparts.

These results indicate that a multifunctional molecular system can exhibit high PR performance if it is carefully designed. This work is further expanded to gain a better understanding of the relationship between the PR properties and structural parameters, such as the effect of the 3-alkyl chain substituents, as well as the degree of conjugation of the oligothiophene-based system. Therefore, a homologous series of fully functionalized regioregular oligo(3-alkylthiophenes) **1** to **3** bearing NLO chromophore is synthesized [92]. The optical properties of the series of the three materials are summarized in Table 1. Reasonably large net optical gain and diffraction efficiency are achieved in these molecular PR materials. It is found that the gain coefficient increases with the decrease of the grating spacing, and the maximum is reached at about 1 μm. It then decreases steeply at a small grating space range (Fig. 7). A diffraction efficiency of 19.8% at an applied field of 77 V/μm is achieved in films of 130-μm thickness for compound **2**.

The phase-shift of the index grating for compound **1** is found to increase monotonically with the applied field starting from almost zero at a low field (<30 V/μm) to about 78° at 77 V/μm (Fig. 8). The typical transmitted intensities of the two coupling beams during the translation of the grating are shown as an inset in Fig. 8. Both the energy transfer and phase-shift originate from the nonlocalization of the PR grating. The grating can be completely erased and rewritten with good reproducibility.

Table 1 Physical Properties of Compounds **1** to **3**

	Oligomers[a]		
	1	**2**	**3**
α (cm^{-1})[b]	9.74	10.3	25.7
Γ_p (cm^{-1})[c]	76.9	94.7	30.7
Φ $(°)$[d]	78	88	72
η_p $(\%)$[e]	18.9	25.1	10.0
Δn^{fi} (10^{-4})[f]	8.06	7.57	4.02
Δn^{sc} (10^{-4})[g]	8.07	9.40	4.32
r_n[h]	1.00	1.24	1.17

[a] All PR data were determined at a light wavelength of 632.8 nm and applied field of 78 V $\mu\mathrm{m}^{-1}$.
[b] α is the absorption coefficient.
[c] Γ_p is the PR gain coefficient for two p-polarized beam couplings.
[d] Φ is the phase-shift between the index and the intensity gratings.
[e] η_p is the diffraction efficiency for a p-polarized light.
[f] Δn^{fi} is the applied field-induced birefringence.
[g] Δn^{sc} is the index modulation by the space-charge field.
[h] r_n is defined as $\Delta n^{\mathrm{sc}}/\Delta n^{\mathrm{fi}}$ (see text).

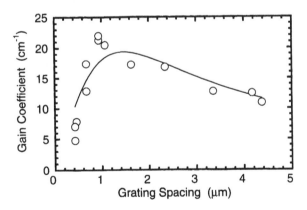

Figure 7 Gain coefficient of PR material of compound **1** as a function of grating spacing measured at an applied field of 66 V/μm. The circles are the experimental measurements. The solid line is a theoretical fitting using Eqs. (4), (19), (20), (38), and (39), with $E_q = 102.6$ V/μm. (Reprinted from Ref. 27. Copyright 2000 The American Physical Society.)

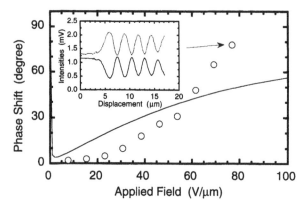

Figure 8 The phase-shift of PR material of compound **1** as a function of the applied field of steady-state refractive index grating with respect to the writing light intensity grating. The inset is a typical intensity of the two beams when the grating is moved about 14 μm in the direction of the grating vector. The solid line is the theoretical fitting using Eq. (20), with $E_q = 99.9$ V/μm. (Reprinted from Ref. 27. Copyright 2000 The American Physical Society.)

In order to compare the strength of the space-charge field among the three materials under the same applied field and geometric conditions, the contribution of the number density of the NLO chromophore for the three oligomers is normalized. The normalization procedure is as follows. Since both Δn^{fi} and Δn^{sc} are proportional to the number density of the NLO chromophore (see Table 1 for definition of Δn^{fi} and Δn^{sc}), the ratio $r_n = \Delta n^{sc}/\Delta n^{fi}$ excludes the influence of the NLO chromophore density for the three oligomers if the external field in all the experiments is the same. It is rationalized that the better performance of **2** over **1** is mainly due to the enhancement of the space-charge field, while a decrease in the number density of the NLO chromophore is responsible for the difference between **2** and **3**. The variation of r_n as a function of the number of thiophene rings in the oligomers **1** to **3** is shown in Table 1. The initial increase in r_n contributes markedly to the increase in PR gain and diffraction efficiency, despite the slight decrease in birefringence Δn^{fi} going from **1** to **3**. On the other hand, as the number of thiophene rings further increases from 8 to 10, both Δn^{sc} and Δn^{fi} drop substantially as a result of the decreasing NLO chromophore density (Table 1). However, also note that although both index modulations Δn^{sc} and Δn^{fi} decrease by 47 and 54%, respectively, the change of the r_n value between **2** and **3** is only 6%, which is within the experimental error. The change in PR properties between oligomers **2** and **3** is attributable mainly to the decrease in the NLO chromophore density.

V. PHOTOREFRACTIVE PROPERTIES OF MOLECULAR MATERIALS BASED ON METHINE DYES

A. Photorefraction and Complementary Grating Competition of Oligothiophenes Functionalized With Methine Dyes

During the investigation of diffraction efficiency of films made from compound **1**, a very interesting competition of complementary holographic gratings is observed. Two transport channels, provided by sexithiophene and methine dye moieties and two trapping centers for photogenerated holes and electrons, respectively, are responsible for the formation of the two complementary gratings [27,92].

Figure 9 shows the diffraction efficiency variation with time measured at a grating spacing of 1.1 μm. At $t = 0$ sec, two writing beams are overlapped inside the sample. The diffraction signal increases rapidly to a maximum value and then decreases gradually until a steady-state diffraction efficiency is reached. At $t = 32$ sec, the sample was illuminated only by a uniform light by blocking one of the writing beams. The diffraction efficiency decays exponentially until it is nearly zero, appears again for a short time, and eventually disappears completely. In another measurement, one of the writing beams is blocked before the maximum

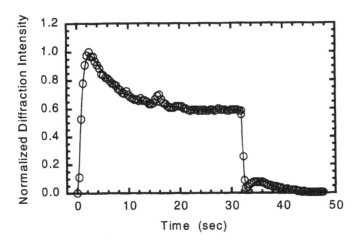

Figure 9 Diffraction efficiency of PR material of compound **1** as a function of time when a grating spacing is 1.1 μm and the applied field is 77 V/μm. The writing beams are turned on at $t = 0$ and one of the writing beams is blocked at $t = 32$ seconds. The solid line is a fitting of the data into biexponential functions [Eqs. (41) and (42)]. (Reprinted from Ref. 27. Copyright 2000 The American Physical Society.)

diffraction efficiency is reached. The grating is exponentially erased as predicted by standard PR theory.

These phenomena are due to the cancellation and revelation of two types of gratings as observed in several inorganic crystals such as $Bi_{12}TiO_{12}$, $Sn_2P_2S_6$, and $Bi_4Ti_3O_{12}$. We propose that the two sets of PR gratings are formed by two types of photoexcited charge carriers, electrons and holes. This is because of the bipolar transport property of this molecular PR material, and the existence of two types of trap centers for the holes and electrons. The sexithiophene moieties provide a transport channel for the hole migration [93], while the methine dye acts as another transport channel for the electron migration because of the presence of a strong electron-withdrawing group. Therefore, the methine dye plays multiple roles as a photogenerator, NLO chromophore, and an electron transporter. The free charge carriers are generated by the photoexcitation of the methine dye. This is because the molecule exhibits an absorption coefficient of 6.42 cm^{-1} at the wavelength of 633 nm mainly from the NLO chromophore.

When writing the gratings, a fast grating is initially built up through trapped holes because of their higher mobility (as shown later). This process results in the initial quick rise in the diffraction efficiency. Thus, the erasure of the grating at this stage shows a single exponential decay. If the grating writing continues for a longer time, the electron traps begin to fill up, thereby creating the slower complementary grating. Since the field built up by the electron traps is in the direction opposite to the field built up by the hole traps, cancellation of the net space-charge field and reduction in the diffraction efficiency are observed. When erasing the gratings in the period of stable diffraction, the space-charge field, as formed by the trapped holes, decays faster than that formed by the trapped electrons. Therefore, the net space-charge field changes signs during the erasure process. The diffraction efficiency first decreases to zero and then increases again. A revelation of the diffraction signal is thus observed. Measurement of the electron mobility by a TOF technique confirms this assumption. A typical transient current signal for holes and electron transport is shown in Fig. 10. The electron transit time is much longer and the amplitude of the current is also smaller than that observed in hole transport. This strongly suggests the existence of an electron transport channel in our molecular system.

Cyclic voltammetry studies showed the ionization potential and electron affinity of each component of the molecule in solution. The HOMO and LUMO energy levels were estimated from the equations: $E_{HOMO} = E_{ox}^0 + 4.4$ eV and $E_{LUMO} = E_{re}^0 + 4.4$ eV, where E_{ox}^0 and E_{re}^0 were oxidation and reduction potentials with respect to the standard hydrogen electrode (SHE) and the value of 4.4 is the ionization potential for hydrogen in eV [94,95]. The HOMO and LUMO energy levels of the methine dye (compound **6**) (Scheme 13) were determined to be -5.82 and -3.48 eV, respectively, with respect to the vacuum level from

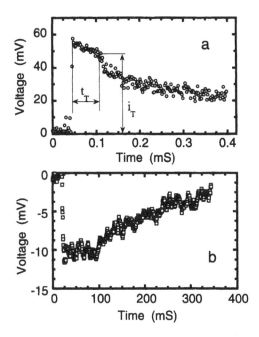

Figure 10 Two typical time-of-flight signals for (a) hole and (b) electron transport. The t_T is transit time. (Reprinted from Ref. 27. Copyright 2000 The American Physical Society.)

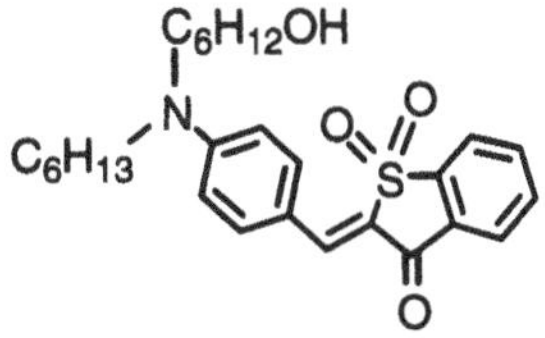

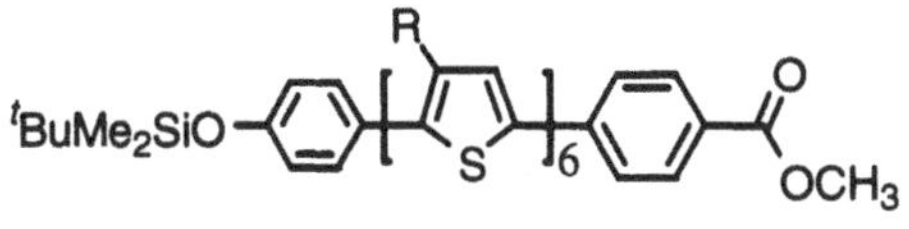

Scheme 13

its lowest oxidation and reduction potentials. The band-gap $\Delta E_{electrochemical}$ was estimated to be 2.34 eV. This value is in good agreement with the spectroscopic estimate of the band-gap $\Delta E_{optical}$, 2.30 eV. The HOMO energy of a sexithiophene molecule alone (compound **7**) was estimated to be −5.48 eV. Since the reduction potential of sexithiophene is out of the solvent window, the LUMO energy level (−3.00 eV) was deduced from the band-gap of the sexithiophene backbone which was estimated from the photoabsorption edge (2.48 eV). If one assumes that there is no ground-state intramolecular interaction between the sexithiophene backbone and the methine dye, the energy levels in our PR molecule should remain identical to their individual components. Indeed, the HOMO and LUMO energy levels for the PR molecules (compound **1**) were electrochemically determined to be −5.42 eV and −3.48 eV, respectively. These are close to those of the individual compounds. A potential energy diagram can be constructed as shown in Fig. 11a. A photogeneration of charge carriers occurs upon absorption of a photon by the dye. An electron can be transported away along the LUMO of the methine dye, and the hole tends to be transferred to the sexithiophene HOMO under the action of an external field. The holes and electrons are further drifted away by sequential hopping to neighboring sexithiophene backbones and methine dyes and can be fixed by individual trapping centers, respectively (Fig. 11b).

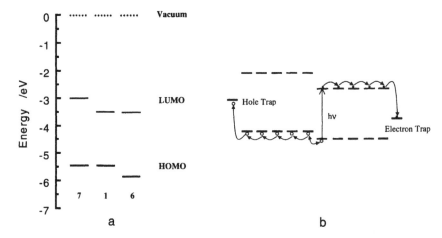

Figure 11 (a) Estimated energy levels of HOMO and LUMO of a photorefractive molecule (compound **1**), the methine dye (compound **6**), and the sexithiophene derivative (compound **7**) with respect to the vacuum level. (b) Proposed charge carrier transport channels. (Reprinted from Ref. 27. Copyright 2000 The American Physical Society.)

With the bipolar transport nature and channels being clarified, a further question is whether the slow electrons would neutralize electric positive dark regions charged by the trapped fast holes. Or would the electrons be immobilized separately by their own individual traps? This question physically corresponds to the applicability of the bipolar two-trap or single-trap model. The bipolar two-trap model is based on the assumption that two types of active centers are involved in simultaneous electron-hole transport [96–99]. There are two gratings that are 180 degrees out of phase with each other, known as complementary gratings. One of the gratings is set up by the redistribution of electrons and another is set up by the redistribution of holes. Since the total space-charge field is the sum of the fields created by each type of carrier caught by different traps, the net amplitude of the space-charge field is smaller than that of the contribution from the principal set of space charges. If their characteristic time constants are different, then the individual gratings can be revealed during writing and erasure. The expression for the writing and erasing diffraction efficiency is [96–98]:

$$\eta(t) \propto \left| E_e \left[1 - \exp\left(-\frac{t}{\tau_e} \right) \right] - E_h \left[1 - \exp\left(-\frac{t}{\tau_h} \right) \right] \right|^2, \tag{41}$$

and

$$\eta(t) \propto \left| E_e \exp\left(-\frac{t}{\tau_e} \right) - E_h \exp\left(-\frac{t}{\tau_h} \right) \right|^2, \tag{42}$$

where E_e and E_h are the steady-state space-charge fields built up by the trapped electrons and holes, respectively, and τ_e and τ_h are the time constants for the electron- and hole-transport systems.

The bipolar single-trap model assumes that both electrons and holes share identical trap centers. Since sequential trappings of the electrons and holes by the identical centers mean the neutralization of the electric charge, the effective space-charge field will depend on the relative power (i.e., the mobilities) of electron and hole transports. The expressions for the writing and erasing diffraction efficiency are [100]:

$$\eta(t) \propto \left| E_{sc} \left\{ 1 - \exp\left[-\left(\frac{1}{\tau_e} + \frac{1}{\tau_h} \right) t \right] \right\} \right|^2 \tag{43}$$

$$\eta(t) \propto \left| E_{sc} \exp\left[-\left(\frac{1}{\tau_e} + \frac{1}{\tau_h} \right) t \right] \right|^2 \tag{44}$$

Equations (43) and (44) reveal that the bipolar single-trap model predicts that the formation and erasure of the grating follow one effective time constant that consists of uncoupled response times of the electron- and hole-transport processes (τ_h and τ_e).

The dynamic behaviors of the diffraction efficiency in Fig. 9 support the bipolar two-trap model. Since the characteristic time constants of complementary gratings are different, the sign of the space-charge field, and hence the direction of energy transfer in the erasing process changes after the dip [$\eta = 0$ for Eq. (42)]. The time for the dip is delayed when the grating spacing increases as shown in Fig. 12, where dips appear at about 2.5, 7.3, and 20.0 sec of erasure time for the gratings with spacings of 1.1, 1.9, and 4.4 μm, respectively. The sign of the space-charge field does not change in the writing process for two reasons: the space-charge field contributed from the hole is dominated, and the buildup time constant for the hole grating is faster than that for the electron grating. The space-charge field is completely canceled at the dip, a zero diffraction efficiency dip during erasure due to the 180° phase difference of the complementary gratings (Figs. 9 and 12).

Although similar grating cancellation and revelation behaviors have been observed in other PR polymeric composites, they are attributed either to the trap's intercommunication or to the residual ionic motion [101–103]. However, the complementary gratings in the present study are formed by the space-charge field of two types of *photogenerated charge carriers*.

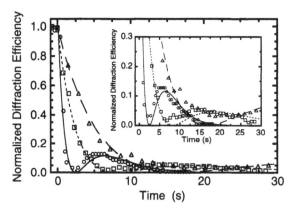

Figure 12 Normalized diffraction efficiency measured from a PR material of compound **1** as a function of time during the grating erasure. The circles, squares, and triangles are experimental measurements with the grating spacings of 1.1, 1.9, and 4.4 μm, respectively, at an applied field of 77 V/μm. The lines are theoretical fittings using a biexponential function Eq. (42). The inset is its enlarged plot. (Reprinted from Ref. 27. Copyright 2000 The American Physical Society.)

B. Molecular PR Materials Based on Methine Dyes

Methine dyes are well studied as near-infrared dyes [104] and for their second-order nonlinear optical effects [89,105]. The studies on the complementary grating competition in compound **1** revealed that these methine dyes are electronic photoconductive materials [27]. Therefore, methine dye molecules possess the four functionalities necessary for manifesting the PR effect. In order to study their PR effect, compounds **8** and **9** (Scheme 14) were designed and synthesized [92]. A net PR gain of 215 cm^{-1} was obtained for samples of compound **9**. The diffraction efficiency of 87.6% at a semiconductor laser wavelength of 780 nm was observed [14]. These molecular materials possess high optical quality, long-term stability, and low absorption loss. More importantly, interesting phenomena associated with their large index modulation can be observed, such as the pulsation behavior in the self-defocusing and the optical limiting effect. To ensure that the materials obtained will be amorphous, two branched alkyl chains bearing chiral carbons were attached to the amino group. Films exhibiting excellent optical quantity can be easily prepared from these materials and their optical transparency is stable for a long term. Wide-angle X-ray diffraction experiments confirmed the amorphous nature of the materials. The T_g for the compounds **8** and **9** was *ca.* 6 and 26°C, respectively, determined with a differential scanning calorimeter. The low-T_g value benefitted the photoinduced index modulation through the orientational enhancement [38]. Solution electrochemistry measurements indicated that the molecules have low reduction potentials (0.83 and 0.41V versus SHE for compounds **8** and **9**, respectively) and act as an electronic photoconductor. Photocurrent measurements indicated that these materials are indeed photoconductive (Fig. 13).

Figure 14 shows the dependence of the gain coefficients on the external field for samples made of compounds **8** and **9**. Large gain values of 117.7 and 221.4 cm^{-1} were observed for materials of **8** and **9**, respectively, at an external

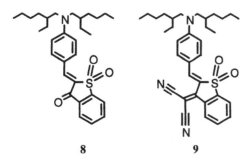

8   9

Scheme 14

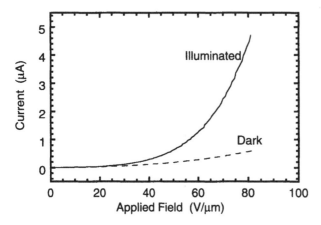

Figure 13 A typical photo-/dark current dependence on the applied field on the sample of molecule **9** (27 μm sandwiched between an ITO glass and an aluminum plate). The light has a wavelength of 780 nm with an irradiation intensity of 128 mW/cm². (Reprinted from Ref. 14. Copyright 2001 American Institute of Physics.)

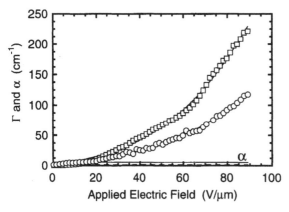

Figure 14 The optical gain and absorption coefficients as a function of the applied field for samples of molecules **8** (circles) and **9** (squares). The solid lines are theoretical fittings using Eqs. (4), (19), (20), (38), and (39). (Reprinted from Ref. 14. Copyright 2001 American Institute of Physics.)

field of 89 V/μm. Since the absorption coefficients of the two samples at the working wavelength (780 and 633 nm) are 1.64 and 5.54 cm^{-1}, net optical gains of 116.1 and 215.9 cm^{-1} were achieved. The latter value is one of the highest gain coefficients reported for organic PR materials. The diffraction efficiencies reached the maximum of 74.3% for material of compound **8** at 633 nm at the applied field of 53 V/μm ($E_{\pi/2}$) and 87.6% for materials of compound **9** at 780 nm and 44 V/μm (Fig. 15). The solid lines in the figure represent the theoretical fittings of the field-dependence of the diffraction efficiency with the standard PR model. Figure 16 is a plot of index modulations as a function of the applied field calculated from the diffraction efficiency. Calculated from Eq. (9), the maximum index modulations Δn_p of 0.0056 and 0.0104 were obtained for samples of compounds **8** and **9**, respectively.

Because of the large photoinduced index modulations and the high optical qualities, these materials are readily used as the medium for holographic information storage. The grating is reversible and can be completely erased and rewritten with good reproducibility. It is also found that the dynamics of the grating formation does not follow a single-exponential function; rather it follows a biexponential one. For compound **9** at an external field of 84 V/μm, 52% of the index change is formed with a time constant of 16.6 ms and the remaining 48% has a time constant of 1.2 sec.

The large index modulation also leads to a large self-defocusing of light beams. The intensity distribution and the propagation dispersion of the output

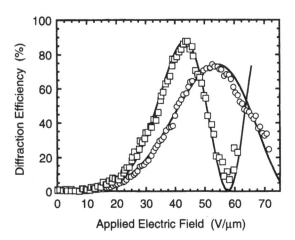

Figure 15 The diffraction efficiency of the samples of molecules **8** (circles) and **9** (squares), respectively, as a function of the applied field at wavelengths of 633 nm and 780 nm. The solid lines are theoretical fittings using Eqs. (9), (19), (38), and (39). (Reprinted from Ref. 14. Copyright 2001 American Institute of Physics.)

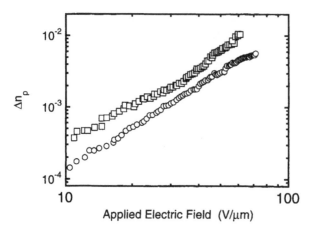

Figure 16 The index modulations of the samples of molecules **8** (circles) and **9** (squares), respectively, as a function of the applied field, calculated according to Eq. (9). (Reprinted from Ref. 14. Copyright 2001 American Institute of Physics.)

beam can be changed dramatically. The near field pattern of the output beam is characterized as a series of bright and dark rings. With the increase of the applied field, the size of the first bright ring increases monotonously. When the applied field surpasses a threshold, which is a function of light intensity, the self-defocusing undergoes a periodical pulsation. The size of the rings expands with time in an accelerating fashion until a dramatic expansion occurs, followed by a collapse to its original (no defocusing) pattern. A typical sequence of the image of one pulsation is shown in Fig. 17a. Figure 17b shows the four cycles of the field angle of the first ring as a function of time. The self-defocusing of light beams is the consequence of the self-modulation of the refractive index by the optical field via the nonlocal PR effect. The external field aligns the dipolar molecules in the direction of film normal. The photoexcited mobile electrons diffuse and drift downward along the intensity gradient to the outside of the beam spot and form a space-charge field pointing radically outward. This space-charge field can reorient the molecule in the radial axis and change the refractive index accordingly. The on-axis index remains a minimum and a doughnut-shaped index distribution is formed. The beam is thus defocused due to diffraction by the light-induced index ring. If the phase difference between the center and wing portions of the beam is about 2π, the first intensity ring appears. A rough estimation gives the index modulation at the first ring, $\Delta n_1 = \lambda \cos \theta / d$, where d is the thickness of the film, and θ is the inner angle of the beam with respect to the film normal. The two rings that we observe correspond to a refractive index modulation $\Delta n \sim 0.0093$. The field component perpendicular to the beam favors

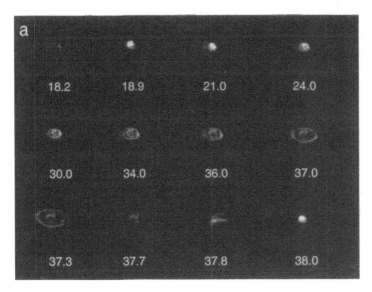

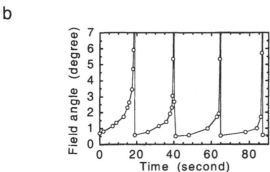

Figure 17 (a) A sequence of the near-field pattern variation. Parallel 780 nm light with intensity of 2.87 W/cm² irradiated on the sample of molecule **9** with a spot size of diameter of 850 µm and incident angle of 43°. The patterns are projected onto a screen placed 151 mm away from the sample. The numbers underneath each spot are the time that the counting is started. (b) Time dependence of field angle of the first bright ring in four pulsation circles. (Reprinted from Ref. 14. Copyright 2001 American Institute of Physics.)

the charged carrier drift along one direction because of the tilted geometry, and the near-field pattern is elliptic (Fig. 17a). As the applied field increases, the formation of the index ring becomes faster and the ring size becomes larger because the PR index modulation is enhanced by the applied field (Figs. 14 and 15). The maximum size of the ring is reached when the charge separation and recombination reach equilibrium. If the field is high enough (about 91 V/μm in this condition), the index ring can grow quickly into a large size. This, in turn, defocuses the light and lowers the light intensity in the beam spot of the sample. The density of the photogenerated electron is reduced and the equilibrium is broken. A sudden disappearance of the space-charge field occurs, and dipolar molecules flip back to the direction of the external field. The transmitted light suddenly returns back to its original distribution and another round of periodic pulsation starts. It can be seen that this instability has limited the gain coefficient. The maximum gain coefficient of a sample of compound **9** shown in Fig. 12 is achieved at a lower field than the unstable threshold.

The large self-defocusing effect suggests interesting applications as the self-focusing, self-trapping, and spatial soliton wave under the designated conditions [106–108]. Optical switching, optical limiting, and transverse optical bistability [109,110] can also be accomplished.

VI. CONCLUSION

Since the disclosure of the first organic photorefractive crystals [7] and later the amorphous polymeric PR materials [8], research on organic photorefractive materials has made very impressive progress. From this review we can see several marked developments. The first net optical gain coefficient was achieved in 1993 [45], and a nearly 100% diffraction efficiency in a 105-μm thin film and over 200 cm^{-1} net optical gain coefficient were realized in 1994 [47] in composite materials. The development of fully functionalized polymeric and molecular materials has kept pace with its polymer composite partner. The first fully functionalized polyurethane showed the PR phenomena in 1992 [69]. A conjugated PR polymer functionalized with NLO chromophore appeared in 1993 [74]. However, net optical gain coefficients in fully functionalized polymers were achieved only recently with conjugated polymers containing transition metal complexes in thin (20 μm) high-T_g [79] and thicker (139 μm) low-T_g [88] films ($\Gamma_{net} = 200$ cm^{-1} and 53 cm^{-1}, respectively). More recently, near 100% diffraction efficiency and 215 cm^{-1} net optical gain coefficient in a 130-μm thin film made of molecular materials were realized [14]. During the exploration of new materials, interesting phenomena have been studied, such as orientational enhancement [38], quasinondestructive read-out [103], deep trap activation [32], local-field effect [85], and the complementary grating competition [27]. Prototype device concepts have also been developed. For example, storage, retrieval, and subsequent erasure of digital data pages with a data density of 0.5 Mbit/cm^2 were demonstrated in 1996 [111]. The spontaneous oscillation and self-pumped phase conjugation were realized in 1997 [67].

These research efforts are driven by the demands on the new technologies for the transport, processing, and storage of large amounts of information. The development of these technologies heavily depends on the emergence of new materials. The ideal characteristics of materials for optical applications include high resolution, better energy sensitivity, broad wavelength sensitivity, real-time capabilities (erasability and speed of operation), good stability and compactness, as well as low cost. Although photorefractive materials are among the most sensitive materials, all of the demands will not be met until some time in the future. Improvements are needed to generate materials exhibiting faster responsive time, requiring lower applied voltage, and operating at the designated wavelengths of telecommunication interest (1.3 and 1.5 μm).

However, further improvements will strongly depend on the understanding of photochemical and photophysical details of the whole photorefractive process. Despite the significant progress, photorefractive mechanisms in organic materials are still not well understood. The reasons may be due to the complexity of the photorefractive effect itself and the complexity of the PR materials. This is especially true in polymeric materials. In these systems a slight change in their functional component or composition may also alter the properties. This is manifested by the dramatic PR property variation due to a small change in the doping ratio in polymer composites.

The molecular PR materials based on the methine dyes are composed of single types of molecules and exhibit the best PR performances. They are chemically pure, structurally well defined, and morphologically stable. They may serve as model materials for detailed photochemical and photophysical studies. Through these model materials, clearer pictures and a deep understanding of the photorefractive effect will emerge. Better mechanistic understanding will surely assist in the search for new PR materials with improved macroscopic properties.

ACKNOWLEDGMENTS

Many students from our own group have made excellent contributions and their names are cited in related references. This work was supported by the National Science Foundation and Air Force Office of Scientific Research. Support from the National Science Foundation Young Investigator program is gratefully acknowledged. This work also benefited from the support of the NSF MRSEC program at the University of Chicago.

REFERENCES

1. A Ashikin, GD Boyd, JM Dziedzic, RG Smith, AA Ballmann, K Nassau. Appl Phys Lett 9:72, 1966.
2. FS Chen. J Appl Phys 38:3418, 1967.
3. FS Chen. J Appl Phys 40:3389, 1969.
4. P Günter, J-P Huignard, eds. Photorefractive Materials and Their Applications. vol 1, 2. New York: Springer, 1988 and 1989.
5. L Solymar, DJ Webb, A Grunnet-Jepsen. In: A Hasegawa, M Lapp, BB Snavely, H Stark, AC Tam, T Wilson, eds. The Physics and Application of Photorefractive Materials, Oxford: Clarendon Press, 1996.
6. F Yu, S Yin, eds. In Photorefractive Optics, Materials, Properties, and Applications. San Diego: Academic, 2000.
7. K Sutter, P Günter. J Opt Soc Am B 7:2274, 1990.
8. S Ducharme, JC Scott, RJ Twieg, WE Moerner. Phys Rev Lett 66:1846, 1991.
9. WE Moerner, SM Silence. Chem Rev 94:127, 1994.
10. WE Moerner, A Grunnet-Jepsen, CL Thompson. Ann Rev Mater Sci 27:585, 1997.
11. LP Yu, WK Chan, ZH Peng, AR Gharavi. Acc Chem Res 29:13, 1996.
12. LP Yu, WK Chan, ZH Peng, WJ Li, A Gharavi. In: HS Nalwa, ed. Handbook of Organic Conductive Molecules and Polymers. vol 4. New York: Wiley, 1997, p 233.
13. Q Wang, LM Wang, LP Yu. Macromol Chem Phys 201:723, 2000.
14. LM Wang, MK Ng, LP Yu. Appl Phys Lett 87:700, 2001.
15. P Yeh. Introduction to Photorefractive Nonlinear Optics. New York: Wiley, 1993.
16. J Feiberg, KR MacDonald. In P Günter, J-P Huignard, eds. vol 2. Photorefractive Materials and Their Applications. New York: Springer, 1989, p 159.
17. CA Wash, WE Moerner. J Opt Soc Am B 9:1642, 1992.
18. M Liphardt, S Ducharme. J Opt Soc Am B 15:2154, 1998.
19. H Kogelink. Bell Syst Tech J 48:2909, 1969.
20. NV Kukhtarev, VB Markov, SG Odulov, MS Soskin, VL Vinetskii. Ferroelectrics 22:949, 1979.
21. NV Kukhtarev, VB Markov, SG Odulov. Opt Commun 23:338, 1977.
22. GC Valley, MB Klein. Opt Eng 22:704, 1983.
23. L Onsager. Phys Rev 54:554, 1938.
24. A Mozumder. J Chem Phys 60:4300, 4305, 1974.
25. PM Borsenberger, AI Ateya. J Applied Phys 49:4035, 1978.
26. PM Borsenberger, LE Contois, DC Hoesterey. J Chem Phys 68:637, 1978.
27. LM Wang, MK Ng, LP Yu, Phys Rev B62:4973, 2000.
28. PM Borsenberger, DS Weiss. In BJ Thompson, ed. Organic Photoreceptors for Xerography. New York: Marcel Dekker, 1998.
29. M Stoka, JF Yanus, DM Pai. J Phys Chem 88:4707, 1984.
30. RH Young, JJ Fitzgerald. J Chem Phys 102:9380, 1995.
31. A Goonesekera, S Ducharme. J Appl Phys 85:6506, 1999.
32. SM Silence, GC Bjorklund, WE Moerner. Opt Lett 19:1822, 1994.

33. A Grunnet-Jepsen, D Wright, B Smith, MS Bratcher, MS DeClue, JS Siegel, WE Moerner. Chem Phys Lett 291:553, 1998.
34. DM Burland, RD Miller, CA Walsh. Chem Rev 94:31, 1994.
35. E Havinga, P van Pelt. Ber Bunsen-Ges Phys Chem 83:813, 1979.
36. JW Wu. J Opt Soc Am B 8:142, 1991.
37. WE Moerner, SM Silence, F Hache, GC Bjorklund. J Opt Soc Am B 11:320, 1994.
38. B Kippelen, K Meerholz, N Peyghambarian. In HS Nalwa, S Miyata, eds. Nonlinear Optics of Organic Molecules and Polymers. Boca Raton, FL: CRC Press, 1997, p 465.
39. R Wortmann, C Poga, RJ Twieg, C Geletneky, CR Moylan, PM Lundquist, RG DeVoe, PM Cotts, H Horn, JE Rice, DM Burland. J Chem Phys 105:10637, 1996.
40. B Kippelen, F Meyers, N Peyghambarian, SR Marder. J Am Chem Soc 119:4559, 1997.
41. CC Teng, HT Man. Appl Phys Lett 56:1734, 1990.
42. F Michelotti, V Taggi, M Bertolotti, T Gabler, HH Hörhold, AJ Bräuer. Appl Phys Lett 83:7886, 1998.
43. GF Harding. In GH Meeten, ed. Optical Properties of Polymers. New York: Elsevier, 1986, p 72.
44. KD Singer, MG Kuzyk, WR Holland, JE Sohn, SJ Lalama, RB Comizolli, HE Katz, ML Schilling. Appl Phys Lett 53:1800, 1988.
45. MCJM Donckers, SM Silence, CA Walsh, F Hache, DM Burland, WE Moerner, RJ Twieg. Opt Lett 18:1044, 1993.
46. B Kippelen, Sandalphon, N Peyghambarian, SR Lyon, AB Padias, HK Hall. Electronics Lett 29:1873, 1993.
47. K Meerholz, BL Volodin, Sandalphon, B Kippelen, N Peyghambarian. Nature 371:497, 1994.
48. Sandalphon, B Kippelen, K Meerholz, N Peyghambarian. Appl Opt 35:2346, 1996.
49. F Wurthner, R Wortmann, R Matschiner, K Lukaszuk, K Meerholz, Y DeNardin, R Bittner, C Brauchle, R Sens. Angew Chem Int Ed Engl 36:2765, 1997.
50. K Meerholz, Y DeNardin, R Bittner, R Wortmann, F Wurthner. Appl Phys Lett 73:4, 1998.
51. B Kippelen, SR Marder, E Hendrickx, JL Maldonado, G Guillemet, BL Volodin, DD Steele, Y Enami, Sandalphon, YJ Yao, JF Wang, H Rockel, L Erskine, N Peyghambarian. Science 279:54, 1998.
52. E Hendrickx, J Herlocker, JL Maldonado, SR Marder, B Kippelen, A Persoons, N Peyghambarian. Appl Phys Lett 72:1679, 1998.
53. BL Volodin, B Kippelen, K Meerholz, B Javidi, N Peyghambarian. Nature 383:58, 1996.
54. SM Silence, MCJM Donckers, CA Walsh, DM Burland, RJ Twieg, WE Moerner. Appl Opt 33:2218, 1994.
55. Y Zhang, Y Cui, PN Prasad. Phys Rev B 46:9900, 1992.
56. ME Orczyk, J Zieba, PN Prasad. J Phys Chem 98:8699, 1994.
57. ME Orczyk, B Swedek, J Zieba, PN Prasad. J Appl Phys 76:4990, 1994.

58. HJ Bolink, VV Krasnikov, GG Malliaras, G Hadziioannou. J Phys Chem 100:16356, 1996.
59. RK Bittner, T Daubler, D Neher, K Meerholz. Adv Mater 11:123, 1996.
60. O Zobel, M Eckl, P Strohriegl, D Haarer. Adv Mater 7:911, 1995.
61. E Hendrickx, BL Volodin, DD Steele, JL Maldonado, JF Wang, B Kippelen, N Peyghambarian. Appl Phys Lett 71:1159, 1997.
62. C Poga, DM Burland, T Hanemann, Y Jia, CR Moylan, JJ Stankus, RJ Twieg, WE Moerner. Proc SPIE 2526: 82, 1995.
63. K Meerholz, R Bittner, Y DeNardin, C Brauchle, E Hendrickx, BL Volodin, B Kippelen, N Peyghambarian. Adv Mater 9:1043, 1997.
64. E Hendrickx, JF Wang, JL Maldonado, BL Volodin, Sandalphon, EA Mash, A Persoons, B Kippelen, N Peyghambarian. Macromolecules 31:734, 1998.
65. AM Cox, RD Blackburn, DP West, TA King, FA Wade, DA Leigh. Appl Phys Lett 68:2801, 1996.
66. A Grunnet-Jepsen, CL Thompson, RJ Twieg, WE Moerner. Appl Phys Lett 70:1515, 1997.
67. A Grunnet-Jepsen, CL Thompson, WE Moerner. Science 277:549, 1997.
68. LP Yu, WK Chan, ZN Bao, S Cao. J Chem Soc Chem Commun 1735, 1992.
69. LP Yu, WK Chan, ZN Bao, S Cao. Macromolecules 26:2216, 1993.
70. YM Chen, ZH Peng, WK Chan, LP Yu. Appl Phys Lett 64:1195, 1994.
71. LP Yu, YM Chen, WK Chan, ZH Peng. Appl Phys Lett 64:2489, 1994.
72. LP Yu, YM Chen, WK Chan. J Phys Chem 99:2797, 1995.
73. WK Chan, YM Chen, ZH Peng, LP Yu. J Am Chem Soc 115:11735, 1993.
74. V Balzani, A Juris, M Venturi, S Campagna, S Serroni. Chem Rev 96:759, 1996.
75. A Juris, V Balzani, F Barigelletti, S Campagna, P Belser, A VonZelewsky. Coord Chem R 84:85, 1988.
76. B Pierre-Alain, M Gratzel. J Am Chem Soc 102:2461, 1980.
77. ZH Peng, A Gharavi, LP Yu. Appl Phys Lett 69:4002, 1996.
78. ZH Peng, A Gharavi, LP Yu. J Am Chem Soc 119:4622, 1997.
79. ZH Peng, ZN Bao, YM Chen, LP Yu. J Am Chem Soc 116:6003, 1994.
80. LP Yu, ZH Peng. Polym Mater Sci Eng 71:441, 1994.
81. MR Wasielewski. Chem Rev 92:435, 1992.
82. K Iida, T Nohara, S Nakamura, G Sawa. Jpn J Appl Phys 28:1390, 1989.
83. ZH Peng, LP Yu. Macromolecules 27:2638, 1994.
84. Q Wang, LM Wang, LP Yu. J Am Chem Soc 120:12860, 1998.
85. LM Wang, Q Wang, LP Yu. Appl Phys Lett 73:2546, 1998.
86. MG Moharam, TK Gaylord, R Magnusson, L Young. J Appl Phys 50:5642, 1979.
87. D Dolphin, ed. The Porphyrin. vol 5. New York: Academic, 1979.
88. Q Wang, LM Wang, JJ Yu, LP Yu. Adv Mater 12:974, 2000.
89. Q Wang, LM Wang, H Saadeh, LP Yu. Chem Commun 17:1689, 1999.
90. WJ Li, A Gharavi, Q Wang, LP Yu. Adv Mater 10:927, 1998.
91. WJ Li, A Gharavi, Q Wang, LP Yu. In: IM Khan, JS Harrison, eds. Field Responsive Polymers: Electroresponsive, Photoresponsive, and Responsive Polymers in Chemistry and Biology, Washington, DC, American Chemical Society, 1999, p 237.

92. MK Ng, LM Wang, LP Yu. Chem Mater 12:2988, 2000.
93. G Horowitz, P Delannoy. In D. Fichou, ed. Handbook of Oligo- and Polythiophenes. Weinheim: Wiley-VCH, 1999, p 283.
94. WC Barrette, HW Johnson Jr, DT Sawyer. Anal Chem 56:1890, 1984.
95. H Reiss, A Heller. J Phys Chem 89:4207, 1985.
96. S Zhivkova, M Miteva. J Appl Phys 68:3099, 1990.
97. MC Bashaw, T-P Ma, RC Baker, S Mroczkowski, RR Dube. J Opt Soc Am B 7:2329, 1990.
98. MC Bashaw, T-P Ma, RC Baker, S Mroczkowski, RR Dube. Phys Rev B 42:5641, 1990.
99. GC Valley. J Appl Phys 59:3363, 1986.
100. MC Bashaw, T-P Ma, RC Baker. J Opt Soc Am B 9:1666, 1992.
101. SP Ducharme, B Jones, JM Takacs, L Zhang. Opt. Lett 18:152, 1993.
102. SM Silence, CA Walsh, JC Scott. TJ Matray, RJ Twieg, F Hache, GC Bjorklund, WE Moerner. Opt Lett 17:1107, 1992.
103. SM Silence, RJ Twieg, GC Bjorklund, WE Moerner. Phys Rev Lett 73:2047, 1994.
104. KA Bello, L Cheng, J Griffiths. J Chem Soc Perkin Trans 11:815, 1987.
105. M Blanchard-Desce, V Alain, PV Bedworth, SR Marder, A Fort, C Runser, M Barzoukas, S Lebus, R Wortmann. Chem Eur J 3:1091, 1997.
106. GC Duree Jr, JL Shultz, GJ Salamo, M Segev, A Yariv, B Crosignani, PD Porto, EJ Sharp, RR Neurgaonkar. Phys Rev Lett 71:533, 1993.
107. GC Duree, M Morin, G Salamo, M Segev, B Crosignani, PD Porto, E Sharp, A Yariv. Phys Rev Lett 74:1978, 1995.
108. MD Iturbe Castillo, JJ Sánchez-Mondragón, SI Stepanov, MB Klein, BA Wechsler. Opt Commun 118:515, 1995.
109. AE Kaplan. Opt Lett 6:360, 1981.
110. IC Khoo, GM Finn, RR Michael, TH Liu. Opt Lett 11:227, 1986.
111. PM Lundquist, R Wortmann, C Geletneky, RJ Twieg, M Jurich, VY Lee, CR Moylan, DM Burland. Science 274:1182, 1996.

7

Dynamic Holography in Photorefractive Liquid Crystals

Gary P. Wiederrecht

Argonne National Laboratory, Argonne, Illinois

I. INTRODUCTION

Photorefractive holography is often proposed as an efficient means to store or process optical images. With the advent of increasingly photosensitive materials, it is now possible to image through highly scattering or absorbing environments. Photorefractivity is initiated by a photoinduced spatial modulation of the charge distribution in a material to create a space-charge field, which in turn alters the index of refraction through electro-optic mechanisms. When developing photorefractive media, the initial charge separation efficiency, charge migration, charge trapping, and electro-optic responses all must be optimized. On top of these technological challenges, cost and processability must also be carefully considered. Thus, although the photorefractive effect was initially observed in inorganic ferroelectric crystals more than 30 years ago, the 1990s saw an explosion in easily processed photorefractive organic materials. These included polymers [1–14], organic crystals [15], glasses [16–18], and liquid crystals [19–30].

The submitted manuscript has been created by the University of Chicago as Operator of Argonne National Laboratory ("Argonne") under Contract No. W-31-109-ENG-38 with the US Department of Energy. The US Government retains for itself, and other acting on its behalf, a paid-up, nonexclusive, irrevocable worldwide license in said article to reproduce, prepare derivative works, distribute copies to the public, and perform publicly and display publicly, by or on behalf of the Government.

Crystalline materials tend to have longer storage times and are generally considered for applications where long-lived holograms are required [31,32]. On the other hand, recent literature frequently cites liquid crystals and polymers for their potential in dynamic holography [33,34]. Media that illustrate large index changes with reasonably fast hologram write and erase times can be used to process optical images in real-time. An example of an application for dynamic holography is "coherence-gated" imaging through highly scattering biological tissue [33,35,36]. Here, the photorefractive medium must act as a sensor that detects only the relatively few photons that travel ballistically through the tissue. Liquid crystals are uniquely positioned in this regard as they can act as photorefractive sensors at extremely low light flux. One group has reported space-charge field-derived index of refraction coefficients of 10 cm^2/W, defined as the magnitude of the index change per unit of optical intensity [34]. This suggests that only 100 $\mu W/cm^2$ is required to induce easily measurable index changes of 10^{-3} and are superior in this regard to photorefractive-like responses in any currently available media.

In order to give the reader a sense of where liquid crystals lie in relation to the field of photorefractivity, it is necessary to briefly review the history of the effect. Photorefractivity was discovered in 1966 as an optical damage mechanism in electro-optic crystals that ultimately destroyed the spatial integrity of laser beams as they passed through the crystals [37]. The researchers deduced that at high optical intensities, two photon absorptions of impurities within the crystals produced excited states that shed their energy in the form of charge separation. The more mobile charge, in this case electron holes, migrated out of the illuminated region due to the photovoltaic effect and was ultimately trapped in the dark regions of the crystal. The inhomogeneity in the charge distribution produced an internal electric field, or space-charge field, that altered the index of refraction though the electro-optic effect and destroyed the optical quality of the laser beam.

The humble beginnings of photorefractivity as a damage mechanism in electro-optic crystals was soon harnessed by scientists seeking to use photorefractivity in holography [38,39]. When two coherent laser beams of identical energy and spatial profile are overlapped in a sample, a sinusoidal interference pattern of light and dark regions is produced as illustrated in Fig. 1. The distance between the peaks of the illuminated regions is called the fringe spacing (Λ). Charge separation occurs in the illuminated regions, followed by preferential charge migration of the more mobile species into the darker regions of the interference pattern. This produces the sinusoidal space-charge field that modulates the refractive index. The refractive index modulation, or grating, forms the basis for the optical signal processing applications of the photorefractive effect. A hologram with read-write-erase capability is produced with this process. It is important to note that only those photons that lie within the coherence length of

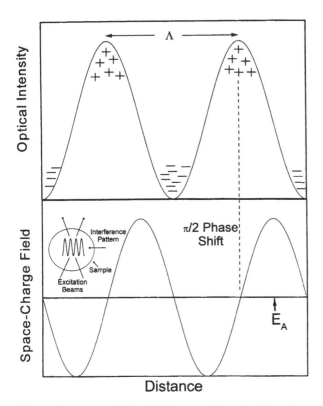

Figure 1 Phase relationship between the optical interference pattern and the space-charge field. For liquid crystals, this example illustrates mobile anions migrating into the nulls of the interference pattern. The application of an applied electric field E_A is usually required to observe a phase-shifted photorefractive grating.

the laser will contribute to the interference pattern. This provides the basis for image sensing through highly scattering media, as scattered photons with longer path lengths pass undetected through the material. Furthermore, as shown in Fig. 1, the photorefractive grating is typically phase-shifted by $\pi/2$ (relative to the interference pattern [39]). This phase-shift is an additional feature of photorefractive holograms that results in asymmetric diffraction, whereby one of the writing beams diffracts light into the other writing beam. This can produce noise-free amplification of the intensity of the laser beam that carries the optical image and is referred to as photorefractive gain.

In the pursuit of improved photorefractive materials, seminal research by a group at IBM led by W. E. Moerner discovered the photorefractive effect in polymers in 1990 [4,40]. Photorefractive polymers are generally composite materials possessing additives that serve to induce the photorefractive effect [1,2,4,40,41]. They initially consisted of photoconductive polymers doped with a high concentration (greater than 30%) of a nonlinear optical (NLO) chromophore, and an electron acceptor that formed a charge transfer complex with the polymer so that mobile charges within the composite could be photogenerated. A notable exception to this strategy incorporated all of the required species for photorefractivity onto a polymer backbone [5]. A few years later it was noted that large increases in the photorefractive sensitivity could be achieved by adding a plasticizer for lowering the glass transition temperature [42]. This permitted reorientation of the NLO chromophores in a viscous polymer, thus producing an additional birefringent contribution to the electro-optic effect. Interestingly, the orientational enhancement effect in some polymeric systems is responsible for the majority of the photorefractive gain [3].

Given the strength of the orientational enhancement effect in polymers, the potential for nematic liquid crystals becomes clearer. In many respects, nematic liquid crystals are ideal for observing the orientational photorefractive effect because nematic liquid crystals are designed to have a strong orientational response, and no nonlinear optical dopant is necessary because the liquid itself is the birefringent component. Thus, essentially 100% of the medium contributes to the birefringence rather than the percentage that is a nonlinear optical dopant. Photorefractive liquid crystals were first reported in 1994 [19,20]. In dye-doped liquid crystals, similar to the polymers, the photorefractive figures of merit rapidly improved in just a few years. These improvements resulted from new liquid crystal mixtures and a better understanding of the photoinduced charge transfer processes in liquid crystals. In this chapter, work is outlined on photoinduced charge transfer reactions in liquid crystals designed to improve photorefractive sensitivity and higher fidelity imaging through better grating resolution. Applications for real-time holographic image detectors and processors are described.

II. INDUCING PHOTOREFRACTIVE HOLOGRAMS IN NEMATIC LIQUID CRYSTALS

A typical experimental apparatus for studying photorefractivity in liquid crystals is illustrated in Fig. 2. Two coherent laser beams from an Ar^+ laser are crossed in the sample, with a total of 5 mW of p-polarized output at 514 nm. The beams are unfocused and have a 1/e diameter of 2.5 mm. The liquid crystal composite is sandwiched between two ITO coated glass slides that are coated with octadecylsilyl groups to induce the liquid crystal director to align perpendicular to the face of the glass slides, that is, homeotropic alignment [43]. The cell thickness is determined by a Teflon spacer that is 12 to 100 μm thick. A small electric field

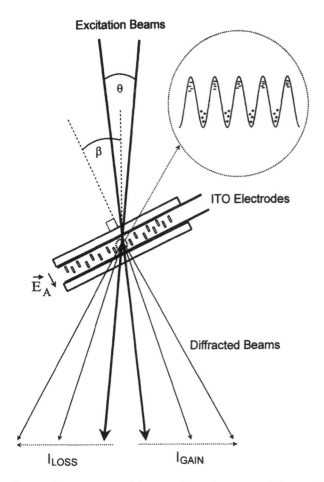

Figure 2 Schematic of the experimental geometry. The sample is tilted at an angle $\beta = 30°$ relative to the bisector of the two beams. This permits charge migration along the grating wavevector resulting in a sinusoidal space-charge field. A phase grating results from the influence of the electric field on the orientational configuration of the birefringent liquid crystal molecules. The beams are polarized in the plane and parallel to the grating wavevector. (Reproduced with permission from Ref. 21. Copyright 1995 American Association for the Advancement of Science.)

E_A of ~0.01 V/μm is then applied. This is required for two reasons: to induce directional charge transport along the wavevector axis, and due to the collective orientational response that produces the index of refraction change. In other words, the applied field, which is greater in magnitude than the internal space-charge field, is required to keep the modulation of the internal electric field greater than zero by limiting reorientation to one direction (Fig. 3). With no applied field, an index grating with a wavevector twice that of the interference pattern will be produced because the reorientation of the director will be in opposing directions at the maxima and minima of the space-charge field. The results is that no diffraction is observed [42].

A means to quantify the magnitude of the photorefractive sensitivity of a material is to measure the *gain coefficient* (Γ). Due to the phase-shift of the index grating relative to the optical interference pattern, asymmetric beam coupling produces a net diffraction of energy from one beam into the other beam. The magnitude of the gain coefficient, for equal intensity beams, is given by the relationship [42]

$$\Gamma = \frac{1}{L} \cdot \left[\ln \frac{I_{12}}{I_1} - \ln \left(2 - \frac{I_{12}}{I_1} \right) \right], \tag{1}$$

where I_1 is the intensity of Beam one in the absence of Beam 2, I_{12} is the intensity of Beam one in the presence of Beam 2, and L is the optical path

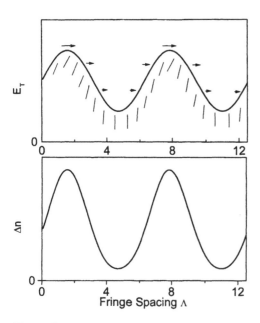

Figure 3 Refractive index change associated with the orientational enhancement effect. The arrows represent the direction of the total field E_T which is the sum of the applied and space-charge fields.

length. The net gain coefficient is cited by subtracting the medium's absorption coefficient (α) and scattering losses. The values for the net gain coefficients in liquid crystals have been reported to be nearly 3000 cm^{-1} [44]. This compares to net gain coefficients for polymers that range upwards of 200 cm^{-1} and inorganic ferroelectric materials with gain coefficients of 10 to 40 cm^{-1} [42].

There is an important caveat to these measurements. Photorefractive gratings in liquid crystals generally operate in the Raman–Nath (thin) grating regime, producing multiple diffraction orders, as illustrated in Fig. 2. This is due to larger fringe spacings in liquid crystals as a result of space-charge field formation through ion diffusion, a process that does not lend itself to small fringe spacings because of ion recombination issues [20]. Also, the collective director orientational response necessitates a larger fringe spacing [45]. It is not necessarily true that photorefractive gain increases exponentially with thickness for a thin grating, as is the case with thick gratings. Therefore, there is some controversy regarding the use of gain coefficients in plane gratings, although some groups do report them for plane gratings in inorganic quantum wells or liquid crystals [44,46]. This can be avoided by simply reporting the beam coupling ratio, I_{12}/I_1 in Eq. (1), as a measure of the beam coupling magnitude.

It must also be verified that the beam coupling is due to a photorefractive mechanism, as other thin grating phenomena can lead to the observation of asymmetric beam coupling in liquid crystals, including thermal, photochromic, order-disorder, and phase change effects [45]. However, these possibilities can be ruled out through diagnostic experiments that are discussed in the literature [20,47]. First, photorefractivity is not observed unless a static electric field is present, suggesting that a space-charge field is present, that leads to director axis reorientation. No effect is observed for an ac applied electric field. Second, the sample must be tilted relative to the writing beams' bisector, as shown in Fig. 2, in order to see the effects. This indicates that a component of the grating wavevector must lie along the direction of the applied electric field in order for directional charge transport to occur along the wavevector. In other words, if the cell plane is perpendicular to the laser beam bisector, the applied field will be perpendicular to the wavevector and no photorefractivity is observed. Third, a grating is only observed when the grating wavevector and the laser beams' polarizations lie in the same axis (extraordinary polarization). This indicates that the index change is a result of reorientation of the liquid crystal molecules in the plane of the two writing beams [20,47]. None of the previously observed alternative effects can be explained by these observations.

Following these control experiments to verify the space-charge field nature of the grating, the holographic experiment can be used to quantify photoinduced charge separation and diffusion properties with a view towards improving sensitivity and grating resolution. With this holographic setup, an optical interference pattern of the form $I = I_o(1 + \cos qx)$ is created, where q is the wavevector of the grating. The result is a modulated space-charge field given by [20,48]

$$E_{sc} = \frac{-mk_BTq}{2e_o} \frac{D^+ - D^-}{D^+ + D^-} \frac{\sigma_{ph}}{\sigma_{ph} + \sigma_d} \sin qx, \tag{2}$$

where the modulation index m is given by

$$m = \frac{2(I_1 I_2)^{1/2}}{I_1 + I_2}. \tag{3}$$

Here, σ_{ph} is the photoconductivity, σ_d is the dark conductivity, k_B is the Boltzmann constant, e_o is the charge of a proton, and D^+ and D^- are the diffusion constants for the cations and anions, respectively. In these experiments, $I_1 \approx I_2$, producing an m value near 1. Thus, the critical factors that determine the magnitude of the space-charge field are the magnitude of the photoconductivity relative to the dark conductivity and the difference in the diffusion coefficients of the cations and anions. These factors allow for one set of charges to preferentially occupy the illuminated regions of the interference pattern and for the opposing charges to occupy the nulls of the interference pattern.

Although the dominant means to create a space-charge field within the interference pattern in liquid crystals is given by Eq. (2), it has been shown that there are other mechanisms to create a space-charge field. One is derived from the conductivity anisotropy and is known as the Carr–Helfrich effect [43,49]:

$$E_{SC}^{\Delta\sigma} = -\left(\frac{\Delta\sigma \sin\theta \cos\theta}{\sigma_\| \sin^2\theta + \sigma_\perp \cos^2\theta}\right) E_A, \tag{4}$$

where σ is the conductivity $\Delta\sigma = (\sigma_\| - \sigma_\perp)$, and θ is the director axis reorientation in response to the applied field E_A. Another mechanism to create a space-charge field is due to dielectric anisotropy, given by [43]

$$E_{SC}^{\Delta\varepsilon} = -\left(\frac{\Delta\varepsilon \sin\theta \cos\theta}{\varepsilon_\| \sin^2\theta + \varepsilon_\perp \cos^2\theta}\right) E_A, \tag{5}$$

where ε is the static dielectric constant and $\Delta\varepsilon = (\varepsilon_\| - \varepsilon_\perp)$. It has been estimated that these terms are approximately 30% of the total relative to the magnitude of E_{SC} given in Eq. (2). The possibility of new mechanisms for inducing space-charge fields in liquid crystals has important ramifications for eventual applications of these crystals. Currently, the resolution of holographic gratings due to charge separation effects in liquid crystals is limited to $\sim 2\ \mu m$, as outlined above. If new mechanisms derived from dielectric anisotropy or the Carr–Helfrich effect can be employed, ultimately these resolutions may be significantly improved. Higher resolution gratings may permit nematic liquid crystals to operate in the Bragg regime for the first time. Currently, the low resolutions permit only a plane grating response. Bragg behavior would permit multiplexing of numerous images at the same site in the sample, an important feature for high-bandwidth applications.

A. Charge Generation in Nematic Liquid Crystals

In order to generate the mobile ions that can produce a space-charge field in liquid crystals, the composites are doped with easily oxidized and reduced moieties that undergo photoinduced charge separation. The scheme for ion generation is illustrated in Fig. 4. Photoexcitation of the donor molecule (D) is followed by collisional interactions with the neutral electron acceptor (A) molecules. This is followed by charge separation and the generation of an ion pair. The critical final step is for a fraction of the ion pairs to form solvent separated ions, with rate constant k_{SEP}. Only the solvent separated ions can diffuse through the sample to create an inhomogeneous spatial charge distribution over macroscopic length scales.

Straightforward charge separation theory can be used to optimize the magnitude of the photorefractive effect through improved generation of mobile charge. Marcus theory predicts that the rate of charge separation (k_{CS}) will be the greatest when the exothermicity of the reaction is equal to the sum of the reorganization energy of the solvent (λ_o) and the internal vibrational reorganization energy of the ions (λ_i) [50–52]. For free energies of charge separation that are

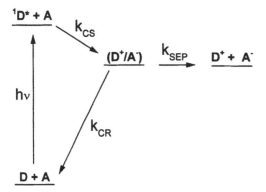

Figure 4 Energy-level diagram illustrating the mechanism for production of the photoinduced solvent separated ions in liquid crystals, where D is the electron donor and A is the electron acceptor.

more or less than the total reorganization energy, the rates of charge separation will be slower. This can be restated by the following equations [50–52].

$$k = Z \exp\left(\frac{-E_a}{kT}\right), \tag{6}$$

where

$$E_a = \frac{(\Delta G_{CS} + \lambda)^2}{4\lambda} \tag{7}$$

$$\lambda = \lambda_0 + \lambda_i \tag{8}$$

$$\lambda_0 = e_o^2 \left(\frac{1}{2r_1} + \frac{1}{2r_2} + \frac{1}{r_{12}}\right)\left(\frac{1}{\varepsilon_\infty} - \frac{1}{\varepsilon_s}\right). \tag{9}$$

Here, Z is the bimolecular collision rate of the uncharged species, ΔG_{CS} is the free energy for charge separation, e_o is the charge of an electron, λ is the total reorganizational energy of the solvent (λ_0) and of the internal intramolecular bonds (λ_i), r_1 and r_2 are the radii of the reactants, and r_{12} is the center-to-center distance between the two reactants. The free energy for charge separation (ΔG_{CS}) was calculated using the relation:

$$\Delta G_{CS} = E_{OX} - E_{RED} - \frac{e_o^2}{\varepsilon_s r_{12}} - E_S, \tag{10}$$

where E_{OX} is the oxidation potential of the donor, E_{RED} is the reduction potential of the acceptor, E_S is the first excited singlet state of the donor, and ε_s is the static dielectric constant.

From Fig. 4, it can be seen that the longer the initial ion pair is present, the more likely it is that solvent separated ions will form. In order to accomplish this, the value of ΔG_{CS} can be gradually changed by using different donors and acceptors with a range of redox values. By monitoring the changes in the photorefractive beam coupling, the generation of solvent separated ions can be optimized.

Further quantitation can be experimentally determined through the relation of these quantities to the diffraction efficiency η of a Raman–Nath orientational grating [20,48]:

$$\eta = \left(\frac{Lmk_bT}{\lambda_{opt}n_eKqe_o} \frac{E_A\varepsilon_s\varepsilon_\infty \sin\beta}{1 + \frac{\varepsilon_s E_A}{2\pi Kq^2}} \nu \frac{\sigma_{ph}}{\sigma_{ph} + \sigma_d}\right)^2, \tag{11}$$

where

$$\nu = \frac{D^+ - D^-}{D^+ + D^-}, \tag{12}$$

where L is the optical path length, n_e is the index of refraction along the extraordinary axis, K is the single constant approximation of the Frank elastic constant [20,48], λ_{opt} is the laser wavelength, and ε_{∞} is the high-frequency dielectric constant. Fortunately, the only variables that are a function of the dopants are the diffusion ration (ν) and conductivity ratio. Rudenko and Sukhov showed that the conductivity term in Eq. (11) saturates at higher light intensities, permitting ν to be determined [20,53].

B. Initial Studies

The first photorefractive experiments in a nematic liquid crystal used the liquid crystalline material 4′-pentyl-4-biphenylcarbonitrile (5CB) with small amounts of the laser dye R6G [19,20]. R6G functioned as the charge generator through heterolytic cleavage. Utilizing the typical crossed laser beam technique to induce a sinusoidal light interference pattern, a photorefractive grating was observed due to the large birefringence of the liquid crystalline material and a gain coefficient of approximately 25 cm^{-1} was reported [19]. These results were very promising, although charge generation and transport of the dyes in 5CB were not efficient. For example, a value for ν of 0.02 was reported, which indicated a very small difference in the diffusion coefficients of the positive and negative ions. Such a result implied that a space-charge field nearly two orders of magnitude higher could be formed within the liquid crystal if appropriate cations and anions were utilized.

We reported a large increase of the orientational photorefractive effect in liquid crystalline materials through the addition of organic electron donors and acceptors [21]. The samples consist of homeotropically aligned mixtures of 35% 4′-(octyloxy)-4-biphenylcarbonitrile (8OCB) and 65% 5CB through treatment of the ITO slides with octadecyltrichlorosilane [43]. This mixture lowers the liquid crystalline to solid phase transition from 24°C for pure 5CB to 5°C [49]. This mixture has improved photorefractive performance due to a greater reorientation angle of the molecules as a result of the lower orientational viscosity of the liquid crystal mixture relative to 5CB. The birefringence of 8OCB is also slightly higher than that of 5CB [54]. In addition, neither component possesses any visible absorption, reducing unwanted competition with the charge generator for photons. The samples utilized were either 88-μm or 37-μm thick, as determined by a Teflon spacer. Thinner samples permitted higher electric fields to be applied due to reduced hydrodynamic turbulence resulting from charged particle motion between the two ITO plates [49]. A battery and low voltage power supply, respectively, were utilized to apply voltages of up to 1.5V (88 μm) and 2.5 V (37 μm) to the sample. This resulted in an electric field of up to 0.01 V/μm for the 37-μm thick sample versus 0.02 V/μm for the 88 μm-thick sample.

Perylene was chosen as the dopant that functions as the charge generator for several reasons. First, perylene is a much more ideal solute in the 8OCB/5CB mixture than is R6G, with solubility unlimited for the purpose of these experiments. Second, it possesses a strong visible absorption ($\varepsilon = 40,000$ M^{-1} cm^{-1} at 442 nm) which has a weak tail that extends farther to the red. At 514 nm, the absorption is 1.5% ($\varepsilon = 610$ M^{-1} cm^{-1}) of the value at 442 nm. Third, perylene is easily oxidized, with an electrochemical oxidation potential of 0.8 V versus SCE [55].

Further increases in the conductivity of the material were achieved by doping the 8OCB/5CB mixture with the electron acceptor N,N'-dioctyl-1,4:5,8-naphthalenediimide (NI) [56]. NI is easily reduced, with an electrochemical reduction potential of -0.5 V versus SCE compared to -1.9 V versus SCE for 5CB. It has no visible absorption in its ground state, but its anion does have a strong absorption ($\varepsilon = 30,000$ M^{-1} cm^{-1}) at 480 nm. It also has high solubility in the 8OCB/5CB mixture. In addition to enhancing the solubility, the octyl tails and the cylindrical shape of this molecule aid its alignment within the liquid crystal. The extinction coefficient of the anion at 514 nm is approximately 7000 M^{-1} cm^{-1}. We determined that an excess concentration of NI versus perylene resulted in the best photoconductivity results, as illustrated in Fig. 5. The photoconductivity of these composites, normalized for absorption, was found to be 50 times greater than the R6G/5CB mixtures.

Photocurrent enhancements per unit absorption of more than an order of magnitude were observed relative to Rhodamine 6G. Beam coupling experiments, shown in Fig. 6, for the 37-μm thick cells resulted in beam amplification (loss) of 88% and photorefractive gain coefficients of 640 cm^{-1}. Photorefractive rise times as short as 40 msec were observed, although at the expense of photorefractive gain. This was accomplished with very low light intensities (100 mW/cm^2) and low applied electric fields (0.04 V/μm) that required only a low voltage (1.5 V) battery. In addition, the samples showed no decomposition over a one-year period.

C. Optimization of Intermolecular Charge Transfer Dopants

As discussed earlier, the photoconductivity and the difference in diffusion coefficients for the cations and anions (ν) need to improved in order to increase the magnitude of the space-charge field in liquid crystals. In order to explore these possibilities, we performed studies with four different intermolecular charge transfer dopant combinations [22] (see Scheme 1). The redox potentials and driving forces of the molecules for charge separation and charge return are given in Table 1. The absorption of the ANI chromophore peaks at 395 nm and extends well into the visible region, permitting the use of the 457-nm line

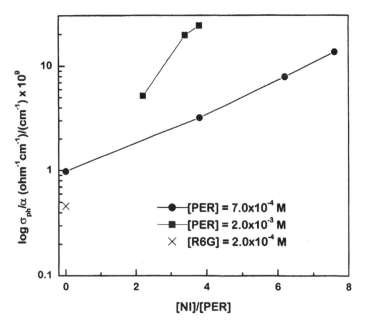

Figure 5 Superior photoconductivity, normalized to an absorption of 1 cm^{-1}, in the perylene and NI doped liquid crystal versus R6G doped sample.

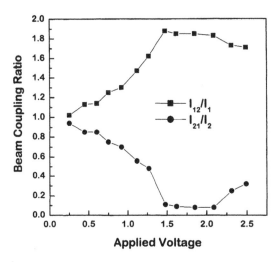

Figure 6 Plot of the beam coupling ratio versus applied voltage for the beam that increases in intensity (■) and for the beam that decreases in intensity (●).

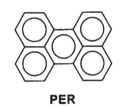

PER

ANI

NI

PI

Scheme 1

Table 1 Lowest Excited Singlet State, Redox Potentials Versus Saturated Calomel Electrode (SCE), and Free Energies for Charge Separation and Return

Molecule	E_S (eV)	E_{OX} (V)	E_{RED} (V)	ΔG_{CS} (eV)	ΔG_{CR} (eV)
PER, NI	2.8	0.8	−0.5	−1.5	−1.3
PER, PI	2.8	0.8	−0.8	−1.2	−1.6
ANI, NI	2.8	1.0	−0.5	−1.3	−1.5
ANI, PI	2.8	1.0	−0.8	−1.0	−1.8

Source: Ref. 22.

of an Ar^+ laser. The 514-nm line was used for the PER chromophores. The concentration of the donors in all cases was 2.0×10^{-3} M.

The results of the conductivity saturation studies that give the value of ν are shown in Fig. 7, and the values for saturation diffraction efficiency (η_{SAT}), q, and ν are shown in Table 2. Values for the constants $E_a = 400$ V/cm, $L = 37$ µm, $K = 7 \times 10^{-7}$ dyne,[1] $m = 1$, ε_∞ of 2.25, ε_s of 15, and $298K$ for T were used to determine ν from Eq. (11). Time-of-flight measurements were performed to determine the diffusion coefficients for D^- of NI^- and D^+ of ANI^+, which can be utilized in conjunction with Eq. (12) to determine the remaining diffusion coefficients.

Clearly, the data indicate that there is a large difference in the values of ν. For the ANI/NI system, the time-of-flight measurements show that the anion (NI^-) is more mobile than the cation (ANI^+). The results shown in Fig. 7

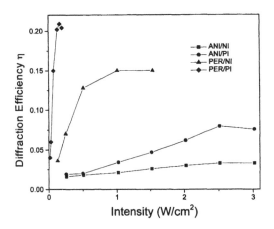

Figure 7 Plot of the diffraction efficiencies versus incident laser power of composite material containing ANI/NI, ANI/PI, PER/PI, and PER/NI.

Table 2 Saturation Diffraction Efficiencies (η_{SAT}), Wavevectors (q), ν, and Diffusion Coefficients for Each Molecular Component

Molecule	η_{SAT}	$q \times 10^{-3}$ cm^{-1}	ν	$D^+ \times 10^8$ cm^2/sec	$D^- \times 10^8$ cm^2/sec
PER, NI	0.15	4.7	0.17	2.5	3.4
PER, PI	0.21	4.7	0.21	2.5	3.8
ANI, NI	0.033	4.2	0.11	2.7	3.4
ANI, PI	0.080	4.2	0.17	2.7	3.8

further support this conclusion, since the smaller PI has increased mobility that produces a larger value for ν than NI doped samples. It follows from Fig. 7 that PER is less mobile than ANI chromophores, because samples doped with PER have superior saturation diffraction efficiencies for a given acceptor. Thus, the best sample is the PER/PI combination which has a value for ν of 0.21, more than 10 times better than R6G/5CB samples.

The PER/PI sample also saturates at much lower intensities (100 mW/cm^2) than the other composites. This means that the efficiency of mobile charge generation is much greater in this sample. This seems counterintuitive since the more easily reduced NI sample would suggest a better charge separation efficiency. However, as discussed for Eq. (7), the interdependence of ΔG_{CS} and λ in the quadratic exponent indicate that the rate for charge separation is maximized for the PER/PI system in which ΔG_{CS} is -1.2 eV. The larger driving force for the NI/PER composite actually slows the rate of charge separation. Following previous precedent, these values ignore the Coulomb term which is small in polar environments [52,57]. Although the optimal ΔG_{CS} value at first glance appears to be rather high, it has been established that the solvent reorganization energies for solvent separated ion pairs, as opposed to tight ion pairs, are higher [52,57].

In addition to optimizing ΔG_{CS}, the free energy for charge return in the PER/PI system is greater than that for PER/NI. This places ΔG_{CR} farther out in the Marcus inverted regime for the PER/PI system, resulting in slower rates for charge return and therefore increasing the efficiency of mobile charge generation [51,52,58]. Thus, the PER/PI liquid crystal composite is the best of both worlds: it has the largest value for ν, which maximizes E_{SC}, and also has the best efficiency of mobile charge generation.

D. Intramolecular Charge Transfer Dopants

We have reported further improvements in the photorefractive sensitivity of liquid crystals by doping with chromophores that undergo efficient photoinduced intramolecular charge transfer [24]. The magnitude of the observed photorefractivity is found to increase with the lifetime of the intramolecular charge separated state. We further show that intramolecular charge transfer dopants represent a superior mechanistic method to produce photorefractivity. This allows for lower concentrations and reduced absorption losses for these materials.

Four compounds (**1** to **4**) were utilized for the intramolecular charge transfer dopants (see Scheme 2). Their synthesis is described elsewhere [59]. They were chosen for the following reasons:

(1) The lifetimes of the charge separated state in degassed toluene for **1** through **4** are 5, 104, 150, and 300 ns, respectively, as determined by time-resolved transient absorption experiments [60–62]. As Table 3 illustrates, this trend is preserved in the liquid crystalline environment, with the caveat that the ion pair lifetimes are increased by at least one order of magnitude. This variation allows for the study of photorefractivity as a function of the lifetime of the intramolecular charge separated state.

(2) They undergo intramolecular charge separation with nearly 100% quantum yield.

(3) Photoinduced electron transfer occurs through excitation of the 400-nm absorption bands of the donor chromophores based on the aminonapthalene–dicarboximide derivatives. The tails of the dopants' absorption bands extend to at least 500 nm, which allows for the use of an Ar^+ laser. Figure 8 illustrates the ground-state absorption spectra of the donor and acceptor for both the intramolecular and intermolecular charge transfer dopants in toluene. The spectra are similar for all of the dopants, with the exception of **2**, which has a 50-nm red-shifted absorption band. The inset illustrates the broadened spectra in the liquid crystalline environment. The extinction coefficient at 457 nm varies from approximately 1000 M^{-1} cm^{-1} for **4**, 2000 M^{-1} cm^{-1} for **1**, 5000 M^{-1} cm^{-1} for **3**, and 10,000 M^{-1} cm^{-1} for **2**.

(4) They are well characterized in liquid crystalline environments [62].

Samples containing **1** through **4** were compared to liquid crystalline composites containing only the donor chromophore **5**(ANI) and to composites containing both donor molecule **5** and the acceptor **NI**. This permitted a direct comparison of photorefractivity for systems in which the initial charge separation occurred through either a intermolecular or intramolecular mechanism. **3** consists of donor and acceptor chromophores that are identical to **5/NI**. The excited states and free energies for charge separation are illustrated in Table 4.

For quantitative comparison of the grating strengths in the different liquid crystal composites, the first-order diffraction efficiency measurements of the Raman–Nath gratings are more amenable to analysis than the beam coupling ratio. Several concentrations for each of the dopants were utilized and Fig. 9 illustrates the highest diffraction efficiency values versus applied voltage for the samples with the optimal concentration of each dopant. A wavevector value of $q = 1 \times 10^3$ cm^{-1} was again utilized. The first clearly noticeable fact is

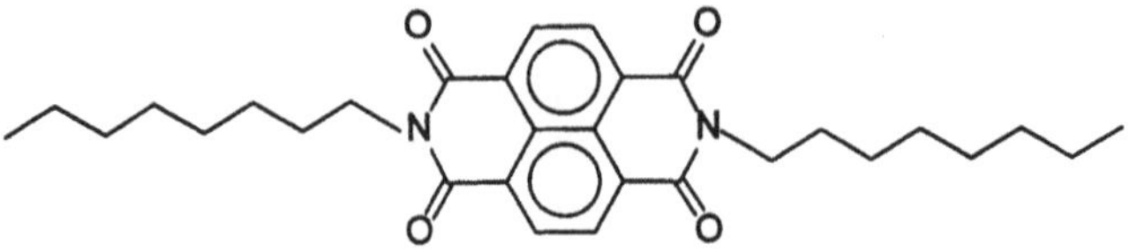

NI

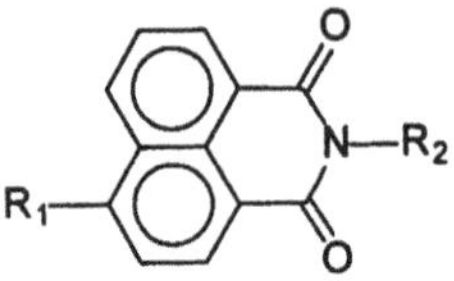

A= 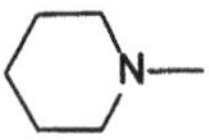   B=

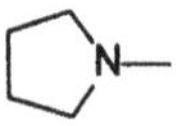

C=    D=

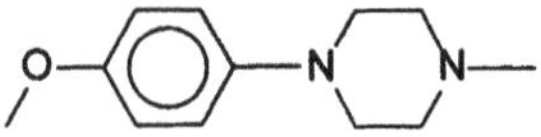

E=

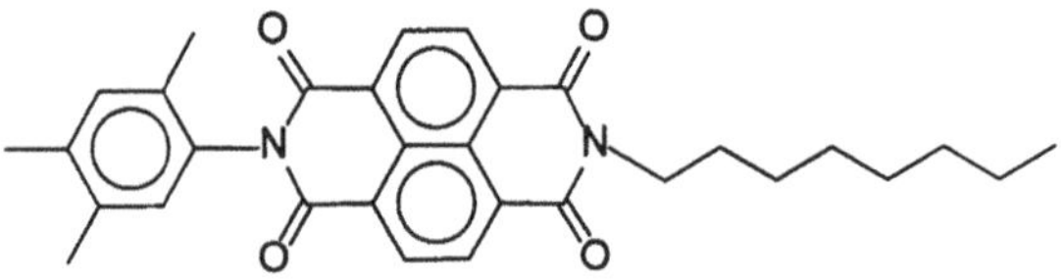

Compound	R₁	R₂
1	D	C
2	B	E
3	A	E
4	D	E
5	A	C

Scheme 2

Table 3 Charge Separation and Charge Return Constants in Toluene and Liquid Crystal

Molecule	τ_{CS} Toluene (ns)	τ_{CR} Toluene (ns)	τ_{CR} LC (μs)
1	0.011	5.3	0.530
2	0.45	104	0.770
3	1.4	150	4.4
4	0.008	300	3.2

Source: Ref. 24.

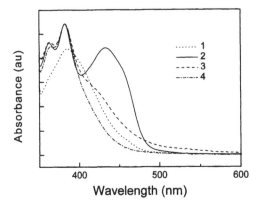

Figure 8 Ground-state absorption spectra of **1** to **4** in toluene. (Reproduced with permission from Ref. 24.)

Table 4 Free Energy for Charge Separation and Charge Return in Polar Liquids[a]

Mol.	E_S (eV)	r_{12} (Å)	E_{OX} (eV)	E_{RED} (eV)	ΔG_{CS} (eV)	ΔG_{CR} (eV)
1	2.87	8.7	0.6	−1.54	−0.80	−2.07
2	2.48	15.3	0.85	−0.7	−0.99	−1.49
3	2.80	15.3	1.02	−0.7	−1.12	−1.68
4	2.87	20.0	0.6	−0.7	−1.60	−1.17

[a] Calculations are made assuming a value for ε_s of 10.5.

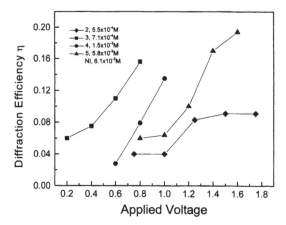

Figure 9 Diffraction efficiency of photorefractive grating in the composite systems. Note that high diffraction efficiency for the composites containing the intramolecular charge transfer dopants **3** and **4** occurs at lower applied voltages than those for the intermolecular charge transfer dopants.

that the intramolecular charge transfer molecules **3** and **4** are superior to the intermolecular charge transfer dopants for inducing photorefractivity. For these dopants, larger photorefractive gain is achieved at lower applied fields. Photorefractivity for **3** is observed for an applied voltage as low as 0.2 V, corresponding to an applied field of only 50 V/cm. For applications purposes, **3** has superior chemical stability in the liquid crystals relative to **4**. We have noted that the methoxy group of **4** leads to long-term degradation and a loss of photorefractivity over a few-day period. The voltages were not increased to higher values for these samples because the measurements were not found to be reliable for very strong gratings with numerous (>5) diffracted beams. Also, theories relating the diffraction efficiency of Raman–Nath gratings to various physical parameters are not valid for diffraction efficiencies above 0.2.

Figure 10 illustrates a plot of η as a function of intensity for samples with dopant **3** (7.1×10^{-4} M) and **5/NI** (5.8×10^{-3} M). The values of η in the saturation limit are different by a factor of five, indicating that ν is a factor of 2.2 larger for the sample doped with **3** relative to that doped with **5/NI**. By using values for $K = 7 \times 10^{-7}$ dyne,[1] ε_∞ of 2.25, ε_s of 10.5, and 298 K for T, we obtain for the composite containing **3** a value for $\nu = 0.29$. *This is the largest value for ν reported for photorefractive liquid crystals.*

Figure 9 further illustrates that the magnitude of photorefractivity for samples with intramolecular dopants **1** through **4** is dependent upon the lifetime of the charge separated state. **1** has no measurable photorefractivity and a charge separated lifetime of only 530 ns in the liquid crystal. **2** has a charge separated

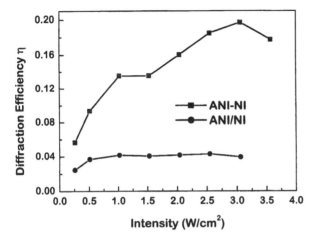

Figure 10 Diffraction efficiency of photorefractive grating as a function of optical intensity for composites containing either **3** or **5/NI**. The composite containing **3** saturates at higher diffraction efficiencies than the composite containing **5/NI**.

lifetime of 770 ns and has photorefractive gain values comparable to those for samples doped with **5**. However, **3** and **4** have dramatically improved photorefractive sensitivities over other materials and ion pair lifetimes of 4.4 and 3.2 μs, respectively.

Figures 9 and 10 can be understood in terms of charge hopping as a mechanism for charge transport, where the likelihood that charges will hop to neighboring molecules increases with the lifetime of the charge separated state. Figure 11 shows the mechanism for bulk charge separation for composites containing intramolecular charge transfer dopants, and can be compared to Fig. 4 for intermolecular dopants. Charge hopping in liquids has long been discussed as an enhancement mechanism for conduction in addition to diffusion. For example, electron-exchange mechanisms were first discussed by Levich and Dahms [63,64]. Ruff and Friedrich generalized these theories and named the process "transfer diffusion" [65]. This process has also been discussed in the framework of Marcus theory [66].

Quantitation of the impact of longer ion pair lifetimes on intermolecular charge separation can be made considering the collision frequency (Z) of the ion pairs and neutral dopants in the liquid crystal. The relations $D_{DA} = k_B T/6\pi\eta a$ and $Z = 4k_B T N_A/3\eta$ can be utilized to obtain the relation $Z = 8D\pi a N_A$. Here, D_{DA} is the diffusion constant of the neutral dopant, N_A is the number density of the dopant, a is the radius of the dopant, and η is the viscosity [64]. D_{DA} can be estimated to be equal to the self-diffusion constant of the liquid crystal, which is 3×10^{-6} cm²/sec, $a \approx 10$Å, and the optimal dopant concentration for **3** is 4.3×10^{17} molecules/cm³. This gives a collision frequency of 3.2×10^6 sec⁻¹, or

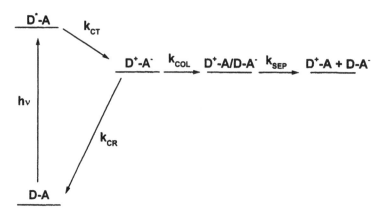

Figure 11 Energy-level diagram illustrating the mechanism to produce photoinduced solvent separated ions through a covalently bound ion pair.

a collision every 300 ns. Given that the lifetimes of the charge separated states of **3** and **4** in a liquid crystalline environment are several microseconds (Table 3), approximately 10 collisions occurring during the lifetime of the intramolecular charge separation are expected. This compares to an excited state lifetime for **5** of 3.5 ns and a collision frequency of 350 ns. Therefore, the number of collisions during the lifetime of the charge transfer/excited state is 1000 times greater for samples containing **3** versus **5/NI**.

We can follow the framework of Suga and Aoyagui, who utilized the collision frequency and Marcus theory to analyze intermolecular electron transfer rates for a reaction scheme of $R_1^- + R_2 \rightarrow R_1 + R_2^-$ [51,66–68]. For the first step, ΔG_{CS} changes only by the Coulomb attraction term, which results in a $\Delta G_{CS} = 0.04$ eV. For intermolecular electron transfer, λ_i may be neglected in the estimate of λ, leaving λ_o to be estimated [61]. Although Eq. (9) predicts a value for λ_o of approximately 0.3 eV, Suga and Aoyagui found that their observed reaction rates implied a value for λ_o of approximately one third the calculated value. This may be due to the fact that the solvent molecules in the neighborhood of the collision are already reorganized to solvate an ion [59]. A value for $\lambda_o - 0.1$ eV produces an initial charge separation reaction rate that is 14% of the collision frequency. Given that the collision frequency allows for approximately 10 collisions during the lifetime of the charge separated state, this suggests that the intermolecular charge separation should occur with near unity quantum yield.

For the second step of the bulk charge separation, charge migration can occur through either diffusion or transfer diffusion. The rate of the transfer diffusion can be easily calculated for the reaction $DA^- + DA \rightarrow DA + DA^-$;

where ΔG_{CS} has a value of zero [65]. This produces a rate constant for charge separation of:

$$k = Z \exp\left(\frac{-\lambda}{4kT}\right). \tag{13}$$

Again utilizing $\lambda_o = 0.1$ eV, this produces a transfer diffusion rate that is 40% of the collision frequency. Thus, this framework illustrates that efficient bulk charge separation can occur within systems that possess long-lived intramolecular ion pairs.

A third observation is that the concentrations of the intramolecular dopants to achieve similar diffraction efficiencies are less than the intermolecular charge transfer dopants. For example, the ideal concentration of the intramolecular dopant **3** is a factor of six less than that for the intermolecular case of **5/NI**, despite the fact that the identical chromophore is optically excited. As a result, reduced absorption for the intramolecular ion pair dopants is achieved with superior diffraction efficiencies. Furthermore, the photorefractivity of the **5/NI** composites increases monotonically with dopant concentration and is limited only by solubility. This is in contrast to the composites containing **3**, where photorefractivity was optimized for concentrations of 7×10^{-4} M. These observations reveal that there is an important mechanistic difference between the intramolecular and intermolecular dopants for producing bulk charge separation. As illustrated in Fig. 4, in order for mobile charge generation to occur in samples containing **5/NI**, an optically excited **5** must undergo a collision with **NI**. Therefore, the rate of charge transfer is ultimately limited by the concentration of **NI**, which alters the reaction rate as $k_{CT} = 1/[\mathbf{NI}]\tau_{CT}$. Given the short excited state lifetime of **5**, this relation shows that the **5/NI** system is limited by the bimolecular collision rate and explains the monotonic increase of photorefractivity with an increase in the concentration of dopants for this system. This contrasts with the intramolecular charge transfer dopants discussed above, where the long lifetime of the initial charge separated state results in near unity quantum yield of mobile charge carriers.

The diffusion coefficients of the ions can be determined through time-of-flight techniques. For a liquid crystalline material between two ITO coated plates with an initial voltage V_1, a sudden increase in the voltage to V_2 allows the ionic mean transit time τ_T and the ion mobility to be determined by the relations [69]:

$$i(t) = i_s \left(1 + \left(\frac{V_2}{V_1} - 1\right) e^{(-2t/\tau_T)}\right) \tag{14}$$

$$\tau_T = \frac{2L}{E_A(\mu_+ + \mu_-)}, \tag{15}$$

where i_s is the saturation current at long times. In order to determine the mobility of the cation, we first studied a sample doped with only **5**. This gave a value for τ_T of 17 sec across a 37-μm sample thickness. Since μ_- is negligible in the absence of a strong acceptor for this sample, a value for $\mu_+ = 6.0 \times 10^{-7}$ cm^2/V sec is obtained. Utilizing $D = \mu k_B T / e_o z$, where z is the charge number, gives a value for $D^+ = 2.7 \times 10^{-8}$ cm^2/sec [66]. This value compares similarly to previously determined diffusion coefficients of ions of approximately equal size in liquid crystals [69]. The same experiment for samples doped with **3** gives $\tau_T = 30 \pm 2$ sec and a value for $D^+ = 1.5 \times 10^{-8}$ cm^2/sec. Thus, the cation of **3** is less mobile than the cation of **5**. However, given the fact that samples of **3** are more photorefractive than **5**, the anions must be more mobile. In other words, the slower mobility of the cation of **3** will produce a greater value of v, and therefore greater photorefractivity, if negatively charged ions are more mobile. We can obtain from Eq. (12) a value for D^- in **3** of 2.7×10^{-8} cm^2/sec and $D^-(\text{NI}^-)$ in **5/NI** of 3.4×10^{-8} cm^2/sec. We also attempted the time-of-flight measurements on samples doped with **NI** alone, in order to determine the mobility for **NI**$^-$. However, lack of electrochemistry at the electrodes resulted in a saturation current that was too low to make an accurate measurement. The diffusion constants for the negative ions are small compared to the self-diffusion coefficient of 5CB, which is 3×10^{-6} cm^2/sec and represents an area of possible improvement [70].

III. LIQUID CRYSTAL/POLYMER COMPOSITES

It would be desirable to combine the very large reorientational effects found in LCs with the charge transport characteristics of polymers to produce higher resolution or potentially longer-lived index gratings. Polymers routinely operate with fringe spacings of approximately 1 μm, while liquid crystals operate with fringe spacings an order of magnitude larger. Along these lines, liquid crystal polymer dispersions are already of great interest for their electro-optic applications. The dispersions are usually studied with a higher concentration of polymer and a smaller concentration of LC molecules which separate into droplets. They are opaque due to the random orientations of the LC directors of the droplets until an electric field is applied. The resulting alignment of the LC directors within each droplet along the electric field produces a transparent material. Such materials are of interest for optical shutters. Orientational photorefractivity has recently been reported in them, albeit at much higher applied fields and with lower photorefractive sensitivity [71–75]. However, an entirely different type of material results if a small amount (1 to 2 weight %) of a polymeric monomer is dissolved in an aligned LC and subsequently photopolymerized with UV radiation in the presence of a photoinitiator. The resulting material is comparable to an anisotropic gel [76–78]. Such materials are referred to as polymer-stabilized liquid crystals (PSLCs) and have been discussed most often in reference to the development of shock-resistant LC displays. The birefringence of the nematic LCs is by and large preserved following polymerization because the monomers are rod-like acrylates that align well with the LC director.

We have worked towards the development of a polymer-stabilized liquid crystal as a new type of organic photorefractive material [79]. These materials are designed to maintain the high birefringence and reorientation characteristics typical of liquid crystals, but permit smaller photorefractive grating fringe spacings by altering the charge transport properties to include trapping effects. The smaller fringe spacing produces a photorefractive grating in liquid crystalline media that operates in the thick grating (Bragg diffraction) regime.

The nematic LC is the original eutectic mixture of 35% (weight %) $4'(n$-octyloxy)-4-cyanobiphenyl (8OCB) and 65% $4'$-(n-pentyl)-4-cyanobiphenyl (5CB). The sample is doped with perylene and 2% (mol %) of NIAC, which is acrylate monomer containing the easily reduced 1,4:5,8-naphthalenediimide moiety (see Scheme 3). Finally, 0.5% (mol %) of benzoin methyl ether (BME) is added to photoinitiate polymerization of the NIAC.

The improved photorefractive grating resolution due to polymer stabilization is illustrated by the asymmetric beam coupling measurements shown in Fig. 12. The unpolymerized samples do not show any measurable beam coupling for fringe spacings (Λ) below 8 μm; whereas the polymerized samples exhibit beam coupling down to $\Lambda = 2.5$ μm. We found that the samples polymerized for two minutes exhibited the best photorefractivity at small fringe spacings. Those samples polymerized longer exhibited reduced beam coupling at all fringe spacings, whereas those polymerized for only one minute showed high beam coupling at larger fringe spacings, but no beam coupling at shorter fringe spacings in a manner similar to the unpolymerized samples. Figure 13 illustrates the kinetics of beam coupling for the two-minute polymerized samples for $\Lambda = 4.8$ μm. The inset of Fig. 13 shows the beam coupling at the smallest fringe spacing of 2.5 μm.

In order to determine whether the grating is a thin (plane) or volume grating, the following well-known quality parameter (Q) can be used [80]:

$$Q = \frac{2\pi L \pi}{\Lambda^2 n},$$ (16)

where λ is the wavelength of the light, n is the index of refraction, and L is the thickness of the grating ($L = d / \cos \beta$, where d is the cell thickness). For $Q \ll 1$, the grating is considered to be a plane grating, and for $Q \gg 1$, a volume grating is created. Although the literature does not appear to specifically designate an exact value for Q in which the thick grating regime is reached, the most rigorous treatments designate that Q values of 10 are required to produce a true volume

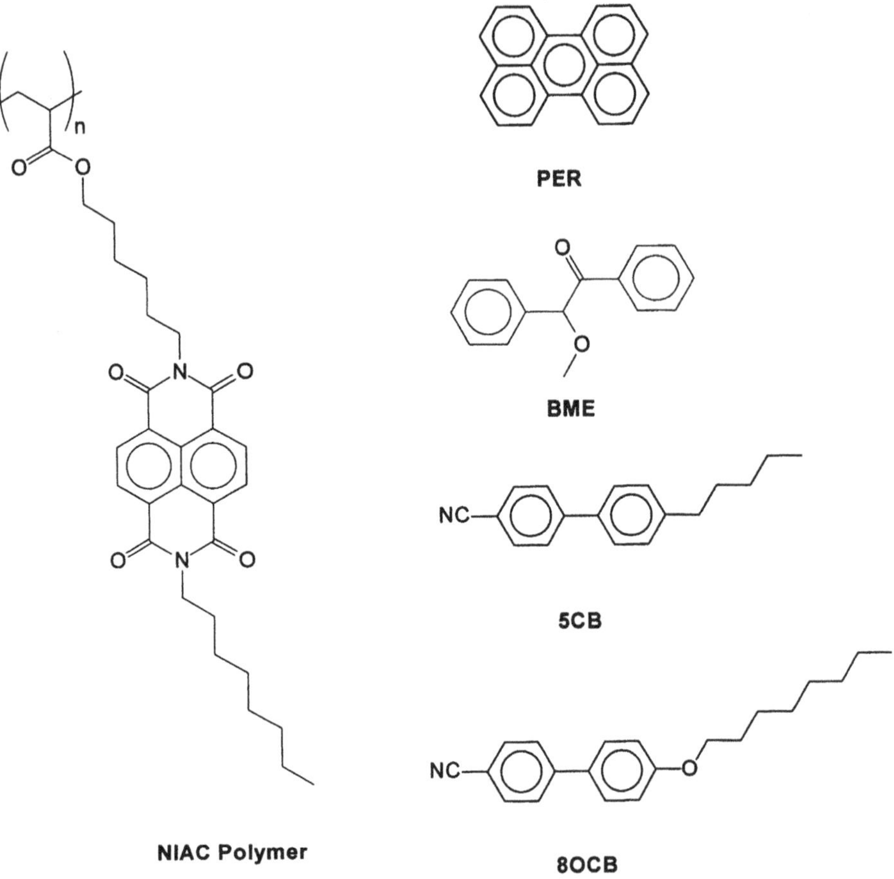

Scheme 3

hologram [81]. For our samples, $Q = 10$ is achieved for $\Lambda = 2.6\ \mu$m, when $L = 26\ \mu$m$/\cos 30° = 30\ \mu$m, and $n = 1.5$. At this fringe spacing, our samples still exhibit a small amount of beam coupling (about $\pm 1\%$), so the PSLCs can be considered to be operative in the thick grating regime. This compares favorably to the best-case scenario for the LCs, where beam coupling is observed for fringe spacings no smaller than $\Lambda = 8\ \mu$m, giving $Q = 1$.

The absorbance of the sample at 514 nm is 0.003, resulting in values for α of only 1 cm^{-1}. Therefore, for $\Lambda = 2.5\ \mu$m a net photorefractive gain of $\Gamma = 5$ cm^{-1} results. For slightly less rigorous treatments of the value of Q, higher photorefractive gains can be achieved. For example, some researchers indicate only that $\Lambda^2 \approx D\lambda$ is sufficient for the thick grating regime to be reached, so

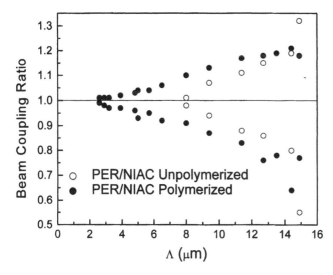

Figure 12 Beam coupling for PSLC composite and unpolymerized LC. Beam coupling in PSLC is observed down to 2.5 μm for applied voltage of 1.8 V. (Reproduced with permission from Ref. 79.)

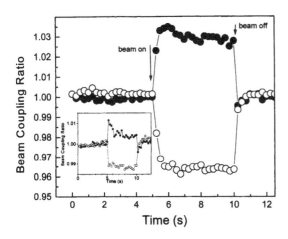

Figure 13 Kinetics of asymmetric beam coupling within PSLC for $\Lambda = 4.8$ μm. The inset shows the beam coupling kinetics for $\Lambda = 2.5$ μm. Beam 1: ●, Beam 2: ○.

that for $\Lambda = 4$ μm ($Q = 4$), the net photorefractive gain is 15 cm^{-1} [71]. It should be noted that impressive photorefractive gains as high as 2890 cm^{-1} are currently being reported in LCs, but these materials do not operate in the thick grating regime [44].

The photorefractive rise (τ_{pr}) and decay (τ_{pd}) times versus Λ, as measured by the four-wave mixing experiments, are shown in Fig. 14a and b for the polymerized and unpolymerized samples, respectively. The decay times are

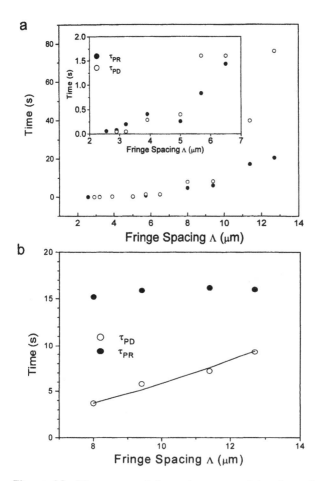

Figure 14 Rise, τ_{PR}, and decay times τ_{PD} of the photorefractive grating versus Λ for (a) PSLC and (b) unpolymerized LC. The unpolymerized LC decay times exhibit a quadratic dependence on fringe spacing, consistent with an ion diffusion, whereas the PSLCs show a steeper dependence versus Λ.

measured following the blockage of one beam. Note that the rise times of the photorefractive gratings in the polymerized samples are always faster than the decay times when one beam is blocked for all but the smallest three fringe spacings. Conversely, for the unpolymerized samples, the decay times are always faster than the rise times even for the large fringe spacings studied. Therefore, the polymerized samples produce a more stable photorefractive grating and, as a result, smaller fringe spacings can be reached.

The time-resolved photoconductivity measurements shown in Fig. 15 give further support for a difference in the photoinduced charge transport in the polymerized samples versus the unpolymerized samples. For the incident laser of 100 mW/cm^2 and a spot size of 2.5 mm, the decay time of the photoconductivity for the unpolymerized samples is 7.4 sec, whereas the photoconductivity of the polymerized samples does not significantly drop over a 30 sec period. Also, the photoconductivity of the polymerized sample is nearly twice that of the unpolymerized samples even at the peak of the unpolymerized photoconductive response. The unnormalized values for the dark conductivity in both samples is 1.7×10^{-10} S cm^{-1}. The photoconductivity is 5.8×10^{-11} S cm^{-1} for the unpolymerized sample and 1.1×10^{-10} S cm^{-1} for the PSLC at an optical intensity of 2 W cm^{-2}.

The beam coupling, four-wave mixing, and photoconductivity data indicate that the mechanism for charge transport within the PSLCs is dramatically different than for the unpolymerized samples. The unpolymerized samples should exhibit a grating decay time that decreases quadratically with fringe spacing. This is supported by the data in Fig. 14b [47,80]. However, the data for the

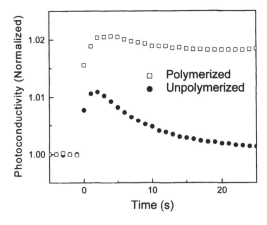

Figure 15   Improved photoconductivity of PSLC relative to LC.

polymerized samples clearly indicate a decay time dependence versus Λ that is far greater than quadratic.

For the PSLCs, reduction of immobile NIAC on the polymer matrix must be followed by charge migration through a different mechanism than within the unpolymerized samples in order to achieve bulk charge separation. In other polymeric materials, holes are generally the mobile species, and a series of shallow traps that empty thermally provide a mechanism through which holes can hop until a deeper trap is encountered. Thus, a mechanism is present for bulk separation to occur in the polymers. If the oxidized deep traps absorb the incident light, charge separation is further encouraged to occur along the shallow traps. A similar process appears to be present in our PSLC, but the mobile charges are now electrons that are transported along the polymer network. We believe that this is the only mechanism that can explain the increased photoconductivity of the PSLCs because clearly the diffusion coefficients of the cations and anions should decrease relative to pure LCs. Charge migration can occur through repeated optical excitation of the reduced trapped species, NIAC$^-$, followed by charge hopping along the polymer chain. As Fig. 16 illustrates, NIAC$^-$ has a large absorption that peaks at 480 nm (30,000 M^{-1} cm^{-1}) and has a considerable extinction coefficient at 514 nm.

As a test of the charge hopping theory, we also observed the decay of the photorefractive grating with both beams blocked and compared it with the data for only one beam blocked. Typical four-wave mixing data are given in Fig. 17. Clearly, the photorefractive grating decays very quickly for the case when both writing beams are blocked as opposed to the case when only one beam is blocked. In fact, the lifetimes of the gratings are enhanced more than an order of magnitude when one beam is incident on the sample. Furthermore, the

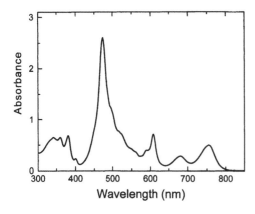

Figure 16 Absorption spectrum of NIAC$^-$.

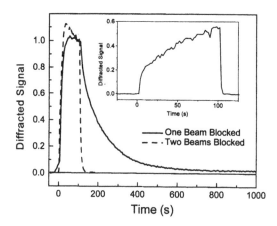

Figure 17 Difference in decay kinetics for one beam blocked versus two beams blocked in PSLC. The lifetime of the grating is enhanced when one beam is incident on the sample. The inset shows the decay kinetics for the unpolymerized LC when one beam is incident on the sample. No enhancement of the grating lifetime is observed. Both beams are incident on the sample at $t = 0$, and either one or both beams are blocked as specified above at $t = 100$ s.

PSLCs exhibit a weak dc signal when one beam is incident on the sample that decays over a much longer time scale (a few hours). No such lifetime increase is observed for photorefractive gratings in unpolymerized LCs, which is illustrated by the inset of Fig. 17. These experiments show that a photoinduced process occurs in the PSLCs that encourages charge separation, but also discourages charge recombination.

One possible explanation for these observations is suggested by focusing on the electron transfer equilibrium that occurs between monomeric $NIAC^-$ and the polymerized NIAC traps, as optimal photorefractivity occurs for incomplete polymerization under short UV illumination times. After the grating is formed, removal of both laser beams from the sample results in rapid charge recombination between monomeric $NIAC^-$ and PER^+ because of the greater mobility of monomeric $NIAC^-$ relative to that of polymerized $NIAC^-$. The rapid depletion of monomeric $NIAC^-$ in the PSLCs results in a shift in the charge equilibrium between monomeric NIAC and polymerized $NIAC^-$ to move charge onto the more mobile monomeric NIAC, which in turn transfers the electron back to PER^+. This process most likely occurs in parallel with direct electron transfer from polymerized $NIAC^-$ to PER^+. On the other hand, when a single laser beam remains incident on the grating, PER^+ and $NIAC^-$ are generated throughout the existing grating. This laser beam generates excess $NIAC^-$ which shifts the equilibrium for electron transfer between monomeric NIAC and polymerized $NIAC^-$

to favor retention of charge on polymerized NIAC. Thus, the previously generated spatial grating due to electron trapping on polymerized NIAC is preserved for a longer period of time.

A. Through Bond Charge Transport

As shown previously, nematic liquid crystals reorient easily in weak electric fields and their high birefringence provides an efficient electro-optic mechanism that makes them excellent candidates for photorefractive materials. However, charge transport relies on the generation of mobile anions or cations. These mobile charges obey the current density (J) equations given by [82,83]

$$J = J^+ + J^- \tag{17}$$

$$J^\pm = e_o \mu^\pm n^\pm(x,t)(E_{SC}(x,t) - E_A) \mp e_o D^\pm \frac{\partial n^\pm[x,t]}{\partial x}, \tag{18}$$

where $\mu^\pm$ is the mobility of the cations and anions, x is the grating wavevector axis, $n^\pm(x,t)$ is the ion density, $E_{SC}(x,t)$ is the magnitude of the space-charge field, E_A is the magnitude of the applied field, and $D^\pm$ is the diffusion constant of the cations and anions, respectively. The first term on the right-hand side of Eq. (18) describes charge drift, and the second term describes ion diffusion. Previous studies of photorefractivity in liquid crystals indicate that ion diffusion is the charge transport mechanism that creates the space-charge field, a process that limits the efficiency and speed of the effect. The charge drift mechanism has not been a factor because of the short ionic drift length L_E, given by [82]

$$L_E = \frac{\mu^\pm \tau^\pm V}{d}, \tag{19}$$

where $\tau^\pm$ is the carrier lifetime, V is applied potential, and d is the cell thickness. Typically, $2\pi L_E \ll \Lambda$ in liquid crystals, while the drift mechanism can only contribute to a photorefractive grating is $2\pi L_E \geq \Lambda$ [82].

The conjugated polymer poly(2,5-bis(2'-ethylhexyloxy)-1,4-phenylene-vinylene), BEH-PPV shows high through bond electron and hole mobility [84] (see Scheme 4). We incorporated BEH-PPV into a nematic liquid crystal mixture in order to explore the possibility of observing through bond charge transport in liquid crystal composites. A eutectic mixture of 35% (weight %) 4'-(n-octyloxy)-4-cyanobiphenyl, 8OCB and 65% 4'-(n-pentyl)-4-cyanobiphenyl, 5CB was doped with 10^{-5} M BEH-PPV (200 kD by GPC), as the electron donor [85]. The molecular weight of the BEH-PPV polymer implies that 500 repeat units of the monomer are present with an extended chain length of 0.35 μm. N,N'dioctyl-1,4:5,8-naphthalenediimide, NI, 8×10^{-3} M, was added as the electron acceptor [59]. Two other liquid crystal composites were also studied as controls. The first control composite contained the five-unit phenylenevinylene oligomer, 5PV

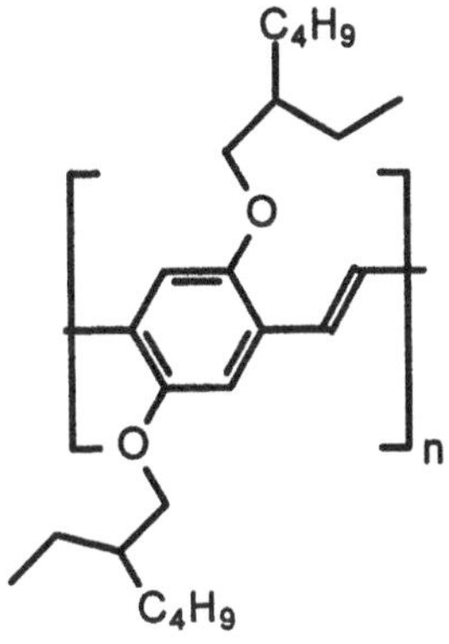

BEH-PPV

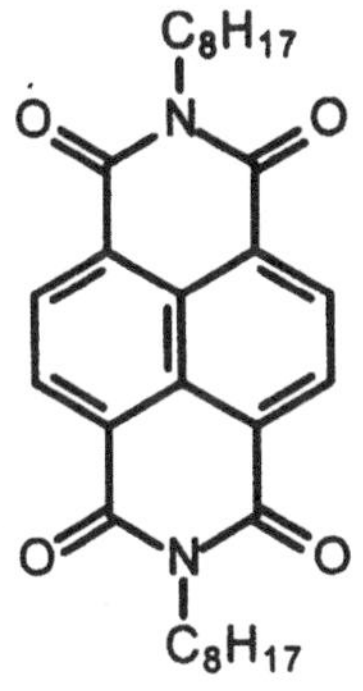

NI

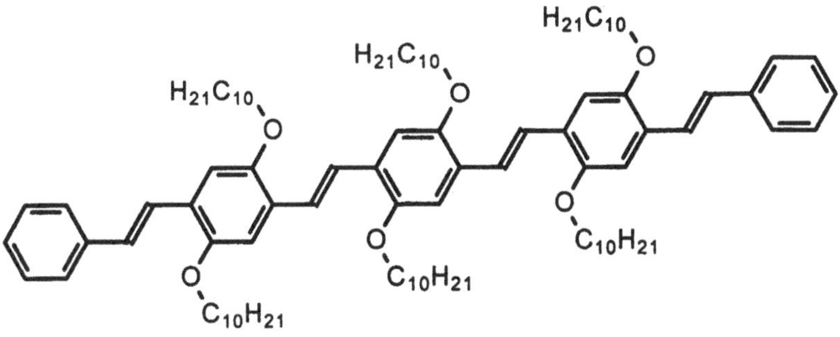

5PV

Scheme 4

as the electron donor (4×10^{-5} M) along with 8×10^{-3} M NI acceptor. The second control composite contained 10^{-5} M BEH-PPV polymer with no NI present. The free energy change for the photoinduced electron transfer reaction ^{1*}(BEH-PPV) + NI $\rightarrow$ (BEH-PPV)$^{+}$ + NI^{-} was -1.0 eV. The free energy change for the photoinduced electron transfer reaction ^{1*}5PV + NI $\rightarrow$ 5PV* + NI^{-} was -1.3 eV, comparable to that for the BEH-PPV/NI pair. These free energy values were derived from Eq. (10) and the oxidation potentials (E_{OX}) for BEH-PPV and 5PV that are both 0.9 V versus SCE, respectively. Their lowest excited state energies E_S were 2.4 eV and 2.7 eV, respectively. The reduction potential E_{RED} of NI was -0.5 V versus SCE.

Figures 18a and b show the kinetics of beam coupling for two different values of Λ in the BEH-PPV/NI composite and the control composite containing 5PV/NI. Comparing the data illustrated in Fig. 18a with that 18b, the direction of beam coupling at smaller Λ is the same for both composites, while it is opposite for the two composites at larger Λ. Comparing Fig. 19a and b for $\Lambda < 5.5$ μm, the direction of beam coupling in the BEH-PPV/NI composite is the same as that of the 5PV/NI control composite. A further observation is that beam coupling is observed in the BEH-PPV/NI composite at the high resolution $\Lambda = 1.5$ μm, whereas the smallest Λ achieved with the 5PV/NI composite is 3 μm. These

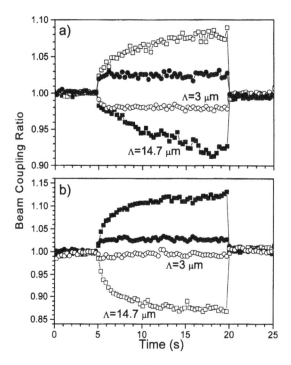

Figure 18 (a) Kinetics of beam coupling in the BEH-PPV/NI liquid crystal composite for $\Lambda = 14.7$ μm ($\square$ and $\blacksquare$) and 3.0 μm ($\bigcirc$ and $\bullet$). (b) Kinetics of beam coupling in the 5PV/NI liquid crystal composite for $\Lambda = 14.7$ μm ($\square$ and $\blacksquare$) and 3.0 μm ($\bigcirc$ and $\bullet$). Open symbols indicate the same beam, while solid symbols indicate the other beam. For each curve a photodiode monitors the intensity of one beam, while the other beam is incident on the sample at 5 sec and blocked at 20 sec. (Reproduced with permission from Ref. 85.)

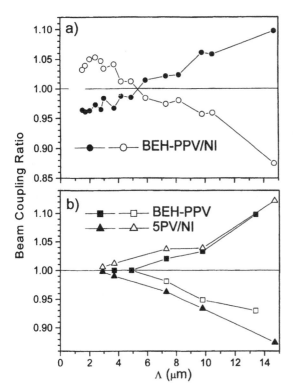

Figure 19   (a) Magnitude and direction of beam coupling in the BEH-PPV/NI liquid crystal composite. (b) Beam coupling ratio versus Λ for control composites containing only BEH-PPV or 5PV/NI. Open symbols indicate the same beam for all three composites, while solid symbols indicate the other beam.

observations are consistent with a space-charge field created through an ion diffusion mechanism. Within this model, the magnitude of the space charge field in liquid crystals increases as the difference between the diffusion coefficients for the cation and anion increases. NI^- has a larger diffusion coefficient than oxidized electron donors of comparable size, including $5PV^+$. Since BEH-PPV$^+$ polymer cation has a much smaller diffusion coefficient than $5PV^+$, the space-charge fields derived from diffusion within both the BEH-PPV/NI and 5PV/NI composites should have the same polarity and, therefore, the same asymmetric energy exchange direction. The larger size and smaller diffusion coefficient of BEH-PPV$^+$ relative to that of $5PV^+$ results in a higher resolution grating attained with the BEH-PPV/NI composite because the difference in diffusion coefficients between BEH/PPV$^+$ and NI^- is larger than that between $5PV^+$ and NI^-.

The sign of the asymmetric beam coupling inverts for the BEH-PPV/NI composite at approximately $\Lambda = 5.5$ µm, while neither control composite shows this inversion. This suggests that the polarity of the space-charge field inverts at this point [39]. Comparing Fig. 19a and 19b for $\Lambda > 5.5$ µm, the direction of beam coupling in the BEH-PPV/NI composite is the same as that with the control composite containing BEH-PPV alone. This indicates that the sign of the mobile charge carrier that contributes to space-charge field formation is the same for $\Lambda > 5.5$ µm in these two composites, and opposite to that for the control composite containing 5PV/NI. Therefore, since negative charges carried by diffusing NI^- are the most mobile charges that contribute to a phase-shifted space-charge field in the BEH-PPV/NI composite for $\Lambda < 5.5$ µm, positive charges are the most mobile charges that contribute to a phase-shifted space-charge field at $\Lambda > 5.5$ µm.

Since the high 200-kD molecular weight of $BEH-PPV^+$ precludes rapid ion diffusion, a charge migration mechanism other than diffusion is producing the space-charge field at larger values of Λ. Solutions of alkoxylated PPVs display very high intrachain mobilities for both holes and electrons [84,86]. Fast hole transport along the BEH-PPV chain, coupled with hole hopping to other BEH-PPV chains can lead to long-distance hole migration in the composite. This mechanism of charge transport can be considered a charge drift mechanism in the trap density limited regime. This means that the charge drift length L_E is larger in the BEH-PPV doped liquid crystals relative to the 5PV doped liquid crystals. In the context of Eq. (19), it is likely that $\tau^\pm$ and $\mu^\pm$ are significantly enhanced in the BEH-PPV polymer due to the delocalization of charge and fast intrachain hole mobilities. The increase in $\tau^\pm$ and $\mu^\pm$ through this mechanism is not possible for smaller cations and anions such as $5PV^+$ and NI^-. Furthermore, the magnitude of the current density due to drift should be larger [Eq. (18)] in BEH-PPV doped composites because $n^\pm$ is proportional to the carrier lifetime [82]. Thus, L_E and $n^\pm$ are large enough in the BEH-PPV composite to observe charge drift as a contributor to space-charge field formation for the first time in liquid crystals.

The fringe spacing limit in which each charge transport mechanism dominates is consistent with our analysis. The contribution to the magnitude of the space-charge field derived from ion diffusion is inversely proportional to Λ [20,47]. However, the contribution to the space-charge field magnitude from charge drift, in the trap density limited model, is independent of Λ [82]. The space-charge field due to drift can contribute to the photorefractive grating as long as $2\pi L_E \geq \Lambda$ so that the space-charge field is phase-shifted from the optical interference pattern. Therefore, in the BEH-PPV polymer/liquid crystal composites where L_E is enhanced, the charge drift mechanism should dominate space-charge field formation at the larger value of Λ explored in our experiments, but the ion diffusion mechanism should dominate at smaller Λ (Fig. 19).

Furthermore, the fact that the control composite containing only BEH-PPV does not show any beam coupling below 5.5 μm is consistent with the availability of only one charge transport mechanism, charge drift, because the absence of NI$^-$ precludes diffusional space-charge fields at lower values of Λ. This discussion only applies to which charge transport mechanism will dominate space-charge field formation at a given fringe spacing. The magnitude of the space-charge field does not correlate with the beam coupling magnitude, because the orientational response of liquid crystals decreases at smaller fringe spacings for $\Lambda < 2d$ [47].

We have shown that the BEH-PPV conjugated polymer provides a mechanism to enhance the photorefractive effect in each fringe spacing limit. In the ion diffusion limit, the much lower mobility of BEH-PPV$^+$ relative to that of NI$^-$ decreases the fringe spacing in which photorefractive beam coupling can be observed. Furthermore, the use of BEH-PPV enhances the charge drift length so that a new space-charge field generation mechanism in liquid crystals becomes possible. The versatility of conjugated polymers in enhancing the photorefractive effect in liquid crystal composites may lead to the development of new optical devices that use photorefractivity for information processing and storage.

IV. POTENTIAL APPLICATIONS

The photorefractive effect can be used as a method for optical image storage and retrieval. For this application, one of the hologram "writing" beams contains the optical image and the other beam is the "reference" beam. A third "probe" beam is sent through the sample anticollinear with the reference beam. A phase conjugate diffracted beam will result that travels along the image beam path and contains the image stored in the material.

The phase conjugate nature of the image retrieval process permits aberration-(noise) free data storage. If an image is generated at a given source, it may travel through imperfect optics that distort the stored image. However, if a phase conjugate beam is created in the retrieval process, then as the beam travels back through the optical path, the original image is retrieved noise free. This technique has been found to work perfectly within experimental signal-to-noise [39]. Other experimenters have found that photorefractive phase conjugation can be used for image amplification of several orders of magnitude [87], image retrieval through dynamic distorters [88], optical tracking of time-dependent holographic images [89], and so on. Another particularly useful application is that of a phase conjugate oscillator, where regular mirrors are replaced with phase conjugate mirrors. Such a technique allows a laser to run with any longitudinal mode desired, without the cavity length requirements presently associated with conventional oscillators [38].

In order to perform real-time imaging through highly scattering media, the holographic material must be highly photosensitive and be capable of fast hologram write and erasure times. The high photosensitivity is essential because of the low photon flux through the sample. Ease of hologram erasure is important for real-time monitoring, so that faster erasure times lead to faster frame rates. One of the most promising classes of materials that satisfy these requirements are organic photorefractive materials. In particular, photorefractive liquid crystals, which possess a higher index of refraction changes (Δn) within the hologram and faster grating erasure times than virtually any other photorefractive material, present an excellent opportunity to develop a direct, real-time optical imaging device.

Optical imaging through highly scattering media has been proposed for a variety of applications, such as medical diagnostics and battlefield imaging [33,35,36]. In these applications, either biological tissue or haze can lead to optical scattering that interferes with currently available optical signal-processing techniques. Dynamic holography is frequently suggested as a solution to optical imaging through scattering media, if a suitably photosensitive holographic material can be found. The concept of dynamic holography is illustrated in Fig. 20. Three laser beams are split from the same laser, usually an inexpensive near-infrared diode laser. A hologram is formed from the interference pattern between the reference beam and the weak beam of ballistic photons that travel through the scattering material to be imaged. The sensitivity of the technique relies on the fact that an interference pattern is formed only for those photons

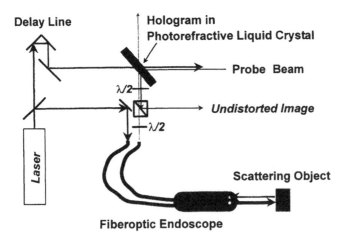

Figure 20 Dynamic holography apparatus for imaging through scattering media such as biological tissue.

in the two beams that lie within the coherence length of the laser. Since the ballistic photons travel faster through the scattering object than the scattered photons, a delay line is used to set the delay of the reference beam to that of the ballistic photons. Thus, a background-free hologram is written in the material without expensive and complicated time-gating technology. Lasers that have a short (~1 mm) coherence length, such as semiconductor diode lasers, increase the likelihood of selecting only the ballistic photons to write a noise-free hologram. Once the hologram is written, a third probe beam is used to diffract off the hologram, and the noise-free image can be recovered with simple frame grabbing technology.

Using medical imaging as an example, optical imaging of biological tissue has several advantages over other imaging technologies. These include the fact that optical radiation is nonionizing, the spatial resolution is potentially higher, and the costs of current magnetic and X-ray imaging devices are prohibitive. Optical imaging of biological tissues is made possible due to the relative transparency of these materials to near-infrared light (NIR) between 700 nm and 1100 nm. Holograms have been written under such conditions with photorefractive polymers and inorganic crystals with optical densities of the scattering medium up to four, but liquid crystals should produce superior image reconstruction due to their higher photorefractive sensitivity as measured through Δn modulation in the hologram. Furthermore, liquid crystals require only a tiny applied field (E_A) on the order of 0.1 V/μm. This compares favorably to other organic materials that require applied fields of up to 100 V/μm. Further more, the erasure time of liquid crystal holograms is less than 1 sec, which will permit improved real-time imaging capabilities, as the object beam is moved through the biological tissue or other scattering medium. Finally, the resolution of the technique should also be ultimately limited only by the fringe spacing of the hologram, which is on the order of 10 μm.

Photorefractive media may be a strong competitor to current optical coherence tomography (OCT) techniques for sensitive optical imaging of biological tissue [90,91]. OCT uses an interferogram that is Fourier transformed to reproduce an optical image. Thus, both OCT and photorefractive technology rely on coherent interference patterns, with photorefractive materials acting to reconstruct the image rather than Fourier transforms. Photorefractive holography is attractive for this reason because the image is instantly recovered, and no manipulation of the interference pattern is required.

A great deal of recent interest is related to the Doppler capabilities of OCT for resolving blood flow [92,93]. In principle, there is no reason why OCT or photorefractive imaging cannot perform similarly in this regard. The similar capabilities are a result of the fact that both techniques require an optical interference pattern to be created between two overlapping beams whose path lengths lie within the coherence length of the laser. The only difference between the two techniques is how the Doppler shift ($\Delta\omega$) manifests itself in the interference pattern. For the scanning systems of OCT, a beating pattern at the Doppler frequency of $2\Delta\omega v/c$ will occur, where v is the blood flow velocity (corrected for the flow velocity angle relative to the probe beam) and c is the speed of light. This produces a small but detectable change in the frequency of the interference pattern in the scanning interferometer.

For the interference pattern in photorefractive media, where both beams are stationary, a Doppler shift in one arm will producing a spatially moving, rather than stationary, interference pattern that will move one fringe spacing in a period of $c/\Delta\omega v$. Based on the literature of optical Doppler tomography, the value of v is on the order of 1000 μm/s for blood flow. This is a small value that should produce a moving period on the order of tens of seconds in the interference pattern of stationary beams. One possibility for monitoring the moving period is to measure the changing phase-shift of the optical interference pattern relative to the index change in the photorefractive medium. As the interference pattern moves, the index grating produced by the spatial inhomogeneity of charge in the photorefractive liquid crystal will lag behind. Methods to measure the phase-shift of an index grating relative to the interference pattern have been reported [94]. Another possibility is to monitor the beam coupling change, which is actually known to increase if a moving grating is present. Thus, the moving interference pattern could enable us to obtain increased sensitivity and Doppler information [95].

V. PERSPECTIVES

Organic photorefractive materials continue to progress rapidly towards real-time dynamic holography that can process optical images of unprecedented low light flux. Among organics, this field is focused primarily on photorefractive polymers and liquid crystals. Interestingly, in many respects related to dynamic holography, the two media are converging. In both types of materials, orientational birefringence is the most effective means to induce a large index of refraction change. Furthermore, many of the most promising new materials are composites containing both polymers, which tend to have better charge transport properties, and liquid crystals, which tend to have superior orientational birefringence. Thus, liquid crystal/polymer composites are among the most likely candidates to eventually produce a sensitive and fast index of refraction responses to effectively perform dynamic holography under low light flux. Further important challenges will be to develop materials that are sensitive over broader wavelength ranges and are stable over larger temperature fluctuations. As important advances continue to be made in these area, the future for dynamic holography in organic materials is bright.

ACKNOWLEDGMENTS

Support is gratefully acknowledged from the Technology Research Division, Office of Advanced Scientific Computing Research, U.S. Department of Energy, under contract W-31-109-ENG-38.

REFERENCES

1. WE Moerner, A Grunnet-Jepsen, CL Thompson. Annu Rev Mater Sci 27:585, 1997.
2. Y Zhang, Y Cui, PN Prasad. Phys Rev B 46:9900, 1992.
3. BL Volodin, B Kippelen, K Meerholz, B Javidi, N Peyghambarian. Nature 383:58, 1996.
4. S Ducharme, JC Scott, RJ Twieg, WE Moerner. Phys Rev Lett 66:1846, 1991.
5. L Yu, WK Chan, A Peng, A Gharavi. Acc Chem Res 29:13, 1996.
6. R Bittner, TK Daeubler, K Meerholz. Adv Mater 11:123, 1999.
7. MA Diaz-Garcia, D Wright, JD Casperson, B Smith, E Glazer, WE Moerner, LI Sukhomlinova, RJ Twieg. Chem Mater 11:1784, 1999.
8. A Grunnet-Jepsen, CL Thompson, RJ Twieg, WE Moerner. Appl Phys Lett 70:1515, 1997.
9. M Liphardt, A Goonesekera, BE Jonees, S Ducharme, JM Takacs, L Zhang. Science 263:367, 1994.
10. SR Marder, B Kippelen, AK-Y Jen, N Peyghammbarian. Nature (London) 388:845, 1997.
11. K Meerholz, BL Volodin, Sandalphon, B Kippelen, N Peyghambarian. Nature 371:497, 1994.
12. Z Peng, AR Gharavi, L Yu. J Am Chem Soc 119:4622, 1997.
13. Q Wang, LM Wang, LP Yu. J Am Chem Soc 120:12860, 1998.
14. JG Winiarz, L Zhang, M Lal, CS Friend, PN Prasad. J Am Chem Soc 121:5287, 1999.
15. K Sutter, J Hulliger, R Schlesser, P Gunter. Opt Lett 18:778, 1993.
16. PM Lundquist, R Wortmann, C Geletneky, RJ Twieg, M Jurich, VY Lee, CR Moylan, DM Burland. Science 274:1182, 1996.
17. U Hofmann, M Grasruck, A Leopold, A Schreiber, A Schloter, C Hohle, P Strohriegl, D Haarer, SJ Zilker. J Phys Chem B 104:3887, 2000.
18. SJ Zilker, M Grasruck, J Wolff, S Schloter, A Leopold, MA Kolchenko, U Hofmann, A Schreiber, P Strohriegl, C Hohle, D Haarer. Chem Phys Lett 306:285, 1999.
19. IC Khoo, H Li, Y Liang. Opt Lett 19:1723, 1994.
20. EV Rudenko, AV Sukhov. JETP Lett 59:142, 1994.
21. GP Wiederrecht, BA Yoon, MR Wasielewski. Science 27:1794, 1995.
22. GP Wiederrecht, BA Yoon, MR Wasielewski. Adv Mat 8:535, 1996.
23. GP Wiederrecht, BA Yoon, MR Wasielewski. Synth Met 84:901, 1997.
24. GP Wiederrecht, BA Yoon, WA Svec, MR Wasielewski. J Am Chem Soc 119:3358, 1997.

25. H Ono, I Saito, N Kawatsuki. Appl Phys Lett 72:1942, 1998.
26. A Miniewicz, S Bartkiewicz. Opt Commun 149:89, 1998.
27. S Bartkiewicz, A Miniewicz, F Kajzar, M Zagorska. Appl Opt 37:6871, 1998.
28. T Sasaki, M Goto, Y Ishikawa, T Yoshimi. J Phys Chem B 103:1925, 1999.
29. NV Tabiryan, C Umeton. J Opt Soc Am B 15:1912, 1998.
30. H Ono, T Kawamura, NM Frias, K Kitamura, N Kawatsuki, H Norisada. Adv Mater 12:143, 2000.
31. D Psaltis, F Mok. Scientific American November:70, 1995.
32. L Solymar, DJ Webb, A Grunnet-Jepsen. The Physics and Applications of Photoreactive Materials. Oxford: Clarendon Press, 1996.
33. B Kippelen, SR Marder, E Hendrickx, JL Maldonado, G Guillemet, BL Volodin, DD Steele, Y Enami, Sandalphon, YJ Yao, JF Wang, H Rockel, L Erskine, N Peyghambarian. Science 269:54, 1998.
34. IC Khoo, MV Wood, MY Shih, PH Chen. Opt Exp 4:432, 1999.
35. SCW Hyde, NP Barry, R Jones, JC Dainty, PMW French, MB Klein, BA Wechsler. Opt Lett 20:1331, 1995.
36. SCW Hyde, NP Barry R Jones, JC Dainty, PMW French. Opt Commun 122:111, 1996.
37. A Ashkin, GD Boyd, JM Dziedzic, RG Smith, AA Ballman, JJ Lexinstein, K Nassau. Appl Phys Lett 9:72, 1966.
38. CR Giuliano. Physics Today April:27, 1981.
39. P Gunter, JP Huignard. Photorefractive Materials and Their Applications 1: Fundamental Phenomena. Berlin: Springer-Verlag, 1998.
40. S Ducharme, JC Scott, RJ Tweig, WE Moerner. Postdeadline Paper OSA Annual Meeting. Boston, 1990.
41. We Moerner, SM Silence. Chem Rev 94:127, 1994.
42. WE Moerner, SM Silence, F Hache, GC Bjorklund. J Opt Soc Am B 11:320, 1994.
43. IC Khoo. Liquid crystals: Physical Properties and Nonlinear Optical Phenomena. New York: Wiley, 1995.
44. IC Khoo, BD Guenther, MV Wood, P Chen, M-Y Shih. Opt Lett 22:1229, 1997.
45. IC Khoo. Opt Lett 20:2137, 1995.
46. Q Wang, RM Brubaker, DD Nolte, MR Melloch. J Opt Soc Am B 9:1626, 1992.
47. IC Khoo. IEEE J Quant Elec 32:525, 1996.
48. NV Tabiryan, AV Sukhov, BY Zeldovich. Mol Cryst Liq Cryst 135:1, 1986.
49. PJ Collings. Liquid Crystals: Nature's Delicate Phase of Matter. Princeton: Princeton University Press, 1990.
50. MR Wasielewski. Chem Rev 92:435, 1992.
51. RA Marcus. J Chem Phys 24:966, 1956.
52. IR Gould, D Ege, JE Moser, S Farid. J Am Chem Soc 112:4290, 1990.
53. EV Rudenko. AV Sukhov. JETP 78:875, 1994.
54. S Sen, P Brahma, SK Roy, DK Mukherhee, SB Roy. Mol Cryst Liq Cryst 100:327, 1983.
55. MR Wasielewski, RL Smith, AG Kostka. J Am Chem Soc 102:6923, 1980.
56. GP Wiederecht, WA Svec, MP Niemczyk, MR Wasielewski. J Phys Chem 99:8918, 1995.

57. IR Gould, D Noukakis, L Gomez-Jahn, RH Young, JL Goodman, S Farid. Chem Phys 176:439, 1993.
58. R Marcus. J Chem Phys 43:679, 1965.
59. SR Greenfield, WA Svec, D Gosztola, MR Wasielewski. J Am Chem Soc 118:6767, 1996.
60. K Hasharoni, H Levanon, SR Greenfield, DJ Gosztola, WA Svec, MR Wasielewski. J Am Chem Soc 118:10228, 1996.
61. GP Wiederrecht, S Watanabe, MR Wasielewski. Chem Phys 176:601, 1993.
62. K Hasharoni, H Levanon, SR Greenfield, DJ Gosztola, WA Svec, MR Wasielewski, J Am Chem Soc 117:8055, 195.
63. VG Levich. Adv Electrochem Eng 4:814, 1966.
64. H Dahms. J Phys. Chem 72:362, 1968.
65. I Ruff, VJ Friedrich. J Phys Chem 75:3297, 1971.
66. K Suga, S Aoyagui. Bull Chem Soc Jpn 46:755, 1973.
67. MR Wasielewski, GLG III, MP O'Neil, WA Svec, MP Niemczyk, L Prodi, D Gosztola. In: N Mataga, T Okada, H Masuhara, eds. Dynamics and Mechanisms of Photoinduced Electron Transfer and Related Phenomena. New York: Elsevier, 1992.
68. M Assel, T Hofer, A Laubereau, W Kaiser. Chem Phys Lett 234:151, 1995.
69. G Briere, R Herino, F Mondon. Mol Cryst Liquid Cryst 19:157, 1972.
70. A Lomax, R Hirasawa, AJ Bard. J Electrochem Soc 119:1679, 1972.
71. H Ono, N Kawatsuki. Opt Lett 22:1144, 1997.
72. A Golemme, BL Volodin, B Kippelen, N Peyghambarian. Opt Lett 22:1226, 1997.
73. G Cipparrone, A Mazzulla, F Simoni. Opt Lett 23:1505, 1998.
74. A Golemme, B Kippelen, N Peyghambarian. Appl Phys Lett 73:2408, 1998.
75. A Golemme, B Kippelen, N Peyghambarian. Chem Phys Lett 319:655, 2000.
76. CA Guymon, EN Hoggan, DM Walba, NA Clark, CN Bowman. Liq Cryst 19:719, 1995.
77. RAM Hikmet. J Appl Phys 68:4406, 1990.
78. CV Rajaram, SD Hudson, LC Chien. Chem Mater 7:2300, 1995.
79. GP Wiederrecht, MR Wasielewski. J Am Chem Soc 120:3231, 1998.
80. HJ Eichler, P Gunter, DW Pohl. Laser-Induced Dynamic Gratings. Berlin: Springer-Verlag, 1986.
81. YH Ja. Elect Lett 17:488, 1981.
82. P Gunter. Phys Rep 93:199, 1982.
83. B Maximus, ED Ley, AD Meyere, H Pauwels. Ferroelectrics 121:103, 1991.
84. RJOM Hoofman, MPd Haas, LDA Siebbeles, JM Warman. Nature 392:54, 1998.
85. GP Wiederrecht, MP Niemczyk, WA Svec, MR Wasielewski. Chem Mater 11:1409, 1999.
86. WB Davis, WA Svec, MA Ratner, MR Wasielewski. Nature 396:60, 1998.
87. J Feinberg, RW Hellwarth. Opt Lett 5:519, 1980.
88. J Shamir, HJ Caulfield, BM Hendrickson. Appl Opt 27:2912, 1988.
89. DZ Anderson, DM Lininger, J Feinberg. Opt Lett 12:123, 1987.
90. GJ Tearney, ME Brezinski, BE Bouma, SA Boppart, C Pitris, JF Southern, JG Fujimoto. Science 276:2037, 1997.

91. S Boppart, GJ Tearney, BE Bouma, JF Southern, ME Brezinski, JG Fujimoto. Proc Natl Acad Sci 94:4256, 1997.
92. Z Chen, TE Milner, X Wang, S Srinivas, JS Nelson. Photochem Photobiol 67:56, 1998.
93. Z Chen, Y Zhao, SM Srinivas, JS Nelson, N Prakash, RD Frostig. IEEE J Sel Top Quantum Electron 5:1134, 1999.
94. M Liphardt, S Ducharme. J Opt Soc Am B 15:2154, 1998.
95. CL Thompson, WE Moerner. Opt Commun 145:145, 1998.

8

Hologram Switching and Erasing Strategies With Liquid Crystals

Michael B. Sponsler
Syracuse University, Syracuse, New York

I. INTRODUCTION

You hand your credit card to the cashier, and the human touch on the card starts the hologram in motion: the bird takes flight or the globes spin. This is just one of many varied applications that might rely on switchable holograms. Each thin hologram in a multilayer stack, switched on in turn, can provide a frame of a three-dimensional movie, giving life to the static holograms that have become so numerous around us.

The last decade has brought numerous advances in holographic recording media, and some of the most exciting offer various capabilities in the new area of switchable holograms and holographic gratings [1]. Liquid crystals (LCs), whose controllable optical properties have been exploited for over a quarter century in applications such as liquid crystal displays (LCDs), have not yet found widespread application in holographic recording media. The pace of research in this area, however, has accelerated by approximately 10 times over the last 10 years, as judged by the published literature. The purpose of this chapter is to review the recent literature on switchable and erasable LC-containing media for holography.

A holographic grating (or hologram), once written, might be switched or modulated, meaning that the intensity of the diffracted beam (or image) can be varied—ideally from zero to a high value. Such switching phenomena can be divided into several categories relating to the means of control and repeatability (switchability versus erasability). The switching might be accomplished under electrical, optical, or mechanical control. While erasability and switchability are different properties of a hologram, erasing of the hologram in rewritable media can be considered as switching of a different sort. Erasing requires that the hologram be rewritten each time it is switched on, but it also allows a different hologram to be written each time. This chapter includes only persistent media and excludes a large body of transient holographic effects in liquid crystals [2]. Also excluded from this chapter are nonholographic means for generating switchable or erasable holograms, such as electrically addressed spatial light modulators [3], electrochemically switched gratings with patterned electrodes [4], nonholographic patterning of alignment layers [5], electric field-induced grating formation [6], and cholesteric-templated gratings [7].

Many applications of switchable holographic gratings and holograms, in addition to moving images [8], have been described and explored, including beam steering optics, refocusable lenses, reconfigurable waveguide interconnects, and fiber-optic communication switches [9]. For information on these and other applications, the reader is directed elsewhere; the focus herein is on the materials and recording strategies.

II. SWITCHABLE HOLOGRAPHIC GRATINGS

Although our everyday experience with holograms involves three-dimensional images, the great majority of the work on switchable holograms has concerned holographic gratings, the simplest of holograms. A switchable grating is an optic that can be used to steer a laser beam: when switched off, the beam is transmitted unchanged, and when switched on, the beam is diffracted in a direction determined by the grating orientation and spacing. By stacking several such gratings, the beam can be steered into a number of different directions.

A holographic grating is the hologram one obtains when no object image is recorded, that is, when two collimated laser beams are directed onto a recording material (Fig. 1). The interference pattern generated by the two intersecting beams is a grating with a sinusoidal intensity profile. The planes of the grating are parallel to the plane that bisects the two beams, and the spacing depends on the wavelength of the light and the angle between the beams, with larger angles giving smaller spacings as described by Bragg's Law. Using visible light, grating spacings of about 0.2 to 15 μm are easily attainable. The light intensity grating is written into the film through photochemical processes that occur to a greater extent in the bright regions of constructive interference. The diffraction efficiency (percentage of light that is diffracted from a single beam) of the written grating then depends on the photochemically induced differences in refractive index

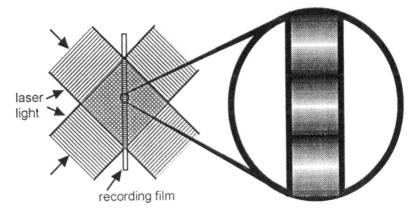

Figure 1 Formation of a holographic grating. Crossed laser beams create a sinusoidal intensity pattern recorded in the film through photochemical processes.

and absorbance between the bright and dark regions. Gratings that rely solely on refractive index differences, called *phase* gratings, are generally preferable to absorbance gratings.

LCs are anisotropic materials, typically containing rod-shaped molecules, that show very pronounced refractive index effects. They are strongly birefringent, meaning that the refractive index measured with the electric field vector of the light parallel to the LC director (extraordinary index n_e) is very different from that in any of the perpendicular orientations (ordinary index n_o). Macroscopic LC samples may be uniformly aligned with appropriate surface layers in directions either parallel to the substrates (planar orientation) or perpendicular to the substrates (homeotropic orientation). Application of electric or magnetic fields can overcome the surface effects, causing the LC molecules to be reoriented and changing the refractive index by up to 0.3 or more (ca. 20%). Since many holographic materials operate with index changes of far less than 1%, diffractive effects are potentially much greater with LC materials and comparable effects can be achieved with much thinner films. Also, since the refractive index is controllable by application and removal of fields, the gratings in LC materials are potentially switchable. The differences in the various recording strategies lie in the photochemical processes that distinguish the bright and dark regions, as well as the means of control over the LC refractive index. In addition to orientation changes, variations in degree of order also lead to large index changes. Therefore, another basic recording strategy is to photochemically reduce the LC anisotropy, for example, by introducing disorder or elevating the temperature.

III. SWITCHABLE, NONERASABLE HOLOGRAPHIC GRATINGS

The control of switching in most cases described below is electrical, and this is also the most convenient for most applications. Electrical switching requires the use of transparent electrodes and a voltage source, but ITO-plated glass is fairly inexpensive, as is electricity, as long as the voltages required are not too high. Optical (photoinduced heating) and mechanical switching have also been demonstrated, and each of these also has potential for application. Thermal switching is a potential option for several of the materials below, but it is not discussed specifically unless photoinduced heating results have been reported.

A. Strategy 1: Photopolymerization of a LC Monomer

A recording strategy that makes full use of an LC's birefringence would be one that uses the LC as the only component and affects a 90° rotation of the LC molecules in either the bright or dark regions. The trick is to enforce this rotation in one region while preventing it in the adjacent ones. One of only two reported strategies for switchable holograms that accomplishes this is that of Sponsler and coworkers (Fig. 2) [10]. A liquid crystalline monomer is used, such that photopolymerization of the LC in the bright regions prevents reorientation, allowing selective reorientation of the dark regions with the application of small electric fields. It was indeed shown that the refractive index in the two regions can be varied continuously from zero to the birefringence of the LC.

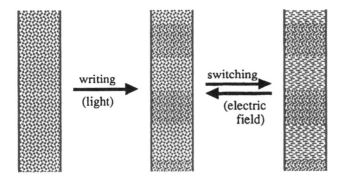

Figure 2 Strategy 1 for recording switchable holograms. An LC monomer mixture is photopolymerized in the bright regions, creating a grating of aligned polymer alternating with monomer. The refractive index modulation is very small in this grating, but can be reversibly switched to a very high value with an electric field, which reorients the monomeric regions.

The recording and switching processes are depicted in Fig. 2. A 6-μm cell is constructed from glass plates with transparent electrode (indium-tin oxide, ITO) coatings, over which are coated polyimide alignment layers. The alignment layers, rubbed on cloth, enforce a uniform planar alignment of the LC sample in the absence of applied fields. The LC sample, which contains 80 to 90% of a liquid crystalline diacrylate (Fig. 3) [11], 10 to 20% of a dielectric dopant (a component that helps the mixture respond to an applied electric field), and 0.1% of a free radical initiator, is capillary filled into the cell at 80 to 100°C (in the nematic range of the mixture). Irradiation with crossed beams creates a grating with polymer in the bright regions, into which the liquid crystalline order has been locked. The use of a diacrylate ensures that the polymer will be highly crosslinked, such that it will be unresponsive to electric fields even at low degrees of polymerization. Therefore, application of a uniform ac field across the grating results in selective reorientation of the dark regions. The grating can then be switched off and on indefinitely, or it can be made permanent, if desired, through further irradiation when it is switched on.

Using this scheme with a grating spacing of 10 μm, an optimum voltage of 3 V was found, giving the theoretical maximum diffraction efficiency (34%) for an optically thin grating (Fig. 4). At 3 V, the LC molecules in the dark regions were rotated about 58°, corresponding to a refractive index change of 46% of the birefringence [12]. At higher voltages, the refractive index continued to increase up to the birefringence of the LC (at 90° rotation), and the diffraction efficiency from the overmodulated, optically thin grating dropped off. Gratings were also made to switch in the opposite mode, that is, switched on by turning off the field. This was accomplished simply by writing the grating with the field on.

The results with this three-component LC mixture were sharply dependent upon exposure. If the exposure was too low, then too little polymer was formed and reorientation was not selective. If the exposure was too high, then the grating did not respond to the applied field, indicating that polymerization occurred in both the bright and dark regions. Polymerization in the dark regions is more problematic as the grating spacing is decreased, and weaker diffraction was observed with a spacing of 5 μm. Since the intensity profile is sinusoidal, the "dark" regions are not completely dark allowing some initiation there, and mi-

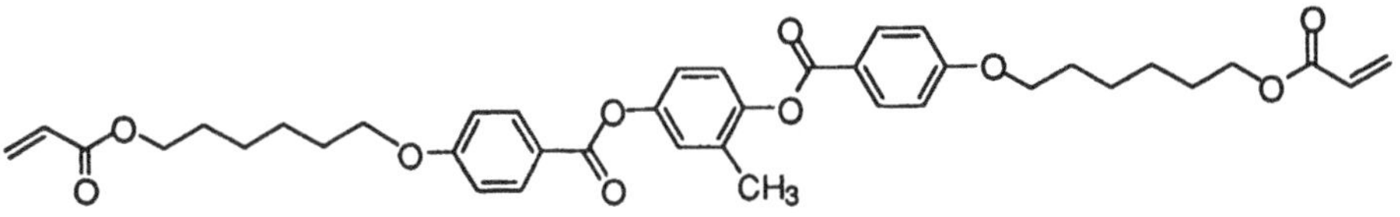

Figure 3 The LC diacrylate monomer used for Strategy 1, as well as some variations of Strategy 2.

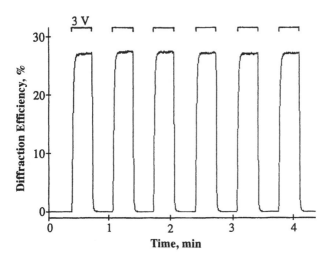

Figure 4   Switching in Strategy 1. Efficiency of the first-order diffraction is repeatedly switched from essentially zero to values near the theoretical limit for an optically thin grating (34%) by application of a 3-V electric field.

gration of free radicals from the bright regions can also occur either by diffusion or polymer chain growth. By adding small amounts of a radical inhibitor to the mixture (0.01%), the response at small grating spacings greatly improved, and strong diffraction (>25%) was observed for switchable gratings with spacings as low as 2 μm.

Good points about this strategy include high sensitivity (20 mJ/cm^2), high efficiency, clean on/off switching ($I_{on}/I_{off} > 100$), and latency of the recorded hologram. A latent hologram is one that does not diffract during the writing stage, and this is desirable since it prevents distortion in the recording. The prime drawbacks are the requirement of elevated temperature and the fact that the switchable gratings remain sensitive to the writing wavelength. In order to produce room temperature, light-stable switchable gratings, efforts were made to remove the unpolymerized LC with solvent treatments and replace it with a room temperature nematic LC. These experiments were only marginally successful, as the polymeric regions were found to be very fragile due to the low degree of polymerization.

B.   Strategy 2: Formation of Holographic Polymer-Dispersed Liquid Crystals (H-PDLCs)

Another photopolymerization strategy towards electrically switchable gratings is the crossed-beam irradiation of a mixture of a non-LC monomer (typically with two or more acrylate groups to ensure high levels of crosslinking), a nonreactive LC (10 to 50%), and a radical photoinitiator [13]. Polymerization (depletion) of monomer in the bright regions leads to diffusion of additional monomer from the dark regions. Concurrently, the LC diffuses into the dark regions, where it phase-separates from the solidifying polymer into droplets. After complete polymerization, most of the LC may be confined to the dark regions, where it can be reoriented with an electric field (Fig. 5). With appropriate index-matching between the polymer and the LC (n_o), a nondiffracting "off" state can be achieved when the field is applied.

Of all the strategies for switchable holographic gratings, this one has received the most attention. Sutherland and coworkers have done much careful work [14], and a number of other groups have also studied fundamental aspects, variations, and applications [15]. Exposures are typically 0.1 to 1 J/cm^2, and the gratings have been shown to exhibit fast (<20 to 150 μs), high contrast switching, with index modulations up to about 25% of the LC birefringence [16]. This strategy has been used successfully in the production of switchable images [17] as well as reflection-mode gratings [13]. Simultaneous recording of multiple gratings (two reflection and one transmission) has also been accomplished, allowing switchable diffraction of a beam to multiple directions, albeit with a loss of efficiency [18]. The use of near-IR photoinitiators has also allowed H-PDLC production with light from 800 to 855 nm [19].

Scanning electron microscopic (SEM) images, reported by several groups, have confirmed the PDLC structure and have helped to clarify the effects of many experimental variables, relating both to composition and recording procedure. For example, the morphology of the final structure is very sensitive to grating spacing, since diffusion processes between bright and dark regions are important. Depending on experimental variables, the bright region composition can vary

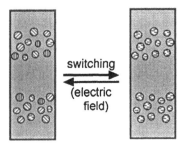

Figure 5   Switching in Strategy 2. An H-PDLC, whose droplets of LC form primarily in the dark regions during the writing step, can be reversibly switched by application of an electric field. The field reorients the LC such that the refractive index matches that of the surrounding polymer and the diffraction is therefore switched off.

from a single phase of polymer to a PDLC structure. Even if LC droplets exist in the bright regions, a switchable grating may still exist due to a nonuniform LC distribution.

A variety of studies has helped to map out the characteristics of H-PDLCs, and detailed diffusion models have been developed and correlated with experiments [20]. Various mixtures containing di- to hexafunctional monomers have been shown to give very different results, with the optimum functionality being about 4.5 [21]. The addition of surfactants has been shown to reduce the electric field threshold for switching from about 10 V/μm to about 5 V/μm [22].

1. Variations on Strategy 2

Interestingly, Kitzerow and coworkers showed that a similar H-PDLC can be produced from a dye-containing, pre-cured PDLC that is uniform [23]. With exposures that were long and intense enough (about 1 min with at least 10 W/cm^2), the resulting thermal grating led to a redistribution of LC through diffusion. The H-PDLCs obtained, which had smaller LC droplets in the bright regions (as observed by SEM), were switchable with electric fields. However, the diffraction properties reported were not as good as the H-PDLCs formed by holographic photopolymerization. Similar work to this, however, is discussed below in the section on materials that are both switchable and erasable.

Incorporation of dye into an H-PDLC can also lead to all-optical switchability, as reported by Fuh and coworkers [24]. Using a strong argon-ion laser beam (514 nm, 600 mW), the diffraction of the gratings was increased up to eight-fold in one to two seconds, returning quickly to the weakly diffracting state when the controlling beam was stopped. The authors attributed the switching effect to an alignment of the LC caused by thermal expansion.

The use of a smectic A LC instead of a nematic LC allows for "memory-type" H-PDLCs, as shown by Date et al. [25]. At an appropriate temperature for the LC used, the grating could be switched off with a 10 ms electric field pulse of 30 V/μm. In contrast, nematic H-PDLCs require a continuous application of the field to maintain the off state. The smectic H-PDLCs were turned back on by warming them above a critical temperature.

In an effort to more efficiently use the birefringence of the LC, Kato et al. demonstrated alignment-controlled, reflection-mode H-PDLCs [26]. They made use of a liquid crystalline monomer mixture similar to that used in Strategy 1 (including the diacrylate in Fig. 3), except that the nonreactive LC was the major component (82%). The LC mixture was prealigned by surface effects then completely cured, leading to an H-PDLC in the usual way, except that the LC droplets were all aligned in the same direction by the ordered polymer. The well-ordered and predictable alignment in the initially formed state (without electric field) allowed the authors to accomplish index-matching, making this the "off" state. The grating was then switched on by applying an electric field, giving a reflection-mode efficiency of about 40%. This corresponded to a Δn of about 0.02, or 20% of the birefringence, similar to the best results without alignment control. As expected, the diffraction was very sensitive to polarization, with no diffraction of light polarized perpendicular to the initial alignment direction since Δn is zero in this case.

The same authors reported on a variation of the alignment-controlled H-PDLCs that they called "in-plane" [27]. The strategy was the same, except that the electric field was applied roughly parallel to the substrate (and perpendicular to the LC surface alignment) with the use of 5-μm interdigitated electrodes on one of the glass plates. In principal, such in-plane realignment of the LC should provide a polarization-insensitive switchable grating, and indeed diffraction was observed for light polarized both parallel and perpendicular to the surface alignment direction. However, the parallel-polarized light was diffracted about twice as efficiently, since the electric field was not purely in-plane, and an out-of-plane component also contributed. A drawback to this approach is that the regions directly above the interdigitated electrodes do not experience the electric field and therefore do not diffract in either state. This reduces the overall efficiency by about 50%, significantly negating the birefringence advantage of the alignment control.

Hikmet and Poels have reduced the percentage of difunctional monomer (Fig. 3) in the same type of mixture even further to less than 1% in order to obtain anisotropic gels. Although holographic gratings were not reported, switchable gratings of 100 μm from a mask showed very promising results [28].

Switching of an H-PDLC has also been accomplished mechanically [29]. By using a relatively elastic photopolymer, De Rosa and coworkers were able to switch off the gratings by applying a shear stress to the cell. LC alignment may play a role in the results, however, the authors attributed the effect to a tilting of the grating. In other words, the direction of diffraction should be reversibly tunable by this method.

C. Strategy 3: Postirradiation LC-Filled Holograms

Polymer-based holograms that have open spaces, either on the surface or in the interior, can be filled with an LC and placed between transparent electrodes to make a switchable hologram. There is a clear disadvantage to this method in the need for postirradiation wet processing, but the procedure can be simple, can be done with commercial holographic films, and can produce strongly diffracting, optically thick holograms.

Polymerization processes are often accompanied by material shrinkage, which can lead to surface relief patterns. In fact, other strategies that depend on photopolymerization can sometimes show secondary holograms due to a surface relief structure, especially if the layer and grating spacing are relatively large. Such gratings are also sometimes observed with materials from Strategy 5, below [30], and recent studies have focused on the alignment properties of LC layers on such relief gratings [31]. Switchable, LC-filled relief gratings on dichromated gelatin were reported by Sainov et al. [32]. The diffraction efficiency was electrically modulated in these gratings by about an order of magnitude.

The commercial Polaroid holographic recording material DMP-128 becomes porous after exposure and processing. Therefore, holographic exposure provides regions that differ markedly in porosity. The pores can then be filled with an LC to produce a structure that is conceptually equivalent to an H-PDLC [33]. This strategy therefore represents an alternative to Strategy 2, and the characteristics of the switchable gratings are reasonably similar to the H-PDLCs discussed above. These composites are capable of nearly 100% diffraction with switching times of about 100 μs.

Similar to the all-optical switching mentioned for Strategy 2, incorporation of a dye with the LC led to a similar result in DMP-128. Malcuit and Stone took advantage of the large index changes that occur near the clearing point of a nematic LC to produce strong, light-controlled switching effects [34]. The gratings were thermostated slightly below the clearing point and irradiated with an 8-mW beam from an argon ion laser (514 nm). The switching was much faster than in the example above, occurring in about 1 ms in each direction. Butler and Malcuit investigated these materials further [35].

Another approach to prepare an empty holographic cell that can be filled with an LC is to use a photoresist. In fact, a conceptually simple strategy for making a patterned LC cell in general is to etch the pattern into one of the two electrodes in a sandwich cell. The requirement that makes this impractical for a holographic pattern is that each (micron-sized) region of the patterned electrode must be electrically connected if a field is to be applied in that region. Parker devised a method that circumvents this problem [36]. By placing the patterned electrode in the middle of a three-electrode sandwich, an electric field can then be applied between the outer two (unpatterned) electrodes (Fig. 6). The patterned electrode then serves to modulate the capacitance, and an LC layer between the patterned and one outer electrode will respond differently in the different regions. The holographic patterning of the electrode is done with standard photoresist methods, which allow also for variations in the thickness of the LC layer in the bright and dark regions as well as variations in dielectric materials used and other details. Therefore, different designs have been advanced for optimization of different parameters, such as switching speed and resolution.

D. Strategy 4: Photorefractive LCs

Photorefractive inorganic materials have been studied for some time as holographic media, but advances in photorefractive organic materials are relatively

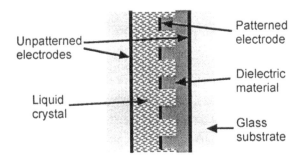

Unpatterned electrodes

Patterned electrode

Dielectric material

Liquid crystal

Glass substrate

Figure 6 Cross-section of a switchable grating from photoresist (Strategy 3). By using standard semiconductor fabrication techniques, a cell can be constructed with a holographically patterned central electrode. After filling with LC, the cell can be switched on and off with an electric field applied between the outer unpatterned electrodes.

recent [37]. High birefringence and susceptibility to electric fields make LCs very attractive as components of photorefractive media. Typical LCs, however, have little capacity for photoinduced charge generation or trapping, so additional components are needed for these roles. DC electric fields are generally applied to induce migration of the charges once they are formed, and the resulting space-charge field causes local changes in the alignment (and therefore, refractive index) of the LC. When the light is shut off, the charges can quickly recombine unless they become trapped. Since this chapter deals with persistent holographic effects, only those media that exhibit charge trapping are mentioned further. While photorefractive effects are inherently erasable or even transient, this strategy is described in this section because it has led in some cases to the production of switchable gratings. (See Chap. 7 for a discussion of transient photorefractive LC materials.)

In LC-containing materials, several photorefractive effects have been identified, making the investigation and understanding of these effects complicated [38]. Refractive index modulations may be induced by the space-charge field, the flow of ions through a region, the flow of LC molecules, or by shear stresses that arise from material flows. Furthermore, space-charge fields are determined not only by charge migration, but also by the LC's dielectric anisotropy, and the charge migration is also affected by anisotropy in the conductivity. Nonetheless, photorefractivity has been observed in a variety of materials, and much progress has been made toward the understanding and rational improvement of the materials.

The photorefractive effect is usually probed by two beam-coupling experiments, in which one beam gains intensity at the expense of the other. This coupling is a characteristic property of the photorefractive effect. Such an asymmetric coupling requires an asymmetric shifting of the refractive index grating

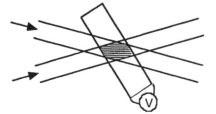

Figure 7 Recording geometry for beam coupling through the photorefractive effect. The LC alignment, enforced by either surface or field effects, is usually homeotropic.

(optimally by π/2, or 1/4 of a grating spacing) with respect to the intensity pattern. In order to allow for an asymmetric shift, the symmetry of the experiment is usually broken by rotating the sample 30° with respect to the light beams (Fig. 7). With this rotation, the grating is tilted 30° with respect to the applied dc field, such that migration of generated charges (either positive or negative) is asymmetric and may occur preferentially from a bright region to one of the two adjacent dark regions (Fig. 8). This results in a space-charge field along the grating axis with the less mobile charges in the bright regions and the more mobile charges in the dark regions. Even if the ultimate space-charge field is symmetric with respect to the grating axis, the effective field, obtained by summing the space-charge and applied fields, is asymmetric. The largest field is then found at one bright/dark boundary (say π/2 from the bright region) and the smallest is at the other boundary (−π/2). These points thus become the peaks and valleys of the index modulation.

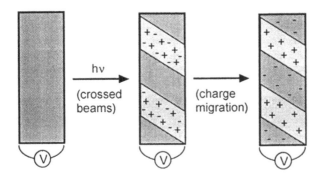

Figure 8 Generation of a space-charge field in a photorefractive material. Photoinduced charges are generated in the bright regions of the interference pattern, and one type of charge (the negative charges, as shown here) preferentially migrate to the dark regions, aided by an applied dc field. The space-charge field causes variations in the LC alignment (not shown) in different regions of the grating.

Comparing photorefractivity results from the literature is also complicated, as pointed out by Wasielewski and Wiederrecht [39], because the beam-coupling parameters depend strongly on the optical thickness of the gratings. However, the results are presented here as given in the original reports.

Khoo and coworkers [40] found that nematic LC films doped with fullerene (C_{60}) for charge generation exhibit multiple photorefractive effects. With exposure to 514 nm light and a dc field of 1 to 2 V, a 25-μm film showed transient grating formation with low exposure (0.1 J/cm^2) and permanent gratings with high exposure (60 J/cm^2). The permanent gratings had efficiencies up to 10% and index modulations up to 0.002 and were found to be electrically switchable, showing no diffraction under a 94-V ac field. The explanation offered for the permanence of the grating was an alteration of the surface alignment layers under the prolonged recording. Gratings were optimized with a spacing of 40 μm, although diffraction was observed as low as 10 μm. Materials of the same type were studied more recently by Zhang et al. [41], who found that for the photorefractive effect, the light exposure and dc field application need not occur simultaneously but can occur sequentially in either order. These authors attributed the photorefractivity to the trapping of charges at the surface of the aligning layer.

Photorefractive experiments have also been done with dye-doped PDLCs. The flow properties of LCs are detrimental to sustained photorefractive effects, particularly at small grating spacings, so immobilizing the phase in a PDLC can increase the attainable resolution. Fullerene was also used as the dopant in PDLCs studied by Ono and Kawatsuki [42]. These gratings were also permanent under some conditions, although no switching experiments were reported. The resolution obtained was similar to that of the bulk LC gratings with index modulations up to 0.008, but the electric fields required are much higher for the PDLC (10 to 40 V/μm). Kippelen et al. [43] studied PDLC materials with different organic dyes and obtained better resolution. A 53-μm film gave diffraction of 56% from a 3-μm grating, corresponding to an index modulation of 0.003. The field was still high, however, at 22 V/μm. No switching experiments were reported with these materials either.

IV. ERASABLE, NONSWITCHABLE HOLOGRAPHIC GRATINGS

An erasable grating is by definition rewritable, so this capability is related to switchability although clearly different. The means of erasing in most cases below is either optical or thermal. Optical erasing, generally preferable due to speed, can be effected through wavelength, intensity, polarization, or a combination of these. Thermal erasing can be done by laser light in some cases.

A. Strategy 4: Photorefractive LCs

This strategy appears in the switchable category and the erasable category, as well as the category for materials that have both capabilities (below). This reflects the fact that photorefractivity is a multifaceted phenomenon with respect to LCs and that different approaches are available that take advantage of different aspects.

Wasielewski and coworkers have made efforts to improve both the resolution and the persistence of photorefractive effects in LC materials. In their use of polymer-stabilized nematic LCs, these workers have explored the incorporation of functionality into the polymer networks. Immobilization of an electron-trapping group on the polymer was shown to greatly improve the attainable resolution, with beam coupling observed down to 2.5 μm, compared to 8 μm for an analogous LC mixture without polymer [39]. With a film thickness of 26 μm, the finest of these gratings are clearly in the Bragg (optically thick) diffraction regime, a rare feat for photorefractive LC materials. Exposures of up to 2 J/cm^2 (for a spacing of 13 μm) were required, with an applied dc electric field of 2 V. For optimum beam coupling, both mobile and immobilized trapping sites were found to be necessary, illustrating the balance of properties that is required. Mobile traps are necessary to provide sufficiently fast charge migration for formation of the space-charge field, but the presence of immobilized traps slows down the decay of the space-charge field. The gratings so produced are not indefinitely persistent but have decay lifetimes up to 80 sec for polymer-stabilized media with grating spacings of about 13 μm, when one of the beams is left on. This work is presented as erasable not because of the inevitable decay, but because the gratings may be erased more than 10 times faster by turning off both beams.

Other accomplishments by Wasielewski et al. include demonstration of Bragg regime gratings for unpolymerized LCs by using very thick samples and magnetic field alignment [44], and the use of a conducting polymer to allow hole migration over long distances [45]. In the latter case, an inversion of the beam-coupling direction with respect to grating spacing was observed due to a change in the mechanism of charge migration. Anion diffusion was the dominant mechanism in fine gratings (less than 5.5 μm), and hole migration on the conducting polymer was dominant with larger spacings.

B. Strategy 5: Thermorecording and Photoinduced Reorientation of LCs

The most studied erasable LC materials for holography are LC phases that contain a dye, most often an azobenzene, either dissolved in the LC or as part of the LC molecule [46]. In these phases, light can cause disordering or reorientation of the LC, either of which is associated with large index changes. In a holographic grating, modulation of either order or orientation is useful, and different mechanisms have been demonstrated for each. Thermorecording and photoinduced reorientation could be handled as separate strategies, but they each require the same type of material: a dye-containing LC. Furthermore, the two processes can occur together to some extent, and the relative importance of each mechanism is not always clear. The general strategy has been demonstrated also in discotic LCs [47] and metallomesogens [48] and with photochromic spyropyran dyes [49].

Light can introduce disorder into a dye-containing LC in a couple of ways. When the dye absorbs light, the local heating can convert the LC from the nematic to the isotropic phase (Fig. 9). If an LC polymer or oligomer is used below (or even somewhat above) its T_g, the isotropic phase remains stable when it cools back down. However, by locally warming the sample and allowing it to anneal, the initial order can be restored. Another mechanism for disordering is the introduction of stresses in the material due to photochemical conformation changes of the dye. For example, a *trans* to *cis* isomerization of an azobenzene results in a change of shape, disturbing the surrounding molecules. In a viscous material, the writing effect, as well as the erasing procedure, is the same.

Light-induced reorientation of a dye-containing LC is also possible, through photoselective rotational diffusion (Fig. 10). If an azobenzene dye molecule absorbs light and undergoes a *trans* to *cis* to *trans* conversion (the second step of which may be either thermal or photochemical), the molecule will likely end up in a different orientation than the one in which it started. If the resulting orientation is one that has the transition moment (roughly the long axis of the molecule) perpendicular to the electric field vector of the light, then the dye molecule is no longer able to absorb the light and the orientation is photostable. Therefore, after sufficient exposure, all of the dye molecules will find themselves

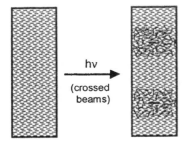

hv

(crossed beams)

Figure 9 Thermorecording. An aligned LC is disordered in the bright regions of the interference pattern, due either to a local increase in temperature or to stresses caused by conformational changes of the dye.

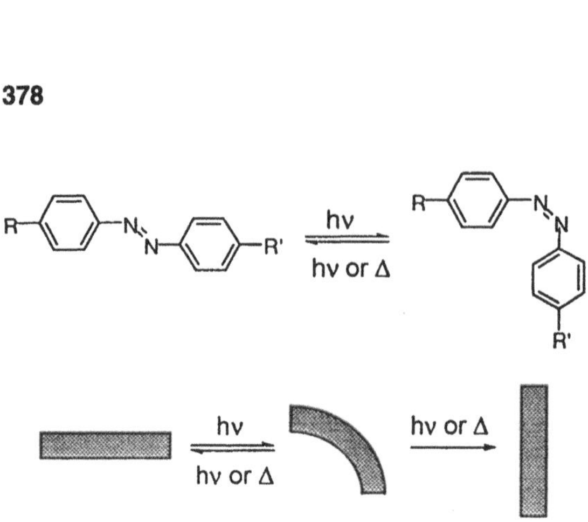

Figure 10 Reorientation of an azobenzene dye by photoisomerization with polarized light. Molecular representations of the *trans* and *cis* forms of an azobenzene are represented by straight and bent shapes to illustrate how a *trans-cis-trans* cycle can effect a reorientation of the *trans*-azobenzene. If the inducing light is polarized in the horizontal direction, the vertically oriented molecule will be unable to absorb the light, and this orientation will therefore be photostable.

rotated in such a way, and the LC director will follow. Again, if the medium is sufficiently viscous, the new orientation may be stable even when the control beams are shut off. Erasure can then be done by heating, as above, or in some cases by changing the polarization of the control beams.

The literature has many examples of azobenzene-containing LC media, some of which operate by introduction of disorder [50], others that make use of rotational diffusion [51], and others whose precise mechanism is unclear from the data presented [52]. In general, the media have the advantages of clean erasability and rewritability, long-term stability of a written hologram, high efficiency (>80%), and high resolution (grating spacings from >20 μm to <0.5 μm). One of the main disadvantages is that the writing and erasing processes are slow (tens of seconds to minutes).

Interestingly, the rotational diffusion strategy can work with either polarized or unpolarized light. If the controlling light is linearly polarized, the dye molecules can rotate such that their transition moment is anywhere in the plane perpendicular to the electric field vector. In practice, rotation in the plane parallel to the substrates is generally observed. In contrast, if the light is unpolarized, the ultimate orientation will be that with the transition moment parallel to the light beam since this is the only orientation that is perpendicular to all of the electric field vectors. The same result is obtained if the control beam is circularly polarized.

As mentioned in Sec. III.C, another mechanism of grating formation has been found to operate in these media: surface relief patterns. Relief gratings are even observed in polarization holograms, in which the intensity of light is uniform during the writing stage, with only the polarization being different in the light and dark regions. No satisfactory mechanism has been brought forward to explain the formation of relief gratings in these media [53].

1. Variations on Strategy 5

A number of reports depart in different ways from the general strategy. Neither thermorecording nor photoinduced reorientation require that the material be liquid crystalline if the material is viscous enough to maintain an ordered or disordered structure. Thus, dye-containing amorphous polymers or oligomers are suitable and even have some advantages relating to optical quality. With LC polymers, care must be taken to achieve monodomain films. The advantage of LC materials is a cooperativity in alignment that leads to amplification. Some workers, however, have shown that LC-like cooperativity can be achieved with non-LC materials.

Zilker and coworkers [54] and Wendorff and coworkers [55] showed that copolymers with both mesogenic and nonmesogenic side chains can give amorphous materials that still benefit from cooperative interactions. In these materials, orientational order can be induced by polarized irradiation. Both groups reported improved properties with respect to other amorphous polymers, and Zilker et al. reported refractive index modulations as high as 0.2, indicating a high degree of light-induced ordering. The exposure required was about 20 J/cm^2.

Recent work by Berg and coworkers represents a thoughtful refinement with the use of azobenzene-containing peptide oligomers, which are not liquid crystalline [56]. When several azo chromophores are linked with a helical peptide backbone and irradiated with polarized light, the helices rotate such that the helix axes lie parallel to the electric field vector in order to position all of the chromophores in nonabsorbing orientations. This provides an ordered arrangement, even though the materials are not liquid crystalline. Best results are obtained with oligoproline backbones, since the ring-constrained amino acid produces more rigid helices. Berg et al. also use circularly polarized beams for both writing and reading, a polarization holography technique that can significantly boost diffraction efficiency when the recording material is sensitive to polarization [57]. With this technique, the interference of the beams produces a modulation of polarization instead of intensity. The results are considerably better than most azobenzene-containing materials, with 80%-efficient gratings written and erased in about one second with a 488-nm exposure of 1 to 3 J/cm^2.

Other variations of Strategy 5 include using an azobenzene-doped LC in porous glass media [58], a spiropyran-doped LC in hydrogel media [59], and small inorganic particles dispersed in a nematic LC [60].

V. SWITCHABLE AND ERASABLE HOLOGRAPHIC GRATINGS

Most applications, it seems, might require only that a recording material be either switchable or erasable. However, media that have both capabilities offer the greatest flexibility for the design of reconfigurable optics. With such a material, one can write a hologram, switch it on and off as desired, then erase the hologram and write a different switchable hologram. For a device that needs to be adapted to different conditions, this capability may be necessary.

Relatively few reports exist at this time describing media that are both switchable and erasable. The general strategy has been to record the information in the form of a reversible surface effect, either on an alignment layer or on the inner surfaces of a PDLC. In this situation, the control of alignment of an LC can be switched between surface and electric field effects, such that the "on" state alignment is provided by the surface control and the "off" state alignment is provided by the electric field.

A. Strategy 4: Photorefractive LCs

In the photorefractive materials described above, the gratings formed are either permanent (though switchable) due to alterations in surface structures that persist in the absence of the photorefractive effect, or subject to decay as the photo-generated charges find and annihilate each other if photorefractivity is solely responsible for the grating. In order to make a material that is both switchable and erasable, reversible surface alterations would be sufficient. Cipparrone and coworkers have found this to be possible in dye-doped PDLCs when sufficient light intensity (≥ 100 W/cm^2 for 2 sec, for an exposure of ≥ 200 J/cm^2) is used to heat the polymer network to its softening point [61]. Under these conditions, the photorefractive realignment and flows noted above are apparently sufficient to alter the PDLC surfaces, and these alterations are frozen in when the irradiation is stopped. If the material is irradiated again, the polymer again softens and a new hologram can be recorded, with efficiencies as high as 30% with a grating spacing of 10 μm. Once stored, the holograms can be switched off by applying an ac electric field (≥ 2 V/μm).

A truly remarkable feature of this work is that the photorefractive recording was done in the absence of any applied electric fields and with the writing beams bisected by the normal to the cell. Nonetheless, the authors claim to have measured a phase-shift of $\pi/2$. They propose that the symmetry is broken by the internal structure of the PDLC and perhaps also by local poling of the polymer by the light gradients during exposure.

B. Strategy 5: Photoinduced Reorientation of LCs

In the late 1980s, Ichimura and coworkers attached azobenzene chromophores onto a quartz substrate and demonstrated that this "command surface" can cause a change in bulk LC alignment under appropriate irradiation [62]. With UV light, the homeotropic LC alignment can be changed to planar as the azobenzene is converted from *trans* to *cis*. Switching to visible light causes the opposite isomerization, and the LC layer returns to homeotropic. In the early 1990s, Gibbons and coworkers showed that azobenzene-containing alignment layers also respond to the polarization of light, producing planar alignments that are stable in the dark, since the *trans*-azobenzenes are simply reoriented by the rotational diffusion mechanism [63]. This mechanism still works the same way as described for LC polymers, except that only the alignment layer needs to be viscous. Thus, a conventional LC can be used in the cell, which may be identical to a common LCD except for the alignment layer. The Gibbons team, at Elsicon, Inc., have put much effort into developing these alignment layers (OptoAlign^{TM}) and have shown that they are suitable for switchable and erasable holographic gratings [64]. The layers also provide for a continuous range of planar alignments from 0 to 90° [65]. Other groups have reported on variations of this strategy [66].

Chen and Brady produced erasable holograms with alignment control by surface-bound azobenzene units, even though the azobenzene in their media was introduced only as a solute in the LC phase [67]. Apparently, in the writing step with 514-nm light, some of the azo dye (methyl red) became anchored to the surface, where it then controlled the alignment of the bulk. Erasure was accomplished with 337-nm light, and 10,000 cycles were demonstrated, although with some degradation. Electrical switching of the gratings was not demonstrated, but could be possible. The authors did show that heating to the clearing point of the LC resulted in a reversible switching off of the diffraction.

A more recent report by Francescangeli and coworkers uses a very similar strategy to that of Chen and Brady, with the main difference being the use of polarization holography [68]. With exposures of about 0.3 J/cm^2, diffraction efficiencies of about 10% were obtained with a broad range of grating spacings, from 50 μm to less than 1 μm. Electrical switching was not demonstrated by these authors either, although it appears that it should work.

VI. COMPARISON OF STRATEGIES AND CONCLUSIONS

The different strategies presented clearly have a variety of advantages and dis-advantages, and some of the important characteristics are compiled in Table 1. For a number of reasons, direct comparisons are difficult to make, so this table

Table 1 Comparison of Properties for Different Strategies[a]

	Strategy 1 photopol. of LC mon.	Strategy 2 formation of H-PDLC	Strategy 4 photoref. LCs	Strategy 5 thermorec., reorient.	Strategy 5 reorient. of surface dyes
Switchable	Y	Y	Y	N	Y
Erasable	N	N	Y	Y	Y
Latent recording	Y	N	N	N	N
Exposure required, J/cm^2	0.02	0.1–1	2–60[b]	0.2–100	0.3
Grating spacing limit, μm	2	0.5	3	0.2	1
(Δn/birefringence) × 100%	100	25	10	100	100[c]
Switching field, V/μm	0.5	3–10	5	—	0.5[d]
Switching time, ms[e]	<33/500	<0.02/0.2	0.2/20	—	20[d]
Temperature, °C	80–100	25	25	25-100	25

[a]The values are given for purposes of rough comparison and should be used with caution (see text).
[b]The exposure required for the switchable and erasable variation is 200 J/cm^2.
[c]Value implied by nonholographic data.
[d]Depends on choice of LC.
[e]The first value corresponds to application of the field and the second is the relaxation time upon removal of the field.

is offered with several caveats. Complete data are not available for all of the strategies, and some of the data were taken from different variations of a given strategy. The optimum value obtainable for one parameter often comes at the expense of a different parameter, so the values shown in the table might not all be achievable simultaneously. The purpose of the table is to present values that are representative of the best that has been reported so that very rough comparisons may be made. In some cases, ranges are given if the optimum value is far from typical. The table includes those methods that do not require any postexposure processing.

The data in Table 1 show large differences in sensitivity, refractive index modulation, and other properties, and if all the variations were included the differences would be much larger. However, each application has its own requirements, so the strategy that is most appropriate depends on all the properties in the table as well as others not listed. Work on all of these strategies will undoubtedly go forward, and continuing improvements in properties can be expected.

Because LCs are strongly birefringent, and because they respond to many aspects of their environment (surfaces, electric and magnetic fields, temperature, flow), there are many possible strategies for using them in switchable and/or erasable holographic media. The five strategies and numerous variations presented in this chapter certainly do not exhaust the possibilities, and additional strategies are likely to emerge. Large-scale commercial application of some of the strategies also seems likely for the near future, perhaps both in the information processing and display markets.

REFERENCES

1. J Zhang, MB Sponsler. In: Molecular and Biomolecular Electronics, RR Birge, ed., ACS Advances in Chemistry Series No. 240; American Chemical Society: Washington, DC, 1994, p 321.
2. (a) B Saad, TV Galstyan, L Dinescu, RP Lemieux, Chem Phys 245:395, 1999; (b) S Bartkiewcz, A Miniewicz, F Kajzar, M Zagorska. Appl Opt 37:6871, 1998; (c) P Rudquist, L Komitov, ST Lagerwall. Liq Cryst 24:329, 1998.
3. S Fukushima. Proc SPIE-Int Soc Opt Eng 2777 (Advanced Materials for Optics and Optoelectronics):85, 1996.
4. K Schanze, TS Bergstedt, BT Hauser, CSP Cavalaheiro. Langmuir 16:795, 2000.
5. PJ Bos, J Chen, JW Doane, B Smith, C Holton, W Glenn. SID 95 Digest 601, 1995.
6. (a) DC Fair, M Tilton, C Hoke. Proc SPIE-Int Soc Opt Eng 1911 (Liquid Crystal Materials, Devices and Applications II):188, 1993; (b) R Williams. J Chem Phys 39:384, 1963.
7. SN Lee, LC Chien, S Sprunt. Appl Phys Lett 72:885, 1998.

8. W Hentschel, W Lauterborn. Opt Eng 24:687, 1985.
9. (a) LH Domash, Y-M Chen, B Gomatam, C Gozewski, RL Sutherland, LV Natarajan, VP Tondiglia, TJ Bunning, WW Adams. Proc SPIE-Int Soc Opt Eng 2689:188, 1996; (b) L Domash, Y-M Chen, C Gozewski, P Haugsjaa, M Oren. Proc SPIE-Int Soc Opt Eng 3010 (Diffractive and Holographic Device Technologies and Applications IV):214, 1997; (c) LH Domash. Proc SPIE-Int Soc Opt Eng 3143 (Liquid Crystals):214, 1997; (d) T Takahashi, H Furue, T Miyama, M Shikada, R Kurihara, S Kobayashi. Mater Res Soc Symp Proc 559 (Liquid Crystal Materials and Devices):57, 1999.
10. (a) J Zhang, CR Carlen, S Palmer, MB Sponsler. J Am Chem Soc 116:7055, 1994; (b) CR Carlen, S Palmer, J Zhang, MB Sponsler. Opt Commun 111:433, 1994.
11. DJ Broer, J Boven, GN Mol, G Challa. Makromol Chem 190:2255, 1989.
12. MB Sponsler. J Phys Chem 99:9430, 1995.
13. RT Pogue, RL Sutherland, MG Schmitt, LV Natarajan, SA Siwecki, VP Tondiglia, TJ Bunning. Appl Spectrosc 54:12A, 2000.
14. (a) RT Pogue, LV Natarajan, SA Siwecki, VP Tondiglia, RL Sutherland, TJ Bunning. Polymer 41:733, 2000; (b) RL Sutherland, LV Natarajan, VP Tondiglia, RT Pogue, SA Siwecki, DM Brandelik, BL Epling, E Berman, C Wendel, MG Schmitt. Proc SPIE-Int Soc Opt Eng 3633 (Diffractive and Holographic Technologies, Systems, and Spatial Light Modulators VI):226, 1999; (c) RL Sutherland, LV Natarajan, VP Tondiglia, TJ Bunning. Chem Mater 5:1533, 1993.
15. (a) D Duca, AV Sukhov, C Umeton. Liq Cryst 26:931, 1999; (b) L Domash, P Haugsjaa, BE Little. Proc SPIE-Int Soc Opt Eng, 3234 (Design and Manufacturing of WDM Devices):146, 1998.
16. RL Sutherland, LV Natarajan, VP Tondiglia, RT Pogue, SA Siwecki, DM Brandelik, BL Epling, E Berman, C Wendel, MG Schmitt. Proc SPIE-Int Soc Opt Eng 3633 (Diffractive and Holographic Technologies, Systems, and Spatial Light Modulators VI):226, 1999.
17. VP Tondiglia, LV Natarajan, RL Sutherland, TJ Bunning, WW Adams. Opt Lett 20:1325, 1995.
18. CC Bowley, AK Fontecchio, GP Crawford, J-J Lin, L Li, S Faris. Appl Phys Lett 76:523, 2000.
19. (a) P Pilot, YB Boiko, TV Galstyan. Proc SPIE-Int Soc Opt Eng 3635 (Liquid Crystal Materials, Devices, and Applications VII):143, 1999; (b) P Pilot, Y Boiko, TV Galstyan. Proc SPIE-Int Soc Opt Eng 3638 (Holographic Materials V):26, 1999.
20. CC Bowley, GP Crawford. Appl Phys Lett 76:2235, 2000.
21. CC Bowley, AK Fontecchio, J-J Lin, H Yuan, GP Crawford. Mater Res Soc Symp Proc 559 (Liquid Crystal Materials and Devices):97, 1999.
22. LV Natarajan, RL Sutherland, VP Tondiglia, TJ Bunning, RM Neal. Proc SPIE-Int Soc Opt Eng 3143:182, 1997.
23. H-S Kitzerow, J Strauss, SC Jain. Proc SPIE-Int Soc Opt Eng 2651:80, 1996.
24. AY-G Fuh, M-S Tsai, L-J Huang, T-C Liu. Appl Phys Lett 74:2572, 1999.
25. M Date, Y Takeuchi, K Kato. J Phys D: Appl Phys 31:2225, 1998.
26. K Kato, T Hisaki, M Date. Jpn J Appl Phys Part 1 38:805, 1999.
27. K Kato, T Hisaki, M Date. Jpn J Appl Phys Part 1 38:1466, 1999.

28. RAM Hikmet, HLP Poels. Liq Cryst 27:17, 2000.
29. ME De Rosa, VP Tondiglia, LV Natarajan. J Appl Polym Sci 68:523, 1998.
30. S Yamaki, M Nakagawa, S Morino, K Ichimura. Appl Phys Lett 76:2520, 2000.
31. (a) A Parfenov, N Tamaoki, S Ohnishi. J Appl Phys 87:2043, 2000; (b) XT Li, A Natansohn, P Rochon. Appl Phys Lett 74:3791, 1999.
32. S Sainov, M Mazakova, M Pantcheva, D Tontchev. Mol Cryst Liq Cryst 152:609, 1987.
33. (a) RT Ingwall, T Adams. Proc SPIE-Int Soc Opt Eng 1555 (Computer and Optically Generated Holographic Optics IV):279, 1991; (b) DH Whitney, RT Ingwall. Proc SPIE-Int Soc Opt Eng 1213 (Photopolymer Device Physics, Chemistry, and Applications):18, 1990.
34. MS Malcuit, TW Stone. Opt Lett 20:1328, 1995.
35. JJ Butler, MS Malcuit. Opt Lett 25:420, 2000.
36. (a) WP Parker. Proc SPIE-Int Soc Opt Eng 2689 (Diffractive and Holographic Optics Technology III):195, 1996; (b) B Parker. Proc SPIE-Int Soc Opt Eng 1914 (Practical Holography VII: Imaging and Materials):176, 1993.
37. G Montemezzani, C Medrano, M Zgonik, P Gunter. Springer Ser Opt Sci 72 (Nonlinear Optical Effects and Materials):301, 2000.
38. IC Khoo, MV Wood, BD Guenther. Mat Res Soc Symp Proc 425:203, 1996.
39. GP Wiederrecht, MR Wasielewski. J Am Chem Soc 120:3231, 1998.
40. (a) IC Khoo. Opt Lett 20:2137, 1995; (b) IC Khoo, H Li, Y Liang, M Lee, B Yarnell, K Wang, M Wood. Proc SPIE-Int Soc Opt Eng 2530:134, 1995.
41. J Zhang, V Ostroverkhov, KD Singer, V Reshetnyak, Y Reznikov. Opt Lett 25:414, 2000.
42. H Ono, N Kawatsuki. Jpn J Appl Phys Part 1 36:6444, 1997.
43. B Kippelen, J Herlocker, JL Maldonado, K Ferrio, E Hendrickx, S Mery, A Golemme, SR Marder, N Peyghambarian. Proc SPIE-Int Soc Opt Eng 3471 (Xerographic Photoreceptors and Organic Photorefractive Materials IV):22, 1998.
44. GP Wiederrecht, MR Wasielewski. Appl Phys Lett 74:3459, 1999.
45. GP Wiederrecht, MP Niemczyk, WA Svec, MR Wasielewski. Chem Mater 11:1409, 1999.
46. (a) FH Kreuzer, C Braeuchle, A Miller, A Petri. In: Polym Electroopt Photoopt Act Media, VP Shibaev, ed; Springer: Berlin, 1996, p 111; (b) K Anderle, JH Wendorff. Mol Cryst Liq Cryst 243:51, 1994; (c) CB McArdle. Pure Appl Chem 68:1389, 1996; (d) S Hvilsted, PS Ramanujam. Curr Trends Polym Sci 1:53, 1996.
47. J Contzen, G Heppke, H-S Kitzerow, D Krueerke, H Schmid. Appl Phys B: Lasers Opt 63:605, 1996.
48. (a) A Stracke, JH Wendroff, D Janietz, S Mahlstedt. Adv Mater 11:667, 1999; (b) J Buey, L Diez, P Espinet, H-S Kitzerow, JA Miguel. Appl Phys B: Lasers Opt 66:355, 1998.
49. LV Natarajan, V Tondiglia, TJ Bunning, RL Crane, WW Adams. Adv Mater Opt Electron 1:293, 1992.
50. T Yamamoto, M Hasegawa, A Kanazawa, T Shiono, T Ikeda. J Mater Chem 10:337, 2000.
51. L Nikolova, T Todorov, M Ivanov, F Andruzzi, S Hvilsted, PS Ramanujam. Appl Opt 35:3835, 1996.

52. N Holme, L Nikolova, T Norris, S Hvilsted, M Pedersen, R Berg, P Rasmussen, P Ramanujam. Macromol Symp 137(Azobenzene-Containing Materials):83, 1999.
53. (a) L Andruzzi, A Altomare, F Ciardelli, R Solaro, S Hvilsted, PS Ramanujam. Macromolecules 32:448, 1999; (b) T Yamamoto, M Hasegawa, A Kanazawa, T Shiono, T Ikeda. J Phys Chem B 103:9873, 1999.
54. SJ Zilker, MR Huber, T Bieringer, D Haarer. Appl Phys B: Lasers Opt 68:893, 1999.
55. T Fuhrmann, M Hosse, I Lieker, J Rubner, A Stracke, JH Wendorff. Liq Cryst 26:779, 1999.
56. (a) PH Rasmussen, PS Ramanujam, S Hvilsted, RH Berg. J Am Chem Soc 121:4738, 1999; (b) RH Berg, S Hvilsted, PS Ramanujam. Nature 383:505, 1996.
57. L Nikolova, T Todorov. Opt Acta 31:579, 1984.
58. O Yaroshchuk, Y Reznikov, G Pelzl. Proc SPIE-Int Soc Opt Eng 2647 (International Conference on Holography and Correlation Optics):243, 1995.
59. RL White, YY Hsu, TM Cooper, JD Greeser, DL Wise, DJ Trantolo. Mater Res Soc Symp Proc 488 (Electrical, Optical, and Magnetic Properties of Organic Solid-State Materials IV):453, 1998.
60. M Kreuzer, T Tschudi, R Eidenschink. Mol Cryst Liq Cryst Sci Technol Sect. A 223:219, 1992.
61. (a) F Simoni, G Cipparrone, A Mazzulla, P Pagliusi. Chem Phys 245:429, 1999; (b) G Cipparrone, A Mazzulla, FP Nicoletta, L Lucchetti, F Simoni. Opt Commun 150:297, 1998; (c) G Cipparrone, A Mazzulla, F Simoni. Mol Cryst Liq Cryst Sci Technol Sect A 299:329, 1997.
62. K Ichimura, Y Suzuki, T Seki, A Hosoki, K Aoki. Langmuir 4:1214, 1988.
63. (a) PJ Shannon, WM Gibbons, ST Sun. Nature 368:532, 1994; (b) WM Gibbons, PJ Shannon, S-T Sun, BJ Swetlin. Nature 351:49, 1991.
64. (a) WM Gibbons, BP McGinnis, PJ Shannon, ST Sun. Proc SPIE-Int Soc Opt Eng 3635 (Liquid Crystal Materials, Devices, and Applications VII):32, 1999; (b) WM Gibbons, PJ Shannon, S-T Sun. Proc SPIE-Int Soc Opt Eng 3143 (Liquid Crystals):102, 1997; (c) WM Gibbons, PJ Shannon, ST Sun. Mol Cryst Liq Cryst 251:191, 1994.
65. WM Gibbons, T Kosa, P Palffy-Muhoray, PJ Shannon, ST Sun. Nature 377:43, 1995.
66. (a) LM Blinov, R Barberi, G Cipparrone, M Iovane, A Checco, VV Lazarev, SP Palto. Liq Cryst 26:427, 1999; (b) H Akiyama, M Momose, K Ichimura, S Yamamura. Macromolecules 28:288, 1995.
67. AG Chen, DJ Brady. Appl Phys Lett 62:2920, 1993.
68. S Slussarenko, O Francescangeli, F Simoni, Y Reznikov. Appl Phys Lett 71:3613, 1997.

9

Novel Molecular Photonics Materials and Devices

Toshihiko Nagamura
Shizuoka University, Hamamatsu, Japan

I. INTRODUCTION

Novel materials, devices, and systems are required for much faster data processing, much higher recording density, and more specific and efficient sensing. Ultrafast switching materials that work in less than one picosecond (ps) are essential for THz communication. Several attempts have been reported for this purpose, which include optical switching by tunneling biquantum well semiconductors or organic nonlinear optical materials [1]. Magnetic "hard" disks and heat-mode optical disks such as phase change or magneto-optical memories have rapidly been increasing their recording density owing to the development of new pickup heads or blue semiconductor lasers. But there is a physical limit in such "plane-type" memories to record data on the surface or in the thin surface layer of recording materials.

Organic molecules have useful optical and electronic functions that can be easily controlled by the structure, substituent, or external fields. Molecular interactions and organized molecular assemblies also can afford much higher functions than isolated or randomly distributed molecules. Photons have many superior properties such as wavelength, polarization, phase, ultrashort pulse, or parallel processability. Through interactions of molecules or molecular assemblies with photons, many properties of photons can be directly converted to changes in physical properties of materials such as fluorescence, absorption, refractive index, conductivity, or optical nonlinearity. Excited-state formation, photochromism, and photoinduced electron transfer are some examples among them. Photon-mode recording or switching based on these changes can therefore achieve ultrafast multiple or three-dimensional recording and parallel process-ability with ultimate resolution at a molecular level. Undoubtedly, molecular photonics based on interactions of molecules and photons has many advantages compared with electric or photoelectric switching, heat-mode, or magnetic recording.

We have been attempting to develop new molecular photonics materials and devices by making various organized molecular systems and by optically controlling their electronic states as schematically shown in Fig. 1. So far we have achieved photoinduced electrochromism (color changes due only to the photoinduced electron transfer and reverse reactions), molecular control of the lifetime and wavelength of colored species over extremely wide ranges, amplified fluorescence quenching in LB films, photo-mode superresolution (PSR) to exceed the diffraction limit of light in optical memory based on transitory photobleaching of phthalocyanine derivatives, ultrafast all-optical two-dimensional control of reflectance, and a parallel optical self-holding switch based on photoinduced complex refractive index changes [2]. A spatial light modulator (SLM) is a device to two-dimensionally control the intensity or the phase of a reading

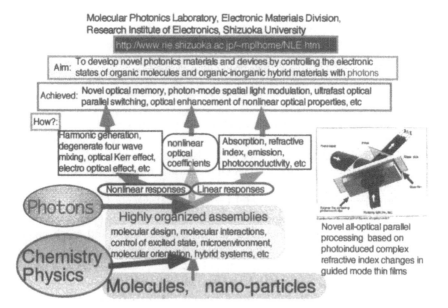

Figure 1 Schematic representation of research at our laboratory.

light by another (writing) light, which plays an essential role in a projection TV and an optical correlator. In the present chapter, some of our recent achievements are discussed.

II. MATERIALS AND METHODS

The structures of typical compounds employed in our study are shown in Fig. 2. 1,2-Dimethoxyethane (DME) and methanol solutions of TFPB$^-$ or iodides (I$^-$) salts of polymeric 4,4'-bipyridinium (PV^{2+}) were used together with polymer films cast from these solutions. The content of 4,4'-bipyridinium ions in a PV^{2+} polymer is 3.3×10^{-4} mol/g. Langmuir–Blodgett (LB) films were prepared by TFPB$^-$ salts of 4,4'-bipyridinium ions with long alkyl chains (HV^{2+}, AV^{2+}), by mixtures of palmitic acid (PA) or arachidic acid (AA) with pure 11-(9-carbazolyl)undecanoic acid (CUA) or commercial 11-(9-carbazolyl)undecanoic acid (MP-CUA) containing a small amount of 5H-benzo[b]carbazolyl derivative (BCZ). Various types of styrylpyridinium (NS$^+$, DCS$^+$, NS$^+$CnNS$^+$) tetraphenylborate (TPB$^-$) salts in DME were also employed to study ultrafast absorption changes in the visible and near-infrared (NIR) region. For nonlinear optical responses poly(3-dodecylthiophene) (PDT) was used in chloroform solutions. Several derivatives of phthalocyanines including hexadeca(2,2,2-trifluoroethoxy) phthalocyanine (H$_2$Pc(TFE)$_{16}$), water soluble copper-phthalocyanine (CuPcS), and zinc-phthalocyanine (ZnPcS) were used for PSR and SLM. Phthalocyanines were used in solutions or in poly(methylmethacrylate) (PMMA) or poly(vinyl alcohol) (PVA) films. A photochromic spiropyran derivative, 1,3,3-trimethylindolino-6'-nitrobenzopyrylospiran (SP) was dispersed in polystyrene, which became a colored photomerocyanine (PM) type upon UV irradiation.

For ultrafast dynamics studies, these dyes were excited in air at room temperature by the second harmonics (400 nm) of a fs Ti:sapphire laser with a regenerative Ti:sapphire amplifier and a double path amplifier pumped with the second harmonics (532 nm) of a Nd:YAG laser. The amplified Ti:sapphire laser delivered pulses with a FWHM of 200 to 250 fs, 10 Hz repetition, a maximum power of 6 mJ/pulse at 800 nm. A probe white light was obtained by focusing the residual 800-nm light into a cell containing a D$_2$O/H$_2$O (2:1) mixture after passing through a BBO crystal to obtain the second harmonics. The transient absorption and the dynamics were observed with a photonic multichannel analyzer (PMA; Hamamatsu Photonics) system using a dual photodiode array (Hamamatsu Photonics C6140) for the UV-visible and an InGaAs multichannel detector (Hamamatsu Photonics C5890-256) for the NIR absorption using an optical delay system. The intensities of the probe light with and without the pump pulses were averaged by 20 times. The block diagram of the fs transient absorption measurement system is shown in Fig. 3.

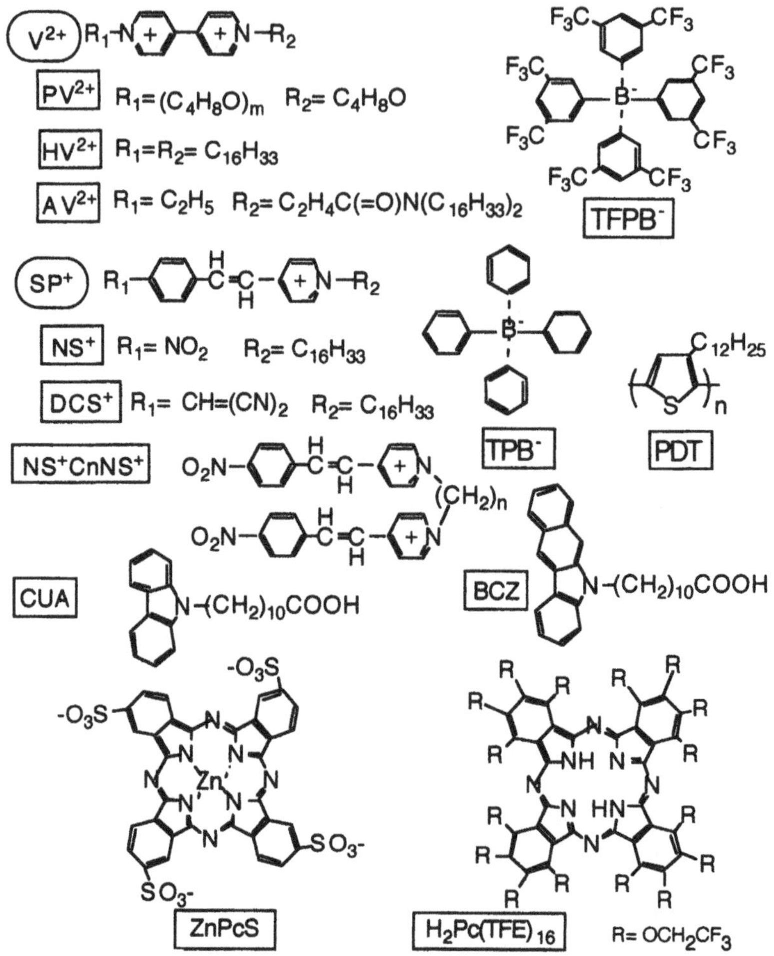

Figure 2 Structures and abbreviations of typical compounds employed.

The experimental setup for measuring basic properties of reflection-type spatial light modulation or parallel optical switching is schematically shown in Fig. 4. The sample plate was index-matched with a BK7 prism, which was set on a computer-controlled rotating stage. The writing beam was a ns OPO laser at 670 nm or the third harmonic (355 nm) of Nd:YAG laser; each with 8 ns pulse width, 0.03 to 2 mJ/pulse, and ca. 0.2 cm^2 beam area. A He–Ne laser (543.5 nm) through a half-wave plate, a polarizer, and a chopper were used as a reading beam. The time dependence of a reflected intensity at a given

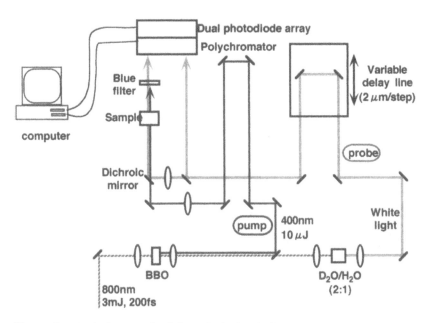

Figure 3 Block diagram of a fs laser flash photolysis system.

incident angle upon ns laser excitation with different powers was detected with a photomultiplier and was recorded with a digital oscilloscope terminated with 50 ohm. The spatial resolution was evaluated by using a USAF Test Target as a mask.

III. VERY SPECIFIC FLUORESCENCE BEHAVIOR IN LB FILMS

The LB deposition is one of the best methods to prepare highly organized molecular systems, in which various molecular parameters such as distance, orientation, extent of chromophore interaction, or redox potential can be controlled in each monolayer. We have been studying photophysical and photochemical properties of LB films in order to construct molecular electronic and photonic devices. The molecular orientation and interactions of redox chromophores are very important in controlling photoresponses at the molecular level. Absorption and fluorescence spectra give important information on them. We have studied photoresponses, specific interactions, and in-plane and out-of-plane orientation of various chromophores in LB films [3–11]. In addition to the change of absorp-

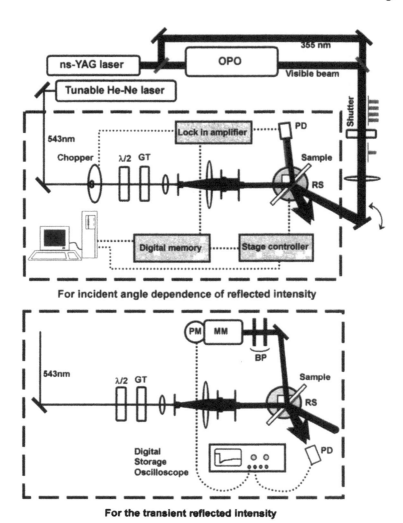

Figure 4 Schematic representation of measurement systems for the incident angle and the time dependences of reflected intensity in guided mode thin films.

tion and fluorescence spectra, the interaction and organization of chromophores in LB films also give various physical properties that cannot be achieved in randomly dispersed systems. Highly amplified fluorescence quenching due to efficient energy transfer was observed in LB films of CUA with and without acceptors [4,12–14]. This result can be applied to highly specific and efficient sensing based on fluorescence.

A. Molar Fraction-Dependent Efficient Energy Transfer in Mixed LB Films

Many studies on the photochemical properties of carbazolyl chromophores in polymers and biscarbazolyl alkanes have been made in relation to the photoconductivity of poly(N-vinylcarbazole) [15], which is one of a few organic materials practically applied in optical information processing. The efficient energy transfer among carbazolyl chromophores has been observed in crystals [16], polymers [17,18], and synthetic bilayer membranes [19] using a small amount of acceptors. The exciton diffusion was shown to be the main mechanism for the electronic energy transfer in molecular crystals of N-isopropyl carbazole or poly(N-vinylcarbazole) [16,17]. The energy transfer from carbazolyl groups in synthetic bilayer membranes of double chain ammonium amphiphiles in aqueous solutions to perylene or anthraquinone derivatives adsorbed on their surface was also analyzed by this mechanism [19]. The fluorescence intensity ratio between acceptors and donors depended linearly on the amount of acceptors in these systems [16–19]. The extent of exciton diffusion reflecting the chromophore aggregation was thus constant in these systems mainly depending on the chemical structure of polymers of amphiphiles. We found the molar fraction dependent efficient energy transfer in mixed LB films of CUA and long chain fatty acids using a small amount of benzocarbazolyl chromophores as an acceptor [4,13]. The molecular orientation and molecular interactions at the ground state of carbazolyl chromophores in mixed LB films of CUA with long chain fatty acids were shown from the changes of absorption and fluorescence spectra [3,4,12].

A commercial MP-CUA containing 470 ppm of 5H-benzo[b]carbazolyl derivative (BCZ) was first used to form mixed LB films with PA or AA. Mixed LB films with PA containing MP-CUA by a molar fraction (f_c) of 0.5, 0.10, and 0.27 were deposited at 15°C and 20 mN m^{-1}. A mixture of MP-CUA ($f_c = 0.20$) and AA was deposited at 20°C and 20 mN m^{-1}. Figure 5 shows the fluorescence spectra (λ_{ex} = 296 nm) of MP-CUA observed under identical conditions for (a) n-hexane solution (4×10^{-7} M) and (b) one monolayer mixed LB film with PA (fc = 0.27). In addition to the red-shift of monomer fluorescence of carbazolyl chromophores observed at 350 and 366 nm in n-hexane by about 6 to 10 nm, additional fluorescence peaks at 417 and 442 nm were observed in LB films. The latter were assigned to the fluorescence of 5H-benzo[b]carbazole [20]. The excitation spectra monitored at 417 and 442 nm corresponded to the absorption spectra of pure CUA. The direct excitation of BCZ at 390 nm in LB films or in dilute n-hexane solution did not give any fluorescence most probably due to negligibly small absorbance ($<1 \times 10^{-6}$). These results indicated that a very small amount of BCZ was excited effectively by the efficient energy transfer from carbazolyl chromophores in LB films which was confirmed by the fluorescence depolarization [4]. Figure 6 shows the fluorescence spectra of

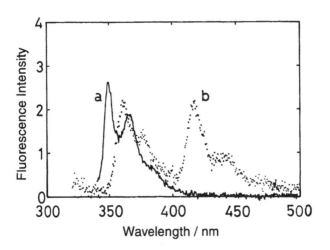

Figure 5 Fluorescence spectra of (a) MP-CUA in n-hexane (0.4 μM, solid line) and (b) one mixed monolayer of MP-CUA ($f_c = 0.27$, dotted line) and PA at 20°C and $\lambda_{ex} = 296$ nm.

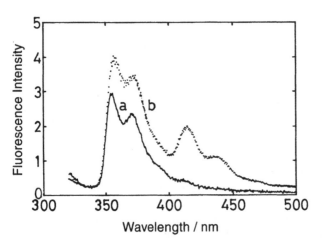

Figure 6 Fluorescence spectra of mixed LB films of MP-CUA and PA at 20°C for (a) 15 monolayers of a 1:19 mixture and (b) 11 monolayers of a 1:9 mixture, $\lambda_{ex} = 296$ nm.

mixed LB films of MP-CUA with PA (a) $f_c = 0.05$ and (b) $f_c = 0.10$. The fluorescence from BCZ is hardly detected for $f_c = 0.05$. A similar fluorescence spectrum was reported for LB films of MP-CUA ($f_c = 0.015$) and stearic acid [20]. An increase of the molar fraction of MP-CUA to 0.10 while keeping the content of acceptors constant (470 ppm) resulted in fairly strong acceptor fluorescence of BCZ as shown in Fig. 6b.

The intensity ratio of the emission of BCZ at 417 nm (I_A) and the monomer emission of carbazolyl groups at 360 nm (I_D) is plotted against f_c values (solid line) in Fig. 7. The result indicates that the I_A/I_D ratio increased steeply with the molar fraction f_c in LB films above about 0.05. At higher f_c values the I_A/I_D ratio increased more gradually. The dotted line shown in Fig. 7 was calculated from Förster energy transfer among randomly distributed donors and acceptors in LB films. It did not explain the observed molar fraction dependence of I_A/I_D ratio in mixed LB films. Then the singlet exciton migration model was applied assuming that carbazolyl chromophores were aggregated in mixed LB films sufficiently to allow the delocalization of excitation energy. This model was originally proposed for the energy transfer in perylene-doped N-isopropyl carbazole crystal or poly(N-vinylcarbazole) [16,17]. According to the hopping model of exciton diffusion, the fluorescence intensity ratio of acceptors and donors is given by $I_A/I_D = nc(1 - F)\Phi_A/\phi_D$, where c is the concentration of

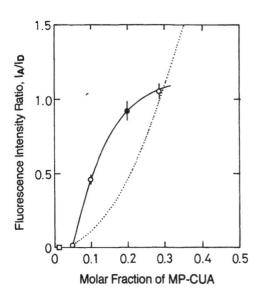

Figure 7 Fluorescence intensity ratio, I_A/I_D of acceptors at 417 nm and donors at 360 nm plotted against the molar fraction of MP-CUA in mixed LB films with (○) PA and (●) AA. The dotted line shows the calculated dependence.

guest molecules (acceptors in this case), n is the number of jumps during the lifetime of donors in the absence of acceptors, and F is the probability that the exciton returns to its starting point during its random walk at least once. Using the values of $\Phi_A = 0.675$ for BCZ, $\Phi_D = 0.32$ for carbazole, and $c = 4.7 \times 10^{-4}$, the value of n was evaluated from the observed fluorescence intensity ratio to be 1600, 1400, 700, and 45 for $f_c = 0.27, 0.20, 0.10,$ and 0.05, respectively [12]. These n-values can be compared with those (about 1000) reported for perylene-doped poly(N-vinylcarbazole) [16,17] or in synthetic bilayer membranes of ammonium amphiphiles containing carbazolyl groups [19]. These results strongly supported the exciton migration model in the present mixed LB films. A much higher value of 4.1×10^4 was found in perylene-doped crystal [16]. The nonstatistical aggregation of chromophores depending on the molar fraction was also shown from these results. The aggregated chromophores in mixed LB films will form domains depending on the molar fraction due to the phase separation between MP-CUA and fatty acids. A schematic representation is given in Fig. 8 for mixed LB films with (a) a higher f_c and (b) a lower f_c value according to these results and the experimentally determined orientation of carbazolyl groups [4]. Both the intra- and interlayer energy transfer will be possible between aggregated carbazolyl chromophores in LB films [13]. The extent of energy migration among aggregated carbazolyl chromophores can be controlled in LB films by molar fractions whereas it is almost constant in other systems reported so far [16–19].

B. Amplified Fluorescence Quenching in LB Films

Amplified photochemical quenching of carbazolyl fluorescence was observed in mixed LB films containing pure CUA and long chain fatty acids [12,14]. A pure CUA was synthesized from 2-nitrobiphenyl and 11-bromoundecanoic acid methyl ester [12,13]. Two monolayers of mixtures of CUA ($f_c = 0.02$ to 0.50) and PA were deposited on five monolayers of cadmium arachidate at 15°C and 20 mN m^{-1} at pH 6.3.

Figure 9 shows the fluorescence spectra of LB films containing CUA with $f_c = 0.25$ during irradiation at 290 nm and 2.5 mW cm^{-2} in an argon atmosphere. It is clearly shown that the fluorescence of CUA decayed considerably and monotonously by irradiation even in argon. Irradiation of similar LB films in air caused much faster decay of fluorescence spectra [12]. No changes were observed in fluorescence spectra and intensities for CUA in solution similarly irradiated or for LB films of CUA stored in the dark. The extent of fluorescence quenching in mixed LB films of CUA and PA during irradiation at 290 nm in air is shown in Fig. 10 as a function of f_c. The fluorescence intensity was normalized by the initial value. These results clearly show very rapid fluorescence quenching upon irradiation for a few minutes and very gradual quenching

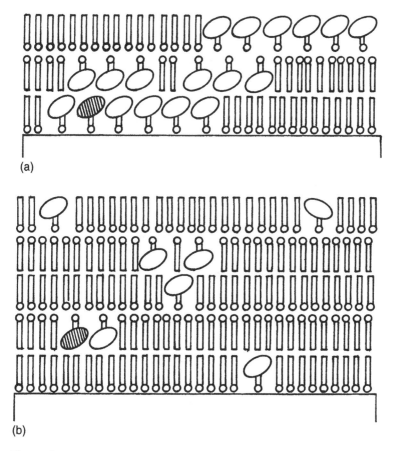

(a)

(b)

Figure 8 Schematic representation of mixed LB films including carbazolyl donors and a small amount of acceptors with (a) higher and (b) lower f_c values.

during later irradiation in LB films with $f_c > 0.05$. Only gradual quenching was observed for a system with $f_c = 0.02$. Similar results were observed for mixed LB films irradiated in argon [14]. The time dependencies are not expressed by a first-order nor a second-order kinetics, which are explained below. No effects of the excitation wavelength on the fluorescence quenching behavior were observed.

No changes in the absorption spectra of LB films were observed in argon during irradiation for up to 225 hr. The decay of fluorescence during irradiation in argon was also found to depend on the temperatures. It became slower at lower temperatures and only 16% decayed at 80°K after 90 min irradiation, which indicated some contribution of thermal process to the fluorescence de-

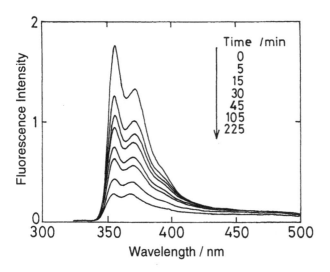

Figure 9 Fluorescence spectra of mixed LB films of pure CUA and PA with $f_c = 0.25$ during irradiation in argon at 290 nm and 2.5 mW cm^{-2}.

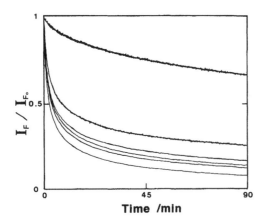

Figure 10 Time dependences of fluorescence intensity at 355 nm for six mixed LB films of pure CUA and PA irradiated at 290 nm and 1.0 mW cm^{-2} in air at 15°C. The f_c values are 0.02, 0.05, 0.15, 0.25, 0.30, and 0.40, respectively, from the top.

cay. From these results, the fluorescence quenching in argon is most probably due to the efficient energy transfer to the nonradiative sites formed by some changes of aggregation structure of carbazolyl chromophores in LB films. Similar fluorescence quenching without changes of absorption spectra was reported in vacuum-deposited films of pyrenecarboxylic acid upon irradiation by a Xe lamp or an excimer laser [21]. It was attributed to the structural changes of aggregates of pyrenyl chromophores to the nonfluorescent aggregate [22].

Meanwhile irradiation of LB films in air caused gradual spectral changes suggesting the photo-oxidation of carbazolyl chromophores [23]. The absorbance at 290 nm, however, decreased only 4% during irradiation for 90 min in air, where about 80% of fluorescence was quenched as shown in Fig. 10. All these results indicated that the fluorescence was quenched to a much higher extent than the changes of absorbance at the excitation wavelength. Such amplified fluorescence quenching in mixed LB films of CUA and PA was most probably caused by the very efficient and molar fraction dependent energy transfer to a trace amount of nonradiative sites formed by photo-oxidation or changes of aggregation structure of carbazolyl chromophores during irradiation.

The efficient energy transfer among carbazolyl chromophores occurred by singlet exciton migration as mentioned above [13]. The mean displacement r_h for randomly hopping excitons corresponding to the number of jumps is 1.9 nm at $f_c = 0.05$ and 20.0 nm at $f_c = 0.27$ [13]. Then the fluorescence from CUA molecules located in a circle with a radius of r_h will be quenched if one CUA molecule within that circle becomes nonradiative by photo-oxidation or changes of aggregation structure. The dependencies of fluorescence quenching on the irradiation time shown in Fig. 10 and on the molar fraction can be explained by the mechanism by which carbazolyl chromophores are distributed or aggregated inhomogeneously in mixed LB films and the extent of their aggregation varied with the molar fraction of CUA. The time dependencies of fluorescence quenching in LB films shown in Fig. 10 can be understood as the result of dispersive electronic excitation transport, that is, a set of single exciton hopping processes with different rates depending on the spatial distribution. As a small number of the energy trap sites are formed during irradiation, the rate of exciton transport slows because the density of carbazolyl chromophores available to accept exciton is reduced. The time dependence of such dispersive processes is known to be expressed by $I(t) = I_0 \cdot \exp(-kt^\alpha)$ $(0 < \alpha \Leftarrow 1)$ with the dispersion parameter α being a measure of the deviation from pure exponential decay. Good linear relationships were obtained from logarithmic plots of $I_F(t)$ versus t^α for fluorescence quenching in air and in argon with $\alpha = 0.20$ except for a small deviation at the very early part of the decay in air [14]. This result strongly suggests that amplified fluorescence quenching is caused by almost the same dispersive energy transfer process among carbazolyl chromophores in both cases. The difference in air and in argon is most probably due to the formation

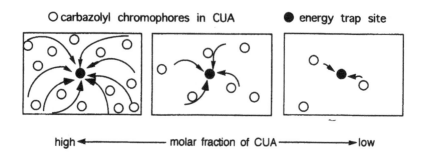

Figure 11 Schematic representation of the energy transfer to a trap site, formed during irradiation, depending on the molar fraction of CUA in mixed LB films with long chain fatty acids which are not shown for simplicity.

rate of nonradiative sites. Similar dispersive electronic excitation transport was observed in polymeric solids, an organic glass, and dye monolayers on crystals [24–47]. Figure 11 shows the schematic representation of the energy transfer to an energy trap site, formed by photo-oxidation or by the changes of aggregation structure, depending on the molar fraction of CUA. The phase separation of CUA and PA occurred in LB films at about $f_c = 0.02$ to 0.05 as mentioned above. The number of aggregated CUA and the extent of aggregation will increase with the molar fraction f_c. The larger the fraction of CUA in mixed LB films, the more fluorescence will then be quenched by the energy transfer to the trap site as shown in Fig. 11. Similar amplified fluorescence quenching was also observed in squarylium dye LB film containing J-aggregates and was successfully used to detect NO_2 gas at as low as a ppb (parts per billion) level [28,29]. The high mobility of excited states in the J aggregate was suggested to cause the very high sensitivity.

IV. PHOTOINDUCED ELECTROCHROMISM OF ION-PAIR CHARGE-TRANSFER COMPLEXES IN VARIOUS MICROENVIRONMENTS

Various photochromic systems employing polymeric thin films or LB films have recently attracted much interest in view of their promising applicability to high-speed and high-density photon-mode optical memory.* The photochromism reported so far involves changes of chemical bonds such as heterolytic cleavage of a pyran ring in spiropyrans, ring opening and closing in diarylethenes and

---

*Please see [2,30–33] and references cited therein.

fulgides, or *cis–trans* isomerization in azobenzenes [2,30–33]. Very recently we have reported novel photoinduced electrochromism as schematically shown in Fig. 12 [34–50]. It is the color change due to photoinduced electron transfer in ion-pair charge transfer (IPCT) complexes of 4,4'-bipyridinium salts with tetrakis[3,5-bis(trifluoromethyl)phenyl]borate [51] (abbreviated to TFPB⁻) and thermal back electron transfer reactions. No changes of chemical structure were involved in photochromism. From steady and laser photolysis results it has been shown that 4,4'-bipyridinium radical cations escaped from the geminate reaction immediately after the photoinduced electron transfer in less than 1 ps [48] upon IPCT excitation became metastable owing to the bulk and chemical stability of TFPB⁻, to the restriction of molecular motion by the microenvironment, and also probably to the very high exothermicity of the reverse reaction in the Marcus inverted region [52,53]. Highly sensitive detection of photoinduced electrochromism in ultrathin LB and polymer films has also been achieved by the optical waveguide method [54–59]. Such photoinduced electrochromism may be applied to ultrafast photon-mode optical memory and to redox sensors. In this section, photoinduced electrochromism and molecular control of orientation of photogenerated radicals in LB films are discussed.

A. Steady Photolysis and Control of Molecular Orientation of Radicals in LB Films

Steady-state photoinduced electrochromism was achieved in organic solutions [34,35], microcrystals [36,37], LB films [7,8,38,39], and polymer films [40–48] which was due only to the photoinduced electron transfer reaction via the excited state of specific IPCT complexes [49,50]. The photochemical coloring and the thermal fading due to the reverse electron transfer were highly reversible in a deaerated atmosphere in all systems [7,8,34–48]. The lifetime of colored (blue)

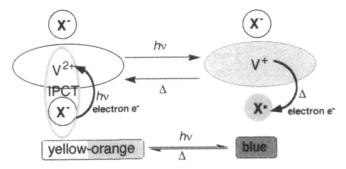

Figure 12 Schematic representation of photoinduced electrochromism in 4,4'-bipyridinium IPCT salts.

state was found to depend markedly on the microenvironments and temperatures. It was 72 hr in polymer films, 4.0 hr in microcrystals, 4.3 hr in LB films, and 26 min in DME solutions at 20°C [43]. Remarkable temperature dependence was observed in polymer films. No decay was observed below 0°C, whereas the colored state was easily erased at 80°C [40,41].

Monolayer properties of several mixtures of AA with TFPB$^-$ salts of HV^{2+} or AV^{2+} as shown in Fig. 2 were studied on an aqueous subphase containing 0.25 mM CdCl$_2$ and 0.05 mM NaHCO$_3$ pH 6.3 at 18°C. LB films were deposited at 20 mN m^{-1} and 18°C on a quartz plate for UV/vis or on a poly(ethyleneterephthalate) film for ESR measurements from 1:1 and 4:1 mixtures of AA and HV^{2+} or AV^{2+}. The deposition ratio was almost unity during 30 deposition cycles for all mixed monolayers. For steady photolysis these samples were irradiated in degassed condition by a Hamamatsu 150-W Xe-Hg lamp equipped with a Toshiba L-39 cutoff filter ($\lambda_{ex} > 365$ nm) and a 10-cm water filter to excite their IPCT absorption band alone. The incident angle dependencies of both s- and p-polarized absorption for 4,4′-bipyridinium radical cations were measured in degassed condition together with the polarization angle dependence at normal incidence.

The π-A isotherms are shown in Fig. 13 for three mixtures of AV^{2+} and AA. The π-A isotherms exhibited several transitions. A similar π-A isotherm

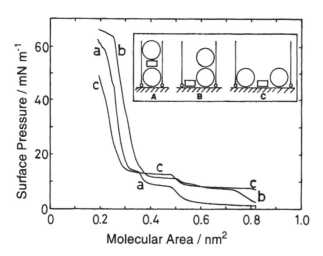

Figure 13 π-A Isotherms for mixtures of AA with AV^{2+} by a molar fraction of (a) 0.10, (b) 0.188, and (c) 0.50 at pH 6.3 and 18°C. The inset shows the schematic representation of surface monolayers during compression processes (C) → (B) → (A). The circle and rectangle represent TFPB$^-$ and the 4,4′-bipyridinium group of AV^{2+}, respectively.

was observed for a mixture of HV^{2+} (18.8%) and AA [39,43]. The apparent limiting area observed at each transition for mixtures of AV^{2+} and AA corresponded well with the calculated values based on the molecular area of $TFPB^-$ (1.4 nm^2) and 4,4′-bipyridinium ion (0.82 nm^2) for a stepwise squeezing out of 4,4′-bipyridinium ion and $TFPB^-$, which does not dissolve in water, as schematically shown in the inset of Fig. 13. From an X-ray analysis on a single crystal of N,N′-dimethyl-4,4′-bipyridinium tetraphenylborate (TPB^-), Moody et al. [60] reported that the 4,4′-bipyridinium ion was sandwiched between two TPB^- ions. IPCT complexes of $TFPB^-$ salts were expected to have similar configurations from several spectroscopic data [34,37,50]. Such a structure of IPCT complexes corresponds well to that schematically shown in inset (a) of Fig. 13 based on the limiting area. AV^{2+} and AA systems showed larger molecular areas than HV^{2+} and AA systems in all corresponding mixtures. This result may reflect the different orientation of 4,4′-bipyridinium ions as mentioned below.

Upon irradiation of an IPCT band in degassed condition ($\lambda_{ex} > 365$ nm), the color of both LB films changed from pale yellow to blue. The UV/vis absorption spectrum after irradiation of LB films with 120 monolayers is shown in Fig. 14, which is characteristic of 4,4′-bipyridinium radical cation monomer [61]. Colored species photogenerated in mixed LB films of AV^{2+}/AA or HV^{2+}/AA systems decayed almost exponentially in the dark *in vacuo* with a lifetime of about 4 hr at 20°C [38,39]. The lifetime of 4,4′-bipyridinium radical cations in LB films was almost the same as that in microcrystalline films [36], which indicated the microenvironment around photogenerated radical cations in LB films

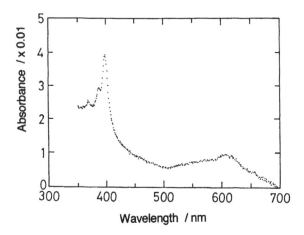

Figure 14 Absorption spectrum of mixed LB films (60 layers × 2) of HV^{2+} and AA (1:4) after excitation ($\lambda_{ex} > 365$ nm) in degassed condition at 20°C for 10 min using nonirradiated LB films as a reference.

is similar to that in microcrystals. Such photochemical coloring and thermal fading was repeated reversibly.

Polarized absorption spectra of photogenerated 4,4'-bipyridinium radical cations were measured *in vacuo* for LB films of HV^{2+}/AA and AV^{2+}/AA systems as a function of polarization angle and incident angle. The thermal decay of radicals during measurements of polarized absorption spectra was corrected by their lifetime [7,8,38,39]. The different optical path length in the incident angle dependence measurements was also corrected from an apparent incident angle dependence of s-polarized absorption in a similar way as for amphiphilic porphyrin [5,9]. No polarization angle dependencies were observed at normal incidence in both LB films. The p-polarized absorption of 4,4'-bipyridinium radical cations at 400 nm, which corresponds to the short-axis transition, are shown in Fig. 15 for (a) $HV^{2+}(18.9\%)/AA$ and (b) $AV^{2+}(18.9\%)/AA$. Figure 15 shows a minimum absorbance in HV^{2+}/AA and a maximum in AV^{2+}/AA at normal incidence. The solid lines in Fig. 15 are calculated by the least square method taking the angle (ϕ) distribution of the transition dipole moments with respect to the surface normal into account. The best fit curves give the following values of ϕ: $45° < \phi < 46°$ for HV^{2+}/AA and $89° < \phi < 90°$ for AV^{2+}/AA systems, respectively. Similar incident angle dependencies were observed at 614 nm which is due to a long-axis transition of 4,4'-bipyridinium radical cations. From these results and the simulation of angular dependencies, it was shown that both the long and short axes of 4,4'-bipyridinium radical cations lay almost flat in LB

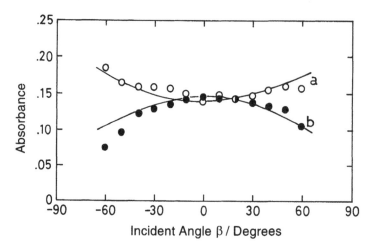

Figure 15 Incident angle dependences of the p-polarized absorbance of 4,4'-bipyridinium radical cations in LB films at 400 nm after correction of the decay and optical path length for photoexcited (a) HV^{2+}/AA and (b) AV^{2+}/AA systems. The solid lines are calculated dependences.

films of AV^{2+}/AA and inclined by about 46° to the substrate surface in LB films of HV^{2+}/AA as schematically shown in Fig. 16. This result can be used to increase memory density by multiple recording based on polarization.

B. Optical Waveguide Detection of Photoinduced Electrochromism in Ultrathin Films

It is very interesting and important to observe color changes in LB films of a single or a few monolayers thick in view of a sensing application with fast response, studying dynamics of photoinduced reactions between organized chromophores, or the easy preparation of good quality LB films. We have applied an optical waveguide (OWG) method for such purposes [54–59]. The electric fields of light propagating through the OWG layer have an exponentially decreasing value as evanescent waves on the surface of the OWG. Evanescent waves have been used to sensitively detect and characterize adsorbates and thin films on the OWG. Thin films (S) deposited on the surface of OWG were degassed by a rotary pump in a small chamber and irradiated with a Xe-Hg lamp through appropriate filters ($\lambda_{ex} > 365$ nm) as shown schematically in Fig. 17. A lin-

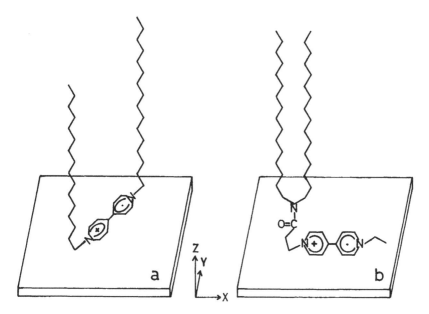

Figure 16 Schematic representation of the orientation of 4,4′-bipyridinium radical cations in (a) HV^{2+}/AA and (b) AV^{2+}/AA LB films. Counteranions (TFPB⁻) and AA are not shown for simplicity.

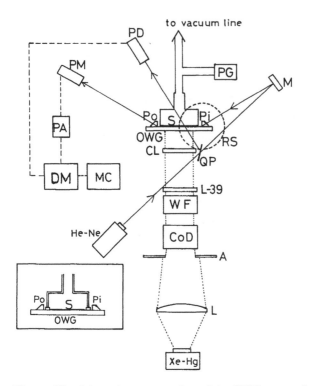

Figure 17 Schematic representation of the OWG system for detecting photoinduced electrochromism of ultrathin films (S) in the evacuation chamber shown in the inset.

early polarized He-Ne laser (632.8 nm) was used as a monitor light. A 150-fold sensitivity of the OWG method as compared with the conventional method was demonstrated from the color change measurement in about 180-nm thick film of $PV^{2+}(TFPB^-)_2$ containing 4,4'-bipyridinium groups as part of the main chain as shown in Fig. 2. The absorbances calculated from the OWG signal, using those before irradiation as a reference, are plotted in Fig. 18 against the irradiation time for $PV^{2+}(TFPB^-)_2$ thin films of various thickness: (a) 10.0, (b) 40.4, (c) 64.9, (d) 95.5, and (e) 179.6 nm. It is clearly seen that the number of photogenerated 4,4'-bipyridinium radical cations increased linearly with irradiation time. The rate of absorbance changes was proportional to the film thickness in the range studied. These results strongly suggested that $PV^{2+}(TFPB^-)_2$ thin films of various thickness are homogeneous and that 4,4'4-bipyridinium groups are distributed randomly throughout the polymer films.

Photoinduced color change in a single-monolayer LB film was successfully detected as shown in Fig. 19a for the HV^{2+}/AA system [55]. Comparison

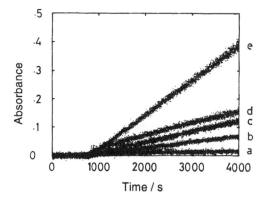

Figure 18 Changes in the OWG absorbance of $PV^{2+}(TFPB^-)_2$ thin films of various thickness during IPCT excitation in degassed condition: (a) 10.0, (b) 40.4, (c) 64.9, (d) 95.5, and (e) 179.6 nm.

of this result with Fig. 14 also demonstrated a more than 120-fold sensitivity of the OWG method. Changes of OWG absorbance are also shown for Y-type LB films deposited on glass slides covered with three monolayers of cadmium arachidate: (b) 2, (c) 4, and (d) 6 monolayers. The absorbance changes increased with the number of monolayers deposited. In contrast with the almost linear increase in absorbance of polymer films as shown in Fig. 18, the absorbance in LB films tended to saturate at longer irradiation times. The "saturated" absorbances increased almost proportionally with the number of monolayers. Similar re-

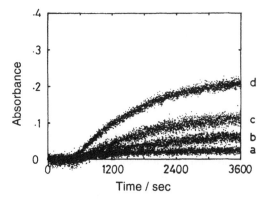

Figure 19 Changes in the OWG absorbance of LB films of HV^{2+}/AA (1:4) with various numbers of monolayers during IPCT excitation in degassed condition: (a) 1, (b) 2, (c) 4, and (d) 6 monolayers.

sults were obtained for LB films of AV^{2+}/AA systems. In LB films the 4,4'-bipyridinium ions are not distributed randomly but are confined to and aligned in a few tenths-nm thick layer periodically distributed in the direction of the surface normal. The long spacing of LB films of HV^{2+}/AA was evaluated as 5.5 nm by small angle X-ray scattering [38,43]. Such structural properties and much smaller thickness of LB films most probably contributed to the "saturation" tendency shown in Fig. 19.

It is possible in principle to determine the orientation of chromophores in a single monolayer on an OWG by the absorption of transverse electric (TE, s-polarized) and transverse magnetic (TM, p-polarized) mode laser. Swalen et al. [62] reported that much stronger absorption was observed for a thin evaporated film of 4-dimethylamino-4'-nitrostilbene with the TM mode and for seven monolayers of cyanine dyes with the TE mode. These results corresponded to the predicted molecular orientation of the two dyes, perpendicular and parallel to the substrate surface, respectively [62]. 4,4'-Bipyridinium radical cations photogenerated in polymer thin films showed the same absorbance for both TE and TM modes. This result corresponded to the random orientation of radical cations, which is consistent with the result from the time dependence of photogeneration mentioned above that 4,4'-bipyridinium groups are randomly distributed throughout the polymer films. In LB films of HV^{2+}/AA with one to six monolayers the OWG signals after photoexcitation displayed anisotropic absorption for TE and TM modes. Both the substituents of the 4,4'-bipyridinium ions and the nature of the substrate surface were found to affect the anisotropic absorption in LB films by the OWG method [57,58]. It is thus strongly suggested that photogenerated 4,4'-bipyridinium radical cations show specific orientation even in a single-monolayer LB film controlled at the molecular level as found in 120 monolayer LB film by a normal incidence method [7,8]. More sensitive detection of photoreaction than such a conventional K^+-doped OWG as mentioned above was recently made by the use of a complex OWG which was composed of a thin layer of higher refractive titanium dioxide on a K^+-doped OWG [59]. The highly sensitive optical detection is useful not only for evaluating photoresponses of ultrathin films needed to construct molecular photonics devices but also for various sensing applications.

C. Ultrafast Color Changes in 4,4'-Bipyridinium Salts

$TFPB^-$ and iodide (I^-) salts of 4,4'-bipyridinium ions showed pale yellow and orange colors, respectively, although each ion is colorless. These new absorption spectra above 350 nm in solutions were attributed to the IPCT complexes with 4,4'-bipyridinium ion as an acceptor and $TFPB^-$ or I^- as a donor. It was thus demonstrated that these ion pairs made electronic interactions at the ground state partially transferring electronic charges from a donor to an acceptor. No color changes were observed with I^- salts by steady photolysis, which is in contrast to TFPB$^-$ salts as mentioned above.

Immediately upon excitation of an IPCT band with a fs laser at 400 nm, transient absorption was observed for both salts in solutions with a peak at about 600 nm, characteristic of 4,4'-bipyridinium radical cations. Figure 20 shows the transient absorption spectra of $PV^{2+}(I^-)_2$ in methanol solution. A marked increase in the absorbance of the 4,4'-bipyridinium radical cations took place within 1 ps after excitation. 4,4'-Bipyridinium radical cations were thus formed in a fs time scale by the photoinduced electron transfer from a donor I^- to an acceptor 4,4'-bipyridinium upon IPCT excitation [48]. The time profiles of transient absorption at 600 nm are shown in Fig. 21 for (a) $PV^{2+}(I^-)_2$ in a film cast from DME and (b) $PV^{2+}(TFPB^-)_2$ in DME solutions. Both of them showed a very rapid rise in about 0.3 ps, which was almost the same as the time resolution of our fs Ti:sapphire laser measurement system with a regenerative amplifier. Similar extremely rapid formation of 4,4'-bipyridinium radical cations was observed for $PV^{2+}(I^-)_2$ salts in methanol and dimethylsulfoxide solutions upon IPCT excitation, respectively. These results demonstrated that the charge separated 4,4'-bipyridinium radical cations were formed directly upon IPCT excitation because of the nature of IPCT absorption bands (that the electrons correlated with the IPCT band are transferred partially at the ground state and completely at the excited state). Such a situation is very different from usual photochromism which is caused by various changes of chemical bonds mainly via the excited singlet state. No transient absorption was observed for $PV^{2+}(I^-)_2$

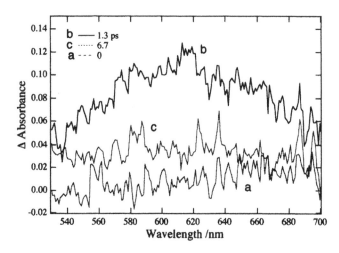

Figure 20 Transient absorption spectra of $PV^{2+}(I^-)_2$ in methanol solution upon fs laser excitation at 400 nm.

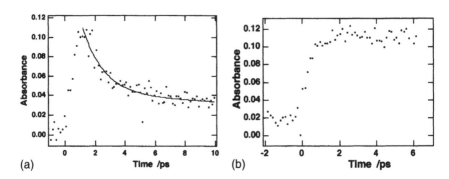

Figure 21   Time profiles of transient absorption at 600 nm for (a) $PV^{2+}(I^-)_2$ in a film cast from DME and (b) $PV^{2+}(TFPB^-)_2$ in DME solutions.

in DME solutions, which was most probably due to the decreased distance between ion pairs in such a less-polar solvent and an appropriate thermodynamic driving force ($-\Delta G°$) for reverse electron transfer reactions [52,53].

The decay behavior due to the reverse electron transfer was found to depend markedly on the microenvironment and the counteranion. The lifetime (τ) and a fraction of a major component for $PV^{2+}(I^-)_2$ was 1.2 ps and 86% in films cast from DME as compared with 4.0 ps and 73% in methanol solutions [48]. Photogenerated 4,4′-bipyridinium radical cations disappeared completely during 10 Hz excitation, which corresponded well with the fact that no steady color changes were observed for $PV^{2+}(I^-)_2$ in solutions and in cast films upon IPCT excitation. No decay was observed in the same time scale as shown in Fig. 21b for $PV^{2+}(TFPB^-)_2$. About one half decayed with $\tau = 71$ ps and the rest survived for an extremely long time corresponding to the reversible and persistent color changes observed by steady photolysis [46]. The lifetime of photogenerated 4,4′-bipyridinium radical cations were thus controlled over a very broad range from about 1 ps to almost infinity by the $-\Delta G°$ value, the polarity of solvents and microenvironments in solid films. The present result of color change in about 0.3 ps with IPCT complexes of PV^{2+} is the fastest response reported so far among materials that show steady photochromism. It will help a great deal to develop novel optical memory and also THz, all optical switching devices using visible light.

D. Charge Resonance Band in the Near-Infrared Region and Ultrafast Dynamics for Styrylpyridinium Tetraphenylborate Salts

Recently we have also reported, for the first time, the charge resonance (CR) band due to dimer radical cation formation as a broad absorption with a peak of 950 to 1700 nm upon steady photoexcitation of styrylpyridinium derivatives such as tetraphenylborate (TPB$^-$) salts of 1-hexadecyl-4-(4-dicyanovinylstyryl) pyridinium (DCS$^+$), 1-hexadecyl-4-(4-nitrostyryl)pyridinium (NS$^+$) or 1,n-bis (4-nitrostyrylpyridinium)alkane (NS$^+$CnNS$^+$) as shown in Fig. 2 in solutions at room temperature [63–70] by steady photolysis. The absorption spectra after irradiation (>365 nm) for TPB$^-$ salts of DCS$^+$ and NS$^+$ in DME are shown in Fig. 22 with respect to those before irradiation. In addition to the absorption spectra in the visible region due to the radical formation, broad specific absorption spectra were observed in the NIR region. The peak wavelength and shape of the latter spectra depended on the substituents. They were assigned to a CR band as schematically shown in Fig. 23 in a dimer radical cation that was formed between a styrylpyridinium cation and a photogenerated styrylpyridinyl radical. We have also observed the CR band with a peak at 1500 to 1700 nm as a charge resonance band for intramolecular dimer radical cations as shown in Fig. 24 for NS$^+$C3NS$^+$, NS$^+$C4NS$^+$, and NS$^+$ in acetonitrile [67]. It is clearly shown that two very strong CR bands with peaks at 950 and 1700 nm were observed only in NS$^+$C3NS$^+$ due to intramolecular dimer radical cations. The CR band energy is twice the stabilization energy of dimer radical cations which is controlled by several factors such as the extent of overlap of two chromophores, their mutual distance, and/or the polarity of solvents. The CR bands at 950 and 1700 nm were assigned to fully and partially overlapped dimer radical cations, respectively.

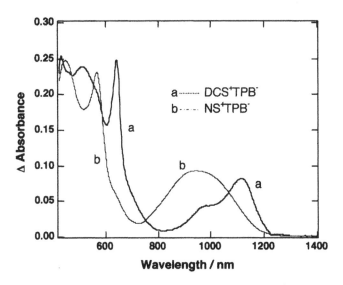

Figure 22 Difference absorption spectra of (a) DCS$^+$TPB$^-$ and (b) NS$^+$TPB$^-$ in DME (0.50 mM) after irradiation at >365 nm in degassed conditions.

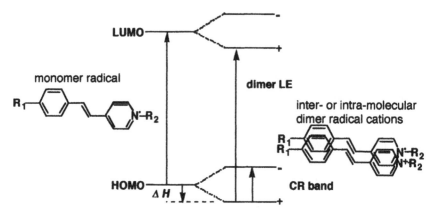

Figure 23 Schematic representation of energy levels for monomer radical and dimer radical cations; ΔH is the stabilization energy of dimer radical cations.

As shown in Fig. 25, the rise of absorption spectra at the visible region due to radical formation and at the near-IR region due to the CR band was observed in less than 1 ps upon a fs laser excitation at 400 nm. These results indicated that the dimer radical cations were formed immediately after the photoinduced electron transfer reaction. The CR band at 900 nm in NS^+TPB^- decayed single exponentially ($\tau = 3.3$ ps) [68]. The transient absorption at 580 nm showed

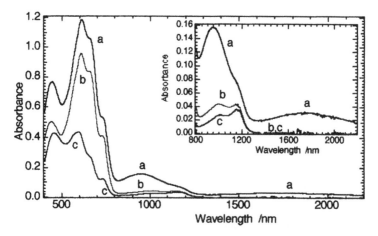

Figure 24 Absorption spectra for TPB^- salts of (a) NS^+C3NS^+, (b) NS^+C4NS^+, (c) NS^+ irradiated in acetonitrile by a Xe-Hg lamp at $\lambda_{ex} > 365$ nm at room temperature in degassed conditions, (a) and (b) 0.25 mM, (c) 0.50 mM.

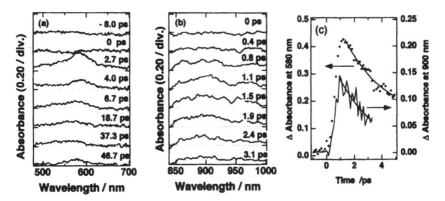

Figure 25 Transient absorption in (a) visible and (b) near-infrared regions together with (c) time dependences at 580 and 900 nm upon fs laser excitation of NS^+TPB^- in DME at room temperature.

double exponential decay with lifetimes of 3.3 and 11.4 ps. Similar results were obtained for DCS^+TPB^-: the decay at 960 nm with $\tau = 3.2$ ps, and that a 650 nm with lifetimes of 3.8 ps for a fast component and 17.4 ps for a slow one [68]. The difference in the decay behavior at the visible and the near-IR region was explained as follows. While the NIR region absorption was attributed to the dimer radical cation alone as the CR band, both the monomer radical and the dimer radical cation contributed to the absorption in the visible region as a HOMO–LUMO transition as schematically shown in Fig. 23. The fast component in the visible region and that in the NIR absorption gave almost the same lifetime. They were attributed to the reverse electron transfer reactions from the dimer radical cation $(SP^+ \cdots SP^\bullet)$ to the $TPB^\bullet$ radical. Then the slow component in the visible region was most probably due to the reverse electron transfer reaction from the monomer radical $(SP^\bullet)$ to the $TPB^\bullet$ radical.

The higher rate of the reverse electron transfer in the dimer radical cations than in the monomer radicals was explained by classical Marcus theory as follows [52,53]. The $-\Delta G^\circ$ values for the reverse electron transfer from $NS^\bullet$ and $DCS^\bullet$ to $TPB^\bullet$ were estimated as 1.69 and 1.61 eV from the redox potentials, respectively. The reduction potential of the dimer radical cation should be less negative by 0.65 and 0.59 V than that of the radical monomer radicals due to the stabilization energy. The $-\Delta G^\circ$ values for the reverse electron transfer from the dimer radical cation to $TPB^\bullet$ were thus estimated to be 1.04 and 1.02 eV for $(NS^+ \cdots NS^\bullet)$ and $(DCS^+ \cdots DCS^\bullet)$, respectively [68].

The total reorganization energy λ in classical Marcus theory was estimated by the method described below. A plot of the maximum of the IPCT band in solution (E_{OP}) against the driving force of electron transfer within the contact ion pair $(-\Delta G_{IP})$ follows the Hush relation $E_{OP} = \lambda + \Delta G_{IP}$ [71]. Since the absorptivity of the IPCT bands is very low and overlaps with the $\pi-\pi^*$ transition of styrylpyridinium chromophore, it is difficult to obtain E_{OP} with sufficient accuracy from absorption spectra. We therefore used the onset of the IPCT band, E_{IPCT} instead of E_{OP}, which is accessible by a simple procedure from the absorption spectra. Thus the correlation of E_{IPCT} and $-\Delta G°$ can be expressed by a modified Hush relation $E_{IPCT} = \lambda + \Delta G°$.

A plot of E_{IPCT} versus $-\Delta G°$ for substituted SP^+TPB^- salts afforded a straight line with a slope of 1.1. This value is very close to the theoretical value of 1.0 as expected from the Hush relation. From the intercept, the reorganization energy λ was evaluated to be 0.90 eV. Similar dependence has been reported for E_{IPCT} evaluated from diffuse reflectance spectra of solid-state IPCT complexes of bipyridinium metal dithiolene salts, which gave a similar value of λ, 0.63 eV [72]. According to classical Marcus theory [52,53], the rate constant for electron transfer will be maximum when $-\Delta G°$ matches the total reorganization energy λ and will decrease if $-\Delta G°$ exceeds λ. The observed values of $-\Delta G°$ in this study for the reverse electron transfer of monomer radicals and dimer radical cations are in the Marcus inverted region [52,53]. The rate constant of the reverse electron transfer reactions from the dimer radical cation to TPB$^{\bullet}$ would become higher due to the smaller of $-\Delta G°$ in the inverted region. This is the reason why the observed decay was faster for dimer radical cations. These results strongly suggest the applicability of the present system to ultrafast optical switching in the NIR region if the appropriate combination of a donor anion and an acceptor cation is used.

V. APPLICATIONS OF ULTRAFAST ABSORPTION CHANGES TO OPTICAL PROCESSING

A. Photon-Mode Superresolution by Transient Bleaching of Phthalocyanines

Upon excitation of molecules the absorption spectrum changes in a femtosecond time scale to cause transient absorption or bleaching. We have proposed photon-mode superresolution (PSR) based on photoinduced transitory bleaching as one of the promising means to increase the recording density of optical disks [73,74]. An organic dye layer showing transitory bleaching at the wavelength of a laser diode is expected to cause the laser beam contraction, since the central part of a Gaussian beam will become "transparent" or less absorptive than the peripheral parts. Then such an organic film on a recording layer will work as PSR material in any type of optical disks, in principle, as schematically shown in Fig. 26(a).

Excitation of phthalocyanines resulted in transient bleaching at the 620 to 600-nm region and transient absorption with a peak at about 530 nm as shown

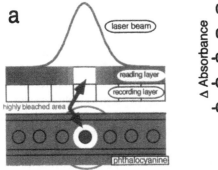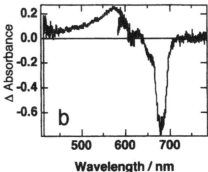

Figure 26 (a) Schematic representation of photon-mode superresolution using transitory photobleaching of phthalocyanine, and (b) absorption changes of a PVA film containing ZnPcS upon photoexcitation with fs laser.

in Fig. 26b for ZnPcS in a PVA film. The transmittance of phthalocyanines in polymer films at the wavelength of laser diodes increased from less than 10% to about 50% depending on the incident power (<5 mW) of the laser diodes [73,74]. The intensity of a probe laser diode increased almost 10 times within 1 ps upon excitation with a fs laser at 400 nm. Instantaneous and power dependent bleaching were thus demonstrated using phthalocyanines. The lifetime of the bleached state can be controlled over a ns–ms range by the dye structures and microenvironment. These results strongly suggest the PSR based on the formation of the photoexcited state of phthalocyanines is very promising for increasing recording density [73,74]. Similar photon-mode superresolution was used to read compact disk (CD) data at higher density by using naphthalocyanine derivatives [75].

B. All-Optical Ultrafast Spatial Light Modulation and Parallel Optical Recording Based on Photoinduced Complex Refractive Index Changes in Guided Wave Geometry Containing Organic Dyes

All-optical data processing has recently attracted much interest especially in the fields of spatial light modulation and optical data storage. Okamoto et al. [76] reported an all-optical photoaddressed spatial light modulator (SLM) using a dye-doped polymer film in a surface plasmon resonance (SPR) configuration. They demonstrated the SLM based on photothermal changes in the refractive index of the methyl orange-doped poly(vinyl alcohol) (PVA) film using an Ar laser of 1 to 6 W cm^{-2} as a writing beam. The rise and fall times at 6 W cm^{-2} laser power were about 10 sec and 2 sec, respectively [76]. Yacoubian and

Aye proposed Fabry–Perot (FP) resonance shifting in attenuated-total-reflection (ATR) geometry using azodye polymers [77]. They reported that their ATR-FP device enhanced optical modulation speed and efficiency as compared with the conventional intensity modulation based on photoinduced birefringence of Disperse Red 1 dye-doped poly(methylmethacrylate) [77]. The response time, 50 to 200 ms, was still relatively slow, although it was improved as compared to the conventional modulation system [77]. Ho et al. proposed a polarization vectorial holographic recording based on birefringent polymeric materials containing a photochromic azobenzene dye. The response time was 80 μs with a 100-MW power (80 μJ/cm^2) of Ar laser [78]. Fichou et al. [79] proposed an incoherent-to-coherent optical converter based on photoinduced absorption of sexithiophene film. No actual properties of such a device including the response time were reported.

We have also proposed a novel all-optical SLM based on complex refractive index changes upon photoexcitation of an organic dye-doped polymer thin film [80–84]. This system is very unique as compared to the previously proposed "all-optical" light modulation systems as mentioned above. In principle fs response can be achieved in this system, because we use resonance condition changes of the guided optical waves (guided mode) in the ATR geometry based on the changes in an imaginary or a real part of the refractive index due to transient absorption or its Kramers–Kronig transformation as schematically shown in Fig. 27. The guided mode "resonance" pattern depends not only on the thickness of a dielectric layer but also on its complex refractive index composed of real and imaginary parts, in general. If the imaginary part increases due to transient absorption as shown in Fig. 26b at about 400 to 600 nm, for example, the reflectance increases as shown in Fig. 27. The change of the real part shifts the resonance as shown in Fig. 27 for the case of the decrease. The intensity of the probe beam can be two-dimensionally controlled by the pump (writing) beam through photoexcitation of a dye. The main advantages using the guided mode are (1) its high sensitivity to small changes in refractive index and thickness, and (2) its sensitivity to both p- and s-polarized light. So far we have achieved repeated light modulation using a pulsed nanosecond (ns) laser and CuPcS and ZnPcS in guided wave-mode geometry. The response time was controlled by the triplet lifetime of phthalocyanines, 30 ns for CuPcS and 0.55 ms for ZnPcS. We are making efforts to achieve much faster responses using a ps or fs laser and appropriate materials.

We have also demonstrated self-held ultrafast parallel optical switching based on the same geometry and using photochromic compounds instead of phthalocyanines [81–84]. Absorption spectra of spiropyran derivative (SP) dispersed in polystyrene with a weight ratio of 1:10 are shown in Fig. 28 before and after UV irradiation for 5 sec. Strong absorption in the visible region due to photomerocyanine (PM) can be held for a long time and be reverted to that of

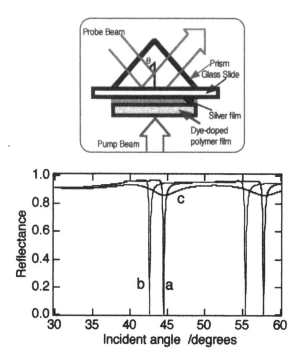

Figure 27 Schematic representation of the all-optical parallel processing in guided mode geometry and the calculated reflectance for a polymer film (1600 nm) on a silver layer (50 nm). The complex refractive index of a polymer layer is (a) 1.60, (b) 1.58, and (c) 1.60 + 0.02i.

SP by visible irradiation. Spectra of extinction coefficient and refractive index changes (Δk and Δn) of polystyrene thin film containing SP upon UV excitation for 5 sec are shown in Fig. 29. The former are based on the observed difference absorption spectra before and after UV irradiation as shown in Fig. 28. The latter, calculated from the extinction coefficient Δk by Kramers–Kronig transformation, are refractive index changes with different signs that can be seen near strong absorption changes. The extinction coefficient and/or refractive index changes over a wide wavelength range from approximately 400 to 800 nm can be utilized to operate a wide range all-optical switch. The guided mode structures are very important for the utilization of ultrafast changes of molecular electronic state upon photoexcitation for practical photonics devices.

The incident angle dependences of measured reflectance of a probe beam at 543 nm are shown in Fig. 30 for a polystyrene film containing SP with a weight ratio of 10:1. Each dip shows the SPR for a silver film (a) and the guided TM wave mode for the composite thin film (b) before and (c,d) after excitation by a

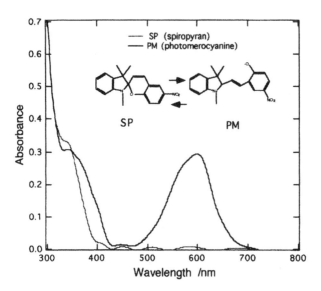

Figure 28 Structures of photochromic SP and PM, and absorption spectra of SP or PM dispersed in a polystyrene thin film with a weight ratio of 1:10 before and after UV irradiation.

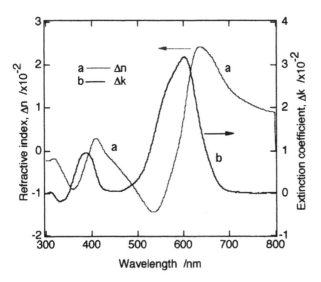

Figure 29 Spectra of (a) refractive index and (b) extinction coefficient of polystyrene thin film containing SP and PM. The former is based on the observed difference absorption spectra shown in Fig. 28 before and after UV irradiation. The former was calculated from (b) by Kramers–Kronig transformation.

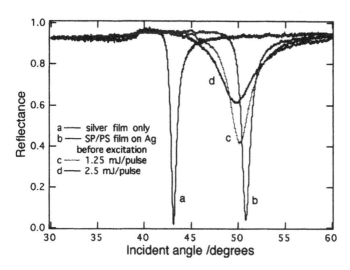

Figure 30 Incident-angle dependences of observed reflectance of a glass slide covered with (a) a silver film (50 nm) alone, and the SP in polystyrene (1:10) film (264-nm thick), (b) before, and (c) and (d) after ns laser excitation (1.25 and 2.5 mJ/pulse, respectively) at 355 nm.

pulsed Nd-YAG laser at 355 nm. The power of the pulsed laser at 355 nm was set to 1.25 or 2.5 mJ/pulse. Photochromism induced by transformation from SP to PM increased the reflectance and slightly shifted the dip to lower incident angles as shown in Fig. 30. The simulation gave an almost perfect reproduction of the observed results, from which the complex refractive indices before and after excitation were determined. From comparison between the measured and calculated dependences, the thicknesses of a silver film and a polymer film were evaluated as 50 nm and 264 nm, respectively. The reflectance at the incident angle of 50.76 degrees was increased from 0.04 to 0.68 upon excitation as shown in Fig. 30. The reflectance increase and the shift were found to be due to the increase of extinction coefficient and the decrease of refractive index at 543.5 nm of the polystyrene thin film containing SP as shown in Fig. 28 by the formation of the PM form. The changes of refractive index and extinction coefficient were estimated to be -0.015 and $+0.024$ at 2.5 mJ/pulse from comparison between the measured and the calculated dependences. Transmittances of the probe beam before and after excitation of 2.5 mJ/pulse were calculated compared to the present reflectance changes by using the same value for the film thickness and the extinction coefficient changes. The estimated values were 0.99 and 0.86 at 543 nm before and after excitation, respectively. The dynamic range of reflection changes, 17.0, was thus demonstrated to be much better than that of transmission changes, 1.15.

The switching OFF response upon excitation of 600 nm laser is shown in Fig. 31 together with that of switching ON of the 355 nm laser. The power of the pulsed laser at 355 nm and 600 nm was 2.5 mJ/pulse and 6.0 mJ/pulse, respectively. The reflectance at the incident angle of 50.76 degrees increased by 5 to 10 times very rapidly with a rise time of about 20 ns upon pulsed laser excitation at 355 nm depending on its power. This rise time corresponded to the response of a photomultiplier. Much better switching is expected if a picosecond laser and the experimental setup with much better time resolution are used, since the rise time of transient absorption of a polystyrene film containing spiropyran was reported as about 200 ps upon excitation with ps laser [85]. The reflectance after ns pulsed laser excitation at 355 nm was held at a high value without applied powers. The switching OFF was also demonstrated to be very fast with a response time similar to that of switching ON, although its accurate estimation was difficult due to a smaller S/N ratio. These results indicate that this fast reflectance decrease was caused by the reverse photochromic reaction from PM to SP, and not by the thermal reaction. The observed smaller reflectance change as compared with switching ON was due to a lower quantum yield of ring closure of PM. Photochromic dyes, which have a higher quantum yield for reverse photochromic reaction, will switch from ON to OFF with a larger dynamic range and fast response.

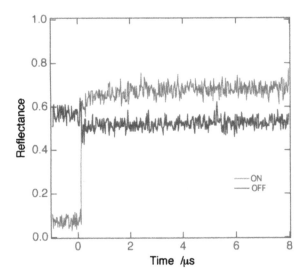

Figure 31 Reflectance changes at the incident angle of 50.76 degrees for switching ON (broken line) and OFF (solid line) responses in a guided mode thin film as in Fig. 30, which were induced by pumping at 355 nm, 2.5 mJ/pulse and at 600 nm, 6.0 mJ/pulse, respectively.

The values of the reflectance after excitation were determined at various probe wavelengths: 0.78 at 543 nm, 0.82 at 594 nm, 0.86 at 632.8 nm, and 0.74 at 694 nm. The reflectance before excitation was as follows: 0.08 at 543 nm, 0.13 at 594 nm, 0.13 at 632.8 nm, and 0.30 at 694 nm. These results gave the extinction ratio, −9.9 dB at 543 nm, −8.0 dB at 594 nm, −8.2 dB at 632.8 nm, and −3.9 dB at 694 nm, respectively. The smaller value of the extinction ratio at 694 nm was due to the fact that a laser diode had a broader emission spectrum than a He-Ne laser. It will be improved by using a thicker polymer film and/or a laser diode that has a narrower emission spectrum. Wavelength dependences of the refractive index and the extinction coefficient changes evaluated from reflectance changes upon pulsed laser excitation in the polystyrene film containing SP corresponded well with the spectra of extinction coefficient and refractive index changes estimated from steady photolysis as shown in Fig. 27. These results confirmed the mechanism responsible for the reflectance changes in guided wave geometry and also demonstrated the wide range of operation wavelength of the present all-optical device.

In addition to very fast photoresponses, it is essential to write and read a two-dimensional image pattern for optical parallel data processing. Pictures of the probe beam at 543.5 nm recorded with a usual camera before and after excitation at 355 nm through a pattern "RIE" of about 1 mm high are shown in Fig. 32. The pattern "RIE" written by the excitation laser can clearly be seen in the reflected beam as shown in Fig. 32b, which was erased completely upon

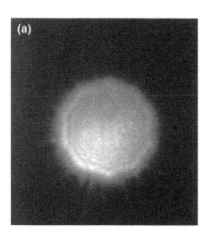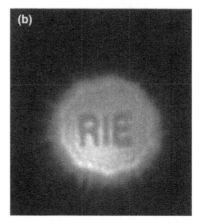

Figure 32 Photographs of a probe He-Ne laser at 543.5 nm reflected from the present device containing SP in the guided mode thin Arton[@] film (a) before and (b) after pulsed laser excitation at 355 nm through a mask patterned "RIE," which stands for Research Institute of Electronics. The diameter of the probe beam is ca. 3 mm, which was expanded by 10 times on a screen to take pictures.

visible excitation. The spatial resolution of the SP-doped polystyrene film in guided wave geometry was evaluated to be better than 128 line pairs (lp) mm^{-1} or about 4 μm by using the USAF test target.

As one of the near-future applications of the present all-optical switch, the architectures of the AND and OR optical parallel processing logic devices were proposed [82]. Optical parallel AND and OR devices are composed of two present switches and one switch only, respectively. They are operated by two parallel data as two input signals, Input 1 and Input 2. An optical parallel NOT device is composed of a polymer film containing a photochromic dye which is colored only during irradiation. Then, the excitation light as an input signal causes the intensity decrease of a probe beam as an output signal. A combination of the present all-optical switch and the photon-mode spatial light modulator will also contribute a great deal to constructing an ultrafast parallel processing optical correlator. Several spatial optical logic devices can be composed simply and easily from the present all-optical switch which employs reflectance increase or decrease by irradiation. An example of possible all-optical parallel correlation based on the present device is schematically shown in Fig. 33.

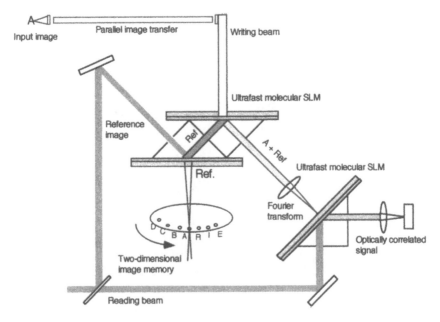

Figure 33 Schematic representation of a possible all-optical parallel correlation system based on the present device.

VI. ULTRAFAST NONLINEAR OPTICAL RESPONSES

Nonlinear optical responses are very important to achieve, for example, wave-length conversion, electro-optical or pure optical control of the refractive index, and all-optical logic. Many organic and inorganic materials have been developed. One of the main problems especially in organic compounds is the small nonlinear optical coefficient. We also have been making efforts to modulate or enhance the second- and third-order nonlinear optical properties by changing the electronic state or the extent of electronic distribution upon photoexcitation [86–93]. Degenerate four-wave mixing (DFWM) signal intensity of PDT in chloroform increased by more than three orders upon excitation at 400 nm as shown in Fig. 34. Enhanced DFWM signal exhibited an instantaneous response that has almost the same temporal profile as a probe pulse at 800 nm. The optical pumping effect on DFWM intensity was independent of the polarization angle of the pumping pulse at 400 nm. These results suggest that the enhancement of DFWM intensity by optical pumping might be ascribed to the gain of the electronic third-order susceptibility. Excitonic states could be produced in a one-dimensional chain backbone of a conjugated polymer like PDT upon photoexcitation. Excitonic states showed an intense transition to higher states in the IR region. These might explain the observed enhancement. We also observed similar enhancement of the second-order optical nonlinearity upon photoexcitation [91,92].

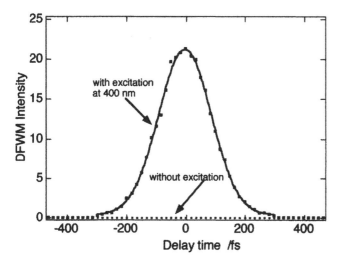

Figure 34 DFWM intensity of poly(3-dodecylthiophene) in chloroform probed at 800 nm without and with excitation by a fs laser at 400 nm.

VII. CONCLUSION

We have developed several molecules and organized molecular assemblies to control the lifetime of photoinduced electrochromism, the complex refractive index, the orientation of photogenerated radicals, the fluorescence, the transmittance, and the nonlinear optical responses by molecular interactions and/or interactions with photons. Application of such materials to novel optical devices based on photoinduced complex refractive index changes was proposed and successfully demonstrated. These results will contribute a great deal to the realization of parallel all-optical ultrafast data processing devices. Considering that our vision was initiated by a simple photoisomerization of 11-*cis* retinal and was processed in a parallel way to achieve extremely high functions, molecular photonics based on an elegant combination of molecules, photons, and appropriate devices is expected to be a very promising information processing method in the near future.

ACKNOWLEDGMENTS

The present work was achieved through collaboration with many researchers and students. The author would like to thank all of them and especially Drs. H. Sakaguchi, K. Sakai, S. Kamata, Y. Isoda, H. Kawai, K. Sasaki, H. Inoue, and A. Harada of Shizuoka University, and Dr. D. Matsunaga of Nippon Kayaku Co. Partial support by the Grant-in-Aids for Scientific Research on Priority Areas "Molecular Superstructures, Design and Creation" (No. 07241102), "Molecular Synchronization for Design of New Materials System" (No. 11167242), "Creation of Novel Delocalized Electronic Systems" (No. 10146219), and Monbusho International Scientific Research Program (Joint Research, No. 08044137, 10044144) from the Ministry of Education, Science, Sports, and Culture, Japan are greatly acknowledged.

REFERENCES

1. T Kamiya, F Saito, W Wada, H Yajima, eds. Femtosecond Technology. In: T Kamiya, B Monemar, H Venghaus, series eds. Springer Series in Photonics 2. Berlin: Springer-Verlag, 1999.
2. T Nagamura. Chapter VI-4. In: T Kamiya, B Monemar, H Venghaus, series eds. Springer Series in Photonics 2. Berlin: Springer-Verlag, 1999, p 376.
3. T Nagamura, K Matano, T Ogawa. Ber Bunsenges Phys Chem 91:759, 1987.
4. T Nagamura, K Kamata, T Ogawa. Nippon Kagaku Kaishi 2090, 1987.
5. T Nagamura, T Koga, T Ogawa. Denki Kagaku 57:1223, 1989.
6. T Nagamura, S Kamata. J Photochem Photobiol A: Chem 55:187, 1990.

7. T Nagamura, Y Isoda, K Saki, T Ogawa. J Chem Soc Chem Commun 703, 1990.
8. T Nagamura, Y Isoda, K Sakai, T Ogawa. Thin Solid Films 210/211:617, 1992.
9. T Nagamura, T Koga, T Ogawa. J Photochem Photobiol A: Chem 66:119, 1992.
10. T Nagamura, T Fujita, M Takasaka, H Takeshita, A Mori, T Nagao. Thin Solid Films 243:602, 1994.
11. T Koga, T Nagamura, T Ogawa. This Solid Films 243:606, 1994.
12. T Nagamura, K Toyozawa, S Kamata, T Ogawa. Thin Solid Films 178:399, 1989.
13. T Nagamura, S Kamata, K Toyozawa, T Ogawa. Ber Bunsenges Phys Chem 94:87, 1990.
14. T Nagamura, K Toyozawa, S Kamata. Colloid Sur 102:31, 1995.
15. F Evero, K Kobs, R Memming, DR Terrell. J Am Chem Soc 105:5988, 1983.
16. W Klöpffer. J Chem Phys 50:1689, 1969.
17. W Klöpffer. J Chem Phys 50:2337, 1969.
18. A Itaya, K Okamoto, S Kusabayashi. Bull Chem Soc Jpn 52:3737, 1979.
19. T Kunitake, M Shimomura, Y Hashiguchi, T Kawanaka. J Chem Soc Chem Commun 833, 1985.
20. N Tamai, T Yamazaki, I Yamazaki. J Phys Chem 91:841, 1987.
21. A Itaya, K Okamoto, S Kusabayashi. Bull Chem Soc Jpn 51:79, 1978.
22. Y Taniguchi, M Mitsuya, N Tamai, I Yamazaki, H Masuhara. Chem Phys Lett 132:516, 1986.
23. G Pfister, DJ Williams. J Chem Phys 61:2416, 1974.
24. F Willig, A Blumen, G Zumofen. Chem Phys Lett. 108:222, 1984.
25. R Richert, B Ries, H Hässler. Phil Mag B49:L25, 1984.
26. R Richert. Chem Phys Lett 118:534, 1985.
27. AD Stein, KA Peterson, MD Fayer. Chem Phys Lett 161:16, 1989.
28. M Furuki, K Ageishi, S Kim, I Ando, LS Pu. Thin Solid Films 180:193, 1989.
29. M Furuki, LS Pu. Thin Solid Films 210/211:471, 1992.
30. H Dürr, H Bouas-Laurent. Photochromism Molecules and Systems. Amsterdam: Elsevier, 1990.
31. M Irie. Chem Rev 100:1685 (and references cited therein), 2000.
32. Y Yokoyama. Chem Rev 100:1717 (and references cited therein), 2000.
33. G Berkovic, C Krongauz, V Weiss. Chem Rev 100:1741 (and references cited therein), 2000.
34. T Nagamura, K Sakai. J Chem Soc Faraday Trans 1 84:3529, 1988.
35. T Nagamura, S Muta, J Photopolym Sci Technol 4:55, 1988.
36. T Nagamura, K Sakai. J Chem Soc Chem Commun 810, 1986.
37. T Nagamura, K Sakai. Ber Bunsenges Phys Chem 93:1432, 1989.
38. T Nagamura, K Sakai, T Ogawa. J Chem Soc Chem Commun 1035, 1988.
39. T Nagamura, K Sakai. Thin Solid Films 179:375, 1989.
40. T Nagamura, Y Isoda. J Chem Soc Chem Commun 71, 1991.
41. T Nagamura. Polym Int 27:125, 1992.
42. T Nagamura, S Muta, K Sakai. J Photopolym Sci Technol 5:561, 1992.
43. T Nagamura. Mol Cryst Liq Cryst 224:75, 1993.
44. T Nagamura, H Sakaguchi, S Muta, T Ito. Appl Phys Lett 63:2762, 1993.
45. T Nagamura, H Sakaguchi, T Ito, S Muta. Mol Cryst Liq Cryst 247:39, 1994.
46. T Nagamura, H Sakaguchi, S Muta. Proc SPIE 2514:241, 1995.

47. T Nagamura. Pure Appl Chem 68:1449, 1996.
48. H Inoue, H Sakaguchi, T Nagamura. Appl Phys Lett 73:10, 1998.
49. T Nagamura, K Sakai. Chem Phys Lett 141:553, 1987.
50. T Nagamura, K Sakai. Ber Bunsenges Phys Chem 92:707, 1988.
51. H Nishida, N Takada, M Yoshimura, T Sonoda, H Kobayashi. Bull Chem Soc Jpn 57:2600, 1984.
52. RA Marcus. J Chem Phys 24:966, 1956.
53. JR Miller, LT Calcaterra, GL Closs. J Am Chem Soc 106:3047, 1984.
54. T Nagamura, H Sakaguchi, K Suzuki, C Moshizuki, K Sasaki. J Photopolym Sci Technol 6:133, 1993.
55. T Nagamura, H Sakaguchi, K Sasaki, C Mochizuki, K Suzuki. This Solid Films 243:660, 1994.
56. T Nagamura, D Kuroyanagi, K Sasaki, H Sakaguichi. Proc SPIE 2547:320, 1995.
57. K Sasaki, T Nagamura. J Photopolym Sci Technol 9:129, 1996.
58. K Sasaki, T Nagamura. Mol Cryst Liq Cryst 294:145, 1997.
59. T Nagamura, T Adachi, K Sasaki, T Matsuura, X-M Chen, K Itoh, M Murabayashi. J Photochem Photobiol A: Chem (submitted).
60. GJ Moody, RK Owusu, AMZ Slawin, N Spencer, JF Stoddart, JDR Thomas, DJ Williams. Angew Chem Int Ed Engl 26:890, 1987.
61. EM Kosower, JL Cotter. J Am Chem Soc 86:5524, 1964.
62. JD Swalen, M Tacke, R Santo, KE Rieckhoff, J Fischer. Hel Chim Acta 61:960, 1978.
63. T Nagamura, A Tanaka, H Kawai, H Sakaguchi. J Chem Soc Chem Commun 599, 1993.
64. T Nagamura, H Kawai, T Ichihara, H Sakaguchi. Synth Metals 71:2069, 1995.
65. H Kawai, T Nagamura. Mol Cryst Liq Cryst 267:235, 1995.
66. T Nagamura, T Ichihara, H Kawai. J Phys Chem 100:9370, 1996.
67. T Nagamura, S Kashihara, H Kawai. Chem Phys Lett 294:167, 1998.
68. H Kawai, T Nagamura. J Chem Soc Faraday Trans 94:3581, 1998.
69. W-S Xia, H Kawai, T Nagamura. J Photochem Photobiol A: Chem 136:35, 2000.
70. H Kawai, T Nagamura. Mol Cryst Liq Cryst 345:209, 2000.
71. NS Hush. Progr Inorg Chem 8:391, 1967.
72. H Kisch. Coord Chem Rev 159:385, 1997.
73. T Nagamura, M Miura, H Fujita, H Sakaguchi, T Sonoda, H Kobayashi. J Photopolym Sci Technol 8:129, 1995.
74. T Nagamura, M Miura, H Fugita, H Sakaguchi, T Sonoda, H Kobayashi, D Matsunaga. Third International Symposium Functional Dyes, P 32. Santa Cruz, CA, 1995.
75. J Seto, S Tamura, N Asai, N Kishi, Y Kijima, N Matsuzawa. Pure Appl Chem 68:1429, 1996.
76. T Okamoto, T Kamiyama, I Yamaguchi. Opt Lett 18:1570, 1993.
77. A Yacoubian, TM Aye. Appl Opt 32:3073, 1993.
78. ZZ Ho, G Savant, J Hirsch, T Jannson. Proc SPIE 1773:433, 1992.
79. D Fichou, J-M Nunzi, F Charra, N Pfeffer. Adv Mater 6:64, 1994.
80. T Nagamura, T Hamada. Appl Phys Lett 69:1191, 1996.
81. K Sasaki, T Nagamura. Appl Phys Lett 71:434, 1997.

82. K Sasaki, T Nagamura. J Appl Phys 83:2894, 1998.
83. T Nagamura, K Sasaki. Proc SPIE 3466:212, 1998.
84. T Nagamura, K Sasaki. Mol Cryst Liq Cryst 344:199, 2000.
85. T Ito, M Hiramatsu, M Hosoda, Y Tsuchiya. Rev Sci Instrum 62:1415, 1991.
86. H Sakaguchi, T Nagamura, T Matsuo. Jpn J Appl Phys 30:L377, 1991.
87. T Nagamura, H Sakaguchi, T Matsuo. Thin Solid Films 210/211:160, 1992.
88. H Sakaguchi, LA Gomez-Jahn, M Prichard, TL Penner, DG Whitten, T Nagamura. J Phys Chem 97:1474, 1993.
89. H Sakaguchi, T Nagamura. Nonlinear Optics 15:73, 1996.
90. H Sakaguchi, T Nagamura. In: PF Barbara, JG Fujimoto, WH Knox, W Zinth eds. Ultrafast Phenomena. Berlin: Springer-Verlag, 1996, pp 62, 209.
91. A Harada, T Nagamura. Mol Cryst Liq. 316:79, 1998.
92. A Harada, T Nagamura. Nonlinear Optics 22:169, 1999.
93. H Sakaguchi, T Nagamura. Nonlinear Optics 22:413, 1999.

10

Molecular Recognition Events Controllable by Photochemical Triggers or Readable by Photochemical Outputs

Seiji Shinkai
Kyushu University, Fukuoka, Japan

Tony D. James
University of Birmingham, Birmingham, United Kingdom

I. INTRODUCTION

A. Why Are Dynamic Functions Indispensable in Molecular Recognition?

In the 1960s the concept of an "enthalpy–entropy compensation relationship" was proposed by Leffler [1]. This concept was very helpful for obtaining insights into a number of thermodynamic data for association and kinetic processes. However, this also implies that high selectivity and high activity cannot appear in one reaction series. Then, how can we create such an exceptional system with high selectivity and high activity that deviate from the enthalpy–entropy compensation relationship? This was the starting point of our chemistry. This query becomes especially important when we mimic the high selectivity and the high activity of enzymes in the field of biomimetic chemistry. The breakthrough was achieved when we realized that as long as one association process or one kinetic process is treated independently, it is restricted by the relationship whereas if two or more systems are coupled reversibly, one may find an exceptional process diverted from the relationship through "switching" between the systems. The original concept has enabled us to create a number of new ion and molecule recognition systems that are combined with switch-functionalized and molecular-assembly systems. In particular, we have noticed that light is a very promising trigger to provide switching functions for functional molecular systems (Fig. 1).

Nature is filled with ions and molecules and we are frequently required to measure the concentration of selected ions and small organic molecules both *in vivo* and *in vitro* processes. One possible approach in this research field is a direct application of biomaterials such as proteins, biomembranes, and the like created by nature after appropriate modifications. This approach has made its mark: for example, one can raise "biosensors" at its representative and successful system. However, systems borrowing ideas from nature are frequently hampered by their inherent boundary conditions such as poor solubility in organic media, instability at high temperature, degradation and denaturation, lack of versatility, and so on because they were created for Nature's own purposes but not for our practical purposes. Thus, the artificial molecular design toward manmade receptors that show high affinity and high selectivity comparable with natural systems has long been a dream for scientists and has recently become a very active area of endeavor. However, one should notice that this concept is only one part in the design of a total sensing system: even though an artificial receptor precisely recognizes a selected ion or molecule, it is still useless unless the guest-binding event can be read out as a convenient physical signal. To exploit the total sensing system, therefore, the binding even must be transduced to some changes in a molecular system and eventually converted to some physical signal; that is, the total sensing system consists of three different components: viz., (1) an ion or molecule recognition site, (2) a signal conversion site, and (3) a signal reading-out site (Fig. 2). To integrate such multicomponent functions

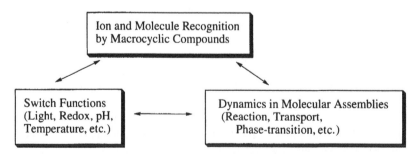

Figure 1 Concept for molecular design of switch-functionalized recognition systems.

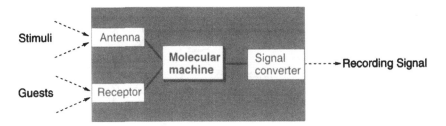

Figure 2   Conceptual scheme for design of a total sensing system.

into one small molecular system seems to be quite difficult. However, there are a considerable number of successful examples in which slight structural changes induced by the ion or molecule binding are efficiently transduced to changes in the subsequent physical signaling processes. These examples teach us that to achieve the precise molecular design, a small molecular system is frequently more advantageous than a polymeric system or a molecular assembly system. Furthermore, to precisely predict the guest-induced structural change is now possible utilizing recently developed computational methods.

In this chapter, we introduce molecular recognition events controllable by photochemical triggers or readable by photochemical outputs, particularly focusing on our own recent research achievements.

II.   THE BIRTH OF PHOTORESPONSIVE CROWN ETHERS

Photoresponsive systems are ubiquitously seen in nature and light is coupled with the subsequent life processes. In these systems, a photoantenna to capture a photon is skillfully combined with a functional group to mediate some subsequent events. It is important that these events are frequently linked with photoinduced structural changes of photoantennas. This suggests that chemical substances which exhibit photoinduced structural changes may serve as potential candidates for the photoantennas. In the past, photochemical reactions such as (E)–(Z) isomerism of azobenzene, dimerization of anthracene, spiropyran–merocyanine interconversion, and the like have been used as practical photoantennas. One can expect that if one of these photoantennas is skillfully combined with a crown ether, many physical and chemical functions of a crown ether family can be controlled by an on–off light switch. This is the basic concept for the design of photoresponsive crown ethers. We believe that this is one of the earliest examples of a "Molecular Machine."

Compound **1** is an early example of a photoresponsive crown ether [2,3]. **1** has a photofunctional azobenzene cap on an N_2O_4 crown ring, so that one can

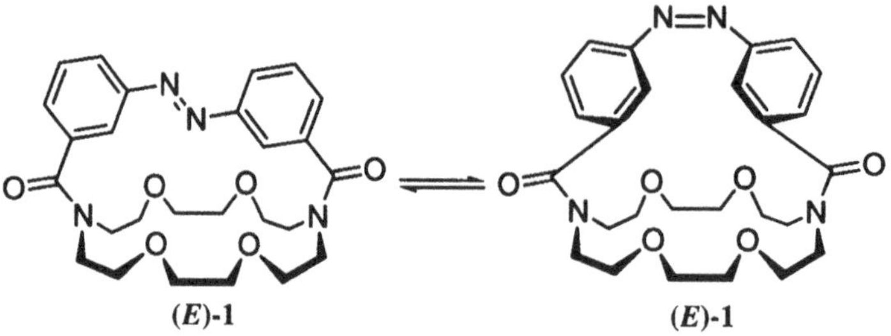

(E)-1   (E)-1

expect that a conformational change in the crown ring occurs in response to the photoinduced configurational change in the azobenzene cap. (E)-1 having the (E)-azobenzene cap selectively binds Na$^+$ while (Z)-1 produced by photoisomerization by UV light irradiation binds K$^+$ more strongly. The finding suggests that the N$_2$O$_4$ ring is apparently expanded by the photoinduced (E)-to-(Z) isomerization. This was confirmed by X-ray crystallographic studies of (E)-2, in

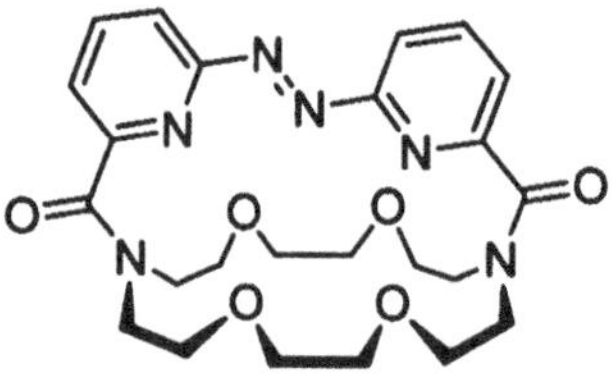

2

contrast to all gauche C–C bonds in 18-crown-6 complexed with K$^+$. The N$_2$O$_4$ crown in 2 has two anti C–C bonds, resulting in an oval-shaped crown ring [4]. Conceivably, this is the reason why (E)-1 favors small Na$^+$ rather than K$^+$.

Photodimerization of anthracene has frequently been cited as a photochemical switch to create photoresponsive crown ethers. Photoirradiation of 3 in the presence of Li$^+$ gives the photocyclo-isomer 4 [5,6]. 4 is fairly stable with Li$^+$ but readily reverts to the open form 3 when Li$^+$ is removed from the ring. In this system, however, intermolecular dimerization may take place competitively

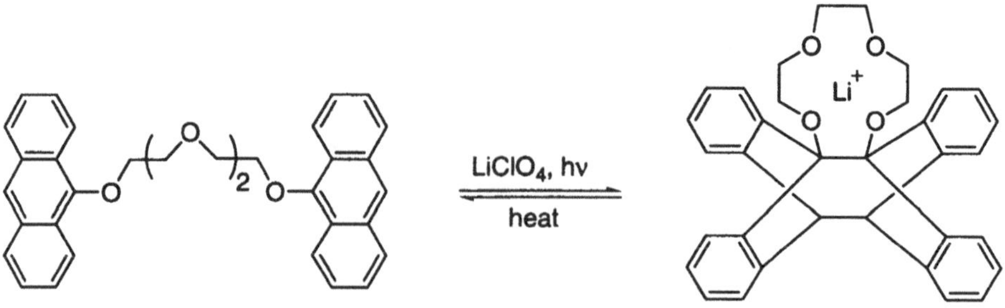

3   4

with intramolecular dimerization. To rule out this possibility, **5** was synthesized and two anthracenes linked by two polyether chains [7]. It was found that intramolecular photodimerization proceeds rapidly in the presence of Na^+ as the template metal cation. Compound **6** has also been synthesized [8]. Although

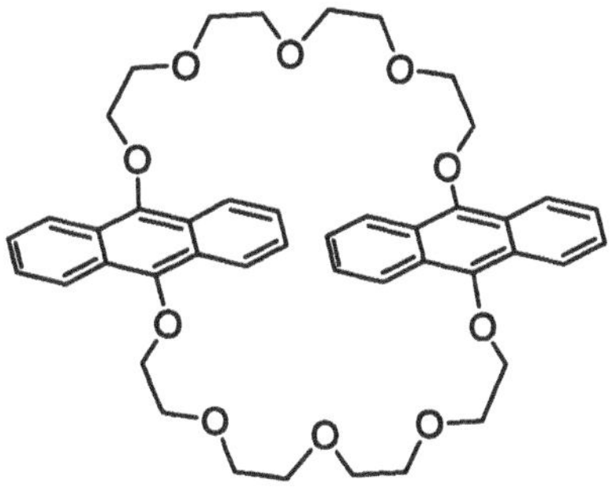

5

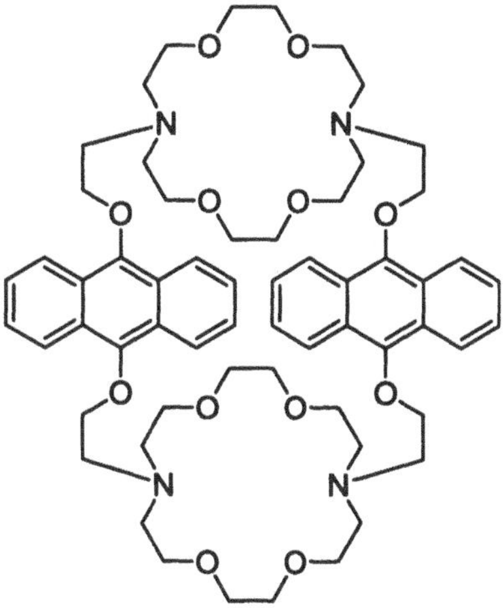

6

this compound was not applied to a light-switch system, it showed a remarkable fluorescence change upon the binding of $RbClO_4$ or $H_3N^+(CH_2)_7NH_3^+$ [8]. Yamashita et al. [9] also synthesized **7** in which intermolecular photodimerization

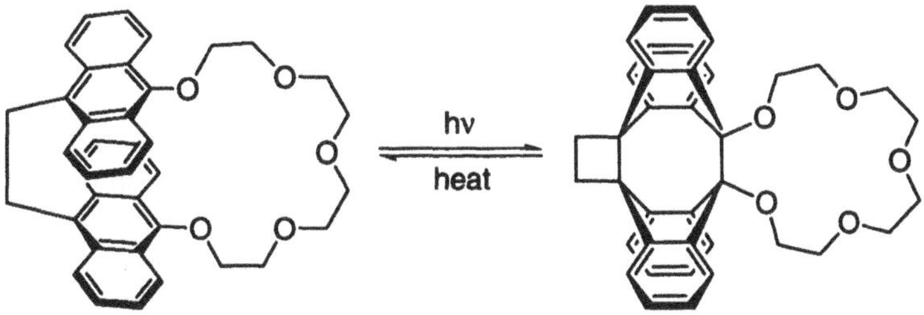

7

8

of anthracene was completely suppressed. The photochemically produced cyclic form [8] showed excellent Na^+ selectivity.

It has been established that alkali metal cations exactly fitting the size of the crown ether ring form a 1:1 complex whereas those that have larger cation radii form a 1:2 sandwich complex. This view was clearly substantiated by using bis(crown ethers). For instance, Kimura et al. [10] reported that the maleate diester of monobenzo-15-crown-5 ((Z)-form) extracts K^+ from the aqueous phase

14 times more efficiently than the fumarate counterpart [(*E*)-form]. The difference stems from the formation of the intramolecular 1:2 complex with the (*Z*) form. If the C=C double bond were replaced by the azolinkage, the resultant bis(crown ethers) would exhibit interesting photoresponsive behaviors. The essence of this section is that the photoinduced change in the spatial distance between two crown rings should be reflected by the change in the ion-binding ability.

A series of azobis(benzocrown ethers) called "butterfly crown ethers" such as **9** and **10** were synthesized [11–15].* The photoresponsive molecular motion is

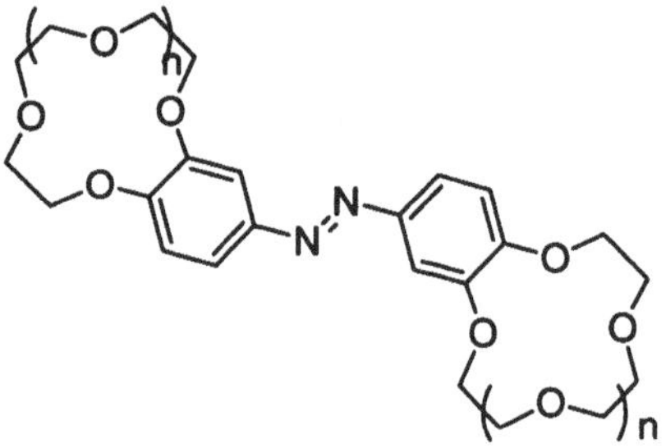

(*E*)-**9**

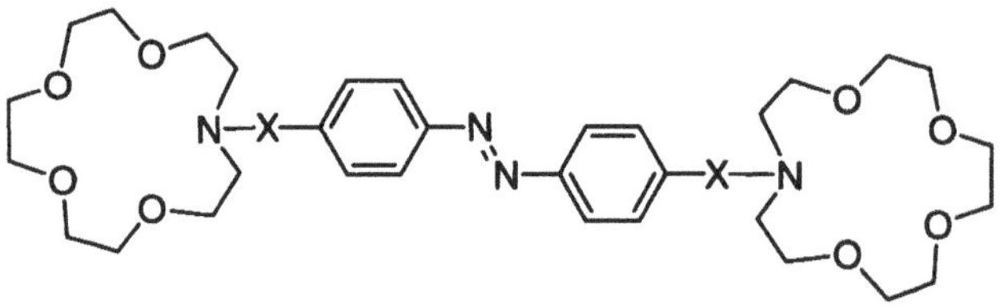

(*E*)-**10**

similar to that of a flying butterfly. It was found that the content of the (*Z*)-forms at the photostationary state grows remarkably with increasing concentration of Rb^+ and Cs^+ which interact with two crown rings in a 1:2 sandwich manner. This is clearly due to the bridge effect of the metal cations with the two crowns. These results support the view that that (*Z*)-forms form an intramolecular 1:2 complex with these metal cations. As expected, the (*Z*)-forms extracted alkali metal cations with large ion radii more efficiently than the corresponding (*E*)-forms. In particular, the photoirradiation effect on **9** is quite remarkable: for example, (*E*)-**9**(n = 2) extracts Na^+ 5.6 times more efficiently than (*Z*)-**9**(n = 2) whereas (*Z*)-**9**(n = 2) extracts K^+ 42.5 times more efficiently than (*E*)-**9**(n = 2) [13].

---

*For a comprehensive review of photoresponsive crown ethers, please see [11].

The solution properties of complexes formed from **9**(n = 3) and poly-methylenediammonium cations, $H_3N^+(CH_2)_mNH_3^+$ have been evaluated in detail [16]. It was found that when the distance between the two ammonium cations is shorter than that between the crown rings in (E)-**9**(n = 3), (e.g., m = 6), they form a polymeric complex (Fig. 3). When the two distances are comparable (e.g., m = 12), they form a 1:1 pseudocyclic complex. Photoisomerized (Z)-**9**(n = 3) showed a different aggregation mode because of the change in distance between the two crown rings: the 1:1 complex for (Z)-**9**(n = 3) + m = 6 diammonium salt and the 2:2 complex for (Z)-**9**(n = 3) + m = 12 diammonium salt. This is a novel example of reversible interconversion between polymers and low molecular-weight pseudomacrocycles. Since the interconversion process is sensitively reflected by electric conductance, one may regard this as the transmission of light energy into an electric signal [16].

Cations are known to be transported through membranes by synthetic macrocyclic polyethers as well as by antibiotics. When the rate-determining step is the ion extraction from the IN aqueous phase to the membrane phase, the transport rate increases with the increasing stability constant. On the other hand, when the rate-determining step is the ion-release from the membrane phase to the OUT aqueous phase, the carrier must reduce the stability constant in order to attain efficient decomplexation. Some polyether antibiotics feature

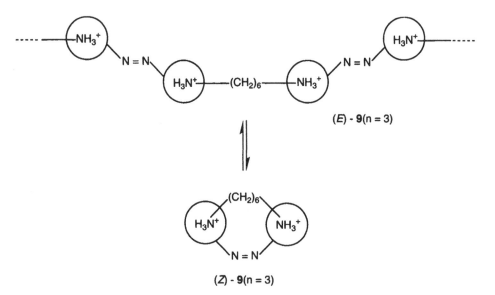

Figure 3 Photoregulation of polymer-pseudomacrocycle interconversion in a **9**(n = 3) + $H_3N^+ - (CH_2)_6 - NH_3^+$ system; the circle indicates benzo-18-crown-6.

an interconversion between the cyclic and noncyclic forms in the membrane, a feature by which the transport system escapes from the limitation of the rate-determining step. In this system, the energy arising from the pH gradient is consumed to compensate for the cyclic–noncyclic interconversion. Here, a new, very attractive idea arises: provided that the ion-binding ability of the carrier at the rate-determining step can be changed by light, the rate of ion-transport can be also changed. This idea should be of particular importance when the ion-release is rate-determining. In K^+ transport with **9**(n = 2) across a liquid membrane, it was found that the rate is accelerated by the UV irradiation which mediates the (E)-to-(Z) isomerization [13]. The rate enhancement is attributed to the increased extraction speed from the IN aqueous phase to the membrane phase. The rate was further enhanced by alternate irradiation by UV and visible [which mediates the (Z)-to-(E) isomerization] light (Fig. 4) [17]. This effect is attributed to the increased release speed from the membrane phase to the OUT aqueous phase. In conclusion, the pH gradient utilized in ion-transport by some polyether antibiotics is replaced by light energy in the present light-driven systems.

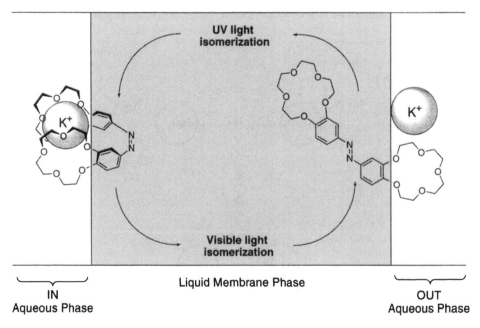

Figure 4 Schematic representation of K^+ transport with **9**(n = 2) accelerated by alternate irradiation of UV and visible light.

III. DYNAMIC ACTIONS OF CALIXARENES IN ION AND MOLECULE RECOGNITION

Calixarenes are $[1_n]$metacyclophanes made up of phenol units linked via alkylidene groups.* They preferably adopt a cone conformation because of the stabilization effect by the intramolecular hydrogen-bonding interaction. Unless bulky substituents are introduced into the OH groups, the rotation of the phenyl units is allowed [18,19]. Hence, the cavity shape that governs the guest-binding properties can be controlled by a change in the phenyl unit rotation. Compound **11** is

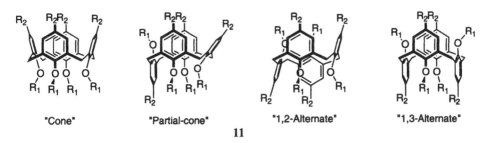

"Cone"   "Partial-cone"   "1,2-Alternate"   "1,3-Alternate"

11

conformationally mobile and four conformers (cone, partial-cone, 1,2-alternate, and 1,3-alternate) are interconverted through the oxygen-through-the-annulus rotation [19,20]. Although the most stable conformer is the partial-cone [19–22], changes in the conformation occur in response to added guests [23]. This conformational freedom characteristic of the calixarene family is very suitable for designing molecular switches. For example, ^{1}H NMR studies established that new peaks appear which are assignable to the cone-**11**($R_2 = Bu^1$) · M^+ complex when $LiClO_4$ or $NaClO_4$ is added [23]. On the other hand, the spectrum was scarcely affected by the addition of $KClO_4$ [23]. The findings suggest that to bind alkali metal cations four oxygens must be arranged in the lower rim of the cone-shaped calix[4]arene and the size of the oxygen-cavity thus composed is comparable with the size of Li^+ or Na^+. In contrast, Ag^+ is efficiently bound to 1,3-alternate-**11**($R_2 = H$) [23–25]. 1,3-Alternate-**11**($R_2 = H$) has two ionophoric cavities at the two sides of the cavity, each of which is composed of two ethereal oxygens and two benzene rings. It is now considered that the binding of Ag^+ is due to the "π-donor participation" characteristic of these cavities [24,25]. The foregoing results indicate that if one can regulate the equilibrium between the cone and 1,3-alternate by some switch function, a change in the metal-binding ability in calix[4]arenes is possible.

---

*For a comprehensive review of calixarene chemistry, please see [18].

It has been shown that calix[4]aryl esters **12** exhibit remarkably high selectivity toward Na^+ [26–30]. This is attributable to the inner size of the ionophoric cavity composed of four $OCH_2C=O$ groups, which is comparable to the ion size of Na^+ [18]. In the absence of guest metals the carbonyls are turned to the exo-annulus direction to reduce electrostatic repulsion whereas in the presence of guest metals they rotate to the endo-annulus direction to coordinate with the

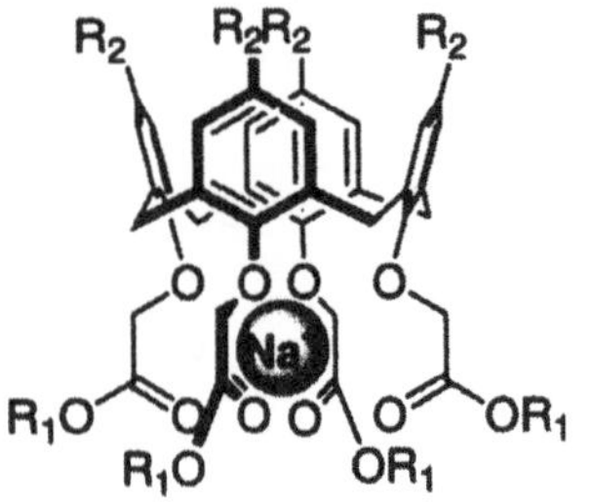

12 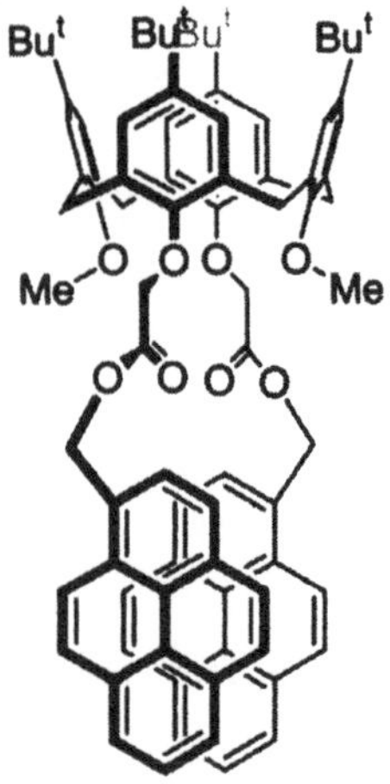   **13**

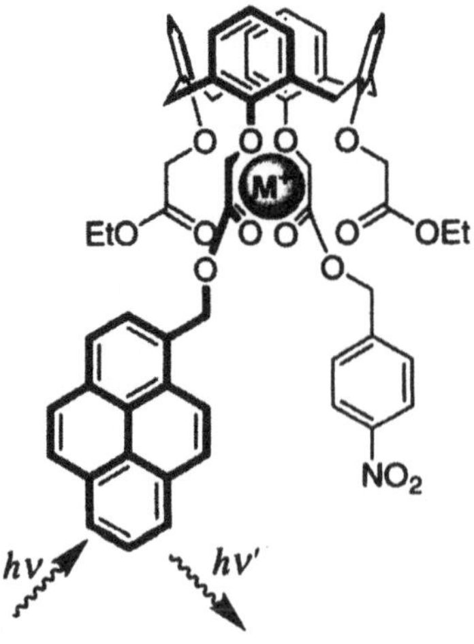

14

bound metal cation. The metal-induced molecular motion of the ester groups has enabled us to design a new fluorogenic calix[4]arene **13**: strong excimer emission (480 nm) was observed in the absence of metal cations because of an approach of two pyrene moieties and with increasing metal concentration (Li^+, Na^+, or K^+) monomer emission increased because of a separation of two pyrene moieties [31]. Thus, one can achieve fluorometric metal sensing over a wide pH range. A similar idea was also reported by Jin et al. [32] and Diamond et al. [33]. In **14** a pyrene fluorophore is combined intramolecularly with a *p*-nitrophenyl quencher [34]. As expected, fluorescence is efficiently quenched in the absence of metal cations while the quenching efficiency becomes low in the presence of Na^+ because of the separation of the fluorophore and the quencher [34].

Hydrogen-bonding interactions have been used extensively in the design of artificial molecular receptors. However, the artificial receptor bearing both hydrogen-bond donors and hydrogen-bond acceptors within a molecule inevitably tends to associate intramolecularly. To avoid such undesired intramolecular association, a hard segment is inserted between the donor and the acceptor so that the two sites cannot form intramolecular hydrogen bonds. This limitation frequently hampers the design of artificial receptors with a structure complementary to the guest molecule. We were thus stimulated to design new artificial receptors in which an "open" form active to the guest is generated from an intramolecularly hydrogen-bonded "closed" form only when it perceives a "stimulus." We already know that in calix[4]aryl esters and amides the four carbonyl groups are turned outward to reduce electrostatic repulsion among carbonyl oxygens whereas bond Na^+ changes the exo-annulus carbonyls to the endo-annulus carbonyls to trap a Na^+ ion [31–34]. We thus considered that the metal-induced structural change can be useful to generate an "open" form from a "closed" form. In chloroform:acetonitrile = 9:1 v/v compound **15** exists as a "closed"

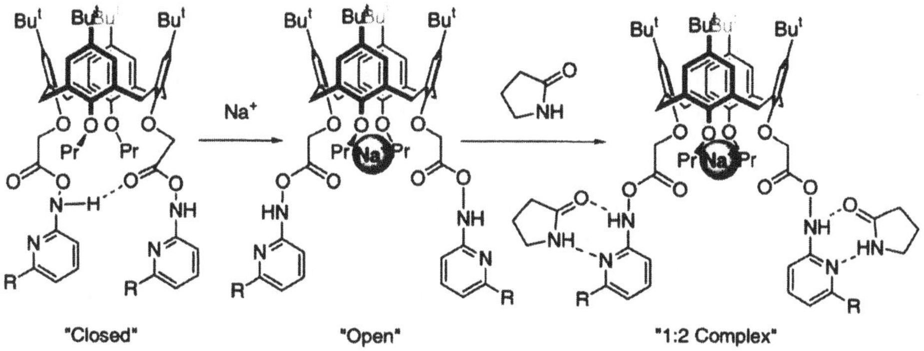

"Closed"   "Open"   "1:2 Complex"

15

form because of the formation of intramolecular hydrogen bonds and cannot bind its complementary guests (e.g., lactams) [35]. On the other hand, Na$^+$ bound to the ionophoric cavity cleaves the intramolecular hydrogen bonds and the exposed receptor sites can bind the guests through intermolecular hydrogen bonds.

The 2,6-diaminopyridine receptor sites in **16** are capable of binding guest molecules with a pteridine moiety. In the absence of metal cations **16** is "closed"

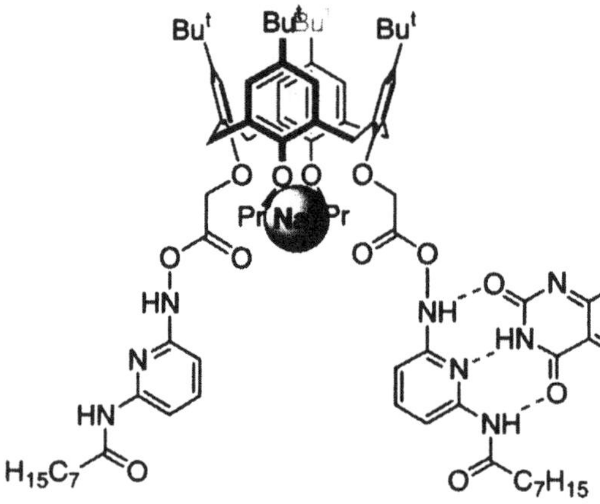

16

because of the intramolecular hydrogen bonds. However, bound Na$^+$ disrupts the intramolecular hydrogen bonds and the receptor sites are associated with the pteridine moiety of a flavin [36a]. Since the flavin is strongly fluorescent, the association process can be conveniently read out by a change in the fluorescence intensity [36a]. Barbituric acid derivatives have two C(=O)−NH−C(=O) binding sites complementary to the 2,6-diamidopyidine group. A 1:1 mixture of "closed" **16** and barbitric acid does not give any new peak in ^{1}H NMR spectroscopy, but when an equimolar amount of NaClO$_4$ is added, the spectral change is induced and the formation of polymeric aggregates is confirmed [36b]. The results indicate that one barbituric acid crosslinks two 2,6-diamidopyridine moieties in "open" **16** to form the polymers. This implies that monomer–polymer interconversion can be attained not only by light [16] but also by metal cations.

To realize a photoregulated ion-binding system in calixarenes we introduced two anthracenes near the metal-binding site of calix[4]arene [37–40]. Compound **17** having a podand-type cavity showed poor ion affinity whereas the photochemically produced isomer **18** with a dimeric anthracene-cap showed much improved ion affinity and sharp Na$^+$ selectivity [38,39]. Interestingly, **17** immobilized in the PVC membrane plasticized with di(2-ethylhexyl)sebacate underwent ring closure to **18** when it was photoirradiated at 381 nm [40]. The

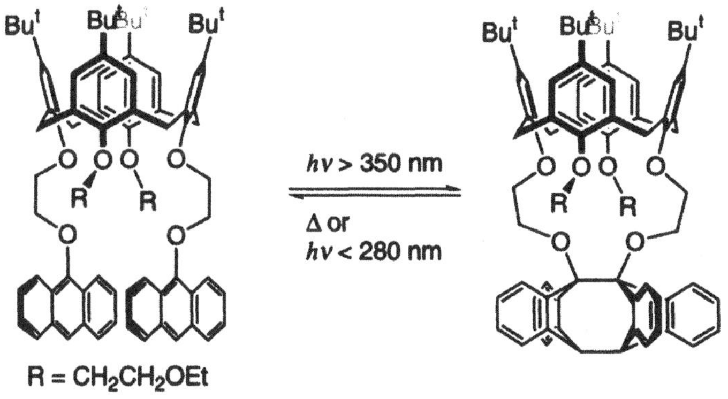

R = CH$_2$CH$_2$OEt

17   18

thermal **18**-to-**17** ring-opening reaction took place slowly [40]. Although the reverse reaction could be accelerated by photoirradiation at 279 nm, this caused serious photodecomposition. We found that in the presence of NaClO$_4$ the thermal reverse reaction was completely inhibited and could be induced only when the membrane was photoirradiated at 279 nm [40]. Added NaClO$_4$ efficiently suppressed the photodecomposition and the reversibility became quite excellent. The results indicate that this system satisfies both the thermal stability and the light stability required for photodevices; that is, photochemically written memories can be stored safely in the presence of NaClO$_4$ and erased by 279 nm irradiation.

IV. MOLECULAR DESIGN OF ARTIFICIAL SUGAR-SENSING SYSTEMS

An overview of past literature teaches us that hydrogen-bonding interactions are versatilely used for recognition of guest molecules.* We have currently been interested in sugar recognition and reading-out of the recognition process [42–45]. Although hydrogen-bonding interactions are also useful for sugar recognition in several systems [46–48], the effect is exerted only in aprotic organic solvents. Hence, hydrogen-bonding interactions are useless for sugar recognition in water while sugars show the significant solubility only in water. Then, how can we "touch" sugars and "recognize" them in water? In an attempt to solve this dilemma we have proposed using a boronic acid that self-associatively forms covalent complexes with a variety of sugar molecules in water (Scheme 1) [42–45,49–51]. Since this covalent-bond formation process is reversible and the rate

---

*For comprehensive reviews, please see [41].

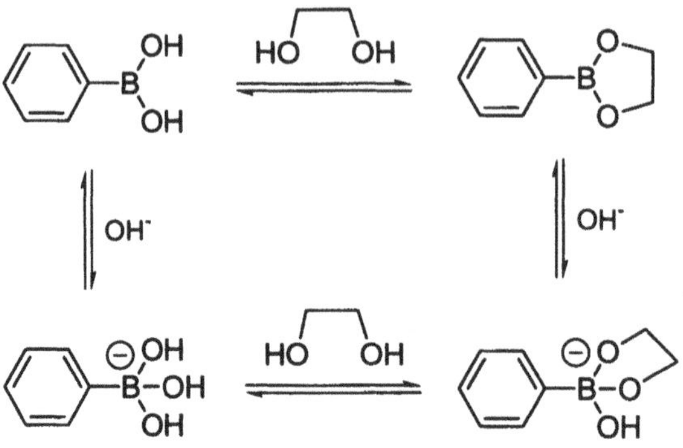

Scheme 1

is much faster than the human time scale, one can treat this system as noncovalent interactions frequently used for molecular recognition. Although this strategy is quite different from that employed by Nature (using hydrogen-bonding interactions) [41,46–48], this is undoubtedly a practical (and probably the sole) way to "touch" sugars in water.

Fluorescence sensors for saccharides are of particular interest in a practical sense. This is in part due to the inherent sensitivity of the fluorescence technique. Only small amounts of a sensor are required (typically 10^{-6} M), offsetting the synthetic costs of such sensors. Also, fluorescence spectrometers are widely available and inexpensive. Fluorescence sensors have also found applications in continuous monitoring using an optical fiber and intracellular mapping using confocal microscopy.

A. PET Receptors

Photoinduced electron transfer (PET) has been widely used as the preferred tool in fluorescent sensor design for atomic and molecular species [52–57]. PET sensors generally consist of a fluorophore and a receptor linked by a short spacer. The changes in the oxidation/reduction potential of the receptor upon guest binding can alter the PET process creating changes in fluorescence.

The first fluorescence PET sensors for saccharides were based on fluorophore boronic acids. Czarnik and Yoon showed that 2- and 9-anthryboronic acid [50] **19** and **20** could be used to detect saccharides. However, the fluorescence change was small [I (in the presence of saccharide)/I_o (in the absence of saccharide) = ca. 0.7]. The pK_a of the fluorophore boronic acids are shifted by saccharide present in the medium. The extent of the effect is in line with the inherent selectivity of phenylboronic acid [49]. The PET from the boronate anion is believed to be the source of the fluorescence quenching. Although

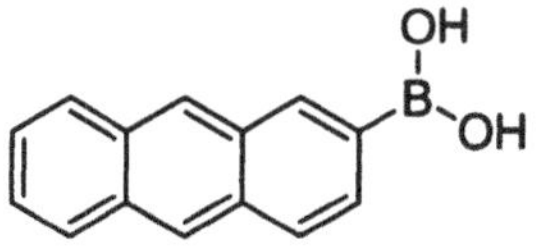

19 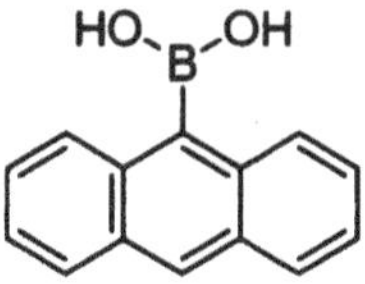   **20**

2-anthrylboronic acid displays only a small fluorescence change, eight aromatic boronic acids have been screened [58,59] and it was determined that **21** and **22** are more suitable candidates for saccharide detection. Aoyama et al. have also shown that 5-indolylboronic acid **23** undergoes fluorescence quenching upon

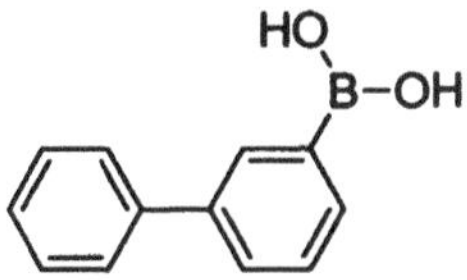

21 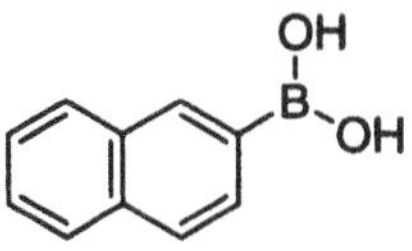   **22** 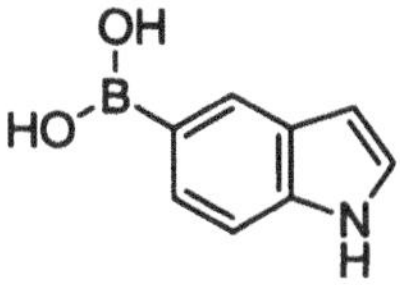   **23**

complexation with oligosaccharides [60]. In the determination of the pK_a of the boronic acid, ^{11}B NMR shifts were consistent with the fluorescence studies. The stability constants of monosaccharides were similar to the inherent selectivity of phenylboronic acid [49]. Oligosaccharides, on the other hand, gave relatively lower stability constants, although higher oligomers showed an increased stabilization relative to lower oligomers due to a secondary interaction with the indole N—H.

With the system outlined above facile boronic acid saccharide complexation only occurs at the high pH required to create a boronate anion. To overcome these disadvantages molecular fluorescence sensors that contain a boronic acid group and an amine group were developed. The boronic acid group is required to bind with and capture sugar molecules in water. The amine group plays two roles in the system: for biological systems the physiological pH is neutral, and boronic acids with a neighboring amine can bind with sugars at neutral pH, but simple boronic acids can only bind with sugars at a high pH; and the fluorescence intensity is controlled by the amine. With no sugar the "free" amine reduces the intensity of the fluorescence (quenching by photoinduced electron transfer). This is the "off" state of the fluorescent sensor. When sugar is added, the amine becomes "bound" to the boron center. The boron-bound amine cannot quench the fluorescence and hence a strong fluorescence is observed. This is the "on" state of the fluorescent sensor. The system described above illustrates the basic concept of an "off-on" fluorescent sensor for sugars useful in the neutral pH region.

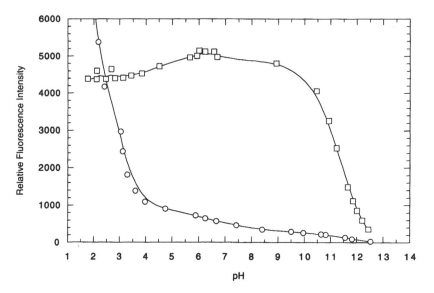

Figure 5   Fluorescence intensity versus pH profile of 1.2×10^{-5} M of **24** in 0.05 M NaCl at 25°C, λ_{ex} 370 nm, λ_{em} 423 nm. (□) 0.05 M [D-glucose], (○) blank.

The first of these fluorescent PET "off-on" saccharide sensors was prepared in 1994 [61,62]. The very large pK_a shift found upon saccharide binding provides a wide pH range for saccharide sensing (Fig. 5). The large shift of the pK_a is due to the interaction found between the boronic acid moiety and the amine group. The boronic acid–amine interaction inhibits the photoinduced electron transfer quenching process in the complex **24b** (Scheme 2). Complete separation of the amine and the acid moiety at very high pH, as in **24c**, quenched the anthracene fluorescence further. However, the fluorescence decrease is not sufficient for the calculation of the pK_a. The introduction of D-glucose remarkably enhances the fluorescence of **24** over a large pH range (Fig. 5). The enhanced interaction between boronic acid and amine, upon saccharide binding, inhibits the electron transfer process giving higher fluorescence (as **24d** in Scheme 2). This

24

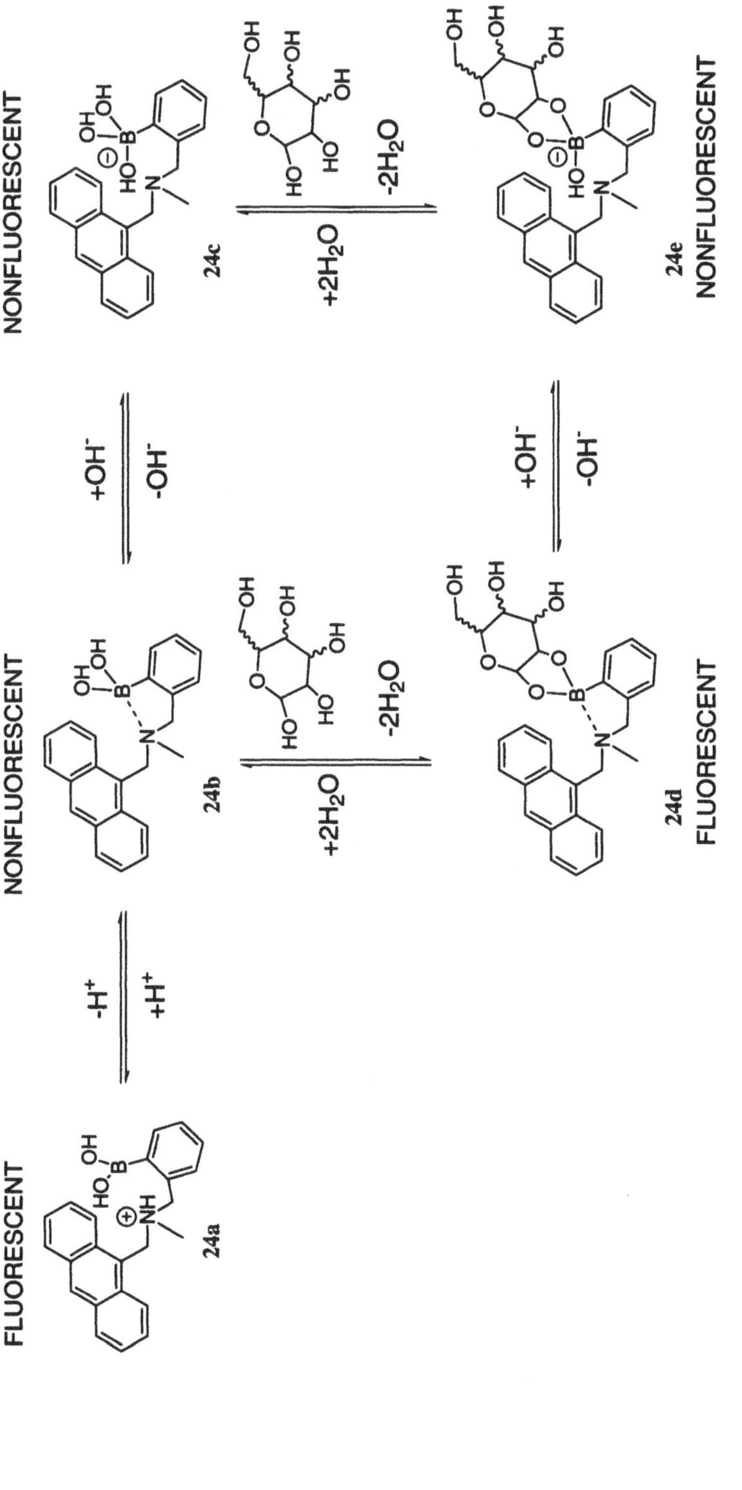

Scheme 2

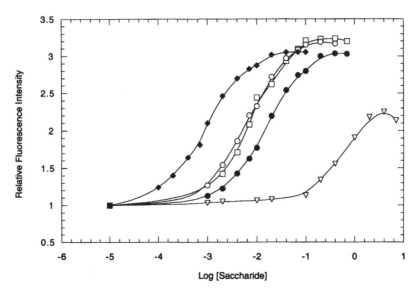

Figure 6 Fluorescence intensity versus log [saccharide or ethylene glycol] profile of
1.0×10^{-5} M of **24** in 33.3% MeOH/H$_2$O pH 7.77 buffer at 25°C, λ_{ex} 370 nm, λ_{em}
423 nm. (◆) D-fructose, (○) D-galactose, (□) D-allose, (●) D-glucose, (▽) ethylene
glycol.

increased interaction would be expected since the saccharide binding to boronic
acid increases its acidity creating a more electron-deficient boron atomic center.
This simple monoboronic acid system **24** shows a selectivity order, which is
inherent to all monoboronic acids (Fig. 6) [49].

The simple "off-on" PET system was improved with the introduction of
a second boronic acid group **25** [62]. With compound **25** similar equilibria

25

to those observed with **14** exist. However, for clarity, only species arising from neutral **25** and saccharide are shown (Scheme 3). For compound **25** two possible saccharide binding modes can inhibit the electron transfer process thus giving higher fluorescence: the 2:1 complex **25a** and 1:1 complex **25b**. Due to fortuitous spacing of the boronic acid groups the diboronic acid was selective for D-glucose over other monosaccharides (Fig. 7).

Norrild et al. have carried out a more detailed investigation of this system in order to confirm the structure of the bound glucose [63]. Norrild and coworkers were interested in the system since the ^{1}H NMR report [62,64] indicated that D-glucose bound to the receptor in its pyranose form. Norrild and Eggert had previously shown that simple boronic acids selectively bind with the furanose form of D-glucose [65]. From ^{1}H NMR observations it was concluded that the diboronic acid initially binds with the pyranose form of D-glucose and over time the bound glucose converts to the furanose form.

Work by Irie et al. on the control of intermolecular chiral 1,1'-binapthyl fluorescence quenching by chiral amines [66] and the use of 1,1'-binaphthyl in the recognition of chiral amines by Cram [67] were the inspiration behind the design of **26** (*R* or *S*). Chiral recognition of saccharides by **26** (*R* or *S*)

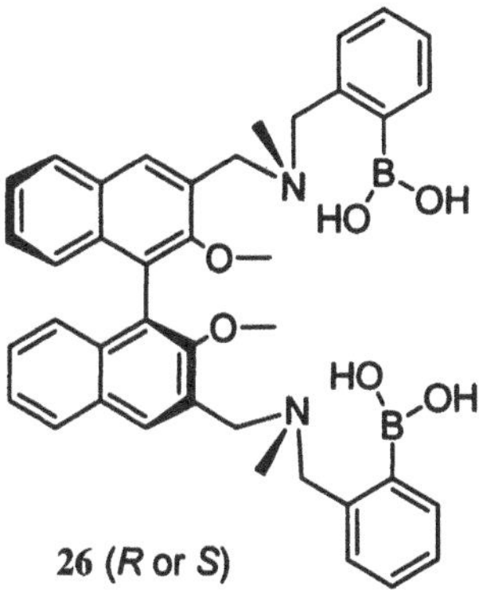

26 (*R* or *S*)

utilizes both steric and electronic factors [68]. The asymmetric immobilization of the amine groups relative to the binaphthyl moiety upon 1:1 complexation of saccharides by D- or L-isomers creates a difference in PET. This difference is manifested in the maximum fluorescence intensity of the complex. Steric factors arising from the chiral binaphthyl building block are chiefly represented by the stability constant of the complex. However, the interdependency of electronic and steric factors upon each other is not excluded. This new molecular cleft, with a longer spacer unit compared to the anthracene-based diboronic acid **25**, gave the best recognition for fructose. Table 1 shows the binding constants for some D- and L-monosaccharides. Discriminative detection of isomers in aqueous media

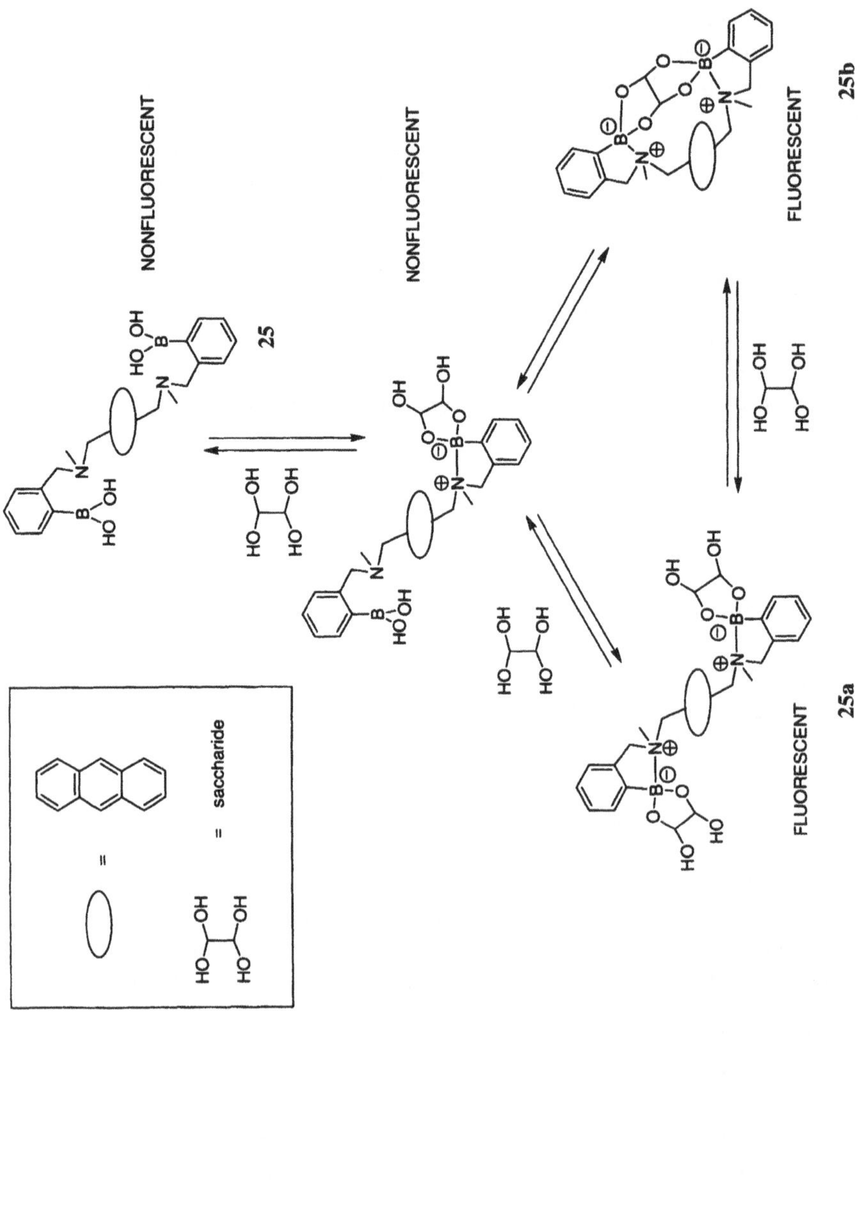

Scheme 3

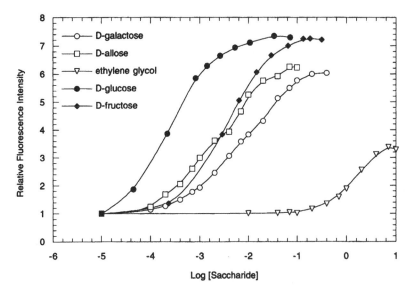

Figure 7 Fluorescence intensity versus log [saccharide or ethylene glycol] profile of 1.0×10^{-5} M of **25** in 33.3% MeOH/H$_2$O pH 7.77 buffer at 25°C, λ_{ex} 370 nm, λ_{em} 423 nm. (◆) D-fructose, (○) D-galactose, (□) D-allose, (●) D-glucose, (▽) ethylene glycol.

by fluorescence, as far as we are aware, had not been achieved before. In this system steric factors and electronic factors bimodally discriminate the chirality of the saccharide. Competitive studies with D- and L-monosaccharides show the possibility of selective detection of saccharide isomers. The availability of both R and S isomers of this particular molecular sensor is an important advantage, since concomitant detection by two probes is possible.

Table 1 Stability Constants and Fluorescence Enhancements[a]

Saccharide	D log K (±0.05)	L log K (±0.05)	D/L fluorescence intensity ratio
Fructose	4.0 (3.7)	3.5 (4.0)	1.47 (0.69)
Glucose	3.3 (3.4)	3.1 (3.5)	1.93 (0.53)
Galactose	3.1	3.3	0.82
Mannose	<2.4	—	—

[a] **17**R (or S) in pH 7.77 (0.05 mol dm^{-3} phosphate buffer, 33% methanolic aqueous solution).

Diboronic acid **27** with a small bite angle has been synthesized and has been shown to be selective for small saccharides such as D-sorbitol [69]. Conversely, diboronic acid **28** with larger spacing of the boronic acid groups loses

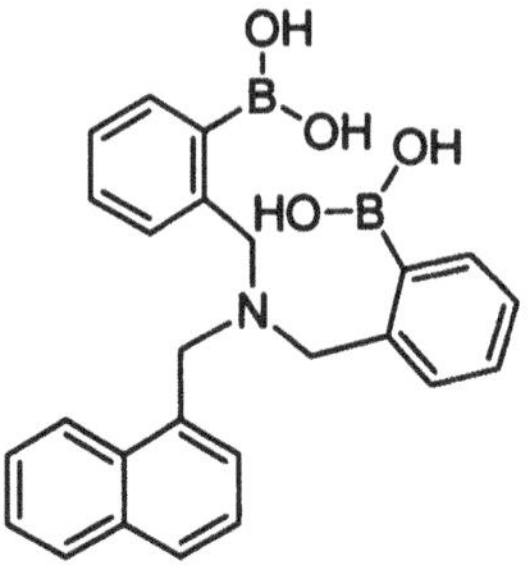

27 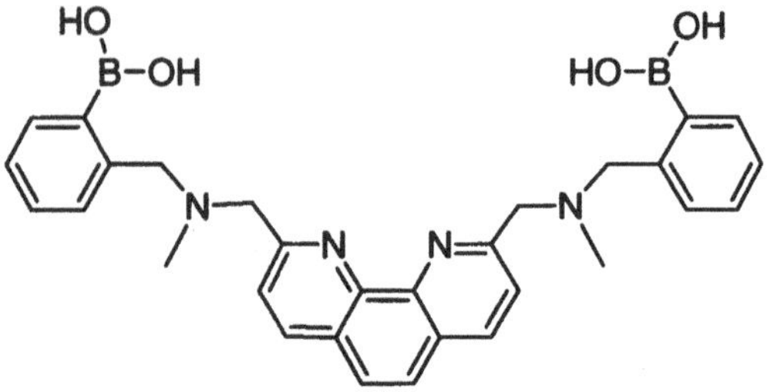   28

selectivity and sensitivity [70]. However, the system may be useful for the detection of saccharides in concentrated solutions, such as those encountered in the brewery and confectionery industries. An allosteric diboronic acid **29** has

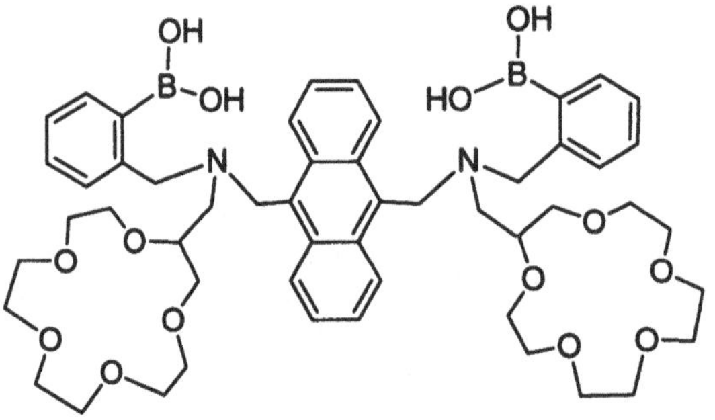

29

been prepared where formation of a metal crown sandwich causes the release of bound saccharide (Scheme 4) [71]. The allosteric concept was also applied in the two-dimensional PET sensor **30**. With the two-dimensional system, the amount of excimer can be directly correlated with the amount of noncyclic saccharide complex formed [72].

More recently the boronic acid PET system has been used in combination with other binding sites. The D-glucosamine selective fluorescent systems **31** and **32** based on a boronic acid and azacrown ether have been explored [73,74]. Sensors **31** and **32** consist of monoaza-18-crown-6 ether or monoaza-15-crown-5 as a binding site for the ammonium terminal of D-glucosamine hydrochloride, while a boronic acid serves as a binding site for the diol (carbohydrate) part

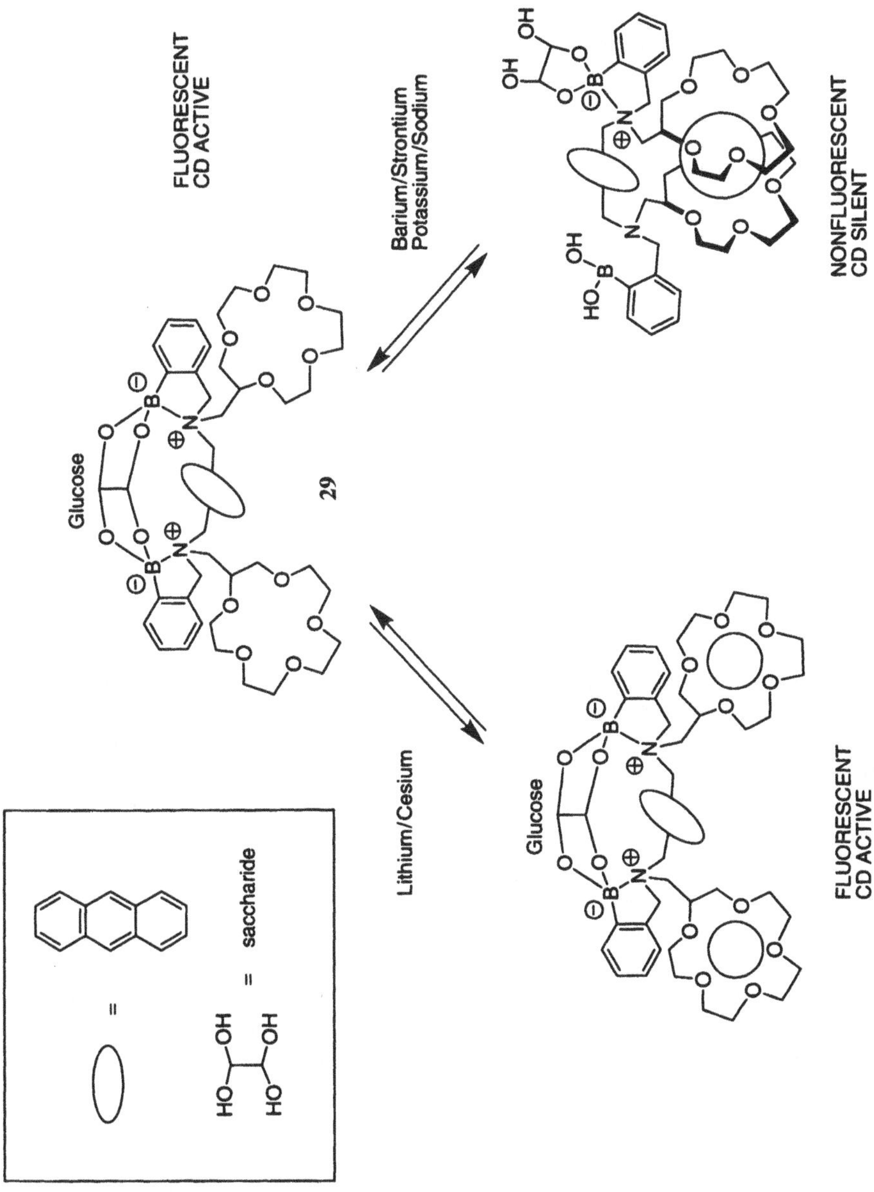
FLUORESCENT
CD ACTIVE
Glucose
29
Barium/Strontium
Potassium/Sodium
NONFLUORESCENT
CD SILENT
Lithium/Cesium
FLUORESCENT
CD ACTIVE
saccharide

Scheme 4

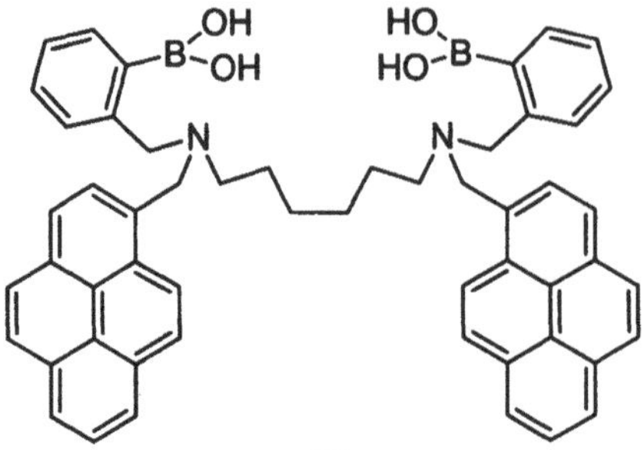

30

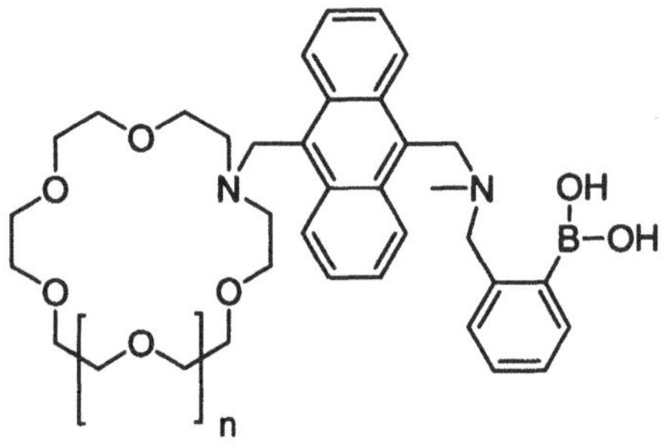

31 (n=0) and 32 (n=1)

of D-glucosamine hydrochloride. The nitrogen of the azacrown ether unit can participate in PET with the anthracene fluorophore; ammonium ion binding can then cause fluorescence recovery. This recovery is due to hydrogen bonding from the ammonium ion to the nitrogen of the azacrown ether. The strength of this hydrogen-bonding interaction modulates the PET from the amine to anthracene. As explained above, the boronic acid unit can also participate in PET with the anthracene fluorophore, and diol binding can also cause fluorescence recovery. The anthracene unit serves as a rigid spacer between the two receptor units, with the appropriate spacing for the glucose moiety. This system behaves like an **AND** logic gate [75,76], in that fluorescence recovery is only observed when two chemical inputs are supplied, for this system the two chemical inputs are an ammonium cation and a diol group. Compounds **33** and **34** do not display any fluorescence enhancement with D-glucose because they have no saccharide-binding site. As expected, compound **24** shows fluorescence enhancement with D-glucose ($K = 67 \pm 3$ M^{-1}) and D-glucosamine hydrochloride ($K = 18 \pm 1$ M^{-1}). With D-glucose the boronic acid has a choice of binding either the 1,2- or 4,6- diols, but with D-glucosamine hydrochloride binding with just the 4,6- diol is possible. The stability constant of **24** with D-glucose is higher than

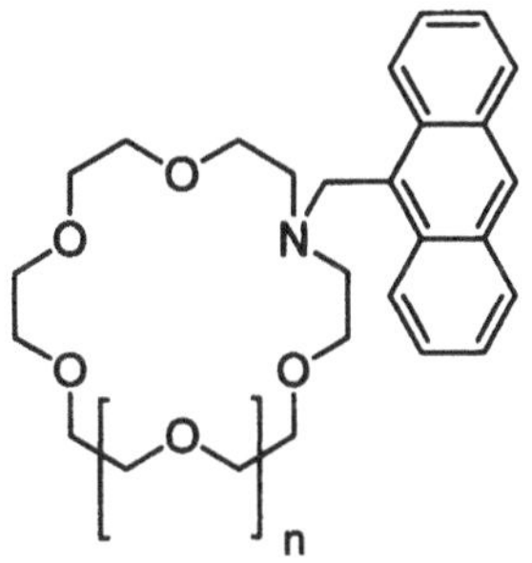

33 (n=0) and 34 (n=1)

that observed with D-glucosamine hydrochloride reflecting the known selectivity of boronic acids for the 1,2-diol of D-glucose. Compounds **31** and **32** show fluorescence increase with D-glucosamine hydrochloride ($K = 18 \pm 2$ M^{-1} and 17 ± 2 M^{-1}, resp.), but no increase with D-glucose. This result clearly demonstrates that for a fluorescent output both a diol and ammonium group must be present in the guest.

In conclusion, the recognition of saccharides by boronic-acid-based molecular receptors has shown tremendous growth during the last few years: from inherent saccharide selectivity with monoboronic acids and controlled selectivity with simple diboronic acids through to the chiral recognition of saccharides. The biggest breakthrough in this study was a combination of the PET sensor concept with the boronic-acid sugar-binding, which has enabled us to solve two difficult problems at one time, sugar-binding at neutral pH region and reading-out of the sugar-binding process.

V. CONCLUSION

This chapter describes two discrete yet related concepts: the control of molecular recognition events using light (or metal ion), and the control of light using molecular recognition events. Coupling molecular recognition events in this way has made it possible to escape the constraints imposed by the enthalpy–entropy compensation relationship. We believe that achievements in molecular design outlined here have been made possible by the accumulation of knowledge in the area of dynamic molecular recognition systems. It is hoped that this collection of our modest achievements will encourage others to explore this new and exciting area of research.

> The World little knows how many thoughts and theories which have passed through the mind of a scientific investigator and have been crushed in silence and secrecy of his own criticism. —Michael Faraday (1791–1867)

REFERENCES

1. JE Leffler, E Grunwald. Rates and Equilibria of Organic Reactions. New York: Wiley, 1963.
2. S Shinkai, T Ogawa, T Nakajima, Y Kusano, O Manabe. Tetrahedron Lett 20:4569, 1979.
3. S Shinkai, T Nakaji, Y Nishida, T Ogawa, O Manabe. J Am Chem Soc 102:5860, 1980.
4. HL Ammon, SK Bhattacharjee, S Shinkai, Y Honda. J Am Chem Soc 106:262, 1984.
5. J-P Desvergne, H Bouas-Laurent. J Chem Soc Chem Commun 403, 1978.
6. H Bouas-Laurent, A Castellan, J-P Desvergne. Pure Appl Chem 52:2633, 1980.
7. H Bouas-Laurent, A Castellan, M Daney, J-P Desvergne, G Guiand, P Marsau, M-H Riffaud. J Am Chem Soc 108:315, 1986.
8. F Fages, J-P Desvergne, H Bouas-Laurent, J-M Lehn, JP Konopelski, P Marsau, Y Barrans. J Chem Soc Chem Commun 655, 1990.
9. I Yamashita, M Fujii, T Kaneda, S Misumi, T Otsubo. Tetrahedron Lett 21:541, 1980.
10. K Kimura, H Tamura, T Tsuchida, T Shono. Chem Lett 611, 1979.
11. S Shinkai, O Manabe. Top Curr Chem 121:67, 1984.
12. S Shinkai, T Ogawa, Y Kusano, O Manabe. Chem Lett 283, 1980.
13. S Shinkai, T Nakaji, T Ogawa, K Shigematsu, O Manabe. J Am Chem Soc 103:111, 1981.
14. S Shinkai, K Shigematsu, Y Kusano, O Manabe. J Chem Soc Perkin Trans 1:3279, 1981.
15. S Shinkai, T Ogawa, Y Kusano, O Manabe, K Kikukawa, T Goto, T Matsuda. J Am Chem Soc 104:1960, 1982.
16. S Shinkai, T Yoshida, O Manabe, F Fuchita. J Chem Soc Perkins Trans 1:1431, 1988.
17. S Shinkai, K Shigematsu, M Sato, O Manabe. J Chem Soc Perkins Trans 1:2735, 1982.
18. CD Gutsche. Calixarenes. Cambridge: Royal Society of Chemistry, 1989.
19. K Iwamoto, K Araki, S Shinkai. J Org Chem 56:4955, 1991.
20. S Shinkai, K Iwamoto, K Araki, T Matsuda. Chem Lett 1263, 1990.
21. PDJ Grootenhuis, PA Kollman, LC Groenen, DN Reinhoudt, GJ van Hummel, F Ugozzoli, GD Andreetti. J Am Chem Soc 122:4165, 1990.
22. T Harada, JM Rudzinski, S Shinkai. J Chem Soc Perkin Trans 2:2109, 1990.
23. K Iwamoto, A Ikeda, K Araki, T Harada, S Shinkai. Tetrahedron 49:9937, 1993.
24. A Ikeda, S Shinkai. Tetrahedron Lett 33:7385, 1992.
25. A Ikeda, S Shinkai. J Am Chem Soc 116:3102, 1994.
26. A Arduini, A Pochini, S Reverberi, R Ungaro. Tetrahedron 42:2089, 1986.
27. S-K Chang, I Cho. J Chem Soc Perkins Trans 1:211, 1986.
28. F Arnard-Neu, EM Collins, M Deasy, G Ferguson, SJ Harris, B Kaitner, AJ Lough, MA McKervey, E Marques, BL Ruhl, MJ Schwing-Weill, EM Seward. J Am Chem Soc 111:8681, 1989.

29. T Arimura, M Kubota, T Matsuda, O Manabe, S Shinkai. Bull Chem Soc Jpn 62:1674, 1989.
30. K Iwamoto, S Shinkai. J Org Chem 57:7066, 1992.
31. I Aoki, Y Kawahara, K Nakashima, S Shinkai. J Chem Soc Chem Commun 1771, 1991.
32. T Jin, K Ichikwa, T Koyama. J Chem Soc Chem Commun 499, 1992.
33. C Perez-Jimenez, SJ Harris, D Diamond. J Chem Soc Chem Commun 480, 1993.
34. I Aoki, T Sakaki, S Shinkai. J Chem Soc Chem Commun 730, 1992.
35. H Murakami, S Shinkai. Tetrahedron Lett 34:4237, 1993.
36. (a) H Murakami, S Shinkai. J Chem Soc Chem Commun 1533, 1993; (b) P Lhotak, S Shinkai. Tetrahedron Lett 36:4829, 1995.
37. G Deng, T Sakaki, Y Kawahara, S Shinkai. Tetrahedron Lett 33:2163, 1992.
38. G Deng, T Sakaki, K Nakashima, S Shinkai. Chem Lett 1287, 1992.
39. G Deng, T Sakaki, Y Kawahara, S Shinkai. Supramol Chem 2:71, 1993.
40. G Deng, T Sakaki, S Shinkai. J Polym Sci Polym Chem 31:1915, 1993.
41. (a) J Rebek, Jr. Angew Chem Int Ed Engl 29:245, 1990; (b) AD Hamilton. Bioorg Chem Front 2:115, 1991.
42. (a) K Tsukagoshi, S Shinkai. J Org Chem 56:4089, 1991; (b) Y Shiomi, M Saisho, K Tsukagoshi, S Shinkai. J Chem Soc Perkin Trans 1:2111, 1993.
43. S Shinkai, K Tsukagoshi, Y Ishikawa, T Kunitake. J Chem Soc Chem Commun 1039, 1991.
44. K Kondo, Y Shiomi, M Saisho, T Harada, S Shinkai. Tetrahedron 48:8239, 1992.
45. TD James, T Harada, S Shinkai. J Chem Soc Chem Commun 857, 1993.
46. K Kano, K Yoshiyasu, S Hashimoto. J Chem Soc Chem Commun 801, 1988.
47. Y Aoyama, Y Tanaka, H Toi, H Ogoshi. J Am Chem Soc 110:634, 1998.
48. Y Kikuchi, K Kobayashi, Y Aoyama. J Am Chem Soc 114:1351, 1992.
49. JP Lorand, JO Edwards. J Org Chem 24:769, 1959.
50. J Yoon, AW Czarnik. J Am Chem Cos 114:5874, 1992.
51. LK Mohler, AW Czarnik, J Am Chem Soc 115:2998, 1993.
52. AP de Silva, T Gunnlaugsson, TE Rice. Analyst (London) 121:1759, 1996.
53. AP de Silva, HQN Gunnaratne, T Gunnlaugsson, PLM Lynch. New J Chem 20:871, 1996.
54. AP de Silva, HQN Gunaratne, T Gunnlaugsson, CP McCoy, PRS Maxwell, JT Rademacher, TE Rice. Pure Appl Chem 68:1443, 1996.
55. AP de Silva, HQN Gunaratne, CP McCoy. J Am Chem Soc 119:7891, 1997.
56. AP De Silva, HQN Gunaratne, T Gunnlaugsson, AJM Huxley, CP McCoy, JT Rademacher, TE Rice. Chem Rev 97:1515, 1997.
57. AP de Silva, T Gunnlaugsson, CP McCoy. J Chem Ed 74:53, 1997.
58. H Suenaga, M Mikami, KRAS Sandanayake, S Shinkai. Tetrahedron Lett 36:4825, 1995.
59. H Suenaga, H Yamamoto, S Shinkai. Pure Appl Chem 68:2179, 1996.
60. Y Nagai, K Kobayashi, H Toi, Y Aoyama. Bull Chem Soc Jpn 66:2965, 1993.
61. TD James, K Sandanayake, S Shinkai. J Chem Soc Chem Commun 477, 1994.
62. TD James, KRAS Sandanayake, R Iguchi, S Shinkai. J Am Chem Soc 117:8992, 1995.

63. M Bielecki, H Eggert, JC Norrild. J Chem Soc Perkin Trans 2:449, 1999.
64. TD James, K Sandanayake, S Shinkai. Angew Chem Int Ed Engl 33:2207, 1994.
65. JC Norrild, H Eggert. J Am Chem Soc 117:1479, 1995.
66. M Irie, T Yorozu, K Hayashi. J Am Chem Soc 100:2236, 1978.
67. DJ Cram. Angew Chem Int Ed Engl 25:1039, 1986.
68. TD James, KRAS Sandanayake, S Shinkai. Nature (London) 374:345, 1995.
69. TD James, H Shinmori, S Shinkai. Chem Commun 71, 1997.
70. P Linnane, TD James, S Imazu, S Shinkai. Tetrahedron Lett 36:8833, 1995.
71. TD James, S Shinkai. J Chem Soc Chem Commun 1483, 1995.
72. KRAS Sandanayake, TD James, S Shinkai. Chem Lett 503, 1995.
73. CR Cooper, TD James. Chem Commun 1419, 1997.
74. CR Cooper, TD James. J Chem Soc Perkin Trans 1:963, 2000.
75. AP de Silva, HQN Gunaratne, CO McCoy. Nature (London) 364:42, 1993.
76. S Iwata, K Tanaka. J Chem Soc Chem Commun 1491, 1995.

11

Probing Nanoenvironments Using Functional Chromophores

Mitsuru Ishikawa and Jing Yong Ye

Joint Research Center for Atom Technology (JRCAT) and
Angstrom Technology Partnership (ATP), Ibaraki, Japan

I. INTRODUCTION

One of the important characteristics of organic molecules is the flexibility of structural frameworks in contrast to metals and semiconductors. In this chapter we discuss the use of flexible organic molecules as molecular probes for nanoenvironments based on detecting fluorescence in the visible spectral region, expecting such molecules to probe nanostructures of condensed matter including polymers and proteins. We consider conjugated molecules whose size is larger than that of benzene but smaller than polymers, and thus refer to them as chromophores. The flexibility, which is closely related to the electronic configurations, will be enhanced especially when such molecules are in the electronic excited states. The intrinsic flexibility of molecular structures and electronic configurations, plus the use of standard visible fluorometry, make it possible and easy to use such molecules as sensitive probes of the immediate local environments around themselves. Especially when such molecules are in the single molecule regime the ability will be exhibited considerably. We will first focus on triphenyl methane (TPM) dyes as a representative for flexible conjugated molecules [1]. We will then discuss two kinds of chromophores other than TPM dyes.

The structure of TPM dyes is characterized by a three-blade propeller-like phenyl rings joined by the central carbon atom (Fig. 1). The photochem-

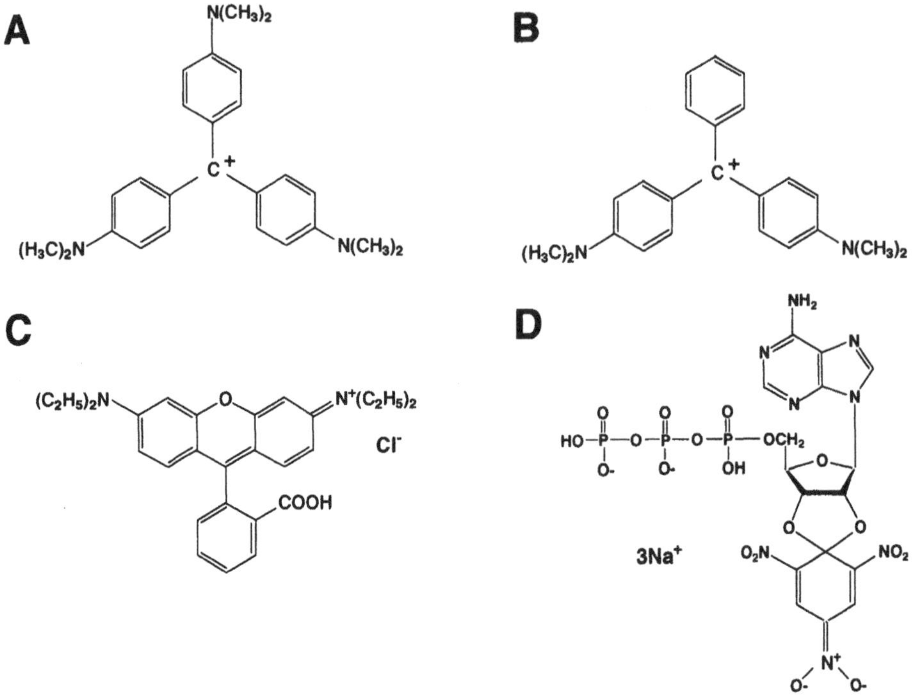

Figure 1 Molecular structures of the chromophores involved in the current work: (A) crystal violet (CV), (B) malachite green (MG), (C) rhodamine B (rhB), and (D) 2′-(or-3′)-O-(2, 4, 6-trinitrophenyl) adenosine 5′-triphosphate (TNP-ATP).

istry and photophysics of TPM dyes in solid and liquid media are described in Ref. [2]. Flexibility of TPM dyes comes from this peculiar structure. Some physical chemists who engage in time-resolved spectroscopy have long been interested in the relaxation mechanisms of the excited states of TPM dyes. Indeed, TPM dyes were studied again and again whenever ultrashort-pulse lasers were newly developed from the picosecond [3] to femtosecond era [4], because the relaxation times of TPM dyes are conveniently adjustable to fit the available pulse width of the lasers used simply by changing solvent viscosity. The fluorescence lifetime (τ_f) of TPM dyes depends strongly on the solvent viscosity from several picoseconds in methanol (0.6 cP) [5–8] at 293 K to ~4.0 ns in alcohols and polymers below 100 K [9–12]. This extreme sensitivity of τ_f to viscosity is the very reason for the use of TPM dyes as molecular probes for viscosity in condensed matter such as liquid, polymers, and glass. Also, oscillating wavelengths of available ultrashort-pulse lasers were suitably fitted with the absorption maximums of TPM dyes (Fig. 2). Recently, we implemented

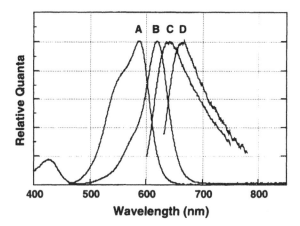

Figure 2   Absorption spectrum of CV (A) and of MG (B) in methanol at 295 K. Fluorescence spectrum of CV (C) and of MG (D) in glycerol at 295 K. (From Ref. 1.)

imaging and spectroscopy of single dye molecules on the basis of fluorescence measurement [13–18] and found a single TPM dye molecule to be useful as a probe to the local viscosity of polymer matrices in which it is embedded [15]. In this chapter we first follow our route to a single-molecule study of a TPM dye starting from femto- and picosecond spectroscopies of TPM dyes in solution. We then present single-molecule studies probing silicon surfaces and enzyme molecules using two kinds of chromophores other than TPM dyes. In the last two studies the role of the flexibility in the chromophores selected is not straightforward compared with TPM dyes; however, these chromophores well exhibited their capabilities as single-molecule proves on the basis of change in the spectrums and lifetimes.

The chapter is divided into five sections. In Sec. III, we discuss how the molecular structure of crystal violet (CV), which is one of TPM dyes selected, is sensitive to its surrounding solvent molecules in solution. Recently, subtle structural differences in the ground state of CV was identified in alcohols by means of a femtosecond spectral hole-burning technique [19–21]. Two key points of these studies are (1) proposal of a novel class of isomers, which are differentiated from one another by solvation and (2) conclusive decision of a long-run dispute over 50 years on whether the ground-state conformational isomers exist.

In Sec. IV we discuss another TPM dye, malachite green (MG), which was used as a molecular probe for glass transition of alcohols and polymers [11,12]. Analysis of the temperature dependence of nonradiative relaxation in MG shed light on the understanding of the mechanism of glass transition. Novel experimental observations are divided into two classes. (1) The critical temperature (T_c) predicted by the mode-coupling theory (MCT) was undoubtedly observed 30–50 K above the calorimetric glass transition temperature (T_g) for monomer alcohols and a polymer without side chains. In this class of alcohols and polymers fluorescence decay curves were expressed using a biexponential function above $T_g + 30$ K, while with a single exponential function below T_g. (2) In polymers having side chains T_c is not found above T_g. However, another transition temperature was observed below T_g in contrast to the monomers and the polymer without side chains. In this class of alcohols and polymers fluorescence decay curves were expressed using a biexponential function both above and below T_g.

In Sec. V we extend the second study to the single-molecule regime using CV as a single-molecule probe [15]. From the observation of biexponential decays of MG fluorescence above and below T_g [12], we assume that polymers with side chains have two sites: one is a liquid-like site which causes fast fluorescence decays and the other is a solid-like site which causes slow fluorescence decays. We succeeded in identifying the above assumption at room temperature (23°C) by examining fluorescence properties of individual CV molecules embedded in poly (methylmethacrylate) (PMMA), for which $T_g = 114$°C [22].

In Sec. VI we examine Si surfaces, focusing our attention on the property of affecting fluorescence characteristics of target molecules adsorbed on the surfaces [16]. The purpose of this work is to explore the possibilities of preparing target molecules at the single-molecule level on Si surfaces for chemical analysis based on fluorescence measurement. The use of Si surfaces is motivated by the fact that Si wafers are widely used as substrates for micro total analysis systems (μ-TAS), where structures in the micrometer regime are defined chemically or photolithographically. A fluorescence single-molecule imaging technique plus time-resolved fluorometry based on far-field light microscopy was used for evaluating position-sensitive fluorescence intensities, spectra, and lifetimes of rhodamine B (rhB) molecules (Fig. 1C) adsorbed on a Si surface covered with an oxide layer. Here, we found a useful guideline for future applications of Si surfaces as plates on which single analyte molecules are prepared.

In Sec. VII we extend the single molecule fluorescence imaging and time-resolved fluorometry from the green to the violet-excitation regime. In fact, previous single-molecule studies were exclusively limited to green excitation and orange-to-red fluorescence measurement [13–17]. Using violet excitation, we observed fluorescent spots from single complexes composed of a nucleotide analogue and the Klenow fragment of DNA polymerase I [18]. We selected 2'-(or 3')-O-(2, 4, 6-trinitrophenyl) adenosine 5'-triphosphate (TNP-TP, Fig. 1D) as a nucleotide analogue. We estimated the fluorescence efficiency (~ 0.24) of TNP-ATP bound to a Klenow fragment from the lifetime (~ 1.2 ns) [23] and a possible value of the radiative lifetime ($\sim 2 \times 10^8$ s^{-1}) evaluated from the absorption maximum. Thus, the fluorescence efficiency increases by a factor of ~ 600 compared with the efficiency in water (0.0002) [24]. This enhancement makes it possible to observe single Klenow fragment-TNP-ATP complexes without separating unbound TNP-ATP molecules before the measurement. TNP-ATP has widely been used in the research of ATP turnover in biochemistry and biophysics. The use of TNP in the single-molecule regime will provide a new possibility in the research involving ATP turnover.

II. EXPERIMENTAL SECTION

A. Light Sources

First, we classified the lasers used in our research in the following manner:

Laser I: A femtosecond dye laser system composed of a femtosecond oscillator and an amplifier, generating tunable (550–640 nm) pulses in 10-Hz and 400-fs fwhm.

Laser II: A femtosecond mode-locked dye laser (Coherent, Satori) synchronously pumped using a cw mode-locked and frequency-doubled Nd:YAB laser (Coherent, Antares), generating pulses in 76 MHz repetition rate and 250-fs fwhm.

Laser III: A picosecond mode-locked and cavity-dumped dye laser (Spectra-Physics, 375B and 344S) synchronously pumped using a cw mode-locked argon ion laser (Spectra-Physics, 2030–18), generating tunable (530–830 nm) pulses in 4-MHz repetition rate and 10-ps fwhm.

Laser IV: A diode-laser-pumped and frequency-doubled (532 nm) cw Nd:YAG laser (Coherent, DPSS 532).

Laser V: A femtosecond laser system composed of a mode-locked Ti: sapphire oscillator (Coherent, Mira 900F) pumped using a cw argon ion laser (Coherent, Innova 425); a regenerative amplifier (Coherent, RegA 9000) generating amplified pulses (800 nm) in 200-kHz repetition rate and 250-fs fwhm; and an optical parametric amplifier (Coherent, OPA 9400) pumped by RegA, generating tunable (500–700 nm) pulses in 200-kHz repetition rate and 150-fs fwhm.

Selection of each laser depends on whether or not the experiment involves time-resolved measurement and what wavelength is suitable to exciting the target molecules. Note, that for single-molecule experiments a laser beam was circularly polarized using a Babinet-Soleil compensator to irradiate uniformly target molecules. All of the experiments otherwise unspecified were carried out at room temperature (23°C) in a clean-air booth (class 1000).

B. Sample Preparation and Measurement

In Sec. III, n-propanol and n-butanol solution (5×10^{-4} M) of CV were used. No spectral change was observed in the absorption spectrum of the solutions from 10^{-3} to 10^{-5} M, thereby showing no aggregation of CV in 5×10^{-4} M solutions. The solutions were circulated through a 0.5-mm thick quartz cell to avoid persistent photobleaching. *Laser I* and the femtosecond pump-and-probe spectrometer were used and more details on the spectrometer are described elsewhere [19].

The effects of a solvent molecule (ethanol) on the molecular structure of CV was examined by molecular orbital calculations. The calculation method used was MNDO-PM3 (modified neglect of diatomic overlap, parametric method 3) [26] using MOPAC Version 6. Optical transition energies of CV were calculated using complete neglect of differential overlap/spectroscopic-included configuration interaction (CNDO/S-CI) [27]. One hundred lowest energies of one-electron-excited configuration were taken into account for the CI calculations. A new g value, $g = e^2/(R_{rs} + ka_{rs})$, was used for two-centered electron repulsion integral [28]. The k-value that reproduced the experimental absorption maximums was 2.75. Even when $k = 1$ or ordinary Nishimoto–Mataga's g-value [28] was used, final results were not affected essentially except the absolute energies of the absorption maximums. MNDO-PM3 is reliable in the calculations of molecular structures in the ground state, but not so in the calculations including hydrogen bonding, transition structures, and molecules containing atoms that are poorly parameterized [29]. The use of CNDO/S-CI is adequate for calculation of optical transition energies and oscillator strength even when the molecules involved are non-planar like ours because both π and σ electrons are considered in this method [27,28]. Alternatively, the use of Pariser-Parr-Pople configuration interaction (PPP-CI) method [30,31] is possible, but PPP-CI is not sufficient for non-planar molecules because this method considers only π electrons.

In Sec. IV, we used *Laser II* oscillating at 642 nm when measuring fast fluorescence decays at temperatures higher than 273 K. The fluorescence was dispersed with a polychromator (Chromex, 250IS) and measured with a synchroscan streak camera (Hamamatsu, C1587). When measuring slow fluorescence decays at temperatures lower than 273 K, we used *Laser III* oscillating at 540 nm. The fluorescence was dispersed with a polychromator (Chromex, 250IS) and measured with a photon-counting streak scope (Hamamatsu, C4334) [14].

In Sec. V for fluorescence single-molecule imaging, we used *Laser IV* in the power density of ~ 6.2 W/cm^2. The emitted fluorescence was determined using an epifluorescence microscope (Nikon, Optiphot XP), equipped with a Nikon BA 580 long-pass filter and a Nikon CF M Plan SLWD 100× objective with a numerical aperture of 0.75. The microscope was coupled with a photon-counting video camera system (Hamamatsu, C2400-40). For single-molecule time-resolved fluorometry, we used 540-nm pulses of OPA in *Laser V*. The fluorescence photons passing through a pinhole placed in the image plane of an optical microscope (Zeiss Axioplan) was focused on the entrance slit of a polychromator (Chromex, 250IS). The polychromator was coupled with a photon-counting streak scope (Hamamatsu, C4334) to determine simultaneously the fluorescence spectrum and lifetime of single CV molecules.

A drop of a methanol solution of CV was spin coated on an ~30-nm-thick PMMA film previously spin coated on a quartz surface. The concentration of a CV solution in methanol was 1 mM for fluorescence lifetime measurements in bulk, and 1.0 and 0.1 nM for single-molecule measurements. The use of nanomolar or lower concentrations of solutions is common for preparing well-separated single molecules on surfaces [13,15,18,31–42].

In Sec. VI, we used an experimental setup as depicted in Fig. 3 and *Laser III* oscillating at 540 nm and the argon ion laser (514.5 nm) in *Laser V*. The colatitude of the incident direction of the laser light was ~65° with respect to the z axis.

Submonolayers of rhB were prepared by first placing a piece of lens-cleaning paper on a (100)-oriented Si wafer in 2-inch diameter covered with a native oxide. Then, a 50-μL drop of a rhB methanol solution was placed on the paper at one end of the wafer. Lastly, the wet paper was gently dragged across the face of the wafer, thus leaving the solvent to naturally evaporate. These procedures (drop and drag treatment) using high-quality methanol or acetone instead of dye solutions are also used as cleaning methods for optics, such as dielectric mirrors. The submonolayers were thus prepared using selected concentrations of rhB-methanol solutions: 1.5×10^{-7}, 10^{-6}, 10^{-5}, and 10^{-4} M.

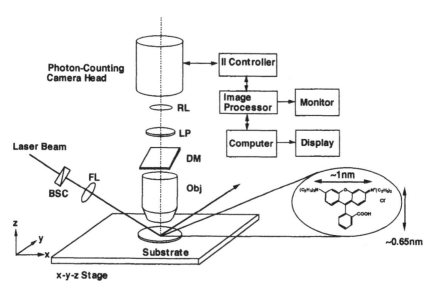

Figure 3 Experimental setup for single-molecule fluorescence imaging. The inset shows the molecular structure and the size of a rhB molecule. (From Ref. 16.)

Using this preparation scheme, almost all of the rhB molecules in the solutions were soaked up by the paper. We estimated the partition ratio of rhB molecules on the paper to those on the wafer to be $\sim 100 : 1$. This high ratio is the very reason why we used a much higher concentration of the rhB solution than that used in previous studies [13,15,18,31–42], in which dye solutions of 10^{-8}–10^{-10} M were dropped or spin-coated on a substrate surface. However, for convenience in our study, we simply refer to the prepared submonolayers as 10^{-7}-M submonolayers, for example, on the basis of the concentration (1.5×10^{-7} M) of the rhB solution used. The partition ratio was estimated by the following procedures. (1) the number of rhB molecules (N_0) before preparing adsorbents was computed from the volume (50 µL) and the concentration (10^{-4} M) of the rhB methanol solution. We kept the rhB solution from spilling from a Si wafer during preparation of a 10^{-4}-M adsorbate. (2) Washing thoroughly the 10^{-4}-M adsorbate with methanol, and then evaluating the concentration of the washed solution by absorption spectroscopy. (3) Again, the number of rhB molecules (N_a) was computed from the volume and the concentration of the washed solution, and then the partition ratio (N_a/N_0) was computed. A similar partition ratio was obtained when using a 10^{-3}-M adsorbate.

Oxide layers of various thickness (260, 182, 98, 62, 43, or 20 Å) were prepared on the Si wafers. At the beginning we prepared an oxide layer thicker than 200 Å by thermal oxidation, and then evaluated the thickness (260 Å). Following that, the 260-Å thick layer was etched down to each thickness using a 1.6%-HF aqueous solution. Each thickness was evaluated with an ellipsometer. Using Auger electron spectroscopy, we determined the thickness of the native oxide layer on the Si wafer to be 11 Å.

Surface irregularities of the Si wafers were evaluated using a NanoScope III AFM (Digital Instruments, Santa Barbara, CA) equipped with at n^+-doped Si tip with a nominal spring constant of 56 N/m. The images were recorded using the tapping mode on an area of 512×512 pixels, and with a scanning line speed of 1 Hz.

In Sec. VII a total internal reflection geometry for irradiating target molecules on a quartz surface was used to reduce scattering background, as demonstrated in Refs. [43–45]. A sample on a quartz substrate was contacted to a 90°-coupling prism with fluorescence-less water (Wako, Japan) as an index-matching fluid. An incident laser beam passing through the prism was reflected at the air/solid interface to produce evanescent field illumination of the sample. The second harmonics (400 nm) of the output of RegA in *Laser V* was used for irradiation of the sample in an average power density from 4 to 80 W/cm^2. The emitted fluorescence was determined using an epifluorescence microscope (Zeiss Axioplan), where a custom-designed dichroic mirror and a dielectric filter for 400-nm excitation and a 100× objective lens (Olympus, LMPlan FI, NA 0.8) were used.

A mixture of TNP-ATP (Molecular Probes) and the Klenow fragment (Stratagene Cloning Systems) was prepared in a buffer solution (2-mM PIPES, pH = 7.5). The concentration of TNP-ATP was 2 nM or 0.4 nM and that of the Klenow fragment was 0.1 μM. The solution was incubated at 40 °C for 1 h to let the enzyme combine with TNP-ATP molecules. A buffer solution only containing 2-nM TNP-ATP without Klenow fragments and a buffer solution of Klenow fragments without TNP-ATP were also prepared for control experiments. A drop of each solution was spin coated at 5000 rpm on a quartz plate, which was pretreated by cleaning with ethanol and Milli-Q water twice, and with an ozone cleaner for 2 h.

III. SOLVATION ISOMERISM OF A TPM DYE IN ALCOHOLS

A. Detective Story Searching for Ground-State Isomers of CV

The visible absorption spectrum of CV in solution appears to be composed of two bands as shown in Fig. 2A. The origin of the two bands was interpreted in three ways: (1) resolution of vibronic structures in one electronic state [46], (2) electronic transition from one ground state to two excited states [47–51], and (3) the existence of two ground-state isomers [5,52–54]. We have an over 50 year long dispute among (1), (2), and (3) over an origin of the twofold absorption spectrum.

In 1942, Lewis and co-workers proposed two ground-state conformational isomers as an origin of the two-band absorption spectrum [52]. One is a propeller structures (D_3 symmetry), in which three phenyl rings is tilted in the same direction. The other is a distorted propeller structure (C_2 symmetry), in which one of the phenyl rings is tilted in the opposite direction. This idea is based on the observation that the shorter wavelength band is reduced, whereas the longer wavelength band is enhanced with decreasing temperature from 296 to 80 K in alcoholic solutions [52]. Similar intensity exchange between the two bands was observed under high pressure and was explained using a model involving chemical equilibrium between two ground-state isomers [53,54].

Several theoretical studies concluded that the two bands originate from electronic transitions from one ground state to two excited states [47–51,55]. The D_3-symmetry conformation was confirmed by an X-ray diffraction study of crystal samples [56], resonance Raman studies [46,51,57], and magnetic circular dichroism measurements [55]. Yet, no C_2-symmetry conformation has been identified yet.

Early pico- and femtosecond pump-probe studies revealed that the bleaching recovery probed on the red side of the bleaching spectrum is faster than that probed on the blue side [3–8,58–68]. This observation was explained by considering either the two ground states [5] or two excited states having different nonradiative decay rates [62]. Ben-Amotz and Harris [7] explained that the apparent faster relaxation at red side of the bleaching spectrum would be due to stimulated emission gains concomitant with the bleaching, however. Thus, the authors concluded that only one ground state and one excited state are adequate for explaining the observed bleaching recovery dynamics.

Many studies concluded that only D_3-symmetry is possible and denied C_2-symmetry structures. However, the authors failed to explain the temperature and pressure dependence of the absorption spectrum as the substantial evidence for two ground states. On the other hand, only the temperature and the pressure dependence of the absorption spectrum evidenced the two ground states. Thus, no one has proposed any models of explaining comprehensively all the previous experimental observations and theoretical calculations. We found a convincing model of CV involving two ground states from experimental observations favored by tunable-excitation femtosecond spectral hole-burning technique and theoretical calculations.

B. Key Observations to Settling the Dispute

In this section we describe experimental observations supporting that the two-band absorption spectrum is composed of two ground states.

Figure 4 shows transient differential absorption (ΔOD) spectra and again a steady-state absorption spectrum of CV in methanol. Note, that the steady-state absorption spectrum in other normal alcohols (data not shown) was similar to that in methanol. The transient ΔOD spectra can roughly be divided into three regions: (1) In the shorter wavelength region than the absorption maximum and shoulder ($\leq$500 nm), where transient absorption was dominantly observed; (2) within the absorption band (500–620 nm), where bleaching was mainly observed; and (3) in the longer wavelength region than the absorption maximum and shoulder ($\geq$620 nm), where stimulated emission gain was dominant. The shape of the bleaching spectrum at 600 fs was almost identical to the steady-state absorption spectrum. Thus, the stimulated emission gain and transient absorption would less overlap on the bleaching spectrum. For all of the pump wavelengths, spectral holes at 600 and 550 nm were observed, each of which corresponds to the absorption maximum and the shoulder, respectively.

Although the location of the holes was independent of the pumping wavelengths, the temporal rise of the holes was dependent on the pumping wavelengths. When pumped at 548 or 558 nm, the two holes appeared simultaneously (Fig. 4A and B). When pumped at 598 nm, however, the hole at 550 nm

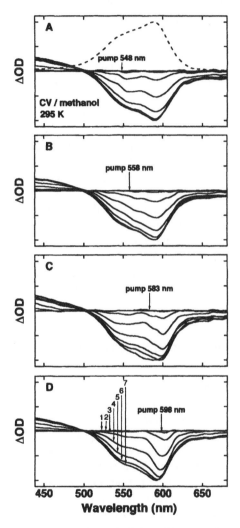

Figure 4 Transient differential absorption (ΔOD) spectra of CV in methanol at 295 K. The probe delay is from -600 (1) to 600 fs (7) at an interval of 200 fs, as shown in the bottom recording. The bleaching maximums of (A), (B), (C), and (D) in optical density (OD) are -1.48, -1.52, -1.01, and -1.23, respectively. (From Refs. 1, 19, 20.)

appeared slower than at 600 nm (Fig. 3D). Figure 5 shows the time evolution
of the bleaching at 550 nm and that at 600 nm when pumped at 548 or 598 nm.
Specifically, when pumped at 598 nm, the bleaching at 550 nm appeared with a
time delay. In Fig. 5C, we introduced a delay time for the rise of the bleaching
τ_d into the fitting function as $1 - \exp(-t/\tau_d)$. The computed delay time was
500 fs. Moreover, we measured delayed bleaching at 550 nm when pumping
at 598 nm in several normal alcohols to find viscosity dependence of the delay
time. The delay time observed was practically constant (500 fs) irrespective of
the alcohols used (data not shown).

   The observed transient ΔOD spectra is free from a coherent effect between
pump and probe pulses and transient Raman gains [19]. Thus, we consider
the temporal rise of the spectral bleaching dependent on the pump and probe
wavelengths as a change in the population of CV in the ground state. If the
ground state of CV is uniform, that is the ground state is common for all
electronic transition, a decrease in the population in the ground state results
in a decrease in all electronic transition intensities. As a result, the bleaching
should rise simultaneously irrespective of the pump-and-probe wavelengths. As
a control experiment we observed transient spectral hole-burning for nile blue
and cresyl violet, both of which are more rigid dye molecules than CV. No
pump-wavelength dependence temporal delay appeared in the transient ΔOD
spectra. Thus, our observations in Figs. 4 and 5 indicate the nonuniformity of
the ground state of CV in alcohols.

C. Decomposition of Two Band Absorption Spectrum Into Each Component

To gain insight into the observed bleaching dynamics, it is necessary to divide
the nonuniform absorption spectrum into uniform components. Figure 6 shows
the absorption spectra of CV in ethanol at several temperatures. With decreasing
temperature the shoulder at 550 nm was reduced, whereas the peak at 590 nm
was enhanced. This observation agreed well with the previous observations [52].
Note, that the shoulder did not disappear completely even at 30 K.

   The procedure for dividing the spectrum is based on the following two
assumptions, previous theoretical studies of the electronic states and the experimental observation (Fig. 6). The first assumption is that the exchange of the
intensity between the shoulder and the peak in the absorption spectrum is due
to the modification of the thermal equilibrium between two ground-state pop-
ulations. The second assumption is that the lower ground state is exclusively
populated at 30 K. Molecular orbital calculations concluded that the lowest singlet excited state (S_1) is two degenerated [47,48,51], and the degeneracy is lifted
by symmetry breaking due to the interaction of a CV molecule with a polar solvent molecule or a counter anion [49–51]. Hence, the absorption spectrum of

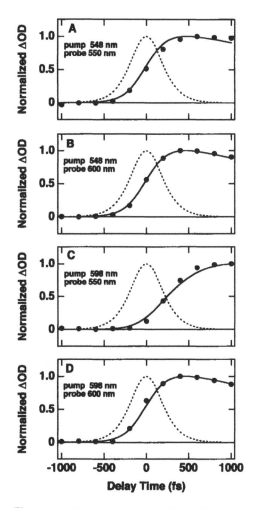

Figure 5 Time evolution of ΔOD of CV in methanol at 25 K. Closed circles show the experimental observations. The dotted lines represent the response function, the shape of which is assumed to be a sech^2 function. The full width of the half maximum of the response function was ~450 fs. The solid lines show theoretical fits. The bleaching recovery times were 1.7 and 6.5 ps. The faster one agrees with the previous observations [5,63]. A flattened top feature was observed near 1000 fs in the time evolution. This feature was analyzed in terms of a relaxation mechanism involving one intermediate state other than the lowest excited singlet state [5,63]. (From Refs. 1, 19, 20.)

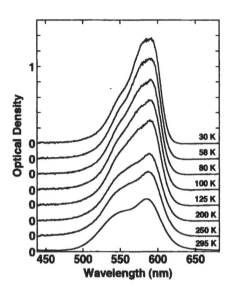

Figure 6 Temperature dependence of the absorption spectrum of CV in ethanol. The enhancement of the OD with decreasing temperature may be due to the change in the refractive index of ethanol. (From Refs. 1, 20.)

CV at 30 K should be composed of two electronic transitions from one ground state to split two S_1 states. In fact, this prediction is consistent with the experimental observation (Fig. 6) that a shoulder remained at a higher-energy side of the absorption maximum. For these reasons, we fit two Gaussian curves on the absorption spectrum at 30 K. On the other hand, two ground states, each of which has S_1 state split into two electronic states, exist together at room temperature. Thus, the absorption spectrum at room temperature should be composed of four Gaussian curves [69].

Figure 7 shows the curve fitting of the absorption spectrum in methanol at room temperature with four Gaussian curves. The longer wavelength component denoted by "L" was well represented by two Gaussian curves whose maximums were 589 and 561 nm, respectively. The shorter wavelength component denoted by "S" was also composed of two Gaussian curves whose maximums are 550 and 524 nm, respectively. The energy difference between the two ground states was evaluated to be ~ 200 cm^{-1} from the Arrhenius plot of the absorption intensity ratio between the two components (data not shown). This value agrees well with that reported by Lewis and co-workers [52]. According to the energy difference, the population ratio between the two ground states is 9.7×10^{-3} from the Boltzmann factor at 30 K. This evaluation shows the validity of the

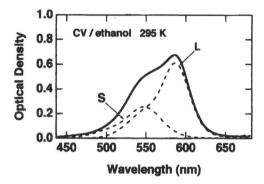

Figure 7 Deconvolution of an absorption spectrum of CV in ethanol at 295 K to four Gaussians. The solid line represents an experimental spectrum. The broken lines show two calculated spectrums, each of which is composed of two Gaussians. The dotted line, which is totally hidden under the solid line, is the sum of the two calculated spectrums. (From Refs. 1, 20.)

assumption that we have little population in the higher energy ground state at 30 K.

D. Summary of the Experimental Observation

The observed bleaching dynamics should be explained using a model involving two ground-state isomers, each of which has two excited states. In Fig. 8 we illustrate possible population-transfer dynamics when pumped at 548 or 598 nm. The 548-nm pumping light excites both ground states because the two absorption components overlap around 550 nm. Consequently, the bleaching is caused all over the absorption spectrum. Similarly, the 558-nm pumping light also excites both isomers. On the other hand, when pumped at 598 nm, only the lower-energy isomer is selectively excited. Thus, only the longer wavelength component L is bleached and therefore the population of the two ground state isomers departs from thermal equilibrium. Note, that the bleaching spectrum at 0 fs (marked with 4) in Fig. 4D closely resembles L component in Fig. 7. If the potential barrier from the higher energy ground state to the lower energy ground state is comparable to kT at room temperature, the population of the higher energy isomers can transfer to the lower ground state within a few picoseconds assuming 10^{12}–10^{13} s^{-1} at the pre-exponential factor of the rate process. The decrease in the population of the higher energy ground state after the population transfer causes bleaching of S component. Thus, the delay time (500 fs) of the rise of the bleaching at 550 nm when pumped at 598 nm is ascribable to the population transfer time on the basis of the two ground state model.

A pump at 548nm

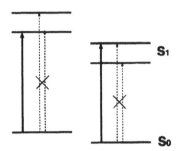

B pump at 598nm

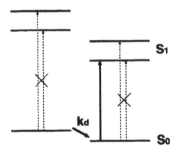

Figure 8 Energy-level schemes and population transitions of CV in alcohols. The arrows with solid lines represent electronic transitions by pump light or the population transition from the higher ground-state to the lower ground-state. The arrows with dotted lines show electronic transitions by probe light. The indicator "x" means the saturation of the transition; k_d is the rate of population transition, which is calculated at $1/500$ $(fs)^{-1}$ by the curve fitting in Fig. 5. (From Refs. 1, 20.)

E. Theoretical Consideration of the Molecular Structure of CV in Alcohols

In this section, solvent effect on the symmetry and the absorption spectrum of CV was examined using molecular orbital (MO) calculation to find what molecular structure and absorption spectrum are possible for CV when solvent molecules interact with CV molecules. For simplicity only one methanol molecule was considered in the current calculation.

We found that a free CV exhibited the D_3-symmetry propeller structure. However, the structure of the CV remained D_3 symmetry (Table 1) even when a methanol molecule was close to a CV molecule. The location of the methanol molecule was calculated on condition that the system of CV plus methanol is

Table 1 Calculated Equilibrium Point of a Methanol Molecule on CV and Structure of CV

Location of methanol	Distance (Å)[a]	Symmetry	Total energy (eV)[b]
—	∞	D_3 (planar)	−4377.72
Central carbon	3.712	D_3 (planar)	−4377.81
Nitrogen	4.101	D_3 (planar)	−4377.81

[a]The distance from oxygen of methanol to CV.
[b]Total energy of CV and a methanol molecule.
Source: Refs. 1, 20.

most stable or in equilibrium. In this calculation a methanol molecule located either 3.712 Å above the central carbon atom or 4.101 Å above one of the nitrogen atoms. Note, that the calculations 3.712 and 4.101 Å are close to the average distance (4.0656 Å) between methanol molecules in neat methanol. This averaged distance was estimated from the specific gravity assuming the uniform distribution of methanol molecules.

When a cation is dissolved in a polar solvent, the solvent molecules will highly solvate the cation. Therefore, it is reasonable to expect that the distance between CV cation and solvent molecules is much shorter than the average methanol-to-methanol distance in neat methanol. Moreover, MNDO calculations tend to underestimate the interaction between two molecules [25]; therefore, the calculations 3.712 and 4.101 Å would be longer than the actual value. For this reason, we calculated total energy and the structure of CV when a methanol molecule was closer to CV than the calculated equilibrium distance. When a methanol molecule is located at 2.0 Å above the central carbon, the structure is stable when the central carbon is 0.157 Å above the original sp^2 plane. On this occasion the CV takes the C_3 pyramidal structure. In contrast, even if a methanol molecule is locate 2.0 Å above one of the nitrogens, the structure of the CV remains D_3 symmetry (Table 2). Note, that the distance 2.0 Å is an arbitrary selection: the distance much shorter than ~4.0 Å is important in these calculations.

Table 2 Calculated Structures of CV When a Methanol Distance is Fixed at 2.0 Å

Location of methanol	Distance (Å)[a]	Symmetry
Central carbon	2.0 (fixed)	C_3 (pyramidal)
Nitrogen	2.0 (fixed)	D_3 (planar)

[a]The distance from oxygen of methanol to CV.
Source: Refs. 1, 20.

When CV has D_3 symmetry, calculated absorption bands are located at 546.2 and 534.5 nm from the lowest energy. When CV is deformed into C_3 symmetry, we obtained red-shifted bands 562.0 and 550.1 nm (Table 3). No effect of a methanol molecule on the structure of a CV molecule other than the point-charge perturbation was considered in the current calculations.

F. Construction of a Model for Ground-State Isomers

Here, we propose a ground state structure of CV other than the D_3-symmetry and the C_2-symmetry based on the experimental observations and MO calculations of our own as well as the previous studies. This proposed structure comprehensively explains all the current and previous experimental observations and theoretical calculations.

In general, a large amplitude vibrational mode with low frequency (≤ 1000 cm^{-1}) will be concerned with the isomerization of organic molecules [70]. From an expectation from the molecular structure of CV, the bending mode of the bonds from the central carbon to the three phenyl rings is a possible large amplitude vibration except for the torsional motion of the phenyl rings. The insensitivity of the rise time to the solvent viscosity shows that the structural change between the two isomers is independent of, or at least not so sensitive to solvent viscosity. In other words, no large structural change such as the torsion of the phenyl rings is appropriate to accounting for the difference between the two isomers of CV. The structural change accompanied by the bending vibration would be less affected by solvent viscosity than the torsional motion of the phenyl rings. Thus, we propose another ground state structure than D_3 and C_2 symmetries whereby the three bonds on the central carbon are bent with D_3 symmetry structure. Such a pyramidal structure has C_3 symmetry. Matsuoka and Yamaoka pointed out the possibility that CV takes the pyramidal structure, although they were not able to experimentally identify any ground-state isomers [71].

Table 3 Calculated Absorption Bands and Oscillator Strength of CV

Symmetry	λ_1[a] (nm)/f[b]	λ_2[a] (nm)/f[b]
D_3 (planar)	546.2/0.88	534.5/0.70
C_3 (pyramidal)	562.0/0.77	550.1/0.71

[a] λ_1, λ_2: absorption maximums.
[b] f: oscillator strength.
Source: Refs. 1, 20.

We now discuss how CV can assume the pyramidal structure. Because only D_3 symmetry was confirmed by the X-ray diffraction study of crystal samples [28], the effects of alcohols should be taken into consideration as the cause of the nonuniformity of the ground state of CV in alcohols. It is well known that liquid alcohol forms a network (monomers, dimers, and higher structures) by hydrogen bonding [72]. A CV molecule in alcohol is surrounded by a network of alcohol molecules. The CV molecule could have two or more solvation forms interacting with alcohol molecules. For simplicity we assume that only a monomer alcohol solvates with CV cation. Of course, we appreciate that the solvation model involving a single solvent molecule is an oversimplified picture. For all this simplicity, however, this model will point out a fundamental picture of the solvation of CV molecules by methanol. The mechanism of solvation was previously discussed by Korppi-Tommola and co-workers [49]. In this mechanism a monomer-dimer equilibrium of alcohols is involved in the solvation equilibrium between monomer alcohol and a CV cation. We, therefore, propose a model of the ground state of CV in Fig. 9. Further discussion of this model is described in Ref. [20].

Our MO calculations shows that when a methanol molecule is located at 2.0 Å above the central carbon, CV has the C_3-symmetry pyramidal structure. Moreover, the calculated absorption bands for both D_3 and C_3 symmetry qualitatively agree with the absorption bands for the components S and L in Fig. 7, respectively. Although many solvent molecules really surround a CV, this consistency strongly supports the possibility of CV having the pyramidal structure in alcohols.

The reason why the C_3-symmetry structure has not been confirmed by the previous resonance Raman studies is probably due to the following reason. The wavelengths of Ar^+ laser used for excitation in the resonance Raman studies [46,51,57] were at the shorter edge of the absorption band. Thus, the excitation resonates only with the D_3-symmetry isomer. In the X-ray diffraction studies [56] the sample was not solution but crystal; therefore, the C_3-symmetry isomer was not confirmed. The MO calculations [51] only denied the C_2-symmetry distorted propeller isomer and show lifting the degeneracy of the S_1 state. On the contrary, the MO calculations support our model involving the split S_1 states. The time resolved bleaching studies, which only concern the bleaching recovery dynamics, were not able to find the isomers. The bleaching recovery times we observed (data not known), however, agree with the previous data [7,64,65], thus showing the validity of our measurements.

G. Concluding Remarks

We propose the solvation isomers whereby CV in alcohols has D_3 or C_3 symmetries in the ground state. Our model successfully explains almost all of the

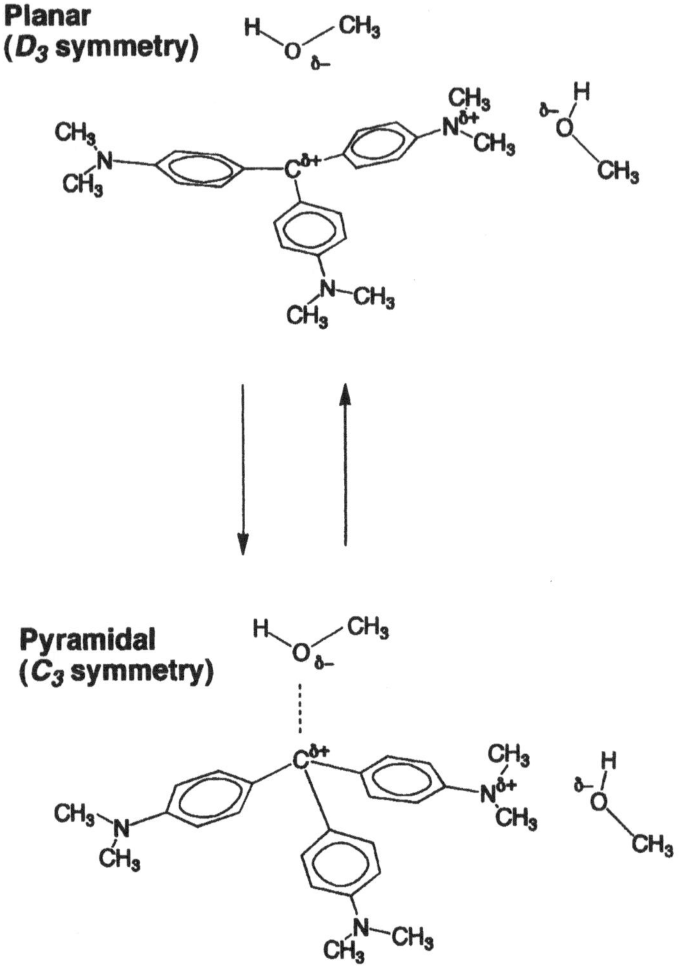

Planar
(D_3 symmetry)

Pyramidal
(C_3 symmetry)

Figure 9 Proposed model of the ground-state structures of CV in alcohols. These two ground states are discriminated by solvation equilibrium between an alcohol molecule and a CV cation. (From Refs. 1, 20.)

previous conflicting observations on the issue of the ground-state isomers. Solvation will occur particularly for polar solute molecules in polar solvents. However, explicit observation of the solvation isomers will depend on the kinds of solute molecules involved. Explicit appearance of the solvation isomers in CV would reflect the extremely flexible nature of TPM dyes even in the ground state. In general, organic molecules such as dye molecules are more flexible in the excited states than in the ground state because an electron in the bonding molecular orbital is promoted to a nonbonding molecular orbital. Thus, we expect fluorescence properties of TPM dyes to be exquisite probes for the immediate environments around them in condensed phase.

IV. PROBING GLASS TRANSITION IN ALCOHOLS AND POLYMERS USING A TPM DYE

A. What is Probed Using a TPM Dye?

In recent years the importance of a supercooled state in liquids occurring well above T_g has widely been recognized in condensed-matter physical chemistry [73]. One of the major theoretical considerations is provided by MCT [74,75]. From the prediction of MCT a dynamical glass transition from ergodicity to nonergodicity occurs at T_c, which locates several tens degrees above T_g. Stimulated by encouraging progress in the theoretical work, much effort has substantially been devoted to the experimental studies of the dynamic of glass-forming materials. The crossover phenomena at T_c were found by neutron scattering [76], light scattering [77], and viscosity measurements [74,78]. However, the observed crossovers are blunt in the previous macroscopic measurements. Thus, ambiguity remains in the precise identification of T_c. The use of molecular probes is important in identifying the crossover point because they are sensitive to the microscopic dynamics involving the dynamical glass transition [11,12].

As stated in Sec. I, nonradiative relaxation of MG, from S_1 to the ground state is strongly influenced by the solvent viscosity. The fluorescence lifetime increases with increasing solvent viscosity from picoseconds to nanoseconds. At cryogenic temperatures (<77 K), the lifetime is exclusively limited by the radiative lifetime; thus, the fluorescence efficiency is close to unity. The viscosity dependent nonradiative process is thought to be due to diffusive rotational motion of the phenyl rings in S_1. Ben-Amotz and Harris [6–8,79] and Abedin and co-workers [9,10] concluded that the potential for the rotational motion of the phenyl rings is barrierless in S_1. The barrierless potential makes the fluorescence lifetime extremely sensitive to changes in solvent viscosity, because the friction from the surrounding solvent molecules is the only impediment to the rotational motion of the phenyl rings. The host matrices selected for MG were glass-forming materials, such as 1-propanol (PR), propylene glycol (PG) and glycerol (GL) as monomers; polybutadiene (PB), poly(vinyl acetate) (PVAc), poly(methyl acrylate) (PMA) and poly(ethyl methacrylate) (PEMA) as polymers. Table 4 lists the melting point (T_m) and the calorimetric glass transition temperature (T_g) for each material. The concentration of MG doped in these glass-forming materials were $\sim 2 \times 10^{-5}$ M. The sample was first cooled down to 10 K at a rate of 2 K/min, and then the fluorescence life-

Table 4   Melting Point (T_m), Critical Temperature (T_c), and Calorimetric
Glass Transition Temperature (T_g) of the Alcohols and Polymers Selected

	$T_m(°C)$	$T_c(°C)$	$T_g(°C)$
1-Propanol (PR)	-127	-143	-173
Propylene glycol (PG)	-59	-68	-101
Glycerol (GL)	17.8	-33	-80
Poly(butadiene) (PB)		-65	-95
Poly(methyl acrylate) (PMA)			7
Poly(vinyl acetate) (PVAc)			30
Poly(ethyl methacrylate) (PEMA)			63
Poly(methyl methacrylate) (PMMA)			114

Source: Modified from Ref. 1.

time was measured from the lowest to the highest temperature step by step. We selected the spectral window for the lifetime measurement from 674 to 692 nm, within which no wavelength dependence of the fluorescence lifetimes was found.

B.   Temperature Dependence of Nonradiative Decays of MG Doped in the Glass-Forming Materials

We measured temperature dependence of MG fluorescence lifetimes in the monomers and polymers in Table 4 from 300 to 10 K. The selected monomers and polymers are divided into two groups on the basis of their effect on the nonradiative decays of MG. The first group consists of the monomers, PR, PG, and GL, and the polymer without side chains PB (Group I). The second group consists of the polymers with side chains, PVAc, PMA, and PEMA (Group II). In temperatures higher than $T_g + 30$ K the fluorescence decays were not fitted to a single exponential function or a stretched exponential function for the every material in Groups I and II, but were well-fitted to a biexponential function. For Group I the fast decay component decreased with decreasing temperature. At temperatures lower than T_g every fluorescence decay curve was well-fitted to a single exponential function; the fluorescence lifetime was close to the radiative lifetime. On the other hand, from Group II the fast decay component survived below T_g down to 100 K. At temperatures lower than 100 K every fluorescence decay curve was well-fitted to a single exponential function; the fluorescence lifetime was close to the radiative lifetime.

When decay curves were analyzed using a biexponential function, the nonradiative decay rate τ_{snr}^{-1} of the slow component was evaluated by subtracting the radiative decay rate from the slow fluorescence decay rate. Figure 10 shows

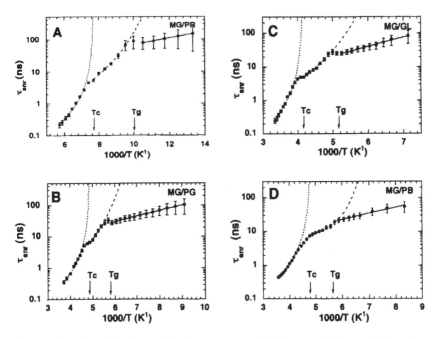

Figure 10 Arrhenius plots of the slow component of biexponential nonradiative decay times of MG in (A) PR, (B) PG, (C) GL, and (D) PB. The dotted, dashed, and solid curves represent fitting experimental observations to the power law, VTF equation, and Arrhenius law in the three temperature regions, respectively. (From Refs. 1, 12.)

the Arrhenius plots of the nonradiative decay time for Group I. One recognizes without doubt three temperature regions in each curve. The lower crossover temperatures are close to each T_g, and the higher crossover temperatures are 30–50 K above each T_g, which are reasonably attributed to T_c predicted by MCT. The reason why the crossover at T_c was successfully identified using MG is attributed to the peculiar molecular structures of TPM dyes. The nonradiative decay time measured is limited by the time required for the diffusive rotation of phenyl rings to reach a sink in the S_1-state potential, from which deactivation occurs to the ground state. Because the S_1-state potential is barrierless and the required rotation for the nonradiative relaxation is small ($\sim 10°$), relaxation dynamics of the surrounding mediums was probed in the range of $10^{-11}–10^{-8}$ s and of few to several angstroms. Both ranges are just relevant to temporal and spatial windows for observing the dynamical glass transition predicted by MCT. This feature of MG is quite different from other molecular probes. The temperature dependence of τ_{snr} was fitted to different equations in each temperature region. Above T_c it was fitted to a power law equation according to MCT, in the middle region between T_c and T_g fitted to the Vogel-Tammann-Fulcher (VTF) equation, and below T_g fitted to the Arrhenius equation. The Arrhenius dependence below T_g is consistent with other studies of the molecular mobility in solid polymers [9,10].

Figure 11 shows Arrhenius plots of the nonradiative decay time of the slow component of Group II. For these branched polymers, we identified no singularities above T_g. For each of the three materials the nonradiative decay time at T_g is shorter than that for the three monomers and PB. This may be due to the fact that even below T_g the side-chain motions are not frozen yet; thus, the microscopic viscosity experienced by MG molecules is not so large as that of the monomers and PB at their T_gs. We missed finding the crossover at T_c, however, the side-chain motions may mask the crossover even if it exists. Moreover, the Arrhenius plot of the nonradiative decay time was not a straight line below T_g, but showed two regions separated by a kink at the temperature several tens degrees below T_g, which can be attributed to the freezing point of the motion of side-chain groups.

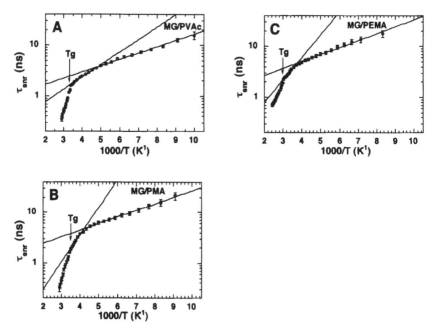

Figure 11 Arrhenius plots of the slow component of nonradiative decay times of MG in (A) PVAc, (B) PMA, and (C) PEMA as functions of temperature. Below T_g the plots show two regions of Arrhenius temperature dependence. (From Refs. 1, 12.)

C. Bimodality in the Dynamic Response of the Glass-Forming Materials

The differences between Groups I and II are due to the following facts. In Group I only diffusive or main-chain motions of the matrix materials are involved in the nonradiative decays; in Group II side-chain motions plus main-chain motions affect the nonradiative decays. By using MG as rotor probes we found the crossover at T_c predicted by MCT 30–50 K above T_g in Group I. In Group II, we observed another kink below T_g which is ascribed to the side-chain motion although we missed any singularities above T_g.

The origin of the biexponential fluorescence decays has not been traced yet, but may be ascribed to possible bimodal sites in the alcohols at the temperatures higher than $T_g + 30$ K and in the polymers with side chains at temperatures from 300 to 100 K. The possible bimodal sites will be characterized by the difference in the viscosity of each site. To conclude whether the biexponential fluorescence decays is due to averaging a heterogeneous distribution of single exponential decays of each molecule or a homogeneous distribution of intrinsic biexponential decays of each molecule, we implemented imaging and time resolved fluorescence measurement of single CV molecules in a PMMA film.

V. SINGLE MOLECULES PROBING BIMODAL DOMAIN STRUCTURES IN A POLYMER FILM

A. What is Revealed Using Single-Molecule Probes?

Ultrafast spectroscopy has rapidly been developed in the femtosecond regime. An example of the femtosecond spectroscopy is described in Sec. III. Also, ultrasensitive measurement has recently reached the single-molecule level [13–18,31–45,80–90]. The characteristics of individual molecules hidden under the ensemble of many molecules, such as spectral diffusion and intensity fluctuations, were revealed using near-field [31–35,38,41,85,86] and far-field microscopies [13,15–18,36,37,39,40,42–45,87–90]. In the previous studies, the selected chromophores were restricted to dye molecules or aromatic hydrocarbon molecules with a rigid molecular structure and high fluorescence efficiencies close to unity. The use of flexible dye molecules, such as TPM dyes, is useful for investigating microscopic dynamics of host matrices as demonstrated in Sec. IV, because flexible dye molecules are more sensitive to the immediate surroundings than rigid molecules. In the current single-molecule study we selected CV instead of MG simply because the wavelength of the excitation light (532 nm) used is near the absorption maximum of CV. A film of PMMA, a polymer having side chains, was selected as a matrix for CV molecules because of its excellent transparency, which is indispensable to single molecule measurement.

In the ensemble-averaged measurements described in Sec. IV, the biexponential fluorescence decays of MG molecules were tentatively ascribed to bimodal structure of the host matrices [11,12]. By virtue of single-molecule fluorescence lifetime measurements we for the first time found evidence for bimodal distribution of sites for individual CV molecules on a thin film of PMMA [15]. Recently, a theoretical study appeared foreseeing bimodality of a supercooled liquid [91].

B. Single Molecules Revealing Ensemble-Averaged Heterogeneity

A fluorescence decay of the ensemble of many CV molecules on a PMMA film is shown in Fig. 12. The decay was not fitted to a single exponential function and a stretched exponential function, but was well fitted to a biexponential function $I(t) = A_f \exp(-t/\tau_f) + A_s \exp(-t/\tau_s)$, where τ_f and τ_s are time constants and A_f and A_s are pre-exponential factors. We obtained $\tau_f = 0.43$ ns and $\tau_s = 1.76$ ns, and the ratio $A_s/A_f = 1.14$. Compared with the excited state lifetime of CV (2–3 ps) in methanol and ethanol [5–8,58–68], the fluorescence lifetime of CV on a PMMA film increased more than two orders of magnitude [9–12]; thus, so did the fluorescence quantum efficiency. The enhancement of the fluorescence efficiency of CV on a PMMA film made it possible to observe single CV molecules. Figure 13 shows fluorescent spots on a PMMA film on which a drop of 1-nM CV in methanol was spin-coated. The number of fluorescent spots in an image linearly increased with increasing concentration of a CV methanol

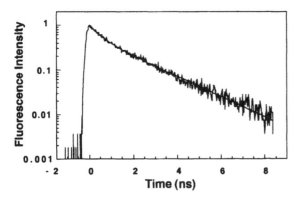

Figure 12 Fluorescence decay curve of 1-μM CV spin-coated on a PMMA film obtained in the bulk measurement. The curve is well fitted to a biexponential function. (From Refs. 1, 15.)

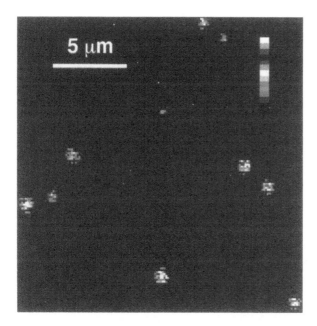

Figure 13 Single-molecule fluorescence image of 1-nM CV spin-coated on a PMMA film. The data accumulation time was 40 s. (From Refs. 1, 15.)

solutions. This observation provides evidence that fluorescence emission comes from individual CV molecules.

Figure 14A shows a histogram of fluorescence photocounts of individual CV molecules. A distinct bimodal distribution is found in this histogram. To examine whether the bimodal distribution is found in this histogram. To examine whether the bimodal distribution is ascribable to the flexible structure of CV, we implemented a control experiment using Texas Red (TR) under the same experimental conditions as those in the CV experiment. Because of the oxygen bridge between two phenyl rings and immobilization of amino groups TR has a rigid molecular structure. Figure 14B shows a histogram of fluorescence photocounts of individual TR molecules. Only one maximum is found in this histogram. The sharp contrast between the two histograms is expected to be due to the flexibility (or rigidity) or the molecular structures. Because of the flexible structure CV is extremely sensitive to the local viscosity of a PMMA film. On the other hand, the fluorescence efficiency of TR is close to unity [92]; thus, less sensitive to the local viscosity. It is not surprising that the photocounts of the maximum of the histogram for TR is smaller than those of the second maximum in the histogram of CV. The long-pass filter used cut a large portion of the fluorescence of TR,

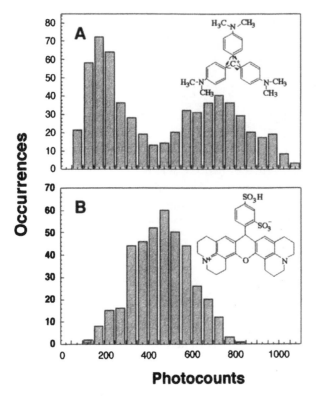

Figure 14   Fluorescence photocounts of individual fluorescent spots versus the number of occurrences of each fluorescent spot (A) for CV and (B) for TR. The bin width is 50 counts. The insets of (A) an (B) show molecular structures of CV and TR, respectively. (From Refs. 1, 15.)

while almost all of the fluorescence from CV passed the filter because of a large strokes shift of the CV fluorescence.

To gain insight into the site-dependent fluorescence characteristics of CV and to provide further evidence for observing single-molecule fluorescence, we implemented time resolved fluorescence measurement of individual fluorescent spots. Figure 15A and B shows representative fluorescence decays corresponding to a strong and a weak fluorescent spot, respectively. Each curve well approximates to a single exponential function, although a slight deviation from the single exponential fitting could be noticed. The slight deviation might be ascribable to the intrinsic decay nature of CV itself [5,11,93,94]. The decays in Fig. 15A and B are clearly different from each other and correspond to the fast and slow decay components observed in the bulk measurement (Fig. 12). For TR

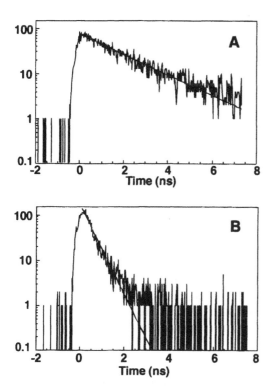

Figure 15 Representative fluorescence decay curves of single CV molecules on a PMMA film. Data accumulation time was 180 s. These curves are fitted to single exponential functions: (A) for a strong fluorescent spot (1.92 ns), and (B) for a weak fluorescent spot (0.44 ns) in the bimodal histogram of Fig. 14A. (From Refs. 1, 15.)

on PMMA film, we found no such distinct difference in fluorescence decays as CV. The fluorescence lifetime of TR varies slightly from molecule to molecule within a range of 3.6–4.4 ns. This observation is rationalized by the fact that the fluorescence lifetime of TR is exclusively determined by the radiative lifetime, whereas that of CV is affected by the internal conversion that is sensitive to the local environment. The observed fluorescence lifetimes varied from molecule to molecule: the fluorescence lifetime of CV distributed in a wide range from 0.37 to 3.11 ns.

In the ensemble of many CV molecules, biexponential fluorescence decays were observed. In individual fluorescent spots, however, single exponential decays were observed. The latter observation is also supporting evidence for observing single-molecule fluorescence in the current work. Thus, the biexponential decays of many CV molecules are composed of single exponential decays of individual CV molecules. Concerning more evidence for observing single-molecule fluorescence, we have evaluated possible observable fluorescence photocounts based on the unitary photocounts or multiples of the unitary photocounts, the stairstep photobleaching, and the calculated photocounts using known experimental parameters [16]. Thus, we can estimate the possible number of single-molecule fluorescence photocounts in the current work.

The observation of a bimodal distribution in the histogram (Fig. 14A) shows that, although the interaction of individual CV molecules with PMMA is different from site to site, the sites of a PMMA matrix are roughly divided into two groups, namely, viscous and less-viscous sites. The less-viscous sites could be attributed to the sites where the side chains of the polymers move to some extent freely, while the less-viscous sites could correspond to the sites where the side-chain motion is suppressed. The background for the above inference that the local viscosity of each site is associated with the degree of side chain motion was addressed in the previous studies [12,95]. These authors assumed that some side-chain motions still occur at temperatures lower than T_g and heterogeneous nature of the side chain motion causes the biexponential fluorescence decays. Our single-molecule measurements were implemented at room temperature. Some side-chain motions causing bimodal local viscosity are expected to be possible in a PMMA film, although the room temperature (296 K) is much lower than $T_g = 387$ K of PMMA [22].

C. Numerical Consistency Between Single-Molecule and Bulk Measurements

Assuming that the bimodal sites are separable at the minimum of the histogram in Fig. 14A, the ratio of the number of high-viscosity sites and that of the low-viscosity sites is estimated to be 0.98. This value is compatible to the amplitude ratio A_s/A_f (= 1.14) obtained in the bulk fluorescence decay measurement (Fig. 12). The mean photocounts of CV molecules at non-viscous (P_N) and viscous (P_V) sites were calculated by averaging the photocounts corresponding to the higher- and lower-photocount maximums of the histogram, respectively. We obtained the ratio $P_N/P_V = 3.73$, which is in good agreement with the ratio of the fluorescence lifetimes $\tau_s/\tau_f = 4.09$ obtained in the ensemble-averaged lifetime measurement within errors. Thus, the nonexponential fluorescence decay of the ensemble of many CV molecules is without doubt the assembly of site-dependent single exponential decays of individual CV molecules.

D. Future Prospect of This Work

The single-molecule experiments illustrated that the interaction between CV molecules and the polymer matrix is strongly site-dependent and bimodal nature.

As a result, the site-dependent nonradiative process of individual C molecules is responsible to biexponential fluorescence decay curves of CV observed in the ensemble-averaged measurement. Our single-molecule study presented in this section will open new possibilities in the experimental study of dynamic response of condensed matter, such as polymers and liquid. We further expect that dye molecules with flexible molecular structures like CV are useful to sensitive local probes for microscopic dynamics of various host mediums.

VI. CHARACTERIZING SILICON SURFACES USING SINGLE-MOLECULE PROBES

A. What is the Appearance of RhB Fluorescence in Submonolayers on Si Surfaces?

Because rhB has extensively been studied not only in solution [96–99] but also in adsorption [100–110], rhB was selected as a probe for evaluating physical and chemical characteristics of Si surfaces. As a future plan to put single-molecule research into practical use, we should consider doing single-molecule chemical analysis particularly at air-substrate interfaces, where single analyte molecules are prepared after being transported. Advantages in the use of solvent-free surfaces include immobilizing target molecules without Brownian motion in solution and reducing possible photobleaching or fluorescence quenching of target molecules, as can be induced by solvents themselves or dissolved material in the solvents used.

To determine the fluorescence distribution of rhB submonolayers on Si surfaces fluorescence images of the submonolayers were observed using fluorescence video microscopy. Circularly polarized light was used for excitation to irradiate uniformly rhB molecules irrespective of their orientations. Representative images (Fig. 16) show three characteristic observations depending on the four rhB concentrations (10^{-7}, 10^{-6}, 10^{-5}, and 10^{-4} M) selected for preparing submonolayers. First, the total photocounts summed over a full area of an image ($66 \times 62 \ \mu m^2$) showed near linear dependence on the rhB concentration (Table 5). Second, the number of bright spots increased with increasing rhB concentration from 10^{-7} to 10^{-5} M, followed by saturation at a concentration between 10^{-5} and 10^{-4} M (Fig. 16C and D). Third, fluorescence not associated with bright spots became strong when the rhB concentration exceeded 10^{-5} M. In the presentation of Fig. 16, however, extra fluorescence not associated with isolated spots is visible only in the 10^{-4}-M submonolayer (Fig. 16D) because of the limited contrast of the presentation.

The first observation provides evidence that the observed photocounts comes from rhB fluorescence. The last two observations motivated us to trace the origin of the non-uniform fluorescence distribution of the submonolayers. For

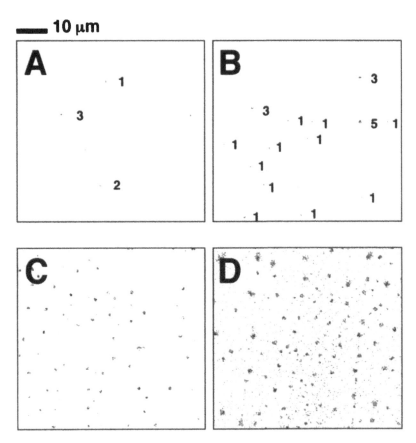

Figure 16 Representative fluorescence images of rhB submonolayers prepared using rhB-methanol solutions in several concentrations: (A) 10^{-7}, (B) 10^{-6}, (C) 10^{-5}, and (D) 10^{-4} M. The number of fluorescent spots for each concentration was (A) 3, (B) 14, (C) 78, and (D) 117. The number of rhB molecules is assigned to the fluorescent spots in (A) and (B) according to the description in subsection B. (From Ref. 16.)

this reason we used position-sensitive time-resolved fluorescence microscopy of rhB submonolayers, plus light scattering microscopy and atomic-force microscopy of Si surfaces.

B. Bright Fluorescent Spots That Hold a Countable Number of RhB Molecules

We also found that unitary photocounts of 100 or multiples of the unitary photocounts occurred in bright fluorescent spots irrespective of rhB concentration

Table 5 Averaged Total Photocounts of an Image (66.0×62.3 μm^2) Versus rhB Concentration Used in Preparing Submonolayers

Conc. (M)	Photocounts/image per 30 s [number of images]
1.5×10^{-4}	524,130 [10]
1.5×10^{-5}	51,426 [15]
1.5×10^{-6}	7,720 [64]
1.5×10^{-7}	976 [59]

The total photocounts for each submonolayer varied among the images, but stayed within a factor of $\sim$3 for each of the 10^{-6}-, 10^{-5}- and 10^{-4}-M submonolayers. In the 10^{-7}-M submonolayers, this scatter increased because the number of fluorescent spots was low in each image. The total photocounts varied by more than a factor of 3, but remained within a factor of 15. The figures in the brackets are the number of images measured. Background photocounts ($\sim$330 counts/image/30 s) are included in the above averages of the total photocounts. Dark-current noise ($\sim$50 counts/image/30 s) that is intrinsic to the photon-counting video camera is included in the background photocounts. *Source*: Ref. 16.

(10^{-7}, 10^{-6}, and 10^{-5} M) using circularly polarized excitation light. We collected the number of occurrences of photocounts from each bright fluorescent spot (Fig. 17). The histograms show several maximums at every $\sim$100 photocounts, although some distribution of photocounts occurred around each maximum. The $\sim$100-count digitization suggest that $\sim$100 photocounts come from a single rhB molecule under the conditions used in this experiment. Two other observations support further our hypothesis that a single rhB molecule produces $\sim$100 photocounts. First, stairstep photobleaching also occurred in a fluorescent spot in units of unitary photocounts of 100 or multiples of the unitary photocounts (data not shown) using the circularly polarized excitation. Second, the unitary photocounts of 100 were similar to the fluorescence photocounts from a single rhB molecule calculated using known experimental parameters, such as absorption cross section and fluorescence quantum efficiency [111].

The number of rhB molecules in bright fluorescent spots (Fig. 16A and B) was determined from the following observations: the digitized histograms, the stairstep photobleaching, and the numerical consistency in the fluorescence photocounts from a single rhB molecule, as described in note 111. The digitized histograms reflect the independent nature of rhB molecules in bright fluorescent spots, within the limits of variability in photocounts around each maximum of

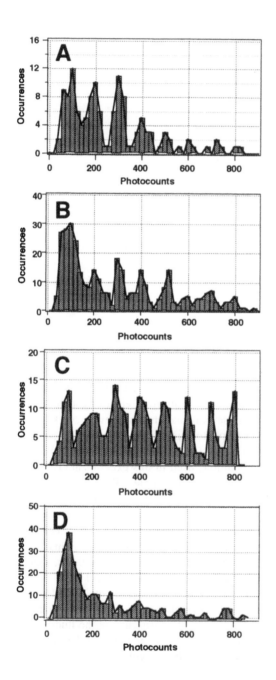

the histograms. In addition to inherent independent nature of each rhB molecule, two technical factors of the current work favor the well-resolved digitized histograms: the use of the off-z-axis and circularly polarized excitation light and the photon-counting technique in the video camera. This excitation is equivalent to simultaneous irradiation of circularly polarized light whose incident directions are along z- and x-axes. Thus, target molecules were equally irradiated using this excitation irrespective of their orientations. Moreover, thanks to the single photon-counting technique, the observed fluorescence intensity is free from fluctuation of the electronics involved in the video microscopy. Thus, the observed distribution of photocounts around each maximum of the histogram is due to the fluctuation of absorbance and fluorescence efficiencies of individual rhB molecules; and the shot-noise-determined fluctuation of the observed photocounts themselves.

To obtain insights into the orientation of rhB molecules entrapped in individual bright spots, excitation-polarization effect on the fluorescence intensity was evaluated. The histogram in Fig. 17D was obtained using p-polarized light whose electric vector was perpendicular to the incident direction and was in the x-z plane (Fig. 3). However, we were not able to make a histogram of photocounts using s-polarized light because of a small number of the photocounts observed (Table 6). (For further details, see Sec. VI.J.)

C. Fluorescence Spectroscopy of RhB Submonolayers on Si Surfaces

To determine if the rhB molecules were monomers in the prepared submonolayers, we measured fluorescence spectra of more than 50 bright fluorescent spots and background regions with no fluorescent spots in the 10^{-5}- and 10^{-4}-M submonolayers. It is widely known that the fluorescence maximum of rhB dimers shifts more than 50 nm from that of monomers [104]. Because the number of fluorescent spots was small in the 10^{-7}- and 10^{-6}-M submonolayers, we selected 10^{-5}- or 10^{-4}-M submonolayers to simultaneously observe as many

---

Figure 17 Fluorescence photocounts of individual fluorescent spots versus the number of occurrences of each fluorescent spot for (A) 10^{-7}-, (B) 10^{-6}-, (C) 10^{-5}-, and (D) 10^{-7}-M submonolayers. Bin width is 20 counts. The histogram of 10^{-4}-M submonolayers was not obtained because of difficulty in identifying fluorescent spots due to an increase in photocounts of background regions without fluorescent spots. The observation in (D) was obtained under the different experimental conditions from those used for the observation in (A), (B), and (C): excitation using linearly-polarized light (se the text in subsection B), the use of 532-nm light, and the reduced excitation power density ($\sim$1.2 W/cm^2) to balance the unitary photocounts ($\sim$100 counts) in (A), (B), and (C) with the unitary photocounts in (D). (From Ref. 16.)

Table 6 Photocounts of Representative 20
Fluorescent Spots Under the Irradiation of Vertically
(V) and Horizontally (H) Polarized Excitation

V (counts/30 s)	H (counts/30 s)	V/H
51	6	8.50
64	10	6.40
73	11	6.64
75	10	7.50
88	10	8.80
95	7	13.6
97	17	5.71
100	8	12.5
101	12	8.42
107	20	5.35
112	16	7.00
112	19	5.89
112	14	8.00
143	14	10.2
154	29	5.31
160	13	12.3
160	16	10.0
204	26	7.85
207	30	6.90
235	29	8.10

Both of the polarized light were perpendicular to the inci-
dent direction: V-polarized light was in the x-z plane and
H-polarized light was perpendicular to the V-polarized light.
Source: Ref. 16.

fluorescent spots as possible within an image. All the spectra from bright spots
were identified as those of rhB monomers (Fig. 18A–E), because the spectral
shape and maximum are similar to those of monomers observed in a dilute rhB
solution (Fig. 18F). For the 10^{-5}-M submonolayers, we were not able to observe
fluorescence spectra in the background region with no fluorescent spots (see the
lower record in Fig. 18E). This inability to observe fluorescence spectra in the
background region is consistent with the fluorescence morphology of a 10^{-5}-M
submonolayer in Fig. 16C, where fluorescence photons in the background region
were thinner than those in a 10^{-4}-M submonolayer. In contrast to the 10^{-5}-M
submonolayers, in the 10^{-4}-M submonolayers, fluorescence spectra were observed from both bright spots and regions with no bright spots, and again were
identified as those of rhB monomers. In the 10^{-3}- and 10^{-2}-M adsorbates, the
coverage of which is mono- to multilayers [104], we observed deeply red-shifted

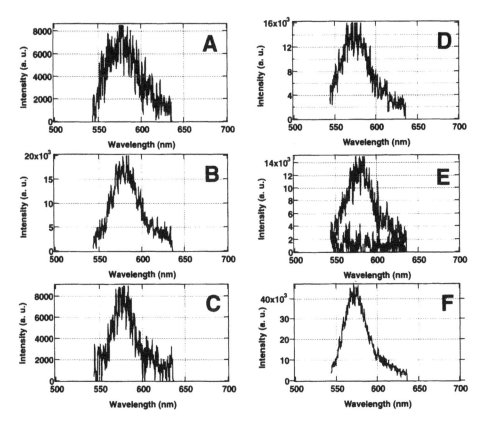

Figure 18 (A)–(E) Fluorescence spectrums of individual fluorescent spots in 10^{-5}-M submonolayers obtained using a streak camera in the focus mode and a CW 514.5-nm excitation. Spectrums from a spot (lower record) and from a background area not containing fluorescent spots (upper record) are displayed in (E). (F) Fluorescence spectrum of rhB in methanol (10^{-6} M) as a reference measured using the same fluorometer used for measuring spectrums (A)–(E). (From Ref. 16.)

($\sim$50 nm) fluorescence spectra only in 10^{-2}-M adsorbates (data not shown). This red-shifted spectra are an indication of rhB dimers (see Sec. VI.I).

D. Fluorescence Lifetimes of RhB in Submonolayers on Si Surfaces

From the near linear dependence of the total fluorescence photocounts on the rhB concentration (Table 5) and the fluorescence spectra of bright spots (Fig. 18A–E and the region with no bright spots (for 10^{-4}-M submonolayers, figure not shown), we conclude that all the fluorescence of rhB submonolayers on a Si surface are ascribed to rhB monomers. To find the origin of the fluorescence in bright spots and in diffused appearance, we measured the fluorescence lifetimes (τ_f) for both bright spots and the region with no bright spots. The 10^{-4}-M submonolayers were selected because both types of fluorescence were measurable (Fig. 16D). The region with no fluorescent spots showed strong quenching ($\tau_f < 50$ ps), as shown in the decay curve (b) in Fig. 19A, whereas individual bright spots showed fluorescence lifetimes equivalent to those of rhB in water and in alcohols, as shown in the decay curve (a) in Fig. 19A. Figure 19B shows a

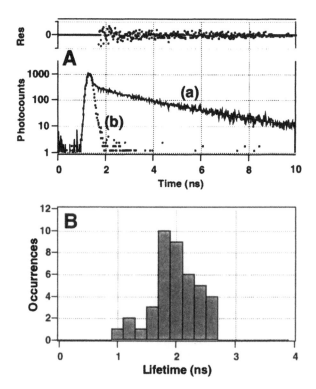

Figure 19 (A) Representative fluorescence decay curves observed on a 10^{-4}-submonolayer. A slow fluorescence decay curve (b), whose lifetime $\tau_f = 2.5$ ns, was observed for an isolated fluorescent spot, whereas a fast decay curve (a), whose lifetime $\tau_f < 50$ ps, was observed in a region not containing isolated fluorescent spots. The upper record shows residuals between the fitting decay curve and the observed slow decay curve. (B) Histogram of observed lifetime τ_f from individual fluorescent spots on 10^{-4}-submonolayers. (From Ref. 16.)

histogram of τ_f for individual fluorescent spots, similar to τ_f previously reported in aqueous ($\tau_f = 1.3$ ns) or alcoholic ($\tau_f = 2.0$–2.7 ns) solutions [97–99]. Thus, our observations identified strong quenching by Si wafers as an important factor in the diffused fluorescence.

E. Correlation Between Light-Scattering Centers and Bright Fluorescent Spots

From the dependence of the number of bright spots on the rhB concentrations (Fig. 16) we inferred that the observed positions of fluorescent spots were associated with intrinsic sites (pits or bumps) dotted on the Si surfaces. One possible reason for this inference is that an increase in the number of bright fluorescent spots in the 10^{-7}-, 10^{-6}-, and 10^{-5}-M submonolayers involves gradual filling of sites on a Si surface by rhB molecules. Thus, the saturation of the number of bright spots and the appearance of the extra fluorescence not associated with the bright spots from the 10^{-5}- to 10^{-4}-M submonolayers might be due to saturation of the sites and the subsequent spillover of rhB molecules not settled within the sites.

To explore the possibility that fluorescent spots are located on the intrinsic sites that scatter excitation light, light-scattering images were contrasted with fluorescent images. For this purpose 10^{-5}-M submonolayers were selected for further experiments of the imaging of light-scattering centers. Because the number of bright spots is nearly saturated in 10^{-5}-M submonolayers (Fig. 16C), we expect a high correlation between the position of fluorescent spots and that of light-scattering centers. On the same field of view, we observed the surfaces of 10^{-5}-M submonolayers under laser irradiation with and without the filters used for the fluorescence imaging experiments. Without the filters, many spots were formed by scattered green laser light; and with the filters, fluorescent spots were observed. The location of light-scattering centers was strongly correlated with that of fluorescent spots, although the intensity of light-scattering centers was weakly correlated with that of fluorescent spots. These observations support our hypothesis that fluorescent spots occur in the position of intrinsic sites (pits or bumps) on the Si wafers. The observed light-scattering centers were isolated diffraction-limited Airy disks. This means that the size of a light-scattering center was smaller than ~0.4 μm in diameter for the microscope we used. Although as much light-scattering contamination as possible is removed, foreign particles from the surroundings may contribute to light-scattering centers. However, when using a Si surface contaminated with foreign particles easily removed by wiping with a piece of lens-cleaning paper, we obtained a lower correlation than the previous comparison between fluorescent spots and light-scattering centers. This observation further supports our hypothesis although the possibility of contributing foreign particles to light-scattering centers cannot be completely eliminated.

F. Physical Properties of Light-Scattering Centers

To verify our hypothesis that fluorescent spots occur in the position of intrinsic sites (pits or bumps) on the Si surface, we evaluated the morphology of possible pits or bumps using an instrument whose resolution is higher than that of light microscope on the same field of view as fluorescent spots were observed. However, this kind of equipment is not available to us. What we can do is observe surface morphology of the Si surfaces independent of the fluorescence measurements and find indirectly the origin of light-scattering centers as described in the last half of this section.

To find qualitatively the mechanical stability of light scattering centers we examined the location of light scattering centers before and after drop-and-drag treatment. The observations of this experiment are divided into three classes. (1) Light-scattering centers that newly appeared after the drop-and-drag treatment. However, we cannot conclude they are particles coming from the paper or surroundings, because they could be fragments of intrinsic bumps collapsed and carried by the drop-and-drag treatment. (2) Light-scattering centers that disappeared after drop-and-drag treatments. They are not pits. If pits, they should still be observed after the drop-and-drag treatment. Nevertheless, we cannot conclude they are particles coming from the paper or surroundings because they could be intrinsic bumps collapsed and swept away by the drop-and-drag treatment. (3) Light-scattering centers that were not removed with repeated drop-and-drag treatment (more than 10 times). They are highly probable intrinsic scattering centers on Si wafers, although we cannot conclude whether they are bumps or pits.

In conclusion, exact evidence was not found that light-scattering centers are pits from the three classes of observations, although the possibility is not entirely eliminated that the light-scattering centers are bumps.

G. Evaluation of Surface Roughness of Si Wafers by Atomic Force Microscopy

To search possible pits or bumps we observed the Si surfaces used for the fluorescence measurement using AFM (Fig. 20A). The surface irregularities of the Si wafers were mostly within ±0.5 nm as shown in Fig. 20B. No pits or bumps that might correspond to the same kind of light-scattering centers were found in the high-resolution (500×500 nm^2) measurement. We again searched Si surfaces for pits or bumps smaller than ~ 0.4 μm in size on larger scanning areas (10×10 μm^2 or 20×20 μm^2) than the high-resolution measurement

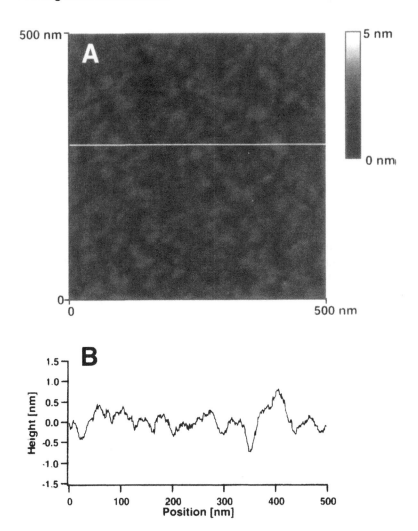

Figure 20 AFM image of a Si wafer used in the fluorescence measurements. No bumps or pits expected from the light-scattering images were observed in the high-resolution (500 × 500 nm^2) scanning. (B) The cross section of the horizontal line in (A). (From Ref. 16.)

did, because they are possible for the light-scattering centers that were observed as diffraction-limited Airy disks. Only bumps (typically 5–20 nm in height and 50–160 nm in diameter) were found. The density of the number of bumps observed were ~40/100 μm^2 at the most. All of the bumps observed were not removed from the surface by repeated scans using the tapping mode of

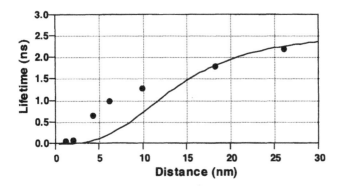

Figure 21 Fluorescence lifetimes τ_f of rhB 10^{-4}-submonolayers on Si wafers covered with an oxide layer in various thickness d (26.0, 18,2, 9.8, 6.2, 4.3, 2.0, and 1.1 nm). Solid line shows the theoretical curve from Eq. (1). Lifetime measurement involved in this figure was carried out in bulk without a microscope; thus, the observed lifetimes were averaged over an area larger than $100 \times 100 \ \mu m^2$. (From Ref. 16.)

AFM measurement. This observation eliminated the possibility that the bumps observed might be liquid or liquid-like materials.

H. Dependence of Fluorescence Lifetimes of RhB Molecules on the Thickness of an Oxide Layer on an Si Surface

Summarizing experimental observations from Secs. VI.A to VI.G, we found that all the fluorescence observed from the rhB submonolayers comes from rhB monomers, bright fluorescent spots are free from strong quenching, the locations of bright fluorescent spots are highly correlated with those of light-scattering centers, and light-scattering centers were likely to be bumps on the Si surfaces.

On the basis of these experimental observations we infer that the fluorescence in bright fluorescent spots, or in light-scattering centers, occurs on bumps and is insulated from strong quenching thanks to the thickness of the bumps. In fact, it seems highly probable that bumps on Si surfaces consist of silicon oxide. Thus, we determined the dependence of τ_f on the thickness (d) of silicon oxide layers using the 10^{-4}-M submonolayers. Figure 21 shows that τ_f increased with increasing d, finally reaching the limiting τ_f (2.6 ns) observed for a submonolayer on a quartz surface. Note that τ_f involved in Fig. 21 was measured in bulk, thus reflecting the averaged nature of 10^{-4}-M submonolayers.

I. No Dimers of RhB Molecules in Submonolayers

The independent nature of individual rhB molecules may be surprising to researchers who are familiar with dye molecules, because dye molecules show self-associative nature, particularly in a concentrated aqueous solution [96]. The independent nature will be lost when dye molecules are strongly associated (e.g., form dimers). Kemnits and co-workers classified chemical species in submono- and monolayers of rhB in the following manner [104]. Species I with fluorescence maximums in the 573–580 nm region are monomers. Species II with maximums in the 590–600 nm region are dimers. According to this classification, no dimers or higher aggregates of rhB molecules were formed in any fluorescence spectra of our measurements involving 10^{-7}- to 10^{-4}-M submonolayers. We observed fluorescence spectra ascribable to dimers only in 10^{-2}-M adsorbates (data not shown). Possible hydrogen bonds and ionic forces on the surfaces may prevent rhB molecules in the submonolayers from forming dimers.

J. Orientations of RhB Molecules in the Bright Spots

The histogram in Fig. 17D shows several maximums at every ~ 100 photocounts, although three maximums near 300, 600, and 800 counts were biased toward lower counts and each maximum was blurred compared with the maximums in Fig. 17. Surprisingly, we found the digitized histogram, although disordered and blurred, using p-polarized excitation. We correlate this observation with the ratio of photocounts determined by the p-polarized excitation to those determined by the s-polarized excitation (p/s in Table 6). Because the value of p/s (5 to 13) is much larger than unity, we consider that the transition dipole of a single rhB molecule is not parallel to the surface assuming that the fluorescence intensity observed is entirely determined by the absorption efficiency. If the transition dipole in a tilted position is completely immobilized on a surface within the acquisition time (30 s), the value of p/s is expected to assume a more sharp fluctuation than the observed fluctuation (within a factor of 3 from spot to spot). Thus, we consider an oversimplified model where the tilted transition dipole can freely rotate within a solid angle during the acquisition time. This model has already been proposed in a previous work [13]. The transition dipole on this model is proportional to $\cos \theta$ along the z-axis and $(1/\sqrt{2}) \sin \theta$ along the y-axis. Note that we canceled azimuth (ϕ) dependence by averaging from 0 to 2π based on the oversimplified model. Using this model plus the known colatitude ($\sim 65°$) of the incident direction of the excitation light, and assuming adsorption efficiency determines exclusively the observed fluorescence intensity, we evaluated possible colatitude of the dipole to be 20–30° to reproduce the values of p/s in Table 6. Transition dipoles not parallel to a surface were concluded in previous studies of dipole orientations in submonolayers using surface second-harmonic generation [101,108] and in a recent single-molecule study [42]. The biased orientation of dipole moments seems to be peculiar to fluorescent spots dotted on silicon surfaces. No such biased orientation is found on quartz surfaces [17].

K. Fluorescence Quenching by Si Wafers

We observed strong fluorescence quenching in the area not containing bright fluorescent spots (Sec. VI.D) and evaluated dependence of τ_f on the thickness of an silicon oxide layer (Sec. VI.H). Here, we take three possible nonradiative processes into consideration to discuss possible mechanisms of this quenching. First, energy transfer among rhB molecules is not important as a mechanism of the strong quenching because of the low coverage of the submonolayers we used. No dimers were identified in the submonolayers. A proposed mechanism of fluorescence quenching is monomer-to-dimer energy transfer, previously characterized by non-single exponential fluorescence decays in rhB mono- to multilayers [100,102,104]. However, single-exponential decays were exclusively observed in our measurements involving 10^{-4}-M submonolayers on quartz and on Si surfaces covered with oxide layers of a various thickness. Second, quenching by electron transfer from rhB molecules to Si wafers is a possible mechanism considering their relative potentials [107,112]. However, a possible electron transfer rate between rhB molecules and a Si wafer, even when rhB molecules come in direct contact with a Si surface without an oxide layer, is slower than the observed quenching rate [113]. Actually, a 1.1-nm-thick oxide layer located between rhB molecules and a S surface even when the shortest τ_f was observed. Third, energy transfer between rhB molecules and Si wafers is also a possible mechanism. The energy transfer is important for the strong quenching we observed, and originally studied for dye-to-metal energy transfer.

Cnossen and co-workers measured τ_f of rhB on an aluminum mirror covered with amylose-acetate ester of a various thickness (1 to 6 nm) [114]. They found Eq. (1) to agree well with the experimental observations:

$$\tau_f(d)^{-1} = \tau_f(\infty)^{-1} \left\{ 1 + \frac{\eta}{8}(dk)^{-3} \left[2\mathrm{Im}\left(\frac{\varepsilon_m(\omega) - \varepsilon_1}{\varepsilon_m(\omega) + \varepsilon_1} \right) \right. \right.$$
$$\left. \left. + 6\xi \frac{1}{k_F d} \frac{\omega}{\omega_p} + 18 \frac{\omega_F}{\omega_p} \frac{\omega}{\omega_p} \frac{1}{k_F d} \right] \right\}, \tag{1}$$

where $\tau_f(d)$ is τ_f at a distance d from the mirror, $\tau_f(\infty)$ is τ_f at an infinite distance from the mirror, k is the magnitude of the wavevector at an emitted frequency ω, $\varepsilon_m(\omega)$ is the complex dielectric constant of the substrate (metal), ε_1 is the dielectric constant of the medium in which the dipole is embedded, and h is an orientation parameter ($h = 3/2$ for a perpendicular dipole and $= 3/4$ for a parallel dipole). Here, k_F is the Fermi wavevector, ω_F the Fermi frequency, ω_p the plasma frequency, and $\xi \approx 1$ is a constant that depends on the electron-gas parameters. Only the first term in the square brackets is important for our analysis, because the last two terms contain electron-gas parameters of metals. The first term represents the bulk contribution to the quenching rate, and is identical to the classical result where the interaction between the transition dipole and the electromagnetic field of its image dipole is considered [114]. We thus computed τ_f using Eq. (1) neglecting the last two terms in the square brackets, and plot the results in Fig. 21 with a smooth curve using $\tau_f(\infty) = 2.6$ ns that was observed on a quartz surface. A large deviation in the experimental observations from the theory is notable below $\sim$15 nm. The reason for the deviation is not yet clear; however, the plot in Fig. 21 is useful for a calibration curve of τ_f versus d of oxide layers. This plot gives an important insight into d where fluorescent spots were observed, assuming that τ_f is determined only by d, and an insight into which d we should select for analyte molecules to be free from quenching in applications of Si wafers to analytical use.

L.   What are the Light-Scattering Centers?

From the correlation between the location of light-scattering centers and that of bright fluorescent spots (Sec. VI.E), physical properties of light-scattering centers (Sec. VI.F), and the AFM measurements (Sec. VI.G), we infer that nanometer-size bumps rather than pits are important as the light scattering centers on which rhB molecules are adsorbed. The light-scattering centers might be SiO_2 bumps on Si surfaces.

Crystal-originated-particles (COP), which are naturally occurring defects on Si wafers, are widely known as light-scattering centers. Finding the origin of COP has been a critical issue in recent silicon manufacturing technologies. Recently, Miyazaki and co-workers revealed the morphology and dimensions of COP [115]. They found the COP on a <100> wafers to be a pit ($0.12 \times 0.12 - 0.3 \times 0.3$ μm^2) in an inverse pyramidal shape surrounded by 4-fold <111> facets, or a pair of such pits. However, we found no pits corresponding to the COPs as pits in our AFM measurement on the large scanning areas. Thus, the COPs as pits are not important for the light scattering centers we observed.

Real aspects of the physical and chemical nature of the light-scattering centers are unknown within the limits of our experiments although the possibility that the bumps observed are liquid or liquid-like materials was denied by the AFM measurements. Furthermore, if the rhB molecules deposited on nanometer-sized bumps show bright spots, we do not know what mechanisms are possible to explain the remarkable uniformity in the fluorescence intensity of the rhB molecules on irradiating p-polarized light. We, however, obtained an important idea for observing single analyte molecules in a high signal-to-background ratio:

preparing analyte molecules on localized insulators dotted on surfaces that are a strong fluorescence quencher.

M. Future Prospects

The single molecule imaging study presented here should be further developed as a key technology, particularly in fluorescence measurement of micro chemical analysis systems (μ-TAS) utilizing chemically or photolithographically defined structures on solid materials. Nowadays, miniaturizing chemical analysis systems is a rapidly developing field in analytical chemistry. An understanding of grouping and quenching of individual dye molecules on a solvent-free surface, as addressed in this chapter, is a good starting point for development of sample preparations, including immobilization and arrangement of single molecules. This work will provide a foundation for developing single-molecule chemical analysis systems on surfaces.

VII. SINGLE-MOLECULE FLUOROMETRY OF AN ANALOG OF NUCLEIC-ACID BASE

A. What is TNP-ATP?

A variety of adenosine $5'$-triphosphate (ATP) fluorescent analogs have been synthesized [116–119] and used to acquire information on ATP turnovers that are crucial to many biochemical processes. Among others a ribose-modified fluorescent nucleotide analog TNP-ATP has widely been used as substrates, inhibitors, and structure probes of many proteins [120–128] because of its unique fluorescence properties and high binding affinity to proteins. The fluorescence efficiency of TNP-ATP is as low as 2×10^{-4} in an aqueous solution [24,120]. However, as described in Sec. I apparent fluorescence intensity increases with increasing viscosity or with decreasing solvent polarity [24,120] and is drastically enhanced when TNP-ATP is bound to proteins [24,120–128]. It is this enhancement of the fluorescence efficiency that makes it possible to observe single complexes of TNP-ATP and protein molecules.

B. Basic Fluorescence Characteristics of TNP-ATP

Despite the wide applicability of TNP-ATP as a useful fluorescent nucleotide analog, the dynamics of its fluorescence is still not fully understood. To find the origin of change in the apparent fluorescence intensity in different environments, it is important to learn how the local environment influences the fluorescence dynamics. Four important fluorescence characteristics of TNP-ATP are summarized in the following manner [23]. First, the fluorescence lifetime of TNP-ATP increases with increasing solvent viscosity, resulting in an increase in quantum efficiency and, therefore, fluorescence intensity. Second, fluorescence intensity increases with increasing pH from 2.4 to 8.1. However, the fluorescence lifetime is essentially independent of the change in pH. The change in fluorescence intensity is not due to variations in fluorescence efficiency but to the change in the absorbance with pH. Third, we examined the interaction of TNP-ATP with the Klenow fragment of DNA polymerase I; and distinguished free TNP-ATP from TNP-ATP bound to two sites of a Klenow fragment by using time-resolved fluorometry. The second binding site with lower affinity was not found in the previous steady-state measurements. Although the fluorescence lifetime of the enzyme-TNP-ATP complex remains unchanged, the fractional fluorescence intensity of the bound TNP-ATP increases, thus leading to an increase in total fluorescence intensity. Lastly, the fluorescence of TNP-ATP in the presence of the enzyme is further enhanced by the addition of Mg^{2+}. This enhancement is due to the increase in the fractional fluorescence intensity of the bound TNP-ATP rather than to the change in fluorescence lifetime. This observation shows that Mg^{2+} raises the affinity of the enzyme for TNP-ATP.

A Klenow fragment is the largest fragment of *Escherichia coli* DNA polymerase I, and also a multifunctional enzyme that carries polymerase and 3′–5′ exonuclease activities on separate catalytic sites of a single polypeptide chain. These complex functions have been the subject of extensive studies using X-ray crystallography [129], NMR [130], and fluorescence microscopy [131]. Interaction between the enzyme and duplex DNA were found to be due to contact of the enzyme with a DNA phosphate backbone [129], while TNP-ATP binds the enzyme with its triphosphate moiety [122]. Thus, TNP-ATP is expected to have the same binding cooperativity to a Klenow fragment as DNA.

C. Single-Molecule Imaging and Time-Resolved Fluorometry of TNP-ATP

In the current single-molecule study, we highlight a complex between TNP-ATP and a Klenow fragment because of the fluorescence efficiency larger than 0.2. Thus, it is highly probable that we succeed in single-molecule imaging and time-resolved fluorometry of the complex. In fact, fluorescence efficiency larger than ~0.1 is a criterion of current single-molecule imaging and time-resolved fluorometry.

TNP-ATP has another merit of single-molecule fluorometry. The large Stokes shift of TNP-ATP (~1800 cm^{-1}) [23] compared with rhodamine or cyanine dyes (~600 cm^{-1}) makes it easy to discriminate fluorescence signal from excitation scattering light in the SMD.

Figure 22A shows well separated bright fluorescent spots of 2-nM TNP-ATP mixed with Klenow fragments spin-coated on a clean quartz surface. Fig-

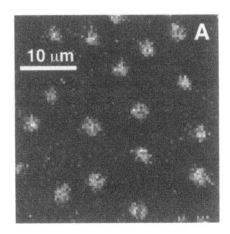

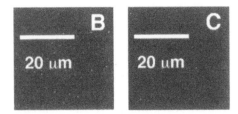

Figure 22 Fluorescence images of (A) 2-nm TNP-ATP mixed with 0.1-mM Klenow fragment, (B) 2-nM TNP-ATP only, and (C) 0.1-mM Klenow fragment only. The data acquisition time was 30 s for each image. No fluorescence spots occurred in (B) and (C), whereas bright spots appeared in (A), which resulted from the fluorescence emission of individual Klenow fragment-TNP-ATP complexes. The signal-to-background ratio was 30:1. (From Ref. 18.)

ure 22B and C show images of 2-nM TNP-ATP and 0.1-μM Klenow fragments spin-coated on a clean quartz surface, respectively. These two images were recorded under the same experimental conditions as those for Fig. 22A. However, no fluorescent spots appear in Fig. 22B or C, in marked contrast with Fig. 22A. These observations indicate that no TNP-ATP molecules form fluorescent spots because of the low fluorescence efficiency. Also Klenow fragments did not generate any fluorescent spots either, because they have no absorption at the excitation wavelength 400 nm (data not shown). Thus, the spots in Fig. 22A result from the fluorescence emission of Klenow fragment-TNP-ATP complexes. Figure 23 shows a histogram of photocounts of individual fluorescent spots. The

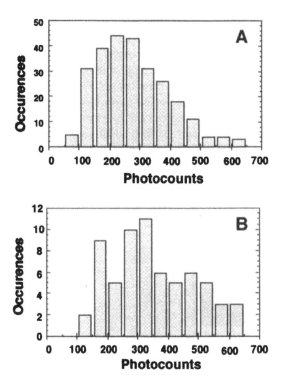

Figure 23 Fluorescence photocounts of individual enzyme-TNP-ATP complexes versus the number of occurrences; the concentration of the complex was (A) 2.0 (from Ref. 18) and (B) 0.4 nm. The bin width is 50 counts.

fluorescence photocounts observed show only one peak of the distribution. This observation implies that the fluorescence owes its origin to one fluorescent component in contrast to the study of individual CV molecules on a PMMA film in Sec. V. The histogram of fluorescence photocounts of individual CV molecules shows a bimodal distribution owing to inhomogeneous interaction between CV and a PMMA film, as shown in Fig. 14A. The broad distribution in Fig. 23A may be ascribed to the different orientation of the transition dipole moment of individual TNP-ATP molecules. Another experiment was also performed using another sample of 0.4-nM TNP-ATP-Klenow-fragment. The histogram of this sample (Fig. 23B) shows a peak of the same fluorescence photocounts as that of the five-fold concentrated sample (Fig. 23A). This observation, plus the observation that the number of fluorescent spots per image linearly decreased with decreasing TNP-ATP concentration, confirmed our observation of single molecule fluorescence. The enzyme-TNP-ATP complex is extremely photostable: no photobleaching occurs more than 1 h under the experimental conditions we used. Thus, we found no stepwise photobleaching of single fluorophores, which is used as a criterion of observing single-molecule fluorescence [16,43].

We further carried out time-resolved fluorescence measurement on a single fluorescent spot of the individual enzyme-TNP-ATP complexes. Figure 24A and B shows a representative fluorescence decay curve and fluorescence spectrums of single enzyme-TNP-ATP complexes. The fluorescence spectrum varied from spot to spot, as an indication of fluorescence from individual complexes. The decay curve cannot be fitted to a single exponential, but was well fitted to a

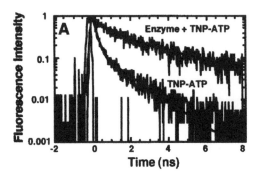

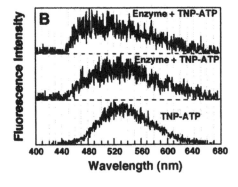

Figure 24 (A) Representative fluorescence decay curve of a single Klenow fragment-TNP-ATP complex obtained by the single-molecule time-resolved spectroscopy, together with a fluorescence decay curve of TNP-ATP obtained by a bulk measurement. Both curves were well fitted to biexponential functions. The instrument-response function in 195-ps fwhm is also displayed. (B) Representative fluorescence spectrums of two individual enzyme-TNP-ATP complexes showing different emission peaks. A fluorescence spectrum of TNP-ATP obtained from a bulk measurement is also displayed for comparison. All spectrums were normalized to unity at their maximum. (From Ref. 18.)

biexponential function $I(t) = A_f \exp(-t/\tau_f) + A_s \exp(-t/\tau_s)$, with two time constants τ_f and τ_s, and two amplitudes A_f and A_s. Figure 25 summarized the fitting parameters for 21 individual enzyme-TNP-ATP complexes. The lifetimes of the slow decay component τ_s and the fast decay component τ_f showed distributions around 2.60 and 0.42 ns, respectively; the ratio of the fast to slow decay component A_f/A_s distributed around 7.73. No significant correlation was found between the fluorescence lifetimes and the fluorescence photocounts. This observation is inconsistent with the above attribution that different orientation of individual molecules is important for the broad distribution of the photocounts in the histogram in Fig. 23A. Although the fluorescence lifetime and the amplitude ratio showed a distribution as a result of the single-molecule measurement, no qualitative difference was found in the decay curves. This observation contrasts sharply with the single-molecule measurement of CV. In Sec. V, a biexponential fluorescence decay in bulk measurements is resolved into two single exponential decays with distinctly different lifetimes corresponding to single CV molecules at two kinds of sites embedded in a PMMA film. In the current TNP-ATP study, however, biexponential fluorescence decays were observed for a single enzyme-TNP-ATP complex. What mechanisms are operative is unknown in the biexponential fluorescence decay. We have, however, two possible explanations for the observation. First, a given molecule has one ground state and two excited states with different decay rates. The molecule in the ground state has a certain possibility to be excited to either of the two excited states, and then the excited states come to equilibrium before emitting fluorescence. Second, a given molecule has two ground states and two excited states of each ground state. The molecule attains equilibrium in the ground states on a time scale much shorter than the time interval of data acquisition, and then each excited state emits fluorescence. In the both cases biexponential decays are possible for

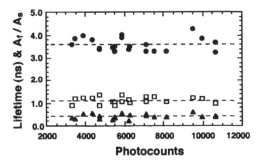

Figure 25 The lifetimes of the slow decay τ_s (●) and the fast decay τ_f (▲) and the amplitude ratio A_f/A_s (□) corresponding to 21 single Klenow fragment-TNP-ATP complexes. The dashed lines indicate the mean values of each parameter. (From Ref. 18.)

a single molecule. A large shoulder at 480 nm, plus the absorption maximum at 408 nm, in the absorption spectrum of TNP-ATP [131] might be an indication of two ground states or two excited states. Similarly, biexponential decays were observed for a single tetramethylrhodamine (TMR) molecule tethered to a large DNA molecule [132], and for an individual tRNA^{Phe}-TMR adduct [133].

Figure 24A also shows a fluorescence decay curve of an ensemble of TNP-ATP molecules in comparison with the decay curve of a single enzyme-TNP-ATP complex. The decay curve was well fitted in with a biexponential function with lifetimes 0.20 and 1.40 ns and the ratio of amplitudes $A_f/A_s = 11$. Compared with the enzyme-TNP-ATP complex, the fast decay component is dominant and the lifetimes are much smaller than those observed in the complex. Assuming that TNP-ATP has the identical radiative lifetime in adsorption and in the complex, we found that the fluorescence efficiency of the enzyme-TNP-ATP complex is 7.3-fold as high as that of TNP-ATP adsorbed on a quartz surface. This fact explains well why fluorescent spots of single TNP-ATP molecules were not observed (Fig. 22B), let alone the fluorescence decay of single TNP-ATP molecules, using a 2-nM TNP-ATP solution.

D. Future Prospects

We extended single molecule fluorescence imaging and time-resolved fluorometry from the green-excitation regime involving standard rhodamine and cyanine dyes to the violet-excitation regime. When TNP-ATP was associated with the enzyme molecule, the fluorescence efficiency of the TNP-ATP was remarkably enhanced to the level on which single-molecule imaging is implemented. This unique property opens up the possibility of allowing TNP-ATP to replace Cy3-conjugated ATP in the study of ATP turnovers at the single-molecule level. In the experiment in which single enzyme molecules are immobilized on a surface in a buffer solution, the use of Cy3-conjugated ATP has thus far been limited to enzymes with which Cy3-conjugated ATP highly associates. Free Cy3-conjugated ATP molecules not associated with enzymes act as background. In contrast, free TNP-ATP makes no contribution to the background fluorescence because of the low fluorescence efficiency in solution.

VIII. CONCLUDING REMARKS

In this chapter we presented the function of TPM dyes and TNP-ATP as probes for environment in the nanometer scale. The function of these chromophores is associated with the extreme sensitivity to their environment, such as viscosity, polarity, and pH. Also, we demonstrated the function of rhB as a probe sensitive to fluorescence quenching, although this function is not specific to rhB but is universal to chromophores. The ability to probe nanostructures will considerably be exhibited especially when functional chromophores are prepared in the single-molecule regime.

ACKNOWLEDGMENTS

This work was supported by New Energy and Industrial Technology Development Organization at the Joint Research Center for Atom Technology.

REFERENCES

1. M Ishikawa, JY Ye, Y Maruyama, H Nakatsuka. J Phys Chem A 103:4319–4331, 1999.
2. DF Duxbury. Chem Rev 93:381, 1993.
3. GE Bush, PM Rentzepis. Science 194:276, 1976.
4. A Migus, A Antonetti, J Etchepare, D Hulin, A Orszag. J Opt Soc Am B2:584, 1985.
5. V Sundstrom, T Gillbro. J Chem Phys 81:3463, 1984.
6. D Ben-Amotz, CB Harris. Chem Phys Lett 119:305, 1985.
7. D Ben-Amotz, CB Harris. J Chem Phys 86:4856, 1987.
8. D Ben-Amotz, R Jeanloz, CB Harris. J Chem Phys 86:6119, 1987.
9. KM Abedin, JY Ye, H Inouye, T Hattori, H Nakatsuka. J Lumin 64:135, 1995.
10. KM Abedin, JY Ye, H Inouye, T Hattori, H Sumi, H Nakatsuka. J Chem Phys 103:6414, 1995.
11. JY Ye, T Hattori, H Inouye, H Ueta, H Nakatsuka, Y Maruyama, M Ishikawa. Phys Rev B53:8349, 1996.
12. JY Ye, T Hattori, H Nakatsuka, Y Maruyama, M Ishikawa. Phys Rev B56:5286, 1997.
13. M Ishikawa, K Hirano, T Hayakawa, S Hosoi, S. Brenner. Jpn J Appl Phys 33:1571, 1994.
14. M Ishikawa, M Watanabe, T Hayakawa, M Koishi. Anal Chem 67:511, 1995.
15. JY Ye, M Ishikawa, O Yogi, T Okada, Y Maruyama. Chem Phys Lett 288:885, 1998.
16. M Ishikawa, O Yogi, JY Ye, T Yasuda, Y Maruyama. Anal Chem 70:5198, 1998.
17. M Yamauchi, JY Ye, O Yogi, M Ishikawa. Chem Lett 735–736, 1999.
18. JY Ye, Y Yamane, M Yamauchi, M Ishikawa. Chem Phys Lett 320:607–612, 2000.
19. M Ishikawa, Y Maruyama. Chem Phys Lett 219:416, 1994.
20. Y Maruyama, M Ishikawa, H Satozono. J Am Chem Soc 118:6257, 1996.
21. Y Maruyama, O Magnin, H Satozono, M Ishikawa. J Phys Chem A 103:5629, 1999.
22. When PMMA was formed in thin films (20 – 40 nm), T_g goes up by $\sim 10°C$ from T_g in bulk. Y Grohens, M Brogly, C Labbe, MO David, J Schultz. Langmuir 14:2929, 1998.

23. JY Ye, M Yamauchi, O Yogi, M Ishikawa. J Phys Chem B 103:2812, 1999.
24. T Hiratsuka. Biochim Biophys Acta 719:509, 1982.
25. JJP Stewart. J Comput Chem 10:209, 221, 1989.
26. JK Bene, HH Jaffe. J Chem Phys 48:807, 4050, 1968.
27. K Nishimoto. Bull Chem Soc Jpn 66:1876, 1993.
28. JB Foresman, A Frisch. Exploring Chemistry with Electronic Structure Methods, 2nd ed: Gaussian, 1996, Ch 6.
29. R Pariser, RG Parr. J Chem Phys 21:466, 767, 1953.
30. JA Pople. Trans Faraday Soc 49:1375, 1953.
31. E Betzig, RJ Chichester. Science 262:1422, 1993.
32. WP Ambrose, PM Goodwin, JC Martin, RA Keller. Phys Rev Lett 72:160, 1994.
33. XS Xie, RC Dunn. Science 265:361, 1994.
34. WP Ambrose, PM Goodwin, JC Martin, RA Keller. Science 265:364, 1994.
35. JK Trautman, JJ Macklin, LE Brus, E Betzig. Nature 269:40, 1994.
36. JK Trautman, JJ Macklin. Chem Phys 205:221, 1996.
37. JJ Macklin, JK Trautman, TD Harris, LE Brus. Science 272:255, 1996.
38. MA Bopp, AJ Meixner, G Tarrach, I Zschokke-Granacher, L Novotny. Chem Phys Lett 263:721, 1996.
39. HP Lu, XS Xie. Nature 385:143, 1997.
40. HP Lu, XS Xie. J Phys Chem B101:2753, 1997.
41. AGT Ruiter, JA Veerman, MF Garcia-Parajo, NJ van Hulst. J Phys Chem A101: 7318, 1997.
42. MA Bopp, Y Jia, G Haran, EA Morlino, RM Hochstrasser. Appl Phys Lett 73:6, 1998.
43. T Funatsu, Y Harada, M Tokunaga, K Saito, T Yanagida. Nature 374:555, 1995.
44. RM Dickson, DJ Norris, TL Tzeng, WE Moerner. Science 274:966, 1996.
45. XH Xu, ES Yeung. Science 275:1106, 1997.
46. L Angeloni, G Smulevich, MP Marzocchi. J Mol Struct 61:331, 1980.
47. CW Looney, WT Simpson. J Am Chem Soc 76:6293, 1954.
48. FC Adam, WT Simpson. J Mol Spectrosc 3:363, 1959.
49. J Korppi-Tommola, RW Yip. Can J Chem 59:191, 1981.
50. J Korppi-Tommola, E Kolehmainen, E Salo, RW Yip. Chem Phys Lett 104:373, 1984.
51. HB Lueck, JL McHale, WD Edwards. J Am Chem Soc 114:2342, 1992.
52. GN Lewis, TT Magel, D Lipkin. J Am Chem Soc 64:1774, 1942.
53. FT Clark, HG Drickamer. J Chem Phys 81:1024, 1984.
54. FT Clark, HG Drickamer. J Phys Chem 90:589, 1986.
55. HPJM Dekkers, ECM Kielman-Van Luyt. Mol Phys 31:1001, 1976.
56. AH Gomes de Mesquita, CH MacGillavry, K Eriks. Acta Crystallogr 18:437, 1965.
57. L Angeloni, G Smulevich, MP Marzocchi. J Raman Spectrosc 8:305, 1979.
58. D Magde, MW Winsor. Chem Phys Lett 24:144, 1974.
59. W Yu, F Pellegrino, M Grant, RR Alfano. J Chem Phys 67:1766, 1977.
60. JM Grzybowski, SE Sugamori, DF Williams, RW Yip. Chem Phys Lett 65:456, 1979.
61. DA Cremers, MW Windsor. Chem Phys Lett 71:27, 1980.
62. R Menzel, CW Hoganson, MW Windsor. Chem Phys Lett 120:29, 1985.

63. A Mokhtari, L Fini, J Chesnoy. J Chem Phys 87:3429, 1987.
64. MM Martin, E Breheret, F Nesa, YH Meyer. Chem Phys 130:279, 1989.
65. MM Martin, P Plaza, YH Meyer. Chem Phys 153:297, 1991.
66. MM Martin, P Plaza, YH Meyer. J Phys Chem 95:9310, 1991.
67. M Vogel, W Rettig. Ber Bunsenges Phys Chem 89:962, 1985.
68. M Vogel, W Rettig. Ber Bunsenges Phys Chem 91:1241, 1987.
69. Because our experiments on a change in the absorption spectrum involved a large number of molecules, we preferred Gaussians. The use of Lorentzians is probably suitable for curve fitting calculation reproducing overlapped spectra when the number of the molecules involved is sparse, such as in single-molecule high-resolution spectroscopy of impurity centers in host crystals at low temperatures (<10 K). WE Moerner. Acc Chem Res 29:563–571, 1996.
70. JM Hollas. Chem Soc Rev 371, 1993.
71. Y Matsuoka, K Yamaoka. Bull Chem Soc Jpn 52:2244, 1979.
72. WC Pierce, DP MacMillan. J Am Chem Soc 60:779, 1938.
73. MD Ediger, CA Angell, SR Nagel. J Phys Chem 100:13200, 1996.
74. W Gotze, L Sjoren. Rep Prog Phys 55:241, 1992.
75. W Gotze. In JP Hansen, D Levesque, J Zinn-Justin, eds. Liquids, Freezing and the Glass Transition. Amsterdam: North-Holland, 1991, p 287.
76. D Richter, B Frick, B Farago. Phys Rev Lett 61:2465, 1988.
77. WM Du, G Li, HZ Cummins, M Fuchs, J Toulouse, LA Knauss. Phys Rev E49:2192, 1994.
78. P Taborek, RN Kleiman, DJ Bishop. Phys Rev B34:1835, 1986.
79. D Ben-Amotz, CB Harris. J Chem Phys 86:5433, 1987.
80. EB Shera, NK Seitinger, LM Davis, RA Keller, SA Soper. Chem Phys Lett 175:553, 1990.
81. T Basche, WE Moerner, M Orrit, T Talon. Phys Rev Lett 69:1516, 1992.
82. M Orrit, M Bernard, RI Personov. J Phys Chem 97:10256, 1993.
83. F Guttler, T Imgartinger, T Plakhotnik, A Renn, UP Wild. Chem Phys Lett 217:393, 1994.
84. S Nie, DT Chiu, RN Zare. Science 266:1018, 1994.
85. RX Bian, RC Dunn, XS Xie, PT Leung. Phys Rev Lett 75:4772, 1995.
86. T Ha, Th Enderle, DF Ogletree, DS Chemla, PR Selvin, S Weiss. Proc Natl Acad Sci USA 93:6264, 1996.
87. T Ha, Th Enderle, DS Chemla, PR Selvin, S Weiss. Phys Rev Lett 77:3979, 1996.
88. S Nie, SR Emory. Science 275:1102, 1997.
89. HP Lu, XS Xie. Nature 385:143, 1997.
90. RM Dickson, AB Cubitt, RY Tsien, WE Moerner. Nature 388:355, 1997.
91. S Bhattachayya, B Bagchi. J Chem Phys 106:7262, 1997.
92. SA Soper, HL Nutter, RA Keller, LM Davis, EB Shera. Photochem Photobiol 57:972, 1993.
93. K Kemnitz, K Yoshihara. J Phys Chem 94:8805, 1990.
94. B Bagchi, GR Fleming, DW Oxtoby. J Chem Phys 78:7375, 1983.
95. K Schmidt-Rohr, AS Kulik, HW Beckham, A Ohlemacher, U Pawelzik, C Boeffel, HW Spiess. Macromolecules 27:4733, 1994.
96. Th Forster, EZZ Konig. Elektrochem 61:344, 1957.

97. IL Arbeloa, KK Rohatgi-Mukherjee. Chem Phys Lett 129:607, 1986.
98. KG Casey, EL Quitevis. J Phys Chem 92:6590, 1988.
99. T-L Chang, HC Cheung. J Phys Chem 96:4874, 1992.
100. N Nakashima, K Yoshihara, F Willig. J Chem Phys 73:3553, 1980.
101. TF Heinz, CK Chen, D Ricard, YR Shen. Phys Rev Lett 48:478, 1982.
102. K Itoh, Y Chiyokawa, M Nakao, K Honda. J Am Chem Soc 106:1620, 1984.
103. Y Liang, PF Moy, JA Poole, AMP Goncalves. J Phys Chem 88:2451, 1984.
104. K Kemnitz, N Tamai, I Yamazaki, N Nakashima, K Yoshihara. J Phys Chem 90:5094, 1986.
105. K Kemnitz, N Tamai, I Yamazaki, N Nakashima, K Yoshihara. J Phys Chem 91:1423, 1987.
106. K Kemnitz, N Nakashima, K Yoshihara. J Phys Chem 92:3915, 1988.
107. K Hashimoto, M Hiramoto, T Sakata. J Phys Chem 92:4272, 1988.
108. ES Peterson, CB Harris. J Chem Phys 91:2683, 1989.
109. T Sakata, K Hashimoto, M Hiramoto. J Phys Chem 94:3040, 1990.
110. MJE Morgenthaler, SR Meech. J Phys Chem 100:3323, 1996.
111. A single rhB molecule absorbs $\sim$1600 photons/s or $\sim$48,000 photons during a 30-s acquisition time. Considering the photocathode efficiency at the fluorescence maximum ($\sim$8%), the NA of the MCP ($\sim$40%), the objective lens collection efficiency ($\sim$17%), the net transparency of the objective lens used at the fluorescence maximum ($\sim$60%),[13] and the transparency of the dichroic mirror used here ($\sim$90%) and the long-pass filter ($\sim$90%), we obtain $\sim$100 fluorescence photons during a 30-s acquisition time assuming $\Phi_f = 1$.
112. D Laser, AJ Bard. J Phys Chem 80:459, 1976.
113. Sakata and co-workers[109] reported an experimental and theoretical relationship between energy gap (D_E) and electron transfer rate (k_{et}) from photoexcited rhB to various semiconductors including Si. The energy gap is the difference between the photoexcited energy level of rhB molecules and the lowest edge of conduction bands of semiconductors. Using this relationship, together with the estimated energy gap $D_E \approx 0.0$ eV, we found that $k_{et} < 10^9$ s^{-1}. Therefore, the minimum fluorescence lifetime τ_f of rhB molecules in direct contact with a Si surface was estimated to be 0.72 ns using $\tau_f = 2.60$ ns obtained on a quartz surface and assuming $k_{et} = 10^9$ s^{-1}. The estimated τ_f(0.72 ns) was longer than the τ_f($<$50 ps) observed on the wafers covered with a 11-Å thick oxide layer.
114. G Cnossen, KE Drabe, DA Wiersma. J Phys Chem 98:5276, 1993.
115. M Miyazaki, S Miyazaki, Y Yanase, T Ochiai, T Shigematsu. Jpn J Appl Phys 34:6303, 1995.
116. JA Secrist, JR Barrio, NJ Leonard. Science 175:646, 1972.
117. NJ Leonard, DIC Scopes, P VanDerLijn, JR Barrio. Biochemistry 17:3677, 1978.
118. DC Ward, E Reich, L Stryer. J Biol Chem 244:1228, 1969.
119. T Hiratsuka, K Uchida. Biochim Biophys Acta 320:635, 1973.
120. EG Moczydlowski, PAG Fortes. J Biol Chem 256:2346, 1981.
121. KE Broglie, M Takahashi. J Biol Chem 258:12940, 1983.
122. GP Mullen, P Shenbagamurthi, AS Mildvan. J Biol Chem 264:19637, 1989.
123. SS Rai, SR Katsuri. Biophys Chem 48:359, 1994.
124. R Liu, FJ Sharom. Biochemistry 36:2836, 1997.

125. J Weber, AE Senior. FEBS Lett 412:169, 1997.
126. J Bandorowicz-Pikula. Mol Cell Biochem 181:11, 1998.
127. C Gatto, AX Wang, JH Kaplan. J Biol Chem 273:10578, 1998.
128. DCA Neville, CR Rozanas, BM Tulk, RR Townsend, AS Verkman. Biochemistry 37:2401, 1998.
129. LS Beese, V Derbyshire, TA Steitz. Science 260:352, 1993.
130. GP Mullen, JB Vaughn Jr., AS Mildvan. Arch Biochem Biophys 301:174, 1993.
131. X Hu, C Aston, DC Schwartz. Biochem Biophys Res Comm 254:466, 1999.
132. L Edman, U Mets, R Rigler. Exp Tech Phys 41:157, 1995.
133. Y Jia, A Sytnik, L Li, S Vladimirov, BS Cooperman, RM Hochstrasser. Proc Natl Acad Sci USA 94:7932, 1997.

Index